PR●BLEMS
3rd Edition

AND

SOLUTIONS

IN

QUANTUM
COMPUTING

AND

QUANTUM
INFORMATION

Willi-Hans Steeb
Yorick Hardy
University of Johannesburg, South Africa

PR⬤BLEMS

3rd Edition

AND

SOLUTIONS

IN

QUANTUM
COMPUTING

AND

QUANTUM
INFORMATION

 World Scientific

NEW JERSEY · LONDON · SINGAPORE · BEIJING · SHANGHAI · HONG KONG · TAIPEI · CHENNAI

Published by

World Scientific Publishing Co. Pte. Ltd.

5 Toh Tuck Link, Singapore 596224

USA office: 27 Warren Street, Suite 401-402, Hackensack, NJ 07601

UK office: 57 Shelton Street, Covent Garden, London WC2H 9HE

British Library Cataloguing-in-Publication Data
A catalogue record for this book is available from the British Library.

PROBLEMS AND SOLUTIONS IN QUANTUM COMPUTING AND
QUANTUM INFORMAITON
(3rd Edition)

ISBN-13 978-981-4366-32-8
ISBN-10 981-4366-32-3

Printed in Singapore.

Preface

The purpose of this book is to supply a collection of problems in quantum computing and quantum information together with their detailed solutions which will prove to be valuable to graduate students as well as to research workers in these fields. All the important concepts and topics such as quantum gates and quantum circuits, entanglement, teleportation, Bell states, Bell inequality, Schmidt decomposition, quantum Fourier transform, magic gate, von Neumann entropy, quantum cryptography, quantum error correction, coherent states, squeezed states, POVM measurement, beam splitter, homodyne detection and Kerr Hamilton operator are included. The topics range in difficulty from elementary to advanced. Almost all problems are solved in detail and most of the problems are self-contained. All relevant definitions are given. Students can learn important principles and strategies required for problem solving. Teachers will also find this text useful as a supplement, since important concepts and techniques are developed in the problems. The book can also be used as a text or a supplement for linear and multilinear algebra or matrix theory. The material was tested in our lectures given around the world.

Any useful suggestions and comments are welcome.

The International School for Scientific Computing (ISSC) provides certificate courses for this subject. Please contact the authors if you want to do this course. More quantum computing exercises can be found on the web page given below.

e-mail addresses of the authors:

whsteeb@uj.ac.za
steeb_wh@yahoo.com
steebwilli@gmail.com
yhardy@uj.ac.za
yorickhardy@gmail.com

Home page of the authors:

http://issc.uj.ac.za

Contents

Notation

$\emptyset$	empty set
$\mathbb{N}$	natural numbers
$\mathbb{Z}$	integers
$\mathbb{Q}$	rational numbers
$\mathbb{R}$	real numbers
$\mathbb{R}^+$	nonnegative real numbers
$\mathbb{C}$	complex numbers
$\mathbb{R}^n$	n-dimensional Euclidean space
$\mathbb{C}^n$	n-dimensional complex linear space
$\mathcal{H}$	Hilbert space
i	$\sqrt{-1}$
$\Re z$	real part of the complex number z
$\Im z$	imaginary part of the complex number z
$A \subset B$	subset A of set B
$A \cap B$	the intersection of the sets A and B
$A \cup B$	the union of the sets A and B
$f \circ g$	composition of two mappings $(f \circ g)(x) = f(g(x))$
t	time variable
$\mathbf{x}$	column vector in $\mathbb{C}^n$
$\mathbf{x}^T$	transpose of $\mathbf{x}$ (row vector)
$\times$	vector product
$\|\cdot\|$	norm
$\mathbf{x} \cdot \mathbf{y} \equiv \mathbf{x}^*\mathbf{y}$	scalar product (inner product) in $\mathbb{C}^n$
$\langle . \,|\, . \rangle$	scalar product in Hilbert space
$\mathbf{x} \times \mathbf{y}$	vector product
$A \otimes B$	Kronecker product of matrices A and B
$f \otimes g$	tensor product of elements f and g of Hilbert spaces
$\det(A)$	determinant of a square matrix A
$\mathrm{tr}(A)$	trace of a square matrix A
$\mathrm{rank}(A)$	rank of matrix A
A^T	transpose of matrix A
$\overline{A}$	conjugate of matrix A
A^*	conjugate transpose of matrix A

$A^\dagger$	conjugate transpose of matrix A (notation used in physics)
I_n	$n \times n$ unit matrix
I	unit operator
$[A, B] := AB - BA$	commutator for square matrices A and B
$[A, B]_+ := AB + BA$	anticommutator for square matrices A and B
δ_{jk}	Kronecker delta with $\delta_{jk} = 1$ for $j = k$ and $\delta_{jk} = 0$ for $j \neq k$
λ	eigenvalue
ϵ	real parameter
H	Hamilton function
$\hat{H}$	Hamilton operator
$\{|0\rangle, |1\rangle, \ldots, |n-1\rangle\}$	arbitrary orthonormal basis for $\mathbb{C}^n$
$\hbar$	$h/2\pi$ with h the Planck constant
ω	frequency
$\mathbf{k}$	wave vector
$\mathbf{r}$	space coordinates
$\mathbf{p}$	momentum
$b, b^\dagger$	Bose annihilation and creation operators
$|n\rangle$	number states, Fock states
$|\beta\rangle$	coherent state, $\beta \in \mathbb{C}$
$|\zeta\rangle$	squeezed state, $\zeta \in \mathbb{C}$
$D(\beta)$	displacement operator, $\beta \in \mathbb{C}$
$S(\zeta)$	one-mode squeezing operator, $\zeta \in \mathbb{C}$
$c, c^\dagger$	Fermi annihilation and creation operators
$\mathbf{E}$	electric field
$\mathbf{B}$	magnetic induction
$\mathbf{P}$	electric polarization

The *Pauli spin matrices* are used extensively in the book. They are given by

$$\sigma_x := \begin{pmatrix} 0 & 1 \\ 1 & 0 \end{pmatrix}, \quad \sigma_y := \begin{pmatrix} 0 & -i \\ i & 0 \end{pmatrix}, \quad \sigma_z := \begin{pmatrix} 1 & 0 \\ 0 & -1 \end{pmatrix}.$$

In some cases we will also use σ_1, σ_2 and σ_3 to denote σ_x, σ_y and σ_z.

We will also use the so-called Dirac notation. Let $\mathcal{H}$ be a Hilbert space and $\mathcal{H}_*$ be the dual space endowed with a multiplication law of the form

$$(c, \phi) = \bar{c}\phi$$

where $c \in \mathbb{C}$ and $\phi \in \mathcal{H}$. The inner product can be viewed as a bilinear form (duality)

$$\langle \cdot | \cdot \rangle \; : \; \mathcal{H}_* \times \mathcal{H} \to \mathbb{C}$$

such that the linear maps

$$\langle \phi | : \psi \to \langle \phi | \psi \rangle, \quad \langle \cdot | : \mathcal{H}_* \to \mathcal{H}'$$

$$| \psi \rangle : \phi \to \langle \phi | \psi \rangle, \quad | \cdot \rangle : \mathcal{H} \to \mathcal{H}'_*$$

where prime denotes the space of linear continuous functionals on the corresponding space, are monomorphisms. The vectors $\langle \phi |$ and $| \psi \rangle$ are called bra and ket vectors, respectively. The ket vector $| \phi \rangle$ is uniquely determined by a vector $\phi \in \mathcal{H}$, therefore we can write $| \phi \rangle \in \mathcal{H}$.

The concept of a Hilbert space will be used throughout the book. A *Hilbert space* is a set, $\mathcal{H}$ of elements, or vectors, $(f, g, h, \ldots)$ which satisfies the following conditions (1) -- (5).

(1) If f and g belong to $\mathcal{H}$, then there is a unique element of $\mathcal{H}$, denoted by $f + g$, the operation of addition $(+)$ being invertible, commutative and associative.

(2) If c is a complex number, then for any f in $\mathcal{H}$, there is an element cf of $\mathcal{H}$; and the multiplication of vectors by complex numbers thereby defined satisfies the distributive conditions

$$c(f + g) = cf + cg, \qquad (c_1 + c_2)f = c_1 f + c_2 f.$$

(3) Hilbert spaces $\mathcal{H}$ possess a zero element, 0, characterized by the property that $0 + f = f$ for all vectors f in $\mathcal{H}$.

(4) For each pair of vectors f, g in $\mathcal{H}$, there is a complex number $\langle f | g \rangle$, termed the inner product or scalar product, of f with g, such that

$$\langle f | g \rangle = \overline{\langle g | f \rangle}$$

$$\langle f | g + h \rangle = \langle f | g \rangle + \langle f | h \rangle$$

$$\langle f | cg \rangle = c \langle f | g \rangle$$

and

$$\langle f | f \rangle \geq 0.$$

Equality in the last formula occurs only if $f = 0$. The scalar product defines the norm $\| f \| = \langle f | f \rangle^{1/2}$.

(5) If $\{ f_n \}$ is a sequence in $\mathcal{H}$ satisfying the Cauchy condition that

$$\| f_m - f_n \| \to 0$$

as m and n tend independently to infinity, then there is a unique element f of $\mathcal{H}$ such that $\| f_n - f \| \to 0$ as $n \to \infty$.

Let $B = \{\, \phi_n \, : \, n \in I \,\}$ be an orthonormal basis in the Hilbert space $\mathcal{H}$. I is the countable index set. Then

$$(1) \qquad \langle \phi_j | \phi_k \rangle = \delta_{jk}$$

$$(2) \qquad \bigwedge_{f \in \mathcal{H}} \quad f = \sum_{j \in I} \langle f | \phi_j \rangle \phi_j$$

$$(3) \qquad \bigwedge_{f,g \in \mathcal{H}} \quad \langle f | g \rangle = \sum_{j \in I} \overline{\langle f | \phi_j \rangle} \langle g | \phi_j \rangle$$

$$(4) \qquad \left(\bigwedge_{\phi_j \in B} \langle f, \phi_j \rangle = 0 \right) \Rightarrow f = 0$$

$$(5) \qquad \bigwedge_{f \in \mathcal{H}} \|f\|^2 = \sum_{j \in I} |\langle f, \phi_j \rangle|^2 .$$

Let $f, g \in \mathcal{H}$. Then we have the inequalities

$$|\langle f, g \rangle| \le \|f\| \cdot \|g\|$$
$$\|f + g\| \le \|f\| + \|g\|$$

and the equality

$$\|f + g\|^2 + \|f - g\|^2 = 2(\|f\|^2 + \|g\|^2) .$$

Part I

Finite-Dimensional Hilbert Spaces

Chapter 1

Qubits

A single *qubit* is a two-state system, such as a two-level atom. The states (kets) $|h\rangle$ and $|v\rangle$ of the horizontal and vertical polarization of a photon can also be considered as a two-state system. Another example is the relative phase and intensity of a single photon in two arms of an interferometer. The underlying Hilbert space for the qubit is $\mathbb{C}^2$. An arbitrary orthonormal basis for $\mathbb{C}^2$ is denoted by $\{\,|0\rangle,\ |1\rangle\,\}$, where (scalar product)

$$\langle 0|0\rangle = \langle 1|1\rangle = 1, \qquad \langle 0|1\rangle = \langle 1|0\rangle = 0\,.$$

Any pure quantum state $|\psi\rangle$ (qubit) of this system can be written, up to a phase, as a *superposition* (linear combination of the states)

$$|\psi\rangle = \alpha|0\rangle + \beta|1\rangle\,, \quad |\alpha|^2 + |\beta|^2 = 1\,, \quad \alpha,\beta \in \mathbb{C}\,.$$

The classical boolean states, 0 and 1, can be represented by a fixed pair of orthonormal states of the qubit.

If the qubit represents a *mixed state* one uses a two-dimensional *density matrix* for its representation. We therefore express one qubit as

$$\rho = \frac{1}{2}(I_2 + \mathbf{n} \cdot \boldsymbol{\sigma}) \equiv \frac{1}{2}(I_2 + n_1\sigma_1 + n_2\sigma_2 + n_3\sigma_3)$$

where $\mathbf{n} \in \mathbb{R}^3$, $\mathbf{n} \cdot \mathbf{n} \equiv n_1^2 + n_2^2 + n_3^2 \le 1$, and $\boldsymbol{\sigma} = (\sigma_1, \sigma_2, \sigma_3)$ denote the Pauli spin matrices. For pure states we have $\mathbf{n} \cdot \mathbf{n} = 1$ and $\rho = |\psi\rangle\langle\psi|$.

Problem 1.   Any state (qubit) in $\mathbb{C}^2$ can be written as

$$\begin{pmatrix} \alpha \\ \beta \end{pmatrix}, \qquad \alpha, \beta \in \mathbb{C}, \quad |\alpha|^2 + |\beta|^2 = 1 \,.$$

Find a parameter representation (i) if the underlying field is the set of real numbers (ii) if the underlying field is the set of complex numbers.

Solution 1.   (i) Using $\alpha = \cos\theta$, $\beta = \sin\theta$ and the identity $\cos^2\theta + \sin^2\theta \equiv 1$ for all $\theta \in \mathbb{R}$ we have

$$\begin{pmatrix} \cos\theta \\ \sin\theta \end{pmatrix} \,.$$

(ii) We have as a representation

$$\begin{pmatrix} e^{i\phi}\cos\theta \\ \sin\theta \end{pmatrix}$$

where $\theta, \phi \in \mathbb{R}$ and $e^{i\phi}e^{-i\phi} = 1$.

Problem 2.   Consider the normalized states ($\theta_1, \theta_2 \in [0, 2\pi)$)

$$\begin{pmatrix} \cos\theta_1 \\ \sin\theta_1 \end{pmatrix}, \qquad \begin{pmatrix} \cos\theta_2 \\ \sin\theta_2 \end{pmatrix} \,.$$

Find the condition on θ_1 and θ_2 such that

$$\begin{pmatrix} \cos\theta_1 \\ \sin\theta_1 \end{pmatrix} + \begin{pmatrix} \cos\theta_2 \\ \sin\theta_2 \end{pmatrix}$$

is normalized.

Solution 2.   From the condition that the vector

$$\begin{pmatrix} \cos\theta_1 + \cos\theta_2 \\ \sin\theta_1 + \sin\theta_2 \end{pmatrix}$$

is normalized it follows that

$$(\sin\theta_1 + \sin\theta_2)^2 + (\cos\theta_1 + \cos\theta_2)^2 = 1 \,.$$

Thus we have

$$\sin\theta_1 \sin\theta_2 + \cos\theta_1 \cos\theta_2 = -\frac{1}{2} \,.$$

It follows that

$$\cos(\theta_1 - \theta_2) = -\frac{1}{2} \,.$$

Therefore, $\theta_1 - \theta_2 = 2\pi/3$ or $\theta_1 - \theta_2 = 4\pi/3$.

Problem 3. Let $\{\,|0\rangle, |1\rangle\,\}$ be an orthonormal basis in the Hilbert space $\mathbb{R}^2$. Let

$$A := |0\rangle\langle 0| + |1\rangle\langle 1|.$$

Consider the three cases

(i) $\quad |0\rangle := \begin{pmatrix} 1 \\ 0 \end{pmatrix}, \qquad |1\rangle := \begin{pmatrix} 0 \\ 1 \end{pmatrix}$

(ii) $\quad |0\rangle := \dfrac{1}{\sqrt{2}} \begin{pmatrix} 1 \\ 1 \end{pmatrix}, \qquad |1\rangle := \dfrac{1}{\sqrt{2}} \begin{pmatrix} 1 \\ -1 \end{pmatrix}$

(iii) $\quad |0\rangle := \begin{pmatrix} \cos\theta \\ \sin\theta \end{pmatrix}, \qquad |1\rangle := \begin{pmatrix} \sin\theta \\ -\cos\theta \end{pmatrix}.$

Find the matrix representation of A in these bases.

Solution 3. We find

(i) $\quad A = \begin{pmatrix} 1 & 0 \\ 0 & 0 \end{pmatrix} + \begin{pmatrix} 0 & 0 \\ 0 & 1 \end{pmatrix} = \begin{pmatrix} 1 & 0 \\ 0 & 1 \end{pmatrix}$

(ii) $\quad A = \dfrac{1}{2} \begin{pmatrix} 1 & 1 \\ 1 & 1 \end{pmatrix} + \dfrac{1}{2} \begin{pmatrix} 1 & -1 \\ -1 & 1 \end{pmatrix} = \begin{pmatrix} 1 & 0 \\ 0 & 1 \end{pmatrix}$

(iii) $\quad A = \begin{pmatrix} \cos^2\theta & \cos\theta\sin\theta \\ \cos\theta\sin\theta & \sin^2\theta \end{pmatrix} + \begin{pmatrix} \sin^2\theta & -\cos\theta\sin\theta \\ -\cos\theta\sin\theta & \cos^2\theta \end{pmatrix}$

$\qquad = \begin{pmatrix} 1 & 0 \\ 0 & 1 \end{pmatrix}.$

For all three cases $A = I_2$, where I_2 is the 2×2 unit matrix. Obviously, the third case contains the first two as special cases. This is the *completeness relation*.

Problem 4. Let $\{\,|0\rangle, |1\rangle\,\}$ be an orthonormal basis in the Hilbert space $\mathbb{C}^2$. The *NOT operation* (unitary operator) is defined as

$$|0\rangle \to |1\rangle, \qquad |1\rangle \to |0\rangle.$$

(i) Find the unitary operator U_{NOT} which implements the NOT operation with respect to the basis $\{\,|0\rangle, |1\rangle\,\}$.
(ii) Let

$$|0\rangle = \begin{pmatrix} 1 \\ 0 \end{pmatrix}, \qquad |1\rangle = \begin{pmatrix} 0 \\ 1 \end{pmatrix}.$$

Find the matrix representation of U_{NOT} for this basis.
(iii) Let

$$|0\rangle = \dfrac{1}{\sqrt{2}} \begin{pmatrix} 1 \\ 1 \end{pmatrix}, \qquad |1\rangle = \dfrac{1}{\sqrt{2}} \begin{pmatrix} 1 \\ -1 \end{pmatrix}$$

be the *Hadamard basis*. Find the matrix representation of U_{NOT} for this basis.

Solution 4.   (i) Obviously,

$$U_{NOT} = |0\rangle\langle 1| + |1\rangle\langle 0|$$

since $\langle 0|0\rangle = \langle 1|1\rangle = 1$ and $\langle 0|1\rangle = \langle 1|0\rangle = 0$.
(ii) For the standard basis we find

$$U_{NOT} = \begin{pmatrix} 0 & 1 \\ 1 & 0 \end{pmatrix}.$$

(iii) For the Hadamard basis we find

$$U_{NOT} = \begin{pmatrix} 1 & 0 \\ 0 & -1 \end{pmatrix}.$$

Thus we see that the respective matrix representations for the two bases are different.

Problem 5.   The *Walsh-Hadamard transform* is a 1-qubit operation, denoted by H, and performs the following linear transform

$$|0\rangle \to \frac{1}{\sqrt{2}}(|0\rangle + |1\rangle), \qquad |1\rangle \to \frac{1}{\sqrt{2}}(|0\rangle - |1\rangle).$$

(i) Find the unitary operator U_H which implements H with respect to the basis $\{\,|0\rangle, |1\rangle\,\}$.
(ii) Find the inverse of this operator.
(iii) Let

$$|0\rangle = \begin{pmatrix} 1 \\ 0 \end{pmatrix}, \qquad |1\rangle = \begin{pmatrix} 0 \\ 1 \end{pmatrix}$$

be the standard basis in $\mathbb{C}^2$. Find the matrix representation of U_H for this basis.
(iv) Let

$$|0\rangle = \frac{1}{\sqrt{2}}\begin{pmatrix} 1 \\ 1 \end{pmatrix}, \qquad |1\rangle = \frac{1}{\sqrt{2}}\begin{pmatrix} 1 \\ -1 \end{pmatrix}$$

be the Hadamard basis in $\mathbb{C}^2$. Find the matrix representation of U_H for this basis.

Solution 5.   (i) Obviously,

$$U_H = \frac{1}{\sqrt{2}}(|0\rangle + |1\rangle)\langle 0| + \frac{1}{\sqrt{2}}(|0\rangle - |1\rangle)\langle 1|$$

$$= \frac{1}{\sqrt{2}}|0\rangle(\langle 0| + \langle 1|) + \frac{1}{\sqrt{2}}|1\rangle(\langle 0| - \langle 1|).$$

(ii) The operator U_H is unitary and the inverse is given by $U_H^{-1} = U_H^* = U_H$, where $*$ denotes the adjoint.

(iii) For the standard basis we find

$$U_H = \frac{1}{\sqrt{2}} \begin{pmatrix} 1 & 1 \\ 1 & -1 \end{pmatrix}.$$

(iv) For the Hadamard basis we find

$$U_H = \frac{1}{\sqrt{2}} \begin{pmatrix} 1 & 1 \\ 1 & -1 \end{pmatrix}.$$

We see that the matrix representations for each of the two bases are the same.

Problem 6. The *Hadamard operator* on one qubit can be written as

$$U_H = \frac{1}{\sqrt{2}}((|0\rangle + |1\rangle)\langle 0| + (|0\rangle - |1\rangle)\langle 1|).$$

(i) Calculate the states $U_H|0\rangle$ and $U_H|1\rangle$.

(ii) Calculate $U_H U_H$.

Solution 6. (i) We obtain the normalized states

$$U_H|0\rangle = \frac{1}{\sqrt{2}}(|0\rangle + |1\rangle), \quad U_H|1\rangle = \frac{1}{\sqrt{2}}(|0\rangle - |1\rangle)$$

(ii) Since $\langle 0|0\rangle = \langle 1|1\rangle = 1$ and $\langle 0|1\rangle = \langle 1|0\rangle = 0$ we obtain

$$U_H U_H = |0\rangle\langle 0| + |1\rangle\langle 1| = I$$

where I is the identity operator (2×2 unit matrix).

Problem 7. Consider the Hilbert space $\mathbb{C}^2$ and the linear operator (2×2 matrix)

$$\Pi(\mathbf{n}) := \frac{1}{2}\left(I_2 + \sum_{j=1}^{3} n_j \sigma_j\right)$$

where $\mathbf{n} := (n_1, n_2, n_3)$ ($n_j \in \mathbb{R}$) is a unit vector, i.e., $n_1^2 + n_2^2 + n_3^2 = 1$. Here σ_1, σ_2, σ_3 are the *Pauli matrices*

$$\sigma_1 = \begin{pmatrix} 0 & 1 \\ 1 & 0 \end{pmatrix}, \quad \sigma_2 = \begin{pmatrix} 0 & -i \\ i & 0 \end{pmatrix}, \quad \sigma_3 = \begin{pmatrix} 1 & 0 \\ 0 & -1 \end{pmatrix}$$

and I_2 is the 2×2 unit matrix.
(i) Describe the properties of $\Pi(\mathbf{n})$, i.e., find $\Pi^\dagger(\mathbf{n})$, $\mathrm{tr}(\Pi(\mathbf{n}))$ and $\Pi^2(\mathbf{n})$.
(ii) Find the vector ($\phi, \theta \in \mathbb{R}$)

$$\Pi(\mathbf{n}) \begin{pmatrix} e^{i\phi} \cos\theta \\ \sin\theta \end{pmatrix}.$$

Discuss.

Solution 7.   (i) For the Pauli matrices we have $\sigma_1^\dagger = \sigma_1$, $\sigma_2^\dagger = \sigma_2$, $\sigma_3^\dagger = \sigma_3$. Thus $\Pi(\mathbf{n}) = \Pi^\dagger(\mathbf{n})$. Since $\mathrm{tr}\sigma_1 = \mathrm{tr}\sigma_2 = \mathrm{tr}\sigma_3 = 0$, $\mathrm{tr}I_2 = 2$, and the trace operation is linear, we obtain $\mathrm{tr}(\Pi(\mathbf{n})) = 1$. Since

$$\sigma_1^2 = \sigma_2^2 = \sigma_3^2 = I_2$$

and

$$[\sigma_1, \sigma_2]_+ = 0, \quad [\sigma_2, \sigma_3]_+ = 0, \quad [\sigma_3, \sigma_1]_+ = 0$$

where $[A, B]_+ := AB + BA$ denotes the *anticommutator*, the expression

$$\Pi^2(\mathbf{n}) = \frac{1}{4}\left(I_2 + \sum_{j=1}^{3} n_j\sigma_j\right)^2 = \frac{1}{4}I_2 + \frac{1}{2}\sum_{j=1}^{3} n_j\sigma_j + \frac{1}{4}\sum_{j=1}^{3}\sum_{k=1}^{3} n_j n_k \sigma_j \sigma_k$$

simplifies to

$$\Pi^2(\mathbf{n}) = \frac{1}{4}I_2 + \frac{1}{2}\sum_{j=1}^{3} n_j\sigma_j + \frac{1}{4}\sum_{j=1}^{3} n_j^2 I_2.$$

Using $n_1^2 + n_2^2 + n_3^2 = 1$ we obtain $\Pi^2(\mathbf{n}) = \Pi(\mathbf{n})$. Thus $\Pi(\mathbf{n})$ is a projection matrix.
(ii) We find

$$\Pi(\mathbf{n}) \begin{pmatrix} e^{i\phi} \cos\theta \\ \sin\theta \end{pmatrix} = \frac{1}{2}\begin{pmatrix} (1 + n_3)e^{i\phi}\cos\theta + (n_1 - in_2)\sin\theta \\ (n_1 + in_2)e^{i\phi}\cos\theta + (1 - n_3)\sin\theta \end{pmatrix}.$$

Problem 8.   The *qubit trine* is defined by the following states

$$|\psi_0\rangle = |0\rangle, \quad |\psi_1\rangle = -\frac{1}{2}|0\rangle - \frac{\sqrt{3}}{2}|1\rangle, \quad |\psi_2\rangle = -\frac{1}{2}|0\rangle + \frac{\sqrt{3}}{2}|1\rangle$$

where $\{|0\rangle, |1\rangle\}$ is an orthonormal basis. Find the probabilities

$$|\langle\psi_0|\psi_1\rangle|^2, \quad |\langle\psi_1|\psi_2\rangle|^2, \quad |\langle\psi_2|\psi_0\rangle|^2.$$

Solution 8.   Using $\langle0|0\rangle = 1$, $\langle1|1\rangle = 1$ and $\langle0|1\rangle = 0$ we find

$$|\langle\psi_0|\psi_1\rangle|^2 = \frac{1}{4}, \quad |\langle\psi_1|\psi_2\rangle|^2 = \frac{1}{4}, \quad |\langle\psi_2|\psi_0\rangle|^2 = \frac{1}{4}.$$

Problem 9. The kets $|h\rangle$ and $|v\rangle$ are states of horizontal and vertical polarization, respectively. Consider the normalized states

$$|\psi_1\rangle = -\frac{1}{2}(|h\rangle + \sqrt{3}|v\rangle)$$

$$|\psi_2\rangle = -\frac{1}{2}(|h\rangle - \sqrt{3}|v\rangle)$$

$$|\psi_3\rangle = |h\rangle$$

$$|\phi_1\rangle = \frac{1}{\sqrt{3}}(-|h\rangle + \sqrt{2}e^{-2\pi i/3}|v\rangle)$$

$$|\phi_2\rangle = \frac{1}{\sqrt{3}}(-|h\rangle + \sqrt{2}e^{+2\pi i/3}|v\rangle)$$

$$|\phi_3\rangle = \frac{1}{\sqrt{3}}(-|h\rangle + \sqrt{2}|v\rangle).$$

Give an interpretation of these states.

Solution 9. Since $\langle h|h\rangle = \langle v|v\rangle = 1$ and $\langle v|h\rangle = \langle h|v\rangle = 0$ we find

$$\langle \psi_1|\psi_2\rangle = -\frac{1}{2}, \qquad \langle \psi_1|\psi_3\rangle = -\frac{1}{2}, \qquad \langle \psi_2|\psi_3\rangle = -\frac{1}{2}.$$

Since the solution to $\cos(\alpha) = -1/2$ is given by $\alpha = 120°$ or $\alpha = 240°$ we find that that the first three states $|\psi_1\rangle$, $|\psi_2\rangle$, $|\psi_3\rangle$ correspond to states of linear polarization separated by $120°$. We find

$$\langle \phi_1|\phi_2\rangle = -\frac{i}{\sqrt{3}}.$$

The states $|\phi_1\rangle$ and $|\phi_2\rangle$ correspond to elliptic polarization and the third state $|\phi_3\rangle$ corresponds to linear polarization.

Problem 10. Let

$$|\psi\rangle = \begin{pmatrix} e^{i\phi}\cos\theta \\ \sin\theta \end{pmatrix}$$

where $\phi, \theta \in \mathbb{R}$.
(i) Find the density matrix $\rho := |\psi\rangle\langle\psi|$.
(ii) Find $\mathrm{tr}\rho$.
(iii) Find ρ^2.

Solution 10. (i) Since

$$\langle \psi| = (e^{-i\phi}\cos\theta, \sin\theta)$$

we obtain the 2×2 density matrix

$$\rho = |\psi\rangle\langle\psi| = \begin{pmatrix} \cos^2\theta & e^{i\phi}\sin\theta\cos\theta \\ e^{-i\phi}\sin\theta\cos\theta & \sin^2\theta \end{pmatrix}.$$

(ii) Since $\cos^2\theta + \sin^2\theta = 1$ we obtain from (i) that $\mathrm{tr}\rho = 1$.
(iii) Since $\langle\psi|\psi\rangle = 1$ we have

$$\rho^2 = (|\psi\rangle\langle\psi|)^2 = |\psi\rangle\langle\psi|\psi\rangle\langle\psi| = |\psi\rangle\langle\psi| = \rho.$$

Problem 11.   Given the Hamilton operator

$$\hat{H} = \hbar\omega\sigma_x.$$

(i) Find the solution

$$|\psi(t)\rangle = e^{-i\hat{H}t/\hbar}|\psi(t=0)\rangle$$

of the *Schrödinger equation*

$$i\hbar\frac{d}{dt}|\psi\rangle = \hat{H}|\psi\rangle$$

with the initial conditions

$$|\psi(t=0)\rangle = \begin{pmatrix} 1 \\ 0 \end{pmatrix}.$$

(ii) Find the probability

$$|\langle\psi(t=0)|\psi(t)\rangle|^2.$$

(iii) The solution of the *Heisenberg equation of motion*

$$i\hbar\frac{d\sigma_z}{dt} = [\sigma_z, \hat{H}](t)$$

is given by

$$\sigma_z(t) = e^{i\hat{H}t/\hbar}\sigma_z e^{-i\hat{H}t/\hbar}$$

where $\sigma_z(t=0) = \sigma_z$. Calculate $\sigma_z(t)$.
(iv) Show that

$$\langle\psi(t=0)|\sigma_z(t)|\psi(t=0)\rangle = \langle\psi(t)|\sigma_z|\psi(t)\rangle.$$

Solution 11.   (i) The solution of the Schrödinger equation is given by

$$|\psi(t)\rangle = \exp(-i\hat{H}t/\hbar)|\psi(t=0)\rangle.$$

Since $\sigma_x^2 = I_2$ we find the unitary matrix

$$\exp(-i\hat{H}t/\hbar) \equiv U(t) = \begin{pmatrix} \cos(\omega t) & -i\sin(\omega t) \\ -i\sin(\omega t) & \cos(\omega t) \end{pmatrix}.$$

Thus the normalized state at time t is

$$|\psi(t)\rangle = U(t)\begin{pmatrix} 1 \\ 0 \end{pmatrix} = \begin{pmatrix} \cos(\omega t) \\ -i\sin(\omega t) \end{pmatrix}.$$

(ii) We find the probability

$$|\langle\psi(t=0)|\psi(t)\rangle|^2 = \cos^2(\omega t).$$

(iii) Since the commutators are given by

$$[\sigma_z, \hat{H}] = \hbar\omega[\sigma_z, \sigma_x] = 2i\hbar\omega\sigma_y, \qquad [\sigma_y, \hat{H}] = \hbar\omega[\sigma_y, \sigma_x] = -2i\hbar\omega\sigma_z$$

we obtain the linear system of matrix-valued differential equations

$$\frac{d\sigma_z}{dt} = 2\omega\sigma_y(t), \qquad \frac{d\sigma_y}{dt} = -2\omega\sigma_z(t)$$

with the initial conditions $\sigma_z(t=0) = \sigma_z$ and $\sigma_y(t=0) = \sigma_y$. Here we used the Heisenberg equation of motion for σ_y to obtain the second differential equation. The solution of this system of matrix-valued linear differential equations is given by

$$\sigma_z(t) = \sigma_z\cos(2\omega t) + \sigma_y\sin(2\omega t)$$

$$\sigma_y(t) = \sigma_y\cos(2\omega t) - \sigma_z\sin(2\omega t).$$

(iv) We find

$$\langle\psi(t=0)|\sigma_z(t)|\psi(t=0)\rangle = \cos(2\omega t)$$

and

$$\langle\psi(t)|\sigma_z|\psi(t)\rangle = \cos^2(\omega t) - \sin^2(\omega t) = \cos(2\omega t).$$

Problem 12. Consider a *Mach-Zehnder interferometer* in which the beam pair spans a two-dimensional Hilbert space with orthonormal basis $\{|0\rangle, |1\rangle\}$. The state vectors $|0\rangle$ and $|1\rangle$ can be considered as orthonormal wave packets that move in two given directions defined by the geometry of the interferometer. We may represent mirrors, beam splitters and relative U_P phase shifts by the unitary matrices

$$U_M = \begin{pmatrix} 0 & 1 \\ 1 & 0 \end{pmatrix}, \quad U_B = \frac{1}{\sqrt{2}}\begin{pmatrix} 1 & 1 \\ 1 & -1 \end{pmatrix}, \quad U_P = \begin{pmatrix} e^{i\chi} & 0 \\ 0 & 1 \end{pmatrix}$$

respectively. Consider the density matrix

$$\rho_{in} = |0\rangle\langle 0|$$

where $\{ |0\rangle, |1\rangle \}$ denotes the standard basis. Using this basis find

$$\rho_{out} = U_B U_M U_P U_B \rho_{in} U_B^\dagger U_P^\dagger U_M^\dagger U_B^\dagger .$$

Give an interpretation of the result.

Solution 12.   Since

$$\rho_{in} = |0\rangle\langle 0| = \begin{pmatrix} 1 \\ 0 \end{pmatrix} (1\ 0) = \begin{pmatrix} 1 & 0 \\ 0 & 0 \end{pmatrix}$$

and

$$U_B U_M U_P U_B = \frac{1}{2} \begin{pmatrix} e^{i\chi} + 1 & e^{i\chi} - 1 \\ -e^{i\chi} + 1 & -e^{i\chi} - 1 \end{pmatrix}$$

we obtain

$$\rho_{out} = \frac{1}{2} \begin{pmatrix} 1 + \cos(\chi) & i\sin(\chi) \\ -i\sin(\chi) & 1 - \cos(\chi) \end{pmatrix} .$$

This yields the intensity along $|0\rangle$ as

$$I \propto 1 + \cos(\chi) .$$

Thus the relative U_P phase χ could be observed in the output signal of the interferometer.

Problem 13.   Let $\{ |0\rangle, |1\rangle \}$ be an orthonormal basis in $\mathbb{C}^2$.
(i) Find the commutator

$$\left[|0\rangle\langle 1|, |1\rangle\langle 0| \right] .$$

(ii) Find the operator $\exp(t|0\rangle\langle 1|)$.
(iii) Find the operator $\exp(t|1\rangle\langle 0|)$.
(iv) Find the operator $\exp(t|0\rangle\langle 1|) \exp(t|1\rangle\langle 0|)$.
(v) Find the operator $\exp(t(|0\rangle\langle 1| + |1\rangle\langle 0|))$.
(vi) Is

$$\exp(t(|0\rangle\langle 1| + |1\rangle\langle 0|)) = \exp(t|0\rangle\langle 1|) \exp(t|1\rangle\langle 0|) ?$$

Solution 13.   (i) We have

$$\left[|0\rangle\langle 1|, |1\rangle\langle 0| \right] = |0\rangle\langle 0| - |1\rangle\langle 1|$$

since $\langle 0|0\rangle = \langle 1|1\rangle = 1$ and $\langle 0|1\rangle = \langle 1|0\rangle = 0$. We see that the commutator is nonzero.

(ii) Since $\langle 0|1\rangle = \langle 1|0\rangle = 0$ we find

$$\exp(t|0\rangle\langle 1|) = \sum_{j=0}^{\infty} \frac{t^j}{j!}(|0\rangle\langle 1|)^j = I_2 + t|0\rangle\langle 1|.$$

(iii) Analogously

$$\exp(t|1\rangle\langle 0|) = \sum_{j=0}^{\infty} \frac{t^j}{j!}(|1\rangle\langle 0|)^j = I_2 + t|1\rangle\langle 0|.$$

(iv) Multiplying the results found above we obtain

$$\exp(t|0\rangle\langle 1|)\exp(t|1\rangle\langle 0|) = I_2 + t(|0\rangle\langle 1| + |1\rangle\langle 0|) + t^2|0\rangle\langle 0|.$$

(v) Since

$$(|0\rangle\langle 1| + |1\rangle\langle 0|)^2 = I_2$$

we obtain

$$\exp(t|0\rangle\langle 1| + t|1\rangle\langle 0|) = \sum_{j=0}^{\infty} \frac{t^{2j}}{(2j)!}I_2 + \sum_{j=0}^{\infty} \frac{t^{2j+1}}{(2j+1)!}(|0\rangle\langle 1| + |1\rangle\langle 0|)$$
$$= \cosh(t)I_2 + \sinh(t)(|0\rangle\langle 1| + |1\rangle\langle 0|).$$

(vi) Clearly when $t \neq 0$ we have

$$\exp(t(|0\rangle\langle 1| + |1\rangle\langle 0|)) \neq \exp(t|0\rangle\langle 1|)\exp(t|1\rangle\langle 0|).$$

Problem 14. Consider the unitary matrix for the NOT gate

$$U_{NOT} = \begin{pmatrix} 0 & 1 \\ 1 & 0 \end{pmatrix}.$$

Show that we can find a unitary matrix V such that $V^2 = U_{NOT}$. Thus V would be the square root NOT gate. What are the eigenvalues of V?

Solution 14. We find the unitary matrix

$$V = \frac{1}{2}\begin{pmatrix} 1+i & 1-i \\ 1-i & 1+i \end{pmatrix}.$$

Obviously $-V$ is also a square root. The eigenvalues of V are 1 and i. The eigenvalues of $-V$ are -1 and $-i$. Note that the eigenvalues of U_{NOT} are 1 and -1.

Problem 15.   Let $\sigma_1, \sigma_2, \sigma_3$ be the Pauli spin matrices. Let **n** be a unit vector in $\mathbb{R}^3$. We define the operator

$$\Sigma := \mathbf{n} \cdot \boldsymbol{\sigma} \equiv n_1\sigma_1 + n_2\sigma_2 + n_3\sigma_3 \, .$$

(i) Calculate Σ^2. From this result and the fact that Σ is hermitian show that Σ is unitary.
(ii) Find the eigenvalues of Σ.
(iii) Let

$$|\psi\rangle = \begin{pmatrix} 1 \\ 0 \end{pmatrix} \, .$$

Calculate the state $\Sigma|\psi\rangle$ and the probability $|\langle\psi|\Sigma|\psi\rangle|^2$.

Solution 15.   (i) Using $n_1^2 + n_2^2 + n_3^2 = 1$, $\sigma_1^2 = \sigma_2^2 = \sigma_3^2 = I_2$ and

$$\sigma_1\sigma_2 + \sigma_2\sigma_1 = 0, \quad \sigma_1\sigma_3 + \sigma_3\sigma_1 = 0, \quad \sigma_2\sigma_3 + \sigma_3\sigma_2 = 0$$

we obtain

$$\begin{aligned}
\Sigma^2 &= (n_1\sigma_1 + n_2\sigma_2 + n_3\sigma_3)^2 \\
&= (n_1^2 + n_2^2 + n_3^2)I_2 \\
&\quad + n_1n_2(\sigma_1\sigma_2 + \sigma_2\sigma_1) + n_1n_3(\sigma_3\sigma_1 + \sigma_1\sigma_3) + n_2n_3(\sigma_2\sigma_3 + \sigma_3\sigma_2) \\
&= I_2 \, .
\end{aligned}$$

Since Σ is hermitian, i.e. $\Sigma = \Sigma^*$ and $\Sigma^2 = I_2$ we find that Σ is a unitary matrix with $\Sigma = \Sigma^{-1}$.
(ii) Since Σ is hermitian and unitary the eigenvalues λ_1, λ_2 can only be ± 1. Since $\mathrm{tr}\Sigma = 0 = \lambda_1 + \lambda_2$ we obtain that the eigenvalues are $+1$ and -1.
(iii) We find

$$\Sigma|\psi\rangle = n_1 \begin{pmatrix} 0 \\ 1 \end{pmatrix} + n_2 \begin{pmatrix} 0 \\ i \end{pmatrix} + n_3 \begin{pmatrix} 1 \\ 0 \end{pmatrix} \, .$$

It follows that

$$|\langle\psi|\Sigma|\psi\rangle|^2 = n_3^2 \, .$$

Problem 16.   Let **n** be a unit vector in $\mathbb{R}^3$, $\boldsymbol{\sigma} = (\sigma_1, \sigma_2, \sigma_3)$ and

$$\mathbf{n} \cdot \boldsymbol{\sigma} := n_1\sigma_1 + n_2\sigma_2 + n_3\sigma_3 \, .$$

(i) Find the unitary matrix $\exp(i\theta\mathbf{n} \cdot \boldsymbol{\sigma})$, where $\theta \in \mathbb{R}$.
(ii) Find the state

$$\exp(i\theta\mathbf{n} \cdot \boldsymbol{\sigma}) \begin{pmatrix} 1 \\ 0 \end{pmatrix} \, .$$

Solution 16. (i) Since

$$\sigma_j \sigma_k = \delta_{jk} I_2 + i \sum_{\ell=1}^{3} \epsilon_{jk\ell} \sigma_\ell$$

where $\epsilon_{123} = \epsilon_{231} = \epsilon_{312} = 1$, $\epsilon_{321} = \epsilon_{213} = \epsilon_{132} = -1$ and 0 otherwise, we obtain

$$\exp(i\theta \mathbf{n} \cdot \boldsymbol{\sigma}) = I_2 \cos\theta + i(\mathbf{n} \cdot \boldsymbol{\sigma}) \sin\theta$$
$$= \begin{pmatrix} \cos\theta + in_3 \sin\theta & i(n_1 - in_2)\sin\theta \\ i(n_1 + in_2)\sin\theta & \cos\theta - in_3\sin\theta \end{pmatrix}.$$

Note that we could also use $(\mathbf{n} \cdot \boldsymbol{\sigma})^2 = I_2$ to find the result.
(ii) Using (i) we find

$$\exp(i\theta \mathbf{n} \cdot \boldsymbol{\sigma}) \begin{pmatrix} 1 \\ 0 \end{pmatrix} = \begin{pmatrix} \cos\theta + in_3\sin\theta \\ i(n_1 + in_2)\sin\theta \end{pmatrix}.$$

Problem 17. Consider the Pauli spin matrix σ_z and the state in $\mathbb{C}^2$

$$|\psi\rangle = \begin{pmatrix} \cos(\theta) \\ \sin(\theta) \end{pmatrix}$$

Calculate the *variance*

$$V_{\sigma_z}(\psi) := \langle\psi|\sigma_z^2|\psi\rangle - (\langle\psi|\sigma_z|\psi\rangle)^2$$

and discuss the dependence on θ.

Solution 17. Using that $\sigma_z^2 = I_2$ we have

$$V_{\sigma_z}(\psi) = \langle\psi|I_2|\psi\rangle - (\langle\psi|\sigma_z|\psi\rangle)^2$$
$$= 1 - (\cos(\theta) \quad \sin(\theta)) \begin{pmatrix} \sin(\theta) \\ \cos(\theta) \end{pmatrix}$$
$$= 1 - 2\cos(\theta)\sin(\theta).$$

For $\theta = 0$ we have $V_{\sigma_z}(\psi) = 1$. The minimum value is 0, for example at $\theta = \pi/4$. The maximum value is 2, for example at $\theta = 3\pi/4$.

Problem 18. Consider the Hamilton operator

$$\hat{H} = \hbar\omega \begin{pmatrix} 0 & \alpha \\ \alpha & 1 \end{pmatrix}$$

where $\alpha \geq 0$. Find α where the energy gap between the two energy levels is the smallest.

Solution 18.   From the eigenvalue equation we find

$$E^2 - \hbar\omega E = \hbar^2\omega^2\alpha^2 .$$

Consequently

$$E_0 = \hbar\omega(1 - \sqrt{1 + \alpha^2}), \qquad E_1 = \hbar\omega(1 + \sqrt{1 + \alpha^2}) .$$

Thus

$$E_1 - E_0 = 2\hbar\omega\sqrt{1 + \alpha^2} .$$

Therefore the shortest energy gap is for $\alpha = 0$.

Problem 19.   Consider the Hamilton operator

$$\hat{H} = \hbar\omega\sigma_z + \Delta\sigma_x$$

where $\Delta \geq 0$.
(i) Find the eigenvalues and the normalized eigenvectors of $\hat{H}$.
(ii) Use the *Cayley-Hamilton theorem* to calculate $\exp(-i\hat{H}t/\hbar)$.

Solution 19.   (i) The Hamilton operator is given by

$$\hat{H} = \begin{pmatrix} \hbar\omega & \Delta \\ \Delta & -\hbar\omega \end{pmatrix} .$$

From $\det(\hat{H} - EI_2) = 0$ we find the two eigenvalues

$$E_\pm = \pm\sqrt{\hbar^2\omega^2 + \Delta^2} .$$

We set $E := \sqrt{\hbar^2\omega^2 + \Delta^2}$. Then from the eigenvalue equation

$$\begin{pmatrix} \hbar\omega & \Delta \\ \Delta & -\hbar\omega \end{pmatrix} \begin{pmatrix} u_1 \\ u_2 \end{pmatrix} = E_+ \begin{pmatrix} u_1 \\ u_2 \end{pmatrix}$$

for the eigenvalue $E_+ = E$ we find $\Delta u_2 = (E - \hbar\omega)u_1$. Thus the eigenvector is given by

$$\begin{pmatrix} \Delta \\ E - \hbar\omega \end{pmatrix} .$$

After normalization we have

$$\frac{1}{\sqrt{\Delta^2 + (E - \hbar\omega)^2}} \begin{pmatrix} \Delta \\ E - \hbar\omega \end{pmatrix} .$$

Analogously we find for the eigenvalue $E_- = -E$ the normalized eigenvector

$$\frac{1}{\sqrt{\Delta^2 + (E + \hbar\omega)^2}} \begin{pmatrix} \Delta \\ -E - \hbar\omega \end{pmatrix}.$$

(ii) Since $E_+ \neq E_-$ and $E_+ = E$, $E_- = -E$ we have to solve the system of equations

$$e^{-iEt/\hbar} = c_0 + c_1 E, \qquad e^{iEt/\hbar} = c_0 - c_1 E$$

for c_0 and c_1. Then

$$e^{-i\hat{H}t/\hbar} = c_0 I_2 + c_1 \hat{H} = \begin{pmatrix} c_0 + c_1\hbar\omega & c_1\Delta \\ c_1\Delta & c_0 - c_1\hbar\omega \end{pmatrix}.$$

The solution of the system of equations is given by

$$c_0 = \cos(Et/\hbar)$$
$$c_1 = \frac{e^{-iEt/\hbar} - e^{iEt/\hbar}}{2E} = \frac{-i\sin(Et/\hbar)}{E}$$

Thus

$$e^{-i\hat{H}t/\hbar} = \begin{pmatrix} \cos(Et/\hbar) - i\sin(Et/\hbar)\hbar\omega/E & -i\sin(Et/\hbar)\Delta/E \\ -i\sin(Et/\hbar)\Delta/E & \cos(Et/\hbar) + i\sin(Et/\hbar)\hbar\omega/E \end{pmatrix}.$$

Obviously, $\exp(-i\hat{H}t/\hbar)$ is a unitary matrix.

Problem 20.   Consider the Pauli spin matrices σ_x, σ_y and σ_z. Can one find an $\alpha \in \mathbb{R}$ such that

$$\exp(i\alpha\sigma_z)\sigma_x \exp(-i\alpha\sigma_z) = \sigma_y ?$$

Solution 20.   We have

$$\exp(i\alpha\sigma_z)\sigma_x \exp(-i\alpha\sigma_z) = \begin{pmatrix} 0 & e^{2i\alpha} \\ e^{-2i\alpha} & 0 \end{pmatrix}.$$

Thus we have to solve

$$\exp(2i\alpha) = -i, \qquad \exp(-2i\alpha) = i.$$

For $\alpha \in [0, 2\pi)$ we obtain $\alpha = 3\pi/4$.

Problem 21.   Consider the *unary gates* (2×2 unitary matrices)

$$N = \begin{pmatrix} 0 & 1 \\ 1 & 0 \end{pmatrix}, \qquad H = \frac{1}{\sqrt{2}} \begin{pmatrix} 1 & 1 \\ 1 & -1 \end{pmatrix},$$

$$V = \begin{pmatrix} 1 & 0 \\ 0 & e^{i\pi/2} \end{pmatrix}, \quad W = \begin{pmatrix} 1 & 0 \\ 0 & e^{i\pi/4} \end{pmatrix}$$

and the normalized state

$$|\psi\rangle = \frac{1}{\sqrt{2}} \begin{pmatrix} 1 \\ 1 \end{pmatrix}.$$

Calculate the state $NHVW|\psi\rangle$ and the expectation value $\langle\psi|NHVW|\psi\rangle$.

Solution 21.   We find the unitary matrix

$$NHVW = \frac{1}{\sqrt{2}} \begin{pmatrix} 1 & -e^{i3\pi/4} \\ 1 & e^{i3\pi/4} \end{pmatrix}.$$

Thus we obtain the state

$$NHVW|\psi\rangle = \frac{1}{2} \begin{pmatrix} 1 - e^{i3\pi/4} \\ 1 + e^{i3\pi/4} \end{pmatrix}.$$

It follows that

$$\langle\psi|NHVW|\psi\rangle = \frac{1}{\sqrt{2}}.$$

Problem 22.   Let $\mathbf{n}$ and $\mathbf{m}$ be a unit vectors in $\mathbb{R}^3$, $\boldsymbol{\sigma} = (\sigma_1, \sigma_2, \sigma_3)$ and

$$\mathbf{n} \cdot \boldsymbol{\sigma} := n_1\sigma_1 + n_2\sigma_2 + n_3\sigma_3.$$

Calculate the commutator $[\mathbf{n} \cdot \boldsymbol{\sigma}, \mathbf{m} \cdot \boldsymbol{\sigma}]$.

Solution 22.   We find

$$\begin{aligned} [\mathbf{n} \cdot \boldsymbol{\sigma}, \mathbf{m} \cdot \boldsymbol{\sigma}] &= 2i((n_2 m_3 - m_2 n_3)\sigma_1 + (n_3 m_1 - m_3 n_1)\sigma_2 \\ &\quad + (n_1 m_2 - m_1 n_2)\sigma_3) \\ &= 2i(\mathbf{n} \times \mathbf{m}) \cdot \boldsymbol{\sigma} \end{aligned}$$

where $\times$ denotes the vector product. Thus the vector $\mathbf{n} \times \mathbf{m}$ is perpendicular to the plane spanned by the vectors $\mathbf{n}$ and $\mathbf{m}$.

Problem 23.   We define a linear bijection, h, between $\mathbb{R}^4$ and $\mathbf{H}(2)$, the set of complex 2×2 hermitian matrices, by

$$(t, x, y, z) \rightarrow \begin{pmatrix} t + x & y - iz \\ y + iz & t - x \end{pmatrix}.$$

We denote the matrix on the right hand side by H.

(i) Show that the matrix can be written as a linear combination of the Pauli spin matrices and the identity matrix I_2.
(ii) Find the inverse map.
(iii) Calculate the determinant of the 2×2 hermitian matrix H. Discuss.

Solution 23. (i) We have

$$H = tI_2 + x\sigma_z + y\sigma_x + z\sigma_y .$$

(ii) Consider $(a, b \in \mathbb{R})$

$$\begin{pmatrix} a & c \\ c^* & b \end{pmatrix} = \begin{pmatrix} t + x & y - iz \\ y + iz & t - x \end{pmatrix} .$$

Comparing the entries we obtain

$$t = \frac{a + b}{2}, \quad x = \frac{a - b}{2}, \quad y = \frac{c + c^*}{2}, \quad z = \frac{c^* - c}{2i} .$$

(iii) We obtain

$$\det H = t^2 - x^2 - y^2 - z^2 .$$

This is the *Lorentz metric*. Let U be a unitary 2×2 matrix. Then $\det(UHU^*) = \det(H)$.

Problem 24. Let $|\psi_1\rangle$ and $|\psi_2\rangle$ be two normalized states in a Hilbert space $\mathcal{H}$. A distance d with $0 \leq d \leq \pi/2$ can be defined as

$$\cos^2 d = |\langle \psi_1 | \psi_2 \rangle|^2 .$$

Let $\mathcal{H} = \mathbb{C}^2$ and consider the states

$$|\psi_1\rangle = \frac{1}{\sqrt{2}} \begin{pmatrix} 1 \\ 1 \end{pmatrix}, \quad |\psi_2\rangle = \frac{1}{\sqrt{2}} \begin{pmatrix} 1 \\ -1 \end{pmatrix} .$$

Find d.

Solution 24. Since $\langle \psi_1 | \psi_2 \rangle = 0$ we have $\cos^2 d = 0$ and therefore $d = \pi/2$.

Problem 25. Let ρ_1 and ρ_2 be density matrices in the same Hilbert space. The *Bures distance* between the two density matrices is defined as

$$D_B(\rho_1, \rho_2) := \sqrt{2(1 - \text{tr}((\rho_1^{1/2} \rho_2 \rho_1^{1/2})^{1/2}))} .$$

Consider the density matrices

$$\rho_1 = \begin{pmatrix} 1 & 0 \\ 0 & 0 \end{pmatrix}, \quad \rho_2 = \begin{pmatrix} 1/2 & 0 \\ 0 & 1/2 \end{pmatrix} .$$

acting in the Hilbert space $\mathbb{C}^2$. Find the Bures distance.

Solution 25. Since
$$\rho_1^{1/2} = \rho_1 = \begin{pmatrix} 1 & 0 \\ 0 & 0 \end{pmatrix}$$
we obtain
$$\rho_1^{1/2} \rho_2 \rho_1^{1/2} = \begin{pmatrix} 1/2 & 0 \\ 0 & 0 \end{pmatrix}.$$
Thus
$$D_B(\rho_1, \rho_2) = \sqrt{2(1 - 1/\sqrt{2})}.$$

Problem 26. Consider the Hilbert space $\mathbb{C}^2$. Show that
$$\Pi_S = \frac{1}{2}\begin{pmatrix} 1 & -i \\ i & 1 \end{pmatrix}, \qquad \Pi_A = \frac{1}{2}\begin{pmatrix} 1 & i \\ -i & 1 \end{pmatrix}$$
are *projection matrices* and decompose the Hilbert space into sub-Hilbert spaces.

Solution 26. We have
$$\Pi_S = \Pi_S^*, \qquad \Pi_S^2 = \Pi_S,$$
$$\Pi_A = \Pi_A^*, \qquad \Pi_A^2 = \Pi_A$$
and
$$\Pi_S + \Pi_A = I_2, \qquad \Pi_S \Pi_A = 0_2.$$
Consider the state
$$|\psi\rangle = \begin{pmatrix} e^{i\phi}\sin\theta \\ \cos\theta \end{pmatrix}.$$
Then
$$\Pi_S|\psi\rangle = \begin{pmatrix} e^{i\phi}\sin\theta - i\cos\theta \\ ie^{i\phi}\sin\theta + \cos\theta \end{pmatrix}, \qquad \Pi_A|\psi\rangle = \begin{pmatrix} e^{i\phi}\sin\theta + i\cos\theta \\ -ie^{i\phi}\sin\theta + \cos\theta \end{pmatrix}$$
with
$$\langle\psi|\Pi_A\Pi_S|\psi\rangle = 0.$$

Chapter 2

Kronecker Product and Tensor Product

Let $\mathcal{H}_1$ and $\mathcal{H}_2$ be two Hilbert spaces and $\mathcal{H}$ be a third Hilbert space defined in terms of $\mathcal{H}_1$ and $\mathcal{H}_2$ with the following specifications. For each pair of vectors f_1, f_2 in $\mathcal{H}_1$, $\mathcal{H}_2$, respectively, there are vectors in $\mathcal{H}$ denoted by $f_1 \otimes f_2$ and $g_1 \otimes g_2$, respectively such that

$$\langle f_1 \otimes f_2 | g_1 \otimes g_2 \rangle = \langle f_1 | g_1 \rangle_{\mathcal{H}_1} \langle f_2 | g_2 \rangle_{\mathcal{H}_2}$$

where $\langle f_1 | g_1 \rangle$ is the scalar product in the Hilbert space $\mathcal{H}_1$. The vector $f_1 \otimes f_2$ is called the tensor product of the vectors f_1 and f_2. The Hilbert space $\mathcal{H}$ consists of the linear combinations of the vectors $f_1 \otimes f_2$ together with the strong limits of their Cauchy sequences. We term $\mathcal{H}$ the *tensor product* of $\mathcal{H}_1$ and $\mathcal{H}_2$ and denote it by $\mathcal{H}_1 \otimes \mathcal{H}_2$. Given a basis $\{\, |\phi_i\rangle \,:\, i \in I \,\}$ in the Hilbert $\mathcal{H}_1$ and a basis $\{\, |\psi_j\rangle \,:\, j \in J \,\}$ in the Hilbert space $\mathcal{H}_2$ we can construct a basis $\{\, |\phi_i\rangle \otimes |\psi_j\rangle \,:\, i \in I,\, j \in J \,\}$ in the product Hilbert space. The tensor product is associative and distributive. If $\hat{A}_1$ and $\hat{A}_2$ are linear operators in $\mathcal{H}_1$ and $\mathcal{H}_2$, respectively, we define the operator $\hat{A}_1 \otimes \hat{A}_2$ in $\mathcal{H}_1 \otimes \mathcal{H}_2$ by the formula

$$(\hat{A}_1 \otimes \hat{A}_2)(f_1 \otimes f_2) = (\hat{A}_1 f_1) \otimes (\hat{A}_2 f_2).$$

$\hat{A}_1 \otimes \hat{A}_2$ is called the tensor product of $\hat{A}_1$ and $\hat{A}_2$. Similarly we can define the tensor product of n Hilbert spaces.

For the finite-dimensional Hilbert spaces $\mathbb{C}^n$ and $\mathbb{R}^n$ the tensor product is realized by the Kronecker product.

Problem 1.   Let $A := (a_{ij})_{ij}$ be an $m \times n$ matrix and B be an $r \times s$ matrix. The *Kronecker product* of A and B is defined as the $(m \cdot r) \times (n \cdot s)$ matrix

$$A \otimes B := \begin{pmatrix} a_{11}B & a_{12}B & \cdots & a_{1n}B \\ a_{21}B & a_{22}B & \cdots & a_{2n}B \\ \vdots & \vdots & \ddots & \vdots \\ a_{m1}B & a_{m2}B & \cdots & a_{mn}B \end{pmatrix}.$$

(i) Let

$$|\phi_1\rangle := \begin{pmatrix} 1 \\ 0 \end{pmatrix}, \qquad |\phi_2\rangle := \begin{pmatrix} 0 \\ 1 \end{pmatrix}.$$

Thus the set $\{ |\phi_1\rangle, |\phi_2\rangle \}$ forms a basis in $\mathbb{C}^2$ (the standard basis). Calculate

$$|\phi_1\rangle \otimes |\phi_1\rangle, \quad |\phi_1\rangle \otimes |\phi_2\rangle, \quad |\phi_2\rangle \otimes |\phi_1\rangle, \quad |\phi_2\rangle \otimes |\phi_2\rangle$$

and interpret the result.

(ii) Consider the Pauli matrices

$$\sigma_x := \begin{pmatrix} 0 & 1 \\ 1 & 0 \end{pmatrix}, \qquad \sigma_z := \begin{pmatrix} 1 & 0 \\ 0 & -1 \end{pmatrix}.$$

Find $\sigma_x \otimes \sigma_z$ and $\sigma_z \otimes \sigma_x$ and discuss. Both σ_x and σ_z are hermitian. Are $\sigma_x \otimes \sigma_z$ and $\sigma_z \otimes \sigma_x$ hermitian? Both σ_x and σ_z are unitary. Is $\sigma_x \otimes \sigma_z$ and $\sigma_z \otimes \sigma_x$ unitary?

Solution 1.   (i) We obtain

$$|\phi_1\rangle \otimes |\phi_1\rangle = \begin{pmatrix} 1 \\ 0 \\ 0 \\ 0 \end{pmatrix}, \qquad |\phi_1\rangle \otimes |\phi_2\rangle = \begin{pmatrix} 0 \\ 1 \\ 0 \\ 0 \end{pmatrix},$$

$$|\phi_2\rangle \otimes |\phi_1\rangle = \begin{pmatrix} 0 \\ 0 \\ 1 \\ 0 \end{pmatrix}, \qquad |\phi_2\rangle \otimes |\phi_2\rangle = \begin{pmatrix} 0 \\ 0 \\ 0 \\ 1 \end{pmatrix}.$$

Thus we find the standard basis in $\mathbb{C}^4$ from the standard basis in $\mathbb{C}^2$.

(ii) We obtain

$$\sigma_x \otimes \sigma_z = \begin{pmatrix} 0 & 0 & 1 & 0 \\ 0 & 0 & 0 & -1 \\ 1 & 0 & 0 & 0 \\ 0 & -1 & 0 & 0 \end{pmatrix}$$

and

$$\sigma_z \otimes \sigma_x = \begin{pmatrix} 0 & 1 & 0 & 0 \\ 1 & 0 & 0 & 0 \\ 0 & 0 & 0 & -1 \\ 0 & 0 & -1 & 0 \end{pmatrix}.$$

We note that $\sigma_x \otimes \sigma_z \neq \sigma_z \otimes \sigma_x$. $\sigma_x \otimes \sigma_z$ and $\sigma_z \otimes \sigma_x$ are hermitian. $\sigma_x \otimes \sigma_z$ and $\sigma_z \otimes \sigma_x$ are unitary.

Problem 2. Given the orthonormal basis

$$|\psi_1\rangle = \begin{pmatrix} e^{i\phi}\cos\theta \\ \sin\theta \end{pmatrix}, \qquad |\psi_2\rangle = \begin{pmatrix} -\sin\theta \\ e^{-i\phi}\cos\theta \end{pmatrix}$$

in the Hilbert space $\mathbb{C}^2$. Use this basis to find a basis in $\mathbb{C}^4$.

Solution 2. A basis in $\mathbb{C}^4$ is given by

$$\{\, |\psi_1\rangle \otimes |\psi_1\rangle, \quad |\psi_1\rangle \otimes |\psi_2\rangle, \quad |\psi_2\rangle \otimes |\psi_1\rangle, \quad |\psi_2\rangle \otimes |\psi_2\rangle \,\}$$

since

$$((\langle\psi_j| \otimes \langle\psi_k|)(|\psi_m\rangle \otimes |\psi_n\rangle)) = \delta_{jm}\delta_{kn}$$

where $j, k, m, n = 1, 2$.

Problem 3. A system of n-qubits can be represented by a finite-dimensional Hilbert space over the complex numbers of dimension 2^n. A state $|\psi\rangle$ of the system is a superposition of the basic states

$$|\psi\rangle = \sum_{j_1, j_2, \dots, j_n = 0}^{1} c_{j_1 j_2 \dots j_n} |j_1\rangle \otimes |j_2\rangle \otimes \cdots \otimes |j_n\rangle.$$

In a short cut notation this state is written as

$$|\psi\rangle = \sum_{j_1, j_2, \dots, j_n = 0}^{1} c_{j_1 j_2 \dots j_n} |j_1 j_2 \dots j_n\rangle.$$

Consider as a special case the state

$$|\psi\rangle = \frac{1}{2}(|0\rangle \otimes |0\rangle + |0\rangle \otimes |1\rangle + |1\rangle \otimes |0\rangle + |1\rangle \otimes |1\rangle) \equiv \frac{1}{2}(|00\rangle + |01\rangle + |10\rangle + |11\rangle)$$

in the Hilbert space $\mathcal{H} = \mathbb{C}^4$ ($n = 2$). Can this state be written as a product state?

Solution 3. Yes, the state can be written as product state. We have

$$\frac{1}{\sqrt{2}}(|0\rangle + |1\rangle) \otimes \frac{1}{\sqrt{2}}(|0\rangle + |1\rangle).$$

Problem 4.   The single-bit *Walsh-Hadamard transform* is the unitary map W given by

$$W|0\rangle = \frac{1}{\sqrt{2}}(|0\rangle + |1\rangle), \qquad W|1\rangle = \frac{1}{\sqrt{2}}(|0\rangle - |1\rangle).$$

The n-bit Walsh-Hadamard transform W_n is defined as

$$W_n := W \otimes W \otimes \cdots \otimes W \qquad (n - \text{times}).$$

Consider $n = 2$. Find the normalized state

$$W_2(|0\rangle \otimes |0\rangle).$$

Solution 4.   We have

$$W_2(|0\rangle \otimes |0\rangle) = (W \otimes W)(|0\rangle \otimes |0\rangle) = W|0\rangle \otimes W|0\rangle.$$

Thus

$$W_2(|0\rangle \otimes |0\rangle) = \frac{1}{2}((|0\rangle + |1\rangle) \otimes (|0\rangle + |1\rangle)).$$

Finally

$$W_2(|0\rangle \otimes |0\rangle) = \frac{1}{2}(|0\rangle \otimes |0\rangle + |0\rangle \otimes |1\rangle + |1\rangle \otimes |0\rangle + |1\rangle \otimes |1\rangle).$$

W_2 generates a linear combination of all states. This also applies to W_n.

Problem 5.   Let A be an $m \times m$ matrix and B be an $n \times n$ matrix. The underlying field is $\mathbb{C}$. Let I_m, I_n be the $m \times m$ and $n \times n$ unit matrix, respectively.
(i) Show that $\text{tr}(A \otimes B) = \text{tr}(A)\text{tr}(B)$.
(ii) Show that

$$\text{tr}(A \otimes I_n + I_m \otimes B) = n\text{tr}(A) + m\text{tr}(B).$$

Solution 5.   (i) The Kronecker product of the matrices $A := (a_{ij})_{ij}$ and $B := (b_{kl})_{kl}$ is defined in problem 1. Thus

$$\text{tr}(A \otimes B) = \sum_{j=1}^{m} \sum_{k=1}^{n} a_{jj}b_{kk} = \text{tr}(A)\text{tr}(B).$$

(ii) Since the trace operation is linear and $\text{tr}I_n = n$ we find

$$\text{tr}(A \otimes I_n + I_m \otimes B) = \text{tr}(A \otimes I_n) + \text{tr}(I_m \otimes B) = n\text{tr}(A) + m\text{tr}(B).$$

Problem 6.   Let A be an arbitrary $n \times n$ matrix over $\mathbb{C}$. Show that

$$\exp(A \otimes I_n) \equiv \exp(A) \otimes I_n . \tag{1}$$

Solution 6.   Using the expansion

$$\exp(A \otimes I_n) = \sum_{k=0}^{\infty} \frac{(A \otimes I_n)^k}{k!}$$

$$= I_n \otimes I_n + \frac{1}{1!}(A \otimes I_n) + \frac{1}{2!}(A \otimes I_n)^2 + \frac{1}{3!}(A \otimes I_n)^3 + \cdots$$

and

$$(A \otimes I_n)^k = A^k \otimes I_n, \qquad k \in \mathbb{N}$$

we find identity (1).

Problem 7.   Let A, B be arbitrary $n \times n$ matrices over $\mathbb{C}$. Let I_n be the $n \times n$ unit matrix. Show that

$$\exp(A \otimes I_n + I_n \otimes B) \equiv \exp(A) \otimes \exp(B) .$$

Solution 7.   The proof of this identity relies on

$$[A \otimes I_n, I_n \otimes B] = 0$$

where $[\,,\,]$ denotes the commutator and

$$(A \otimes I_n)^r (I_n \otimes B)^s \equiv (A^r \otimes I_n)(I_n \otimes B^s) \equiv A^r \otimes B^s, \qquad r, s \in \mathbb{N} .$$

Thus

$$\exp(A \otimes I_n + I_n \otimes B) = \sum_{j=0}^{\infty} \frac{(A \otimes I_n + I_n \otimes B)^j}{j!}$$

$$= \sum_{j=0}^{\infty} \sum_{k=0}^{j} \frac{1}{j!} \binom{j}{k} (A \otimes I_n)^k (I_n \otimes B)^{j-k}$$

$$= \sum_{j=0}^{\infty} \sum_{k=0}^{j} \frac{1}{j!} \binom{j}{k} (A^k \otimes B^{j-k})$$

$$= \left(\sum_{j=0}^{\infty} \frac{A^j}{j!} \right) \otimes \left(\sum_{k=0}^{\infty} \frac{B^k}{k!} \right)$$

$$= \exp(A) \otimes \exp(B) .$$

Problem 8.   Let A and B be arbitrary $n \times n$ matrices over $\mathbb{C}$. Prove or disprove the equation

$$e^{A \otimes B} = e^A \otimes e^B.$$

Solution 8.   Obviously this is not true in general. For example, let $A = B = I_n$. Then

$$e^{A \otimes B} = e^{I_{n^2}}$$

and

$$e^A \otimes e^B = e^{I_n} \otimes e^{I_n} \neq e^{I_{n^2}}.$$

Problem 9.   Let A be an $m \times m$ normal matrix and B be an $n \times n$ normal matrix. The underlying field is $\mathbb{C}$. The eigenvalues and eigenvectors of A are given by $\lambda_1, \lambda_2, \ldots, \lambda_m$ and $\mathbf{u}_1, \mathbf{u}_2, \ldots, \mathbf{u}_m$. The eigenvalues and eigenvectors of B are given by $\mu_1, \mu_2, \ldots, \mu_n$ and $\mathbf{v}_1, \mathbf{v}_2, \ldots, \mathbf{v}_n$. Let ϵ_1, ϵ_2 and ϵ_3 be real parameters. Find the eigenvalues and eigenvectors of the matrix

$$\epsilon_1 A \otimes B + \epsilon_2 A \otimes I_n + \epsilon_3 I_m \otimes B.$$

Solution 9.   Let $\mathbf{x} \in \mathbb{C}^m$ and $\mathbf{y} \in \mathbb{C}^n$. Then we have

$$(A \otimes B)(\mathbf{x} \otimes \mathbf{y}) = (A\mathbf{x}) \otimes (B\mathbf{y}),$$

$$(A \otimes I_n)(\mathbf{x} \otimes \mathbf{y}) = (A\mathbf{x}) \otimes \mathbf{y}, \qquad (I_m \otimes B)(\mathbf{x} \otimes \mathbf{y}) = \mathbf{x} \otimes (B\mathbf{y}).$$

Thus the eigenvectors of the matrix are

$$\mathbf{u}_j \otimes \mathbf{v}_k, \qquad j = 1, 2, \ldots, m \quad k = 1, 2, \ldots, n.$$

The corresponding eigenvalues are given by

$$\epsilon_1 \lambda_j \mu_k + \epsilon_2 \lambda_j + \epsilon_3 \mu_k.$$

Problem 10.   Let A, B be $n \times n$ matrices over $\mathbb{C}$. The $n \times n$ matrices form a vector space. A *scalar product* can be defined as

$$\langle A, B \rangle := \text{tr}(AB^\dagger).$$

This provides a Hilbert space. The scalar product implies a *norm*

$$\|A\|^2 = \langle A, A \rangle = \text{tr}(AA^\dagger).$$

This norm is called the *Hilbert-Schmidt norm*.

(i) Consider the *Dirac matrices*

$$
\gamma_0 := \begin{pmatrix} 1 & 0 & 0 & 0 \\ 0 & 1 & 0 & 0 \\ 0 & 0 & -1 & 0 \\ 0 & 0 & 0 & -1 \end{pmatrix}, \qquad \gamma_1 := \begin{pmatrix} 0 & 0 & 0 & 1 \\ 0 & 0 & 1 & 0 \\ 0 & -1 & 0 & 0 \\ -1 & 0 & 0 & 0 \end{pmatrix}.
$$

Calculate the scalar product $\langle \gamma_0, \gamma_1 \rangle$. Discuss.

(ii) Let U be a unitary $n \times n$ matrix. Find the scalar product $\langle UA, UB \rangle$.

(iii) Let C, D be $m \times m$ matrices over $\mathbb{C}$. Find $\langle A \otimes C, B \otimes D \rangle$.

Solution 10. (i) We find

$$
\langle \gamma_0, \gamma_1 \rangle = \mathrm{tr}(\gamma_0 \gamma_1^\dagger) = 0.
$$

Thus γ_0 and γ_1 are orthogonal to each other.

(ii) Since

$$
\mathrm{tr}(UA(UB)^\dagger) = \mathrm{tr}(UAB^\dagger U^\dagger) = \mathrm{tr}(U^\dagger UAB^\dagger) = \mathrm{tr}(AB^\dagger)
$$

where we used the *cyclic invariance* for matrices, we find that

$$
\langle UA, UB \rangle = \langle A, B \rangle.
$$

Thus the scalar product is invariant under unitary transformations.

(iii) Since

$$
\begin{aligned}
\mathrm{tr}((A \otimes C)(B \otimes D)^\dagger) &= \mathrm{tr}((A \otimes C)(B^\dagger \otimes D^\dagger)) \\
&= \mathrm{tr}((AB^\dagger) \otimes (CD^\dagger)) \\
&= \mathrm{tr}(AB^\dagger)\mathrm{tr}(CD^\dagger)
\end{aligned}
$$

we find

$$
\langle A \otimes C, B \otimes D \rangle = \langle A, B \rangle \langle C, D \rangle.
$$

Problem 11. Let T be the 4×4 matrix

$$
T := \left(I_2 \otimes I_2 + \sum_{j=1}^{3} t_j \sigma_j \otimes \sigma_j \right)
$$

where σ_j, $j = 1, 2, 3$ are the Pauli spin matrices and $-1 \le t_j \le +1$, $j = 1, 2, 3$. Find the matrix T^2.

Solution 11. We have

$$
T^2 = I_2 \otimes I_2 + 2 \sum_{j=1}^{3} t_j \sigma_j \otimes \sigma_j + \sum_{j=1}^{3} \sum_{k=1}^{3} t_j t_k \sigma_j \sigma_k \otimes \sigma_j \sigma_k.
$$

Since

$$\sigma_1\sigma_2 = i\sigma_3, \qquad \sigma_2\sigma_1 = -i\sigma_3$$
$$\sigma_2\sigma_3 = i\sigma_1, \qquad \sigma_3\sigma_2 = -i\sigma_1$$
$$\sigma_3\sigma_1 = i\sigma_2, \qquad \sigma_1\sigma_3 = -i\sigma_2$$

and $\sigma_1^2 = I_2$, $\sigma_2^2 = I_2$, $\sigma_3^2 = I_2$, we find

$$\sum_{j=1}^{3}\sum_{k=1}^{3} t_j t_k \sigma_j \sigma_k \otimes \sigma_j \sigma_k \equiv I_2 \otimes I_2 \sum_{j=1}^{3} t_j^2 - 2(t_1 t_2 \sigma_3 \otimes \sigma_3 + t_2 t_3 \sigma_1 \otimes \sigma_1 + t_3 t_1 \sigma_2 \otimes \sigma_2).$$

Therefore

$$T^2 = (I_2 \otimes I_2)\left(1 + \sum_{j=1}^{3} t_j^2\right)$$
$$+ 2(t_1 - t_2 t_3)\sigma_1 \otimes \sigma_1 + 2(t_2 - t_3 t_1)\sigma_2 \otimes \sigma_2 + 2(t_3 - t_1 t_2)\sigma_3 \otimes \sigma_3.$$

Problem 12. Let $\{\, |0\rangle, |1\rangle, \ldots, |n-1\rangle \,\}$ be an orthonormal basis in the Hilbert space $\mathbb{C}^n$. Is

$$|\psi\rangle = \frac{1}{\sqrt{n}}\left(\sum_{j=0}^{n-2} |j\rangle \otimes |j+1\rangle + |n-1\rangle \otimes |0\rangle\right)$$

independent of the chosen orthonormal basis? Prove or disprove.

Solution 12. Consider the special case $\mathbb{R}^2$. Let

$$|0\rangle = \begin{pmatrix} 1 \\ 0 \end{pmatrix}, \qquad |1\rangle = \begin{pmatrix} 0 \\ 1 \end{pmatrix}.$$

Thus

$$|\psi\rangle = \frac{1}{\sqrt{2}}\begin{pmatrix} 1 \\ 0 \end{pmatrix} \otimes \begin{pmatrix} 0 \\ 1 \end{pmatrix} + \frac{1}{\sqrt{2}}\begin{pmatrix} 0 \\ 1 \end{pmatrix} \otimes \begin{pmatrix} 1 \\ 0 \end{pmatrix} = \frac{1}{\sqrt{2}}\begin{pmatrix} 0 \\ 1 \\ 1 \\ 0 \end{pmatrix}.$$

Now let

$$|0\rangle = \frac{1}{\sqrt{2}}\begin{pmatrix} 1 \\ 1 \end{pmatrix}, \qquad |1\rangle = \frac{1}{\sqrt{2}}\begin{pmatrix} 1 \\ -1 \end{pmatrix}.$$

Then

$$|\psi\rangle = \frac{1}{\sqrt{2}}\frac{1}{2}\begin{pmatrix} 1 \\ 1 \end{pmatrix} \otimes \begin{pmatrix} 1 \\ -1 \end{pmatrix} + \frac{1}{\sqrt{2}}\frac{1}{2}\begin{pmatrix} 1 \\ -1 \end{pmatrix} \otimes \begin{pmatrix} 1 \\ 1 \end{pmatrix} = \frac{1}{\sqrt{2}}\begin{pmatrix} 1 \\ 0 \\ 0 \\ -1 \end{pmatrix}.$$

Thus, $|\psi\rangle$ depends on the chosen basis.

Problem 13. In the product Hilbert space $\mathbb{C}^2 \otimes \mathbb{C}^2 \cong \mathbb{C}^4$ the *Bell states* are given by

$$|\Phi^+\rangle = \frac{1}{\sqrt{2}}(|0\rangle \otimes |0\rangle + |1\rangle \otimes |1\rangle), \qquad |\Phi^-\rangle = \frac{1}{\sqrt{2}}(|0\rangle \otimes |0\rangle - |1\rangle \otimes |1\rangle)$$

$$|\Psi^+\rangle = \frac{1}{\sqrt{2}}(|0\rangle \otimes |1\rangle + |1\rangle \otimes |0\rangle), \qquad |\Psi^-\rangle = \frac{1}{\sqrt{2}}(|0\rangle \otimes |1\rangle - |1\rangle \otimes |0\rangle)$$

and form an orthonormal basis in $\mathbb{C}^4$. Here, $\{|0\rangle, |1\rangle\}$ is an arbitrary orthonormal basis in the Hilbert space $\mathbb{C}^2$. Let

$$|0\rangle = \begin{pmatrix} e^{i\phi}\cos\theta \\ \sin\theta \end{pmatrix}, \qquad |1\rangle = \begin{pmatrix} -e^{i\phi}\sin\theta \\ \cos\theta \end{pmatrix}.$$

(i) Find $|\Phi^+\rangle$, $|\Phi^-\rangle$, $|\Psi^+\rangle$, $|\Psi^-\rangle$ for this basis.
(ii) Consider the special case when $\phi = 0$ and $\theta = 0$.

Solution 13. (i) We obtain

$$|\Phi^+\rangle = \frac{1}{\sqrt{2}} \begin{pmatrix} e^{2i\phi} \\ 0 \\ 0 \\ 1 \end{pmatrix}, \qquad |\Phi^-\rangle = \frac{1}{\sqrt{2}} \begin{pmatrix} e^{2i\phi}\cos(2\theta) \\ e^{i\phi}\sin(2\theta) \\ e^{i\phi}\sin(2\theta) \\ -\cos(2\theta) \end{pmatrix},$$

$$|\Psi^+\rangle = \frac{1}{\sqrt{2}} \begin{pmatrix} -e^{2i\phi}\sin(2\theta) \\ e^{i\phi}\cos(2\theta) \\ e^{i\phi}\cos(2\theta) \\ \sin(2\theta) \end{pmatrix}, \qquad |\Psi^-\rangle = \frac{1}{\sqrt{2}} \begin{pmatrix} 0 \\ e^{i\phi} \\ -e^{i\phi} \\ 0 \end{pmatrix}.$$

(ii) If we choose $\phi = 0$ and $\theta = 0$ which simply means we choose the standard basis for $|0\rangle$ and $|1\rangle$ (i.e., $|0\rangle = (1\ 0)^T$ and $|1\rangle = (0\ 1)^T$), we find that the Bell states take the form

$$|\Phi^+\rangle = \frac{1}{\sqrt{2}} \begin{pmatrix} 1 \\ 0 \\ 0 \\ 1 \end{pmatrix}, \qquad |\Phi^-\rangle = \frac{1}{\sqrt{2}} \begin{pmatrix} 1 \\ 0 \\ 0 \\ -1 \end{pmatrix},$$

$$|\Psi^+\rangle = \frac{1}{\sqrt{2}} \begin{pmatrix} 0 \\ 1 \\ 1 \\ 0 \end{pmatrix}, \qquad |\Psi^-\rangle = \frac{1}{\sqrt{2}} \begin{pmatrix} 0 \\ 1 \\ -1 \\ 0 \end{pmatrix}.$$

Problem 14.   Let $\mathcal{H}_A$ and $\mathcal{H}_B$ be two p-dimensional Hilbert spaces over $\mathbb{C}$, where p is a prime number. Let

$$\{\,|0_A\rangle, |1_A\rangle, \ldots, |(p-1)_A\rangle\,\}, \qquad \{\,|0_B\rangle, |1_B\rangle, \ldots, |(p-1)_B\rangle\,\}$$

be orthonormal bases in these Hilbert spaces. We define the states

$$|\psi(a,b)\rangle := I_p \otimes X^a Z^b \frac{1}{\sqrt{p}} \sum_{j=0}^{p-1} |j_A\rangle \otimes |j_B\rangle$$

in the Hilbert space $\mathcal{H}_A \otimes \mathcal{H}_B$, where $a, b \in \{0, 1, \ldots, p-1\}$. The $p \times p$ matrices X and Z are defined as

$$X|j\rangle = |j+1 \bmod p\rangle, \qquad Z|j\rangle = \omega^j |j\rangle, \quad j = 0, 1, \ldots, p$$

with a complex primitive pth root ω of 1 and $\{\,|0\rangle, |1\rangle, \ldots, |p-1\rangle\,\}$ is the orthonormal basis given above for the Hilbert space $\mathcal{H}_B$. Calculate the states $|\psi(0,0)\rangle$ and $|\psi(1,1)\rangle$.

Solution 14.   Since

$$X^0 = Z^0 = I_p, \quad I_p |j_A\rangle = |j_A\rangle, \quad I_p |j_B\rangle = |j_B\rangle$$

we obtain

$$|\psi(0,0)\rangle = \frac{1}{\sqrt{p}} \sum_{j=0}^{p-1} |j_A\rangle \otimes |j_B\rangle\,.$$

Using

$$\begin{aligned}
(I_p \otimes XZ)(|j_A\rangle \otimes |j_B\rangle) &= |j_A\rangle \otimes (XZ|j_B\rangle) \\
&= |j_A\rangle \otimes \omega^j X |j_B\rangle \\
&= |j_A\rangle \otimes \omega^j |j_B + 1 \bmod p\rangle \\
&= \omega^j |j_A\rangle \otimes |j_B + 1 \bmod p\rangle
\end{aligned}$$

we find

$$|\psi(1,1)\rangle = \frac{1}{\sqrt{p}} \sum_{j=0}^{p-1} \omega^j |j_A\rangle \otimes |j_B + 1 \bmod p\rangle\,.$$

The states $|\psi(a,b)\rangle$ are maximally entangled states in the Hilbert space $\mathcal{H}_A \otimes \mathcal{H}_B$.

Problem 15.   Let σ_x, σ_y, σ_z be the Pauli spin matrices. We have

$$\sigma_x \sigma_y = i\sigma_z, \quad \sigma_y \sigma_z = i\sigma_x, \quad \sigma_z \sigma_x = i\sigma_y$$

and

$$\sigma_y \sigma_x = -i\sigma_z, \quad \sigma_z \sigma_y = -i\sigma_x, \quad \sigma_x \sigma_z = -i\sigma_y.$$

Calculate the commutators

$$[\sigma_x \otimes \sigma_x, \sigma_y \otimes \sigma_y], \quad [\sigma_x \otimes \sigma_x, \sigma_z \otimes \sigma_z].$$

Solution 15. We have

$$\begin{aligned}
[\sigma_x \otimes \sigma_x, \sigma_y \otimes \sigma_y] &= (\sigma_x \otimes \sigma_x)(\sigma_y \otimes \sigma_y) - (\sigma_y \otimes \sigma_y)(\sigma_x \otimes \sigma_x) \\
&= (\sigma_x \sigma_y) \otimes (\sigma_x \sigma_y) - (\sigma_y \sigma_x) \otimes (\sigma_y \sigma_x) \\
&= (i\sigma_z) \otimes (i\sigma_z) - (-i\sigma_z) \otimes (-i\sigma_z) \\
&= 0.
\end{aligned}$$

Analogously $[\sigma_x \otimes \sigma_x, \sigma_z \otimes \sigma_z] = 0$.

Problem 16. Consider the Pauli matrices

$$\sigma_x = \begin{pmatrix} 0 & 1 \\ 1 & 0 \end{pmatrix}, \quad \sigma_y = \begin{pmatrix} 0 & -i \\ i & 0 \end{pmatrix}$$

and the *GHZ state*

$$|\psi\rangle = \frac{1}{\sqrt{2}} \left(\begin{pmatrix} 1 \\ 0 \end{pmatrix} \otimes \begin{pmatrix} 1 \\ 0 \end{pmatrix} \otimes \begin{pmatrix} 1 \\ 0 \end{pmatrix} + \begin{pmatrix} 0 \\ 1 \end{pmatrix} \otimes \begin{pmatrix} 0 \\ 1 \end{pmatrix} \otimes \begin{pmatrix} 0 \\ 1 \end{pmatrix} \right).$$

(i) Show that $|\psi\rangle$ is an eigenvector of the operator $\sigma_y \otimes \sigma_y \otimes \sigma_x$. What is the eigenvalue?
(ii) Is

$$|\phi\rangle = \frac{1}{\sqrt{2}} \left(\begin{pmatrix} 1 \\ 0 \end{pmatrix} \otimes \begin{pmatrix} 1 \\ 0 \end{pmatrix} \otimes \begin{pmatrix} 1 \\ 0 \end{pmatrix} + e^{i\phi} \begin{pmatrix} 0 \\ 1 \end{pmatrix} \otimes \begin{pmatrix} 0 \\ 1 \end{pmatrix} \otimes \begin{pmatrix} 0 \\ 1 \end{pmatrix} \right)$$

an eigenvector of $\sigma_y \otimes \sigma_y \otimes \sigma_x$?

Solution 16. (i) We have

$$\sigma_x \begin{pmatrix} 1 \\ 0 \end{pmatrix} = \begin{pmatrix} 0 \\ 1 \end{pmatrix}, \quad \sigma_x \begin{pmatrix} 0 \\ 1 \end{pmatrix} = \begin{pmatrix} 1 \\ 0 \end{pmatrix}$$

and

$$\sigma_y \begin{pmatrix} 1 \\ 0 \end{pmatrix} = \begin{pmatrix} 0 \\ i \end{pmatrix}, \quad \sigma_y \begin{pmatrix} 0 \\ 1 \end{pmatrix} = \begin{pmatrix} -i \\ 0 \end{pmatrix}.$$

Thus

$$(\sigma_y \otimes \sigma_y \otimes \sigma_x)|\psi\rangle = \frac{1}{\sqrt{2}} \left(\sigma_y \begin{pmatrix} 1 \\ 0 \end{pmatrix} \otimes \sigma_y \begin{pmatrix} 1 \\ 0 \end{pmatrix} \otimes \sigma_x \begin{pmatrix} 1 \\ 0 \end{pmatrix} \right.$$

$$+ \sigma_y \begin{pmatrix} 0 \\ 1 \end{pmatrix} \otimes \sigma_y \begin{pmatrix} 0 \\ 1 \end{pmatrix} \otimes \sigma_x \begin{pmatrix} 0 \\ 1 \end{pmatrix} \Bigg)$$

$$= \frac{1}{\sqrt{2}} \left(\begin{pmatrix} 0 \\ i \end{pmatrix} \otimes \begin{pmatrix} 0 \\ i \end{pmatrix} \otimes \begin{pmatrix} 0 \\ 1 \end{pmatrix} + \begin{pmatrix} -i \\ 0 \end{pmatrix} \otimes \begin{pmatrix} -i \\ 0 \end{pmatrix} \otimes \begin{pmatrix} 1 \\ 0 \end{pmatrix} \right)$$

$$= -\frac{1}{\sqrt{2}} \left(\begin{pmatrix} 0 \\ 1 \end{pmatrix} \otimes \begin{pmatrix} 0 \\ 1 \end{pmatrix} \otimes \begin{pmatrix} 0 \\ 1 \end{pmatrix} + \begin{pmatrix} 1 \\ 0 \end{pmatrix} \otimes \begin{pmatrix} 1 \\ 0 \end{pmatrix} \otimes \begin{pmatrix} 1 \\ 0 \end{pmatrix} \right)$$

$$= -|\psi\rangle .$$

Consequently, the eigenvalue is -1.
(ii) We find using the calculation from (i)

$$(\sigma_y \otimes \sigma_y \otimes \sigma_x)|\phi\rangle = -\frac{1}{\sqrt{2}} \left(e^{i\phi} \begin{pmatrix} 1 \\ 0 \end{pmatrix} \otimes \begin{pmatrix} 1 \\ 0 \end{pmatrix} \otimes \begin{pmatrix} 1 \\ 0 \end{pmatrix} + \begin{pmatrix} 0 \\ 1 \end{pmatrix} \otimes \begin{pmatrix} 0 \\ 1 \end{pmatrix} \otimes \begin{pmatrix} 0 \\ 1 \end{pmatrix} \right).$$

Thus, in general $|\phi\rangle$ is not an eigenvector of $\sigma_y \otimes \sigma_y \otimes \sigma_x$, for example if $\phi = \pi/4$.

Problem 17.   Consider the three-qubit *GHZ state*

$$|\psi\rangle = \frac{1}{\sqrt{2}}(|0\rangle \otimes |0\rangle \otimes |0\rangle + |1\rangle \otimes |1\rangle \otimes |1\rangle)$$

with the standard basis

$$|0\rangle = \begin{pmatrix} 1 \\ 0 \end{pmatrix}, \qquad |1\rangle = \begin{pmatrix} 0 \\ 1 \end{pmatrix}.$$

Let σ_x, σ_y be the Pauli spin matrices.
(i) Calculate the expectation values

$$\langle\psi|(\sigma_y \otimes \sigma_y \otimes \sigma_y)|\psi\rangle, \quad \langle\psi|(\sigma_x \otimes \sigma_x \otimes \sigma_y)|\psi\rangle,$$

$$\langle\psi|(\sigma_x \otimes \sigma_y \otimes \sigma_x)|\psi\rangle, \quad \langle\psi|(\sigma_y \otimes \sigma_x \otimes \sigma_x)|\psi\rangle .$$

(ii) Calculate the expectation values

$$\langle\psi|(\sigma_x \otimes \sigma_y \otimes \sigma_y)|\psi\rangle, \quad \langle\psi|(\sigma_y \otimes \sigma_x \otimes \sigma_y)|\psi\rangle,$$

$$\langle\psi|(\sigma_y \otimes \sigma_y \otimes \sigma_x)|\psi\rangle, \quad \langle\psi|(\sigma_x \otimes \sigma_x \otimes \sigma_x)|\psi\rangle .$$

Solution 17.   (i) We find

$$\langle\psi|(\sigma_y \otimes \sigma_y \otimes \sigma_y)|\psi\rangle = 0, \quad \langle\psi|(\sigma_x \otimes \sigma_x \otimes \sigma_y)|\psi\rangle = 0,$$

$$\langle\psi|(\sigma_x \otimes \sigma_y \otimes \sigma_x)|\psi\rangle = 0, \quad \langle\psi|(\sigma_y \otimes \sigma_x \otimes \sigma_x)|\psi\rangle = 0.$$

(ii) We obtain

$$\langle\psi|(\sigma_x \otimes \sigma_y \otimes \sigma_y)|\psi\rangle = -1, \quad \langle\psi|(\sigma_y \otimes \sigma_x \otimes \sigma_y)|\psi\rangle = -1,$$

$$\langle\psi|(\sigma_y \otimes \sigma_y \otimes \sigma_x)|\psi\rangle = -1, \quad \langle\psi|(\sigma_x \otimes \sigma_x \otimes \sigma_x)|\psi\rangle = 1.$$

Problem 18. Consider the *Bell state*

$$|\Psi^-\rangle = \frac{1}{\sqrt{2}}\left(\begin{pmatrix}1\\0\end{pmatrix}\otimes\begin{pmatrix}0\\1\end{pmatrix} - \begin{pmatrix}0\\1\end{pmatrix}\otimes\begin{pmatrix}1\\0\end{pmatrix}\right).$$

Let **n**, **m** be unit vectors in $\mathbb{R}^3$. Calculate the expectation value

$$E(\mathbf{n}, \mathbf{m}) = \langle\Psi^-|(\mathbf{n}\cdot\boldsymbol{\sigma})\otimes(\mathbf{m}\cdot\boldsymbol{\sigma})|\Psi^-\rangle.$$

Solution 18. Straightforward calculation yields

$$E(\mathbf{n}, \mathbf{m}) = -n_1 m_1 - n_2 m_2 - n_3 m_3 = -\mathbf{n}\cdot\mathbf{m} = -|\mathbf{n}|\cdot|\mathbf{m}|\cos\theta = -\cos\theta_{\mathbf{m},\mathbf{n}}$$

since $|\mathbf{m}| = |\mathbf{n}| = 1$. We write $\cos\theta_{\mathbf{m},\mathbf{n}}$ instead of $\cos\theta$ in order to indicate that $\theta_{\mathbf{n},\mathbf{m}}$ is the angle between the quantization directions **m** and **n**.

Problem 19. Let $A_1, A_2, \ldots, A_m$ and $B_1, B_2, \ldots, B_n$ be two sets of 4×4 matrices over $\mathbb{C}$. Assume that

$$A_j B_k = B_k A_j$$

for all j, k with $j = 1, 2, \ldots, m$ and $k = 1, 2, \ldots, n$. Find two such sets of matrices using the Kronecker product of 2×2 matrices.

Solution 19. If

$$A_j = C_j \otimes I_2, \qquad B_k = I_2 \otimes D_k$$

where C_j, D_k are arbitrary 2×2 matrices we have

$$[A_j, B_k] = [C_j \otimes I_2, I_2 \otimes D_k] = C_j \otimes D_k - C_j \otimes D_k = 0_4$$

for all j, k.

Problem 20. Let A, B be $n \times n$ hermitian matrices over $\mathbb{C}$. Let

$$H = \hbar\omega(A \otimes B)$$

be a Hamilton operator, where $\otimes$ denotes the Kronecker product, $\hbar$ is the Planck constant and ω the frequency. The *Heisenberg equation of motion* for the operator $B \otimes A$ is given by

$$i\hbar \frac{d(B \otimes A)}{dt} = [B \otimes A, H](t).$$

(i) Assume that $[A, B] = 0_n$, where $[\,,\,]$ denotes the commutator and 0_n is the $n \times n$ zero matrix. Simplify the Heisenberg equation of motion using this condition.

(ii) Assume that $[A, B]_+ = 0_n$, where $[\,,\,]_+$ denotes the anticommutator. Simplify the Heisenberg equation of motion using this condition. Give an example for such matrices.

(iii) Assume that $[A, B]_+ = I_n$, where $[\,,\,]_+$ denotes the anticommutator and I_n is the $n \times n$ identity matrix. Simplify the Heisenberg equation of motion using this condition. Give an example for such matrices.

Solution 20.   (i) We have

$$\begin{aligned}
[B \otimes A, \hat{H}](t) &= \hbar\omega[B \otimes A, A \otimes B](t) \\
&= \hbar\omega((B \otimes A)(A \otimes B) - (A \otimes B)(B \otimes A))(t) \\
&= \hbar\omega((BA) \otimes (AB) - (AB) \otimes (BA))(t).
\end{aligned}$$

In general, $(AB) \otimes (BA) \neq (BA) \otimes (AB)$. If $[A, B] = 0_n$ we find

$$i\hbar \frac{d(B \otimes A)}{dt} = 0_{n^2}.$$

Thus $B \otimes A$ is a constant of motion.

(ii) From $[A, B]_+ = 0_n$ we have $AB = -BA$. Thus we also have

$$i\hbar \frac{d(B \otimes A)}{dt} = 0_{n^2}$$

in this case. An example are the Pauli spin matrices. For example, let $A = \sigma_x$ and $B = \sigma_y$. Then $[A, B]_+ = 0_2$. Another example is given by Fermi operators. They have the matrix representation $(j = 1, 2, \ldots, N)$

$$c_j^\dagger = \sigma_z \otimes \cdots \otimes \sigma_z \otimes \left(\frac{1}{2}\sigma_+\right) \otimes I_2 \otimes \cdots \otimes I_2$$

$$c_j = \sigma_z \otimes \cdots \otimes \sigma_z \otimes \left(\frac{1}{2}\sigma_-\right) \otimes I_2 \cdots \otimes I_2$$

where σ_+ and σ_- appears in the j-th place and where $\sigma_+ := \sigma_x + i\sigma_y$, $\sigma_- := \sigma_x - i\sigma_y$. We have $[c_k^\dagger, c_j]_+ = \delta_{jk}I$ and $[c_k^\dagger, c_j^\dagger]_+ = [c_k, c_j]_+ = 0$.

(iii) Since $BA = -AB + I_n$ we find

$$i\hbar \frac{d(B \otimes A)}{dt} = \hbar\omega(I_n \otimes (AB) - (AB) \otimes I_n)(t).$$

Problem 21. The *Bell states* with spin J ($J = 1/2, 1, 3/2, 2, \ldots$) are given by

$$|B_1\rangle = \frac{1}{\sqrt{2J+1}} \sum_{k=0}^{2J} |k\rangle \otimes |k\rangle$$

$$|B_2\rangle = \frac{1}{\sqrt{2J+1}} \sum_{k=0}^{2J} (-1)^k |k\rangle \otimes |k\rangle$$

$$|B_3\rangle = \frac{1}{\sqrt{2J+1}} \sum_{k=0}^{2J} |k\rangle \otimes |2J - k\rangle$$

$$|B_4\rangle = \frac{1}{\sqrt{2J+1}} \sum_{k=0}^{2J} (-1)^k |k\rangle \otimes |2J - k\rangle.$$

Show that for spin $J = 1$ the Bell states are linearly dependent.

Solution 21. For $J = 1$ the Bell states are

$$|B_1\rangle = \frac{1}{\sqrt{3}}(|0\rangle \otimes |0\rangle + |1\rangle \otimes |1\rangle + |2\rangle \otimes |2\rangle)$$

$$|B_2\rangle = \frac{1}{\sqrt{3}}(|0\rangle \otimes |0\rangle - |1\rangle \otimes 1\rangle + |2\rangle \otimes |2\rangle)$$

$$|B_3\rangle = \frac{1}{\sqrt{3}}(|0\rangle \otimes |2\rangle + |1\rangle \otimes |1\rangle + |2\rangle \otimes |0\rangle)$$

$$|B_4\rangle = \frac{1}{\sqrt{3}}(|0\rangle \otimes |2\rangle - |1\rangle \otimes |1\rangle + |2\rangle \otimes |0\rangle).$$

Let $c_1, c_2, c_3, c_4 \in \mathbb{C}$. Then the equation

$$c_1|B_1\rangle + c_2|B_2\rangle + c_3|B_3\rangle + c_4|B_4\rangle = 0$$

provides the solution

$$c_2 = -c_1, \qquad c_3 = -c_1, \qquad c_4 = c_1$$

with c_1 arbitrary. Thus the four states are not linearly independent. We also have

$$\langle B_1|B_2\rangle = \frac{1}{3}, \qquad \langle B_1|B_3\rangle = \frac{1}{3}, \qquad \langle B_1|B_4\rangle = -\frac{1}{3},$$

$$\langle B_2|B_3\rangle = -\frac{1}{3}, \qquad \langle B_2|B_4\rangle = \frac{1}{3}, \qquad \langle B_3|B_4\rangle = \frac{1}{3}.$$

Note that for $J = 1/2$ the four Bell states form an orthonormal basis in $\mathbb{C}^4$.

Problem 22.   Consider the state in $\mathbb{C}^4$

$$|\psi\rangle = \sum_{j_1,j_2=0}^{1} c_{j_1j_2}|j_1\rangle \otimes |j_2\rangle$$

where

$$|0\rangle = \begin{pmatrix} 1 \\ 0 \end{pmatrix}, \qquad |1\rangle = \begin{pmatrix} 0 \\ 1 \end{pmatrix}.$$

Let S, T be 2×2 matrices over $\mathbb{C}$

$$S = \begin{pmatrix} s_{11} & s_{12} \\ s_{21} & s_{22} \end{pmatrix}, \qquad T = \begin{pmatrix} t_{11} & t_{12} \\ t_{21} & t_{22} \end{pmatrix}$$

where $\det S = 1$, $\det T = 1$. This means that S and T are elements of the Lie group $SL(2, \mathbb{C})$. Let

$$\sum_{j_1,j_2=0}^{1} d_{j_1j_2}|j_1\rangle \otimes |j_2\rangle = (S \otimes T) \sum_{j_1,j_2=0}^{1} c_{j_1j_2}|j_1\rangle \otimes |j_2\rangle.$$

Show that

$$d_{00}d_{11} - d_{01}d_{10} = c_{00}c_{11} - c_{01}c_{10}. \tag{1}$$

Owing to (1) the quantity $c_{00}c_{11} - c_{01}c_{10}$ is called an *invariant*.

Solution 22.   Since

$$S \otimes T = \begin{pmatrix} s_{11}t_{11} & s_{11}t_{12} & s_{12}t_{11} & s_{12}t_{12} \\ s_{11}t_{21} & s_{11}t_{22} & s_{12}t_{21} & s_{12}t_{22} \\ s_{21}t_{11} & s_{21}t_{12} & s_{22}t_{11} & s_{22}t_{12} \\ s_{21}t_{21} & s_{21}t_{22} & s_{22}t_{21} & s_{22}t_{22} \end{pmatrix}$$

we have

$$(S \otimes T)|\psi\rangle = c_{00}\begin{pmatrix} s_{11}t_{11} \\ s_{11}t_{21} \\ s_{21}t_{11} \\ s_{21}t_{21} \end{pmatrix} + c_{01}\begin{pmatrix} s_{11}t_{12} \\ s_{11}t_{22} \\ s_{21}t_{12} \\ s_{21}t_{22} \end{pmatrix} + c_{10}\begin{pmatrix} s_{12}t_{11} \\ s_{12}t_{21} \\ s_{22}t_{11} \\ s_{22}t_{21} \end{pmatrix} + c_{11}\begin{pmatrix} s_{12}t_{12} \\ s_{12}t_{22} \\ s_{22}t_{12} \\ s_{22}t_{22} \end{pmatrix}.$$

Thus

$$d_{00} = c_{00}s_{11}t_{11} + c_{01}s_{11}t_{12} + c_{10}s_{12}t_{11} + c_{11}s_{12}t_{12}$$
$$d_{01} = c_{00}s_{11}t_{21} + c_{01}s_{11}t_{22} + c_{10}s_{12}t_{21} + c_{11}s_{12}t_{22}$$

$$d_{10} = c_{00}s_{21}t_{11} + c_{01}s_{21}t_{12} + c_{10}s_{22}t_{11} + c_{11}s_{22}t_{12}$$
$$d_{11} = c_{00}s_{21}t_{21} + c_{01}s_{21}t_{22} + c_{10}s_{22}t_{21} + c_{11}s_{22}t_{22}.$$

From $\det S = 1$ and $\det T = 1$ it follows that $s_{11}s_{22} - s_{12}s_{21} = 1$ and $t_{11}t_{22} - t_{12}t_{21} = 1$. Thus

$$s_{11}s_{22}t_{11}t_{22} + s_{12}s_{21}t_{12}t_{21} - s_{11}s_{22}t_{12}t_{21} - s_{12}s_{21}t_{11}t_{22} = 1.$$

Using this result we obtain (1).

Problem 23. Let I_2 be the 2×2 identity matrix and let σ_x, σ_y and σ_z be the Pauli spin matrices.
(i) Let A be an $m \times m$ matrix over $\mathbb{C}$ and B be an $n \times n$ matrix over $\mathbb{C}$. Find all solutions for m, n, A and B satisfying

$$\frac{1}{2}I_2 \otimes I_2 + \frac{1}{2}\sigma_x \otimes \sigma_x + \frac{1}{2}\sigma_y \otimes \sigma_y + \frac{1}{2}\sigma_z \otimes \sigma_z = A \otimes B.$$

(ii) Let C, D, E, and F be 2×2 matrices over $\mathbb{C}$. Find all solutions for C, D, E and F satisfying

$$\frac{1}{2}I_2 \otimes I_2 + \frac{1}{2}\sigma_x \otimes \sigma_x + \frac{1}{2}\sigma_y \otimes \sigma_y + \frac{1}{2}\sigma_z \otimes \sigma_z = C \otimes D + E \otimes F.$$

(iii) Find the set *stabilized* by

$$S = \{\, \sigma_x \otimes I_2, I_2 \otimes \sigma_x \,\}$$

i.e. find

$$\{\, \mathbf{u} \in \mathbb{C}^4 \mid \forall A \in S: \; A\mathbf{u} = \mathbf{u} \,\}.$$

Solution 23. (i) Straightforward calculation yields

$$\frac{1}{2}I_2 \otimes I_2 + \frac{1}{2}\sigma_x \otimes \sigma_x + \frac{1}{2}\sigma_y \otimes \sigma_y + \frac{1}{2}\sigma_z \otimes \sigma_z = \begin{pmatrix} 1 & 0 & 0 & 0 \\ 0 & 0 & 1 & 0 \\ 0 & 1 & 0 & 0 \\ 0 & 0 & 0 & 1 \end{pmatrix} = A \otimes B.$$

Since $A \otimes B$ is an $mn \times mn$ matrix we have $mn = 4$. For $m = n = 2$ we consider

$$A = \begin{pmatrix} a_1 & a_2 \\ a_3 & a_4 \end{pmatrix}, \qquad B = \begin{pmatrix} b_1 & b_2 \\ b_3 & b_4 \end{pmatrix}.$$

Thus we have $a_1b_1 = 1$, $a_1b_4 = 0$ and $a_4b_4 = 1$. Consequently $a_1 \neq 0$ and $b_4 \neq 0$, so that $a_1b_4 \neq 0$. Thus $m = n = 2$ does not yield a solution.

For $m = 1$ and $n = 4$ we find the solution

$$A = (a), \qquad B = \frac{1}{a}\begin{pmatrix} 1 & 0 & 0 & 0 \\ 0 & 0 & 1 & 0 \\ 0 & 1 & 0 & 0 \\ 0 & 0 & 0 & 1 \end{pmatrix}, \qquad a \in \mathbb{C}/\{0\}.$$

For $m = 4$ and $n = 1$ we find the solution

$$A = \frac{1}{b}\begin{pmatrix} 1 & 0 & 0 & 0 \\ 0 & 0 & 1 & 0 \\ 0 & 1 & 0 & 0 \\ 0 & 0 & 0 & 1 \end{pmatrix}, \qquad B = (b), \qquad b \in \mathbb{C}/\{0\}.$$

(ii) Clearly C and E, and D and F must be linearly independent, otherwise we would have the case $m = n = 2$ discussed in (a) which has no solution. Let

$$C = \begin{pmatrix} c_1 & c_2 \\ c_3 & c_4 \end{pmatrix}, \qquad E = \begin{pmatrix} e_1 & e_2 \\ e_3 & e_4 \end{pmatrix}.$$

We have to satisfy the equations

$$c_1 D + e_1 F = \begin{pmatrix} 1 & 0 \\ 0 & 0 \end{pmatrix}, \qquad c_2 D + e_2 F = \begin{pmatrix} 0 & 0 \\ 1 & 0 \end{pmatrix}$$

$$c_3 D + e_3 F = \begin{pmatrix} 0 & 1 \\ 0 & 0 \end{pmatrix}, \qquad c_4 D + e_4 F = \begin{pmatrix} 0 & 0 \\ 0 & 1 \end{pmatrix}.$$

It follows that $\{D, F\}$ should be a basis for the 2×2 matrices over $\mathbb{C}$, however the span of $\{D, F\}$ is a two dimensional space whereas the 2×2 matrices over $\mathbb{C}$ forms a four dimensional space. Thus we have a contradiction. There are no solutions.

(iii) $\mathbf{u}$ must simultaneously be an eigenvector of $\sigma_x \otimes I_2$ and $I_2 \otimes \sigma_x$ with eigenvalue 1. The eigenspace corresponding to the eigenvalue 1 from $\sigma_x \otimes I_2$ yields

$$\mathbf{u} = \alpha \begin{pmatrix} 1 \\ 1 \end{pmatrix} \otimes \begin{pmatrix} 1 \\ 0 \end{pmatrix} + \beta \begin{pmatrix} 1 \\ 1 \end{pmatrix} \otimes \begin{pmatrix} 0 \\ 1 \end{pmatrix}$$

where $\alpha, \beta \in \mathbb{C}$. We must now satisfy

$$(I_2 \otimes \sigma_x)\mathbf{u} = \alpha(I_2 \otimes \sigma_x)\begin{pmatrix} 1 \\ 1 \end{pmatrix} \otimes \begin{pmatrix} 1 \\ 0 \end{pmatrix} + \beta(I_2 \otimes \sigma_x)\begin{pmatrix} 1 \\ 1 \end{pmatrix} \otimes \begin{pmatrix} 0 \\ 1 \end{pmatrix}$$

$$= \alpha \begin{pmatrix} 1 \\ 1 \end{pmatrix} \otimes \begin{pmatrix} 0 \\ 1 \end{pmatrix} + \beta \begin{pmatrix} 1 \\ 1 \end{pmatrix} \otimes \begin{pmatrix} 1 \\ 0 \end{pmatrix}$$

$$= \mathbf{u}$$

so that $\alpha = \beta$. Thus

$$\mathbf{u} = \alpha \begin{pmatrix} 1 \\ 1 \end{pmatrix} \otimes \begin{pmatrix} 1 \\ 0 \end{pmatrix} + \alpha \begin{pmatrix} 1 \\ 1 \end{pmatrix} \otimes \begin{pmatrix} 0 \\ 1 \end{pmatrix}, \qquad \alpha \in \mathbb{C}.$$

Chapter 3

Matrix Properties

For finite-dimensional quantum systems finding the norm, eigenvalues, eigenvectors, Schmidt rank and inverse (if it exists) of square matrices is important. Let A be an $n \times n$ matrix over $\mathbb{C}$. Then we can define the sup-norm

$$\|A\| := \sup_{\|\mathbf{x}\|=1} \|A\mathbf{x}\|$$

where $\|A\mathbf{x}\|$ denotes the Euclidean norm in $\mathbb{C}^n$. The *Hilbert-Schmidt norm* of a square matrix A is defined as

$$\|A\| := (\operatorname{tr}(AA^*))^{1/2}$$

where tr denotes the trace.

Let A be an $n \times n$ matrix over $\mathbb{C}$. Then the *eigenvalue equation* is defined as $A\mathbf{x} = \lambda\mathbf{x}$, where $\lambda \in \mathbb{C}$ is the eigenvalue and $\mathbf{x} \in \mathbb{C}^n$ with $\mathbf{x} \neq \mathbf{0}$ is a corresponding eigenvector.

Furthermore, decompositions of square matrices such as the singular value decomposition, spectral decomposition, polar decomposition and Schur decomposition are necessary in quantum computing.

The most important function in quantum computing is the exponential function of a square matrix A defined by

$$\exp(A) := \sum_{j=0}^{\infty} \frac{A^j}{j!} = \lim_{k \to \infty} \left(I + \frac{A}{k} \right)^k .$$

Problem 1.   Consider the hermitian 4×4 matrix (Hamilton operator)

$$\hat{H} = \frac{\hbar\omega}{2}(\sigma_x \otimes \sigma_x - \sigma_y \otimes \sigma_y)$$

where ω is the frequency. Find the *norm* of $\hat{H}$, i.e.,

$$\|\hat{H}\| := \sup_{\|\mathbf{x}\|=1} \|\hat{H}\mathbf{x}\|, \qquad \mathbf{x} \in \mathbb{C}^4.$$

Solution 1.   There are two methods to find the norm of $\hat{H}$. In the first method we use the *Lagrange multiplier method* where the constraint $\|\mathbf{x}\| = 1$ can be written as $x_1^2 + x_2^2 + x_3^2 + x_4^2 = 1$. Since

$$\sigma_x \otimes \sigma_x = \begin{pmatrix} 0 & 0 & 0 & 1 \\ 0 & 0 & 1 & 0 \\ 0 & 1 & 0 & 0 \\ 1 & 0 & 0 & 0 \end{pmatrix}, \quad \sigma_y \otimes \sigma_y = \begin{pmatrix} 0 & 0 & 0 & -1 \\ 0 & 0 & 1 & 0 \\ 0 & 1 & 0 & 0 \\ -1 & 0 & 0 & 0 \end{pmatrix}$$

we have

$$\hat{H} = \hbar\omega \begin{pmatrix} 0 & 0 & 0 & 1 \\ 0 & 0 & 0 & 0 \\ 0 & 0 & 0 & 0 \\ 1 & 0 & 0 & 0 \end{pmatrix}.$$

Let $\mathbf{x} = (x_1, x_2, x_3, x_4)^T \in \mathbb{C}^4$. We maximize

$$f(\mathbf{x}) := \|\hat{H}\mathbf{x}\|^2 - \lambda(x_1^2 + x_2^2 + x_3^2 + x_4^2 - 1)$$

where λ is the Lagrange multiplier. To find the extrema we solve the four equations

$$\frac{\partial f}{\partial x_1} = 2\hbar^2\omega^2 x_1 - 2\lambda x_1 = 0$$

$$\frac{\partial f}{\partial x_2} = -2\lambda x_2 = 0$$

$$\frac{\partial f}{\partial x_3} = -2\lambda x_3 = 0$$

$$\frac{\partial f}{\partial x_4} = 2\hbar^2\omega^2 x_4 - 2\lambda x_4 = 0$$

together with the constraint $x_1^2 + x_2^2 + x_3^2 + x_4^2 = 1$. The four equations can be written in the matrix form

$$\begin{pmatrix} \hbar^2\omega^2 - \lambda & 0 & 0 & 0 \\ 0 & -\lambda & 0 & 0 \\ 0 & 0 & -\lambda & 0 \\ 0 & 0 & 0 & \hbar^2\omega^2 - \lambda \end{pmatrix} \begin{pmatrix} x_1 \\ x_2 \\ x_3 \\ x_4 \end{pmatrix} = \begin{pmatrix} 0 \\ 0 \\ 0 \\ 0 \end{pmatrix}.$$

If $\lambda = 0$ then $x_1 = x_4 = 0$ and $\|\hat{H}\mathbf{x}\| = 0$, which is a minimum. If $\lambda \neq 0$ then $x_2 = x_3 = 0$ and $x_1^2 + x_4^2 = 1$ so that $\|\hat{H}\mathbf{x}\| = \hbar\omega$, which is the maximum. Thus we find $\|\hat{H}\| = \hbar\omega$.

In the second method we calculate the positive definite matrix $\hat{H}^*\hat{H}$ and find the square root of the largest eigenvalue of $\hat{H}^*\hat{H}$. Since $\hat{H}^* = \hat{H}$ we find

$$\hat{H}^*\hat{H} = \hbar^2\omega^2 \begin{pmatrix} 1 & 0 & 0 & 0 \\ 0 & 0 & 0 & 0 \\ 0 & 0 & 0 & 0 \\ 0 & 0 & 0 & 1 \end{pmatrix}.$$

Thus the maximum eigenvalue is $\hbar^2\omega^2$ (twice degenerate) and $\|\hat{H}\| = \hbar\omega$.

Problem 2. Let $\hat{H}$ be a hermitian $n \times n$ matrix (Hamilton operator) with eigenvalues $E_0, E_1, \ldots, E_{n-1}$ with corresponding normalized eigenvectors $|\psi_0\rangle$, $\psi_1\rangle$, $\ldots$, $|\psi_{n-1}\rangle$. The *quantum correlation function* of two $n \times n$ hermitian matrices A and B is given by

$$Q_k(t) := \frac{1}{2}\langle\psi_k|(A(t)B - AB(t) + BA(t) - B(t)A)|\psi_k\rangle, \quad k = 0, 1, \ldots, n-1$$

where

$$A(t) = e^{i\hat{H}t/\hbar}Ae^{-i\hat{H}t/\hbar}, \qquad B(t) = e^{i\hat{H}t/\hbar}Be^{-i\hat{H}t/\hbar}.$$

Note that $Q_k(t)$ is real valued. Find $Q_k(t)$ using the properties

$$e^{i\hat{H}t/\hbar}|\psi_k\rangle = e^{iE_kt/\hbar}|\psi_k\rangle, \qquad \sum_{j=0}^{n-1} |\psi_j\rangle\langle\psi_j| = I_n.$$

Solution 2. Since

$$\langle\psi_k|A(t)B|\psi_k\rangle = \sum_{j=0}^{n-1} e^{i(E_k - E_j)t/\hbar}\langle\psi_k|A|\psi_j\rangle\langle\psi_j|B|\psi_k\rangle$$

$$\langle\psi_k|AB(t)|\psi_k\rangle = \sum_{j=0}^{n-1} e^{i(E_j - E_k)t/\hbar}\langle\psi_k|A|\psi_j\rangle\langle\psi_j|B|\psi_k\rangle$$

$$\langle\psi_k|BA(t)|\psi_k\rangle = \sum_{j=0}^{n-1} e^{i(E_j - E_k)t/\hbar}\langle\psi_k|B|\psi_j\rangle\langle\psi_j|A|\psi_k\rangle$$

$$\langle\psi_k|B(t)A|\psi_k\rangle = \sum_{j=0}^{n-1} e^{i(E_k - E_j)t/\hbar}\langle\psi_k|B|\psi_j\rangle\langle\psi_j|A|\psi_k\rangle$$

and utilizing the identity $e^{i\alpha} - e^{-i\alpha} \equiv 2i\sin(\alpha)$ yields

$$Q_k(t) = i\sum_{j=0}^{n-1} \sin((E_j - E_k)t/\hbar)\left(\langle\psi_j|A|\psi_k\rangle\langle\psi_k|B|\psi_j\rangle - \langle\psi_k|A|\psi_j\rangle\langle\psi_j|B|\psi_k\rangle\right).$$

Problem 3.   Let A and B be two $n \times n$ matrices over $\mathbb{C}$. If there exists a non-singular $n \times n$ matrix X such that

$$A = XBX^{-1}$$

then A and B are said to be *similar matrices*. Show that the spectra (eigenvalues) of two similar matrices are equal.

Solution 3.   We have

$$\det(A - \lambda I_n) = \det(XBX^{-1} - X\lambda I_n X^{-1}) = \det(X(B - \lambda I_n)X^{-1})$$
$$= \det(X)\det(B - \lambda I_n)\det(X^{-1}) = \det(B - \lambda I_n).$$

Problem 4.   Let A be an $n \times n$ matrix over $\mathbb{C}$. Then there exists an $n \times n$ unitary matrix Q, such that

$$Q^*AQ = D + N$$

where $D = \text{diag}(\lambda_1, \lambda_2, \ldots, \lambda_n)$ is the diagonal matrix composed of the eigenvalues of A and N is a strictly upper triangular matrix (i.e., N has zero entries on the diagonal). The matrix Q is said to provide a *Schur decomposition* of A. Let

$$A = \begin{pmatrix} 3 & 8 \\ -2 & 3 \end{pmatrix}, \qquad Q = \frac{1}{\sqrt{5}}\begin{pmatrix} 2i & 1 \\ -1 & -2i \end{pmatrix}.$$

Show that Q provides a Schur decomposition of A.

Solution 4.   Obviously, $Q^*Q = QQ^* = I_2$. Now

$$Q^*AQ = \frac{1}{5}\begin{pmatrix} -2i & -1 \\ 1 & 2i \end{pmatrix}\begin{pmatrix} 3 & 8 \\ -2 & 3 \end{pmatrix}\begin{pmatrix} 2i & 1 \\ -1 & -2i \end{pmatrix}$$
$$= \begin{pmatrix} 3+4i & -6 \\ 0 & 3-4i \end{pmatrix} = \begin{pmatrix} 3+4i & 0 \\ 0 & 3-4i \end{pmatrix} + \begin{pmatrix} 0 & -6 \\ 0 & 0 \end{pmatrix}.$$

Consequently, we obtained a Schur decomposition of the matrix A with the given Q.

Problem 5.   Consider a square non-singular matrix A over $\mathbb{C}$. The *polar decomposition theorem* states that A can be written as

$$A = UP$$

where U is a unitary matrix and P is a positive definite matrix. Thus P is hermitian. Show that A has a unique polar decomposition.

Solution 5.   Since A is invertible, so are A^* and A^*A. The positive square root P of A^*A is also invertible. Set $U := AP^{-1}$. Then U is invertible and

$$U^*U = P^{-1}A^*AP^{-1} = P^{-1}P^2P^{-1} = I$$

so that U is unitary. Since P is invertible, it is obvious that AP^{-1} is the only possible choice for U.

Problem 6.   Let A and B be $n \times n$ hermitian matrices. Suppose that

$$A^2 = I_n, \qquad B^2 = I_n \qquad (1)$$

and

$$[A, B]_+ \equiv AB + BA = 0_n \qquad (2)$$

where 0_n is the $n \times n$ zero matrix. Let $\mathbf{x} \in \mathbb{C}^n$ be normalized, i.e., $\|\mathbf{x}\| = 1$. Here $\mathbf{x}$ is considered as a column vector.
(i) Show that

$$(\mathbf{x}^*A\mathbf{x})^2 + (\mathbf{x}^*B\mathbf{x})^2 \le 1. \qquad (3)$$

(ii) Give an example for the matrices A and B.

Solution 6.   (i) Let $a, b \in \mathbb{R}$ and let $r^2 := a^2 + b^2$. The matrix

$$C = aA + bB$$

is again hermitian. Then

$$C^2 = a^2A^2 + abAB + baBA + b^2B^2.$$

Using the properties (1) and (2) we find

$$C^2 = a^2I_n + b^2I_n = r^2I_n.$$

Therefore $(\mathbf{x}^*C^2\mathbf{x}) = r^2$ and

$$-r \le a(\mathbf{x}^*A\mathbf{x}) + b(\mathbf{x}^*B\mathbf{x}) \le r.$$

Let

$$a = \mathbf{x}^*A\mathbf{x}, \qquad b = \mathbf{x}^*B\mathbf{x}$$

then $a^2 + b^2 \le r$ or $r^2 \le r$. This implies $r \le 1$ and $r^2 \le 1$ from which (3) follows.

(ii) An example is $A = \sigma_x$ and $B = \sigma_y$ since $\sigma_x^2 = I_2$, $\sigma_y^2 = I_2$ and $\sigma_x\sigma_y + \sigma_y\sigma_x = 0_2$.

Problem 7. Let K be an $n \times n$ skew-hermitian matrix with eigenvalues $\mu_1, \dots, \mu_n$ (counted according to multiplicity) and the corresponding normalized eigenvectors $\mathbf{u}_1, \dots, \mathbf{u}_n$, where $\mathbf{u}_j^*\mathbf{u}_k = 0$ for $k \ne j$. Then K can be written as

$$K = \sum_{j=1}^{n} \mu_j \mathbf{u}_j \mathbf{u}_j^*$$

and $\mathbf{u}_j\mathbf{u}_j^*\mathbf{u}_k\mathbf{u}_k^* = 0$ for $k \ne j$ and $j, k = 1, 2, \dots, n$. Note that the $n \times n$ matrices $\mathbf{u}_j\mathbf{u}_j^*$ are projection matrices and

$$\sum_{j=1}^{n} \mathbf{u}_j\mathbf{u}_j^* = I_n .$$

(i) Calculate $\exp(K)$.
(ii) Every $n \times n$ unitary matrix can be written as $U = \exp(K)$, where K is a skew-hermitian matrix. Find U from a given K.
(iii) Use the result from (ii) to find for a given U a possible K.
(iv) Apply the result from (ii) and (iii) to the unitary 2×2 matrix

$$U(\theta) = \begin{pmatrix} \cos\theta & \sin\theta \\ -\sin\theta & \cos\theta \end{pmatrix} .$$

(v) Apply the result from (ii) and (iii) to the 2×2 unitary matrix

$$V(\theta, \phi) = \begin{pmatrix} \cos\theta & -e^{i\phi}\sin\theta \\ e^{-i\phi}\sin\theta & \cos\theta \end{pmatrix} .$$

(vi) Every hermitian matrix H can be written as $H = iK$, where K is a skew-hermitian matrix. Find H for the examples given above.

Solution 7. (i) Using the properties of the $n \times n$ matrix $\mathbf{u}_j\mathbf{u}_j^*$ we find

$$\exp(K) = \exp\left(\sum_{j=1}^{n} \mu_j \mathbf{u}_j\mathbf{u}_j^*\right) = \sum_{j=1}^{n} e^{\mu_j}\mathbf{u}_j\mathbf{u}_j^* .$$

(ii) From $U = \exp(K)$ we find

$$U = \sum_{j=1}^{n} e^{\mu_j}\mathbf{u}_j\mathbf{u}_j^*$$

where $\mathbf{u}_j$ $(j = 1, 2, \ldots, n)$ are the normalized eigenvectors of U.
(iii) The matrix K is given by

$$K = \sum_{j=1}^{n} \ln(\lambda_j) \mathbf{u}_j \mathbf{u}_j^*$$

where λ_j $(j = 1, 2, \ldots, n)$ are the eigenvalues of U and $\mathbf{u}_j$ are the normalized eigenvectors of U. Note that the eigenvalues of U are of the form $\exp(i\alpha)$ with $\alpha \in \mathbb{R}$. Thus we have $\ln(e^{i\alpha}) = i\alpha$.
(iv) The eigenvalues of the matrix $U(\theta)$ are $e^{i\theta}$ and $e^{-i\theta}$ with the corresponding normalized eigenvectors

$$\mathbf{u}_1 = \frac{1}{\sqrt{2}} \begin{pmatrix} 1 \\ i \end{pmatrix}, \qquad \mathbf{u}_2 = \frac{1}{\sqrt{2}} \begin{pmatrix} 1 \\ -i \end{pmatrix}.$$

Thus

$$K(\theta) = \ln(e^{i\theta}) \mathbf{u}_1 \mathbf{u}_1^* + \ln(e^{i\theta}) \mathbf{u}_2 \mathbf{u}_2 = \frac{i\theta}{2} \begin{pmatrix} 1 & -i \\ i & 1 \end{pmatrix} - \frac{i\theta}{2} \begin{pmatrix} 1 & i \\ -i & 1 \end{pmatrix} = \begin{pmatrix} 0 & \theta \\ -\theta & 0 \end{pmatrix}.$$

(v) For the matrix $V(\theta, \phi)$ the eigenvalues are $e^{-i\theta}$ and $e^{i\theta}$ with the corresponding normalized eigenvectors

$$\frac{1}{\sqrt{2}} \begin{pmatrix} 1 \\ ie^{-i\phi} \end{pmatrix}, \qquad \frac{1}{\sqrt{2}} \begin{pmatrix} 1 \\ -ie^{-i\phi} \end{pmatrix}.$$

Thus

$$K(\theta, \phi) = \ln(e^{-i\theta}) \mathbf{u}_1 \mathbf{u}_1^* + \ln(e^{i\theta}) \mathbf{u}_2 \mathbf{u}_2^* = \begin{pmatrix} 0 & -\theta e^{i\phi} \\ \theta e^{-i\phi} & 0 \end{pmatrix}.$$

(vi) For $U(\theta)$ we find

$$i\theta \begin{pmatrix} 0 & 1 \\ -1 & 0 \end{pmatrix}.$$

For $V(\theta, \phi)$ we find

$$\begin{pmatrix} 0 & -i\theta e^{i\phi} \\ i\theta e^{-i\phi} & 0 \end{pmatrix}.$$

Problem 8. Let A and B be $n \times n$ hermitian matrices. Suppose that

$$A^2 = A, \qquad B^2 = B \tag{1}$$

and

$$[A, B]_+ \equiv AB + BA = 0_n \tag{2}$$

where 0_n is the $n \times n$ zero matrix. Let $\mathbf{x} \in \mathbb{C}^n$ be normalized, i.e., $\|\mathbf{x}\| = 1$. Here $\mathbf{x}$ is considered as a column vector. Show that

$$(\mathbf{x}^* A \mathbf{x})^2 + (\mathbf{x}^* B \mathbf{x})^2 \leq 1. \tag{3}$$

Solution 8. For an arbitrary $n \times n$ hermitian matrix M we have

$$0 \leq (\mathbf{x}^*(M - (\mathbf{x}^* M \mathbf{x})I_n)^2 \mathbf{x}) = (\mathbf{x}^*(M^2 - 2(\mathbf{x}^* M \mathbf{x})M + (\mathbf{x}^* M \mathbf{x})^2 I_n)\mathbf{x})$$
$$= (\mathbf{x}^* M^2 \mathbf{x}) - 2(\mathbf{x}^* M \mathbf{x})^2 + (\mathbf{x}^* M \mathbf{x})^2 = (\mathbf{x}^* M^2 \mathbf{x}) - (\mathbf{x}^* M \mathbf{x})^2.$$

Thus
$$0 \leq (\mathbf{x}^* M^2 \mathbf{x}) - (\mathbf{x}^* M \mathbf{x})^2 \quad \text{or} \quad (\mathbf{x}^* M \mathbf{x})^2 \leq (\mathbf{x}^* M^2 \mathbf{x}).$$

Thus for $A = M$ we have using (1)

$$(\mathbf{x}^* A \mathbf{x})^2 \leq \mathbf{x}^* A \mathbf{x}$$

and therefore
$$0 \leq (\mathbf{x}^* A \mathbf{x}) \leq 1.$$

Similarly
$$0 \leq (\mathbf{x}^* B \mathbf{x}) \leq 1.$$

Let $a, b \in \mathbb{R}$, $r^2 := a^2 + b^2$ and $C := aA + bB$. Then

$$C^2 = a^2 A^2 + b^2 B^2 + abAB + baBA.$$

Using (1) and (2) we arrive at $C^2 = a^2 A + b^2 B$. Thus

$$(\mathbf{x}^* C \mathbf{x})^2 \leq (\mathbf{x}^* C^2 \mathbf{x}) \leq a^2 + b^2.$$

Let
$$a := (\mathbf{x}^* A \mathbf{x}), \qquad b := (\mathbf{x}^* B \mathbf{x})$$

then
$$(\mathbf{x}^* C \mathbf{x}) = a^2 + b^2 = r^2$$

and therefore $(r^2)^2 \leq r^2$ which implies that $r^2 \leq 1$ and thus (3) follows.

Problem 9. Let $\mathcal{H}_A$ and $\mathcal{H}_B$ be two finite-dimensional Hilbert spaces. The *Schmidt rank* of a linear operator $L : \mathcal{H}_A \otimes \mathcal{H}_B \to \mathcal{H}_A \otimes \mathcal{H}_B$ over $\mathcal{H}_A \otimes \mathcal{H}_B$ is the smallest non-negative integer $\text{Sch}(L, \mathcal{H}_A, \mathcal{H}_B)$ such that L can be written as

$$L = \sum_{j=1}^{\text{Sch}(L, \mathcal{H}_A, \mathcal{H}_B)} L_{j,A} \otimes L_{j,B}$$

where $L_{j,A} : \mathcal{H}_A \to \mathcal{H}_A$ and $L_{j,B} : \mathcal{H}_B \to \mathcal{H}_B$ are linear operators.

Let $\{|0\rangle, |1\rangle\}$ denote an orthonormal basis in $\mathbb{C}^2$. Find the Schmidt rank $\mathrm{Sch}(U_{CNOT}, \mathbb{C}^2, \mathbb{C}^2)$ and $\mathrm{Sch}(U_{SWAP}, \mathbb{C}^2, \mathbb{C}^2)$ where

$$U_{CNOT} = |00\rangle\langle 00| + |01\rangle\langle 01| + |11\rangle\langle 10| + |10\rangle\langle 11|$$
$$U_{SWAP} = |00\rangle\langle 00| + |10\rangle\langle 01| + |01\rangle\langle 10| + |11\rangle\langle 11|.$$

Solution 9. We note that

$$U_{CNOT} = |0\rangle\langle 0| \otimes I_2 + |1\rangle\langle 1| \otimes U_{NOT}$$

where

$$U_{NOT} := |0\rangle\langle 1| + |1\rangle\langle 0|.$$

In other words

$$0 < \mathrm{Sch}(U_{CNOT}, \mathbb{C}^2, \mathbb{C}^2) \le 2.$$

Now suppose U_{CNOT} can be written as the product $A \otimes B$ where

$$A := a_0|0\rangle\langle 0| + a_1|0\rangle\langle 1| + a_2|1\rangle\langle 0| + a_3|1\rangle\langle 1|$$
$$B := b_0|0\rangle\langle 0| + b_1|0\rangle\langle 1| + b_2|1\rangle\langle 0| + b_3|1\rangle\langle 1|.$$

This yields the conditions $a_0 b_0 = 1$, $a_0 b_1 = 0$ and $a_3 b_1 = 1$. These equations are inconsistent, i.e.,

$$\mathrm{Sch}(U_{CNOT}, \mathbb{C}^2, \mathbb{C}^2) \ne 1.$$

Thus

$$\mathrm{Sch}(U_{CNOT}, \mathbb{C}^2, \mathbb{C}^2) = 2.$$

The operator U_{SWAP} has the eigenvalue 1 (three times) with corresponding orthonormal eigenvectors

$$\left\{ |00\rangle, |11\rangle, \frac{1}{\sqrt{2}}(|01\rangle + |10\rangle) \right\}$$

and the eigenvalue -1 with corresponding eigenvector $\frac{1}{\sqrt{2}}(|01\rangle - |10\rangle)$. Defining

$$|\phi_1\rangle := \frac{1}{\sqrt{2}}(|01\rangle + |10\rangle), \qquad |\phi_2\rangle := \frac{1}{\sqrt{2}}(|01\rangle - |10\rangle)$$

we find that

$$U_{SWAP} := |00\rangle\langle 00| + |\phi_1\rangle\langle\phi_1| - |\phi_2\rangle\langle\phi_2| + |11\rangle\langle 11|$$

where $\{|00\rangle, |\phi_1\rangle, |\phi_2\rangle, |11\rangle\}$ forms an orthonormal basis in $\mathbb{C}^4$. In this basis U_{SWAP} is the diagonal matrix

$$
U_{SWAP} = \begin{pmatrix} 1 & 0 & 0 & 0 \\ 0 & 1 & 0 & 0 \\ 0 & 0 & -1 & 0 \\ 0 & 0 & 0 & 1 \end{pmatrix}.
$$

Clearly, the matrices

$$|00\rangle\langle00|, \quad |11\rangle\langle11|, \quad |\phi_1\rangle\langle\phi_1| \text{ and } |\phi_2\rangle\langle\phi_2|$$

are linearly independent. Thus

$$\text{Sch}(U_{SWAP}, \mathbb{C}^2, \mathbb{C}^2) = 4.$$

Problem 10.   The *operator-Schmidt decomposition* of a linear operator Q acting in the product Hilbert space $\mathcal{H} = \mathcal{H}_1 \otimes \mathcal{H}_2$ of two finite-dimensional Hilbert spaces ($\dim\mathcal{H}_1 = m$, $\dim\mathcal{H}_2 = n$) with $\mathcal{H}_1 = \mathbb{C}^m$ and $\mathcal{H}_2 = \mathbb{C}^n$ can be constructed as follows. Let X, Y be $d \times d$ matrices over $\mathbb{C}$. Then we can define a scalar product or inner product $\langle X, Y \rangle := \text{tr}(XY^\dagger)$. Using this inner product we can define an orthonormal set of $d \times d$ matrices $\{ X_j : j = 1, 2, \dots, d^2 \}$ which satisfies the condition

$$\langle X_j, X_k \rangle = \text{tr}(X_j X_k^\dagger) = \delta_{jk}.$$

Thus we can write the matrix Q as

$$
Q = \sum_{j=1}^{m^2} \sum_{k=1}^{n^2} c_{jk} A_j \otimes B_k
$$

where $\{ A_j : j = 1, 2, \dots, m^2 \}$ and $\{ B_k : k = 1, 2, \dots, n^2 \}$ are fixed orthonormal bases of $m \times m$ and $n \times n$ matrices in the Hilbert spaces $\mathbb{C}^m$ and $\mathbb{C}^n$ respectively, and c_{jk} are complex coefficients. Thus $C = (c_{jk})$, with $j = 1, 2, \dots, m^2$ and $k = 1, 2, \dots, n^2$ is an $m^2 \times n^2$ matrix. The *singular value decomposition theorem* states that the matrix C can be written as

$$C = U\Sigma V^\dagger$$

where U is an $m^2 \times m^2$ unitary matrix, V is an $n^2 \times n^2$ unitary matrix and Σ is an $m^2 \times n^2$ diagonal matrix. The matrix Σ is of the form

$$
\Sigma = \begin{pmatrix} s_1 & \cdots & 0 \\ \vdots & \ddots & \vdots \\ 0 & \cdots & s_{n^2} \\ 0 & \cdots & 0 \\ \vdots & \ddots & \vdots \\ 0 & \cdots & 0 \end{pmatrix}.
$$

It is assumed that C, U and V are calculated in orthonormal bases, for example the standard basis. Thus we obtain

$$Q = \sum_{j=1}^{m^2} \sum_{k=1}^{n^2} \sum_{\ell=1}^{n^2} U_{j\ell} s_\ell V_{\ell k} A_j \otimes B_k$$

where s_ℓ is the ℓ-th diagonal entry of the $m^2 \times n^2$ diagonal matrix Σ. Defining

$$H_\ell := \sum_{j=1}^{m^2} U_{j\ell} A_j, \qquad K_\ell := \sum_{k=1}^{n^2} V_{\ell k} B_k$$

where $\ell = 1, 2, \ldots, n^2$ we find the operator-Schmidt decomposition

$$Q = \sum_{\ell=1}^{n^2} s_\ell H_\ell \otimes K_\ell.$$

(i) Consider the *CNOT gate*

$$U_{CNOT} = \begin{pmatrix} 1 & 0 & 0 & 0 \\ 0 & 1 & 0 & 0 \\ 0 & 0 & 0 & 1 \\ 0 & 0 & 1 & 0 \end{pmatrix}.$$

Find the operator-Schmidt decomposition of U_{CNOT}.
(ii) Consider the SWAP operator

$$U_{SWAP} = \begin{pmatrix} 1 & 0 & 0 & 0 \\ 0 & 0 & 1 & 0 \\ 0 & 1 & 0 & 0 \\ 0 & 0 & 0 & 1 \end{pmatrix}.$$

Find the operator-Schmidt decomposition of U_{SWAP}.
(iii) Let

$$Z := \left(\sqrt{1-p} I_2 \otimes I_2 + i\sqrt{p}\sigma_x \otimes \sigma_x \right) \left(\sqrt{1-p} I_2 \otimes I_2 + i\sqrt{p}\sigma_z \otimes \sigma_z \right)$$

where σ_x, σ_y and σ_z are the Pauli spin matrices. Find the operator-Schmidt decomposition of Z.

Solution 10. (i) We have

$$U_{CNOT} = \begin{pmatrix} 1 & 0 \\ 0 & 0 \end{pmatrix} \otimes \begin{pmatrix} 1 & 0 \\ 0 & 1 \end{pmatrix} + \begin{pmatrix} 0 & 0 \\ 0 & 1 \end{pmatrix} \otimes \begin{pmatrix} 0 & 1 \\ 1 & 0 \end{pmatrix}$$

$$= \begin{pmatrix} 1 & 0 \\ 0 & 0 \end{pmatrix} \otimes I_2 + \begin{pmatrix} 0 & 0 \\ 0 & 1 \end{pmatrix} \otimes \sigma_x.$$

(ii) We have

$$U_{SWAP} = \frac{1}{2}(I_2 \otimes I_2 + \sigma_x \otimes \sigma_x + \sigma_y \otimes \sigma_y + \sigma_z \otimes \sigma_z).$$

(iii) We have

$$Z = (1-p)I_2 \otimes I_2 + p\sigma_y \otimes \sigma_y + \sqrt{p(1-p)} \left[\left(e^{i\pi/4}\sigma_x\right) \otimes \sigma_x + \left(e^{i\pi/4}\sigma_z\right) \otimes \sigma_z \right].$$

Problem 11.   Let A, B be $n \times n$ matrices over $\mathbb{C}$. Assume that

$$[A, [A, B]] = [B, [A, B]] = 0. \tag{1}$$

Show that

$$e^{A+B} = e^A e^B e^{-\frac{1}{2}[A,B]} \tag{2a}$$

$$e^{A+B} = e^B e^A e^{+\frac{1}{2}[A,B]}. \tag{2b}$$

Hint. Use the *technique of parameter differentiation*. Consider the matrix-valued function

$$f(\epsilon) = e^{\epsilon A} e^{\epsilon B}$$

where ϵ is a real parameter and calculate the derivative $df/d\epsilon$.

Solution 11.   If we differentiate $f(\epsilon)$ with respect to ϵ we find

$$\frac{df}{d\epsilon} = Ae^{\epsilon A}e^{\epsilon B} + e^{\epsilon A}e^{\epsilon B}B = (A + e^{\epsilon A}Be^{-\epsilon A})f(\epsilon)$$

since $e^{\epsilon A}e^{-\epsilon A} = I_n$. Owing to (1) we have

$$e^{\epsilon A}Be^{-\epsilon A} = B + \epsilon[A, B].$$

Thus we obtain the linear matrix-valued differential equation

$$\frac{df}{d\epsilon} = ((A + B) + \epsilon[A, B])f(\epsilon).$$

Since the matrix $A + B$ commutes with $[A, B]$ we may treat $A + B$ and $[A, B]$ as ordinary commuting variables and integrate this linear differential equation with the initial conditions $f(0) = I_n$. We find

$$f(\epsilon) = e^{\epsilon(A+B)+(\epsilon^2/2)[A,B]} = e^{\epsilon(A+B)}e^{(\epsilon^2/2)[A,B]}$$

since $A + B$ commutes with $[A, B]$. If we set $\epsilon = 1$ and multiply both sides by $e^{-[A,B]/2}$ then (2a) follows. Likewise we can prove the second form of the identity (2b).

Problem 12. Let A be an $n \times n$ matrix. Assume that the inverse matrix of A exists. The inverse matrix can be calculated as follows (*Csanky's algorithm*). Let

$$p(x) := \det(xI_n - A) \tag{1}$$

where I_n is the $n \times n$ unit matrix. The roots are, by definition, the eigenvalues $\lambda_1, \lambda_2, \dots, \lambda_n$ of A. We write

$$p(x) = x^n + c_1 x^{n-1} + \dots + c_{n-1} x + c_n \tag{2}$$

where

$$c_n = (-1)^n \det A.$$

Since A is nonsingular we have $c_n \neq 0$ and vice versa. The *Cayley-Hamilton theorem* states that

$$p(A) = A^n + c_1 A^{n-1} + \dots + c_{n-1} A + c_n I_n = 0_n. \tag{3}$$

Multiplying this equation with A^{-1} we obtain

$$A^{-1} = \frac{1}{-c_n} \left(A^{n-1} + c_1 A^{n-2} + \dots + c_{n-1} I_n \right). \tag{4}$$

If we have the coefficients c_j we can calculate the inverse matrix A. Let

$$s_k := \sum_{j=1}^{n} \lambda_j^k.$$

Then the s_j and c_j satisfy the following $n \times n$ lower triangular system of linear equations

$$\begin{pmatrix} 1 & 0 & 0 & \dots & 0 \\ s_1 & 2 & 0 & \dots & 0 \\ s_2 & s_1 & 3 & \dots & 0 \\ \vdots & \vdots & \vdots & \ddots & \vdots \\ s_{n-1} & s_{n-2} & \dots & s_1 & n \end{pmatrix} \begin{pmatrix} c_1 \\ c_2 \\ c_3 \\ \vdots \\ c_n \end{pmatrix} = \begin{pmatrix} -s_1 \\ -s_2 \\ -s_3 \\ \vdots \\ -s_n \end{pmatrix}.$$

Since

$$\operatorname{tr}(A^k) = \lambda_1^k + \lambda_2^k + \dots + \lambda_n^k = s_k$$

we find s_k for $k = 1, 2, \dots, n$. Thus we can solve the linear equation for c_j. Finally, using (4) we obtain the inverse matrix of A. Apply Csanky's algorithm to the 4×4 permutation matrix

$$U = \begin{pmatrix} 0 & 1 & 0 & 0 \\ 0 & 0 & 1 & 0 \\ 0 & 0 & 0 & 1 \\ 1 & 0 & 0 & 0 \end{pmatrix}.$$

Solution 12.   Since

$$
U^2 = \begin{pmatrix} 0 & 0 & 1 & 0 \\ 0 & 0 & 0 & 1 \\ 1 & 0 & 0 & 0 \\ 0 & 1 & 0 & 0 \end{pmatrix}, \quad U^3 = \begin{pmatrix} 0 & 0 & 0 & 1 \\ 1 & 0 & 0 & 0 \\ 0 & 1 & 0 & 0 \\ 0 & 0 & 1 & 0 \end{pmatrix}
$$

and $U^4 = I_4$ we find

$$\mathrm{tr}U = 0 = s_1, \quad \mathrm{tr}U^2 = 0 = s_2, \quad \mathrm{tr}U^3 = 0 = s_3, \quad \mathrm{tr}U^4 = 4 = s_4.$$

We obtain the system of linear equations

$$
\begin{pmatrix} 1 & 0 & 0 & 0 \\ 0 & 2 & 0 & 0 \\ 0 & 0 & 3 & 0 \\ 0 & 0 & 0 & 4 \end{pmatrix} \begin{pmatrix} c_1 \\ c_2 \\ c_3 \\ c_4 \end{pmatrix} = \begin{pmatrix} 0 \\ 0 \\ 0 \\ -4 \end{pmatrix}
$$

with the solution $c_1 = 0$, $c_2 = 0$, $c_3 = 0$, $c_4 = -1$. Thus the inverse matrix of U is given by

$$
U^{-1} = U^3 = \begin{pmatrix} 0 & 0 & 0 & 1 \\ 1 & 0 & 0 & 0 \\ 0 & 1 & 0 & 0 \\ 0 & 0 & 1 & 0 \end{pmatrix}.
$$

Problem 13.   Let

$$
J^+ := \begin{pmatrix} 0 & 1 \\ 0 & 0 \end{pmatrix}, \quad J^- := \begin{pmatrix} 0 & 0 \\ 1 & 0 \end{pmatrix}, \quad J_3 := \frac{1}{2}\begin{pmatrix} 1 & 0 \\ 0 & -1 \end{pmatrix}.
$$

(i) Let $\epsilon \in \mathbb{R}$. Find

$$e^{\epsilon J^+}, \quad e^{\epsilon J^-}, \quad e^{\epsilon(J^+ + J^-)}.$$

(ii) Let $r \in \mathbb{R}$. Show that

$$e^{r(J^+ + J^-)} \equiv e^{J^- \tanh(r)} e^{2J_3 \ln(\cosh(r))} e^{J^+ \tanh(r)}.$$

Solution 13.   (i) Using the expansion for an $n \times n$ matrix A

$$\exp(\epsilon A) = \sum_{j=0}^{\infty} \frac{\epsilon^j A^j}{j!}$$

we find

$$
e^{\epsilon J^+} = \begin{pmatrix} 1 & 0 \\ 0 & 1 \end{pmatrix} + \epsilon \begin{pmatrix} 0 & 1 \\ 0 & 0 \end{pmatrix}, \quad e^{\epsilon J^-} = \begin{pmatrix} 1 & 0 \\ 0 & 1 \end{pmatrix} + \epsilon \begin{pmatrix} 0 & 0 \\ 1 & 0 \end{pmatrix}
$$

and

$$e^{\epsilon(J^+ + J^-)} = \begin{pmatrix} 1 & 0 \\ 0 & 1 \end{pmatrix} \cosh(\epsilon) + \begin{pmatrix} 0 & 1 \\ 1 & 0 \end{pmatrix} \sinh(\epsilon).$$

(ii) Since

$$e^{2J_3 \ln(\cosh(r))} = \begin{pmatrix} \cosh(r) & 0 \\ 0 & 1/\cosh(r) \end{pmatrix},$$

$1/\cosh(r) + \tanh(r)\sinh(r) \equiv \cosh(r)$ and using the results from (i) we find the identity.

Problem 14. The *Heisenberg commutation relation* ($\hbar = 1$) can be written as

$$[\hat{p}, \hat{q}] = -iI$$

where $\hat{p} := -i\partial/\partial q$ and I is the identity operator. Let $\alpha, \beta \in \mathbb{R}$ and

$$U(\alpha) = \exp(i\alpha\hat{p}), \qquad V(\beta) = \exp(i\beta\hat{q}).$$

Then using the *Baker-Campbell-Hausdorff formula* we find

$$U(\alpha)V(\beta) = \exp(i\alpha\beta)V(\beta)U(\alpha).$$

This is called the *Weyl representation* of Heisenberg's commutation relation. Can we find finite-dimensional $n \times n$ unitary matrices U ($U \neq I_n$) and V ($V \neq I_n$) such that

$$UV = \zeta VU$$

with $\zeta \in \mathbb{C}$, $\zeta^n = 1$?

Solution 14. Such matrices can be found, namely the permutation matrix

$$U := \begin{pmatrix} 0 & 1 & 0 & \dots & 0 \\ 0 & 0 & 1 & \dots & 0 \\ \vdots & \vdots & \vdots & \ddots & \vdots \\ 0 & 0 & 0 & \dots & 1 \\ 1 & 0 & 0 & \dots & 0 \end{pmatrix}$$

and the diagonal matrix

$$V := \begin{pmatrix} 1 & 0 & 0 & \dots & 0 \\ 0 & \zeta & 0 & \dots & 0 \\ 0 & 0 & \zeta^2 & \dots & 0 \\ \vdots & \vdots & \vdots & \ddots & \vdots \\ 0 & 0 & 0 & \dots & \zeta^{n-1} \end{pmatrix}.$$

Problem 15.   Let U be the $n \times n$ unitary matrix

$$
U := \begin{pmatrix} 0 & 1 & 0 & \ldots & 0 \\ 0 & 0 & 1 & \ldots & 0 \\ \vdots & \vdots & \vdots & \ddots & \vdots \\ 0 & 0 & 0 & \ldots & 1 \\ 1 & 0 & 0 & \ldots & 0 \end{pmatrix}
$$

and V be the $n \times n$ unitary diagonal matrix $(\zeta \in \mathbb{C})$

$$
V := \begin{pmatrix} 1 & 0 & 0 & \ldots & 0 \\ 0 & \zeta & 0 & \ldots & 0 \\ 0 & 0 & \zeta^2 & \ldots & 0 \\ \vdots & \vdots & \vdots & \ddots & \vdots \\ 0 & 0 & 0 & \ldots & \zeta^{n-1} \end{pmatrix}
$$

where $\zeta^n = 1$ $(\zeta \neq 1)$. Then the set of matrices

$$
\{ \, U^j V^k \; : \; j, k = 0, 1, 2, \ldots, n-1 \, \}
$$

provide a basis in the Hilbert space for all $n \times n$ matrices with the *scalar product*

$$
\langle A, B \rangle := \frac{1}{n} \mathrm{tr}(AB^*)
$$

for $n \times n$ matrices A and B. Write down the basis for $n = 2$.

Solution 15.   For $n = 2$ we have the combinations

$$
(j, k) \in \{ \, (0,0),\ (0,1),\ (1,0),\ (1,1) \, \} \, .
$$

This yields the orthonormal basis (where $\zeta = -1$)

$$
I_2 = \begin{pmatrix} 1 & 0 \\ 0 & 1 \end{pmatrix}, \quad \sigma_x = \begin{pmatrix} 0 & 1 \\ 1 & 0 \end{pmatrix}, \quad \sigma_z = \begin{pmatrix} 1 & 0 \\ 0 & -1 \end{pmatrix}, \quad -i\sigma_y = \begin{pmatrix} 0 & -1 \\ 1 & 0 \end{pmatrix} .
$$

Problem 16.   Let A and B be $n \times n$ matrices over $\mathbb{C}$. Show that the matrices AB and BA have the same set of eigenvalues.

Solution 16.   Consider first the case that A is invertible. Then we have

$$
AB = A(BA)A^{-1} \, .
$$

Thus AB and BA are similar and therefore have the same set of eigenvalues. If A is singular we apply the *continuity argument*: Consider the matrix

$A + \epsilon I_n$. We choose $\delta > 0$ such that $A + \epsilon I_n$ is invertible for all ϵ, $0 < \epsilon < \delta$. Thus $(A + \epsilon I_n)B$ and $B(A + \epsilon I_n)$ have the same set of eigenvalues for every $\epsilon \in (0, \delta)$. We equate their characteristic polynomials to obtain

$$\det(\lambda I_n - (A + \epsilon I_n)B) = \det(\lambda I_n - B(A + \epsilon I_n)), \quad 0 < \epsilon < \delta.$$

Since both sides are analytic functions of ϵ we find by letting $\epsilon \to 0^+$ that

$$\det(\lambda I_n - AB) = \det(\lambda I_n - BA).$$

Problem 17. The *numerical range*, also known as the *field of values*, of an $n \times n$ matrix A over the complex numbers, is defined as

$$W(A) := \{ \, \mathbf{x}^* A \mathbf{x} \, : \, \|\mathbf{x}\| = 1, \; \mathbf{x} \in \mathbb{C}^n \, \}.$$

(i) Find the numerical range for the matrix

$$A = \begin{pmatrix} 1 & 0 \\ 0 & 0 \end{pmatrix}.$$

(ii) Find the numerical range for the matrix

$$A = \begin{pmatrix} 0 & 0 \\ 1 & 1 \end{pmatrix}.$$

The *Toeplitz-Hausdorff convexity theorem* tells us that the numerical range of a square matrix is a convex compact subset of the complex plane.

Solution 17.   (i) Obviously, the numerical range is the unit interval $[0, 1]$. (ii) The numerical range $W(A)$ is the closed elliptical disc in the complex plane with foci at $(0, 0)$ and $(1, 0)$, minor axis 1, and major axis $\sqrt{2}$.

Problem 18.   An $n \times n$ *circulant matrix* C is given by

$$C := \begin{pmatrix} c_0 & c_1 & c_2 & \cdots & c_{n-1} \\ c_{n-1} & c_0 & c_1 & \cdots & c_{n-2} \\ c_{n-2} & c_{n-1} & c_0 & \cdots & c_{n-3} \\ \vdots & \vdots & \vdots & \ddots & \vdots \\ c_1 & c_2 & c_3 & \cdots & c_0 \end{pmatrix}.$$

For example, the permutation matrix

$$P := \begin{pmatrix} 0 & 1 & 0 & \cdots & 0 \\ 0 & 0 & 1 & \cdots & 0 \\ \vdots & \vdots & \vdots & \ddots & \vdots \\ 0 & 0 & 0 & \cdots & 1 \\ 1 & 0 & 0 & \cdots & 0 \end{pmatrix}$$

is a circulant matrix. It is also called the $n \times n$ *primary permutation matrix*.
(i) Let C and P be the matrices given above. Let

$$f(\lambda) = c_0 + c_1\lambda + \cdots + c_{n-1}\lambda^{n-1}.$$

Show that $C = f(P)$.
(ii) Show that C is a *normal matrix*, that is,

$$C^*C = CC^*.$$

(iii) Show that the eigenvalues of C are $f(\omega^k)$, $k = 0, 1, \ldots, n - 1$, where ω is the nth primitive root of unity.
(iv) Show that

$$\det(C) = f(\omega^0)f(\omega^1) \cdots f(\omega^{n-1}).$$

(v) Show that F^*CF is a diagonal matrix, where F is the unitary matrix with (j, k)-entry equal to

$$\frac{1}{\sqrt{n}}\omega^{(j-1)(k-1)}, \quad j, k = 1, \ldots, n.$$

Solution 18. (i) Direct calculation of

$$f(P) = c_0 I_n + c_1 P + c_2 P^2 + \cdots + c_{n-1}P^{n-1}$$

yields the matrix C, where I_n is the $n \times n$ unit matrix. Notice that P^2, P^3, ..., P^{n-1} are permutation matrices.
(ii) We have $PP^* = P^*P$. If two $n \times n$ matrices A and B commute, then $g(A)$ and $h(B)$ commute, where g and h are polynomials. Thus C is a normal matrix.
(iii) The characteristic polynomial of P is

$$\det(\lambda I_n - P) = \lambda^n - 1 = \prod_{k=0}^{n-1}(\lambda - \omega^k).$$

Thus the eigenvalues of P and P^j are, respectively, ω^k and ω^{jk}, where $k = 0, 1, \ldots, n - 1$. It follows that the eigenvalues of $C = f(P)$ are $f(\omega^k)$, $k = 0, 1, \ldots, n - 1$.
(iv) Using the result from (iii) we find

$$\det(C) = \prod_{k=0}^{n-1} f(\omega^k).$$

(v) For each $k = 0, 1, \ldots, n - 1$, let

$$\mathbf{x}_k = (1, \omega^k, \omega^{2k}, \ldots, \omega^{(n-1)k})^T$$

where ^{T} denotes the transpose. If follows that

$$P\mathbf{x}_k = (\omega^k, \omega^{2k}, \ldots, \omega^{(n-1)k}, 1)^T = \omega^k \mathbf{x}_k$$

and

$$C\mathbf{x}_k = f(P)\mathbf{x}_k = f(\omega^k)\mathbf{x}_k \, .$$

Thus the vectors $\mathbf{x}_k$ are the eigenvectors of P and C corresponding to the respective eigenvalues ω^k and $f(\omega^k)$, $k = 0, 1, \ldots, n - 1$. Since

$$\langle \mathbf{x}_j, \mathbf{x}_k \rangle \equiv \mathbf{x}_j^* \mathbf{x}_k = \sum_{\ell=0}^{n-1} \overline{\omega^{k\ell}} \omega^{j\ell} = \sum_{\ell=0}^{n-1} \omega^{(j-k)\ell} = \begin{cases} 0 & j \neq k \\ n & j = k \end{cases}$$

we find that

$$\left\{ \frac{1}{\sqrt{n}} \mathbf{x}_0, \ \frac{1}{\sqrt{n}} \mathbf{x}_1, \ \ldots, \ \frac{1}{\sqrt{n}} \mathbf{x}_{n-1} \right\}$$

is an orthonormal basis in the Hilbert space $\mathbb{C}^n$. Thus we obtain the unitary matrix

$$F = \frac{1}{\sqrt{n}} \begin{pmatrix} 1 & 1 & 1 & \cdots & 1 \\ 1 & \omega & \omega^2 & \cdots & \omega^{n-1} \\ 1 & \omega^2 & \omega^4 & \cdots & \omega^{2(n-1)} \\ \vdots & \vdots & \vdots & \ddots & \vdots \\ 1 & \omega^{n-1} & \omega^{2(n-1)} & \cdots & \omega^{(n-1)(n-1)} \end{pmatrix}$$

such that

$$F^* C F = \operatorname{diag}(f(\omega^0), f(\omega^1), \ldots, f(\omega^{n-1})) \, .$$

The matrix F is unitary and is called the *Fourier matrix*.

Problem 19. An $n \times n$ matrix A is called a *Hadamard matrix* if each entry of A is 1 or -1 and if the rows or columns of A are orthogonal, i.e.,

$$AA^T = nI_n \quad \text{or} \quad A^T A = nI_n \, .$$

Note that $AA^T = nI_n$ and $A^T A = nI_n$ are equivalent. Hadamard matrices H_n of order 2^n can be generated recursively by defining

$$H_1 = \begin{pmatrix} 1 & 1 \\ 1 & -1 \end{pmatrix}, \qquad H_n = \begin{pmatrix} H_{n-1} & H_{n-1} \\ H_{n-1} & -H_{n-1} \end{pmatrix}$$

for $n \geq 2$. Show that the eigenvalues of H_n are given by $+2^{n/2}$ and $-2^{n/2}$ each of multiplicity 2^{n-1}.

Solution 19. We use induction on n. The case $n = 1$ is obvious. Now for $n \geq 2$ we have

$$\det(\lambda I - H_n) = \begin{vmatrix} \lambda I - H_{n-1} & -H_{n-1} \\ -H_{n-1} & \lambda I + H_{n-1} \end{vmatrix}$$
$$= \det((\lambda I - H_{n-1})(\lambda I + H_{n-1}) - H_{n-1}^2).$$

Thus

$$\det(\lambda I - H_n) = \det(\lambda^2 I - 2H_{n-1}^2)$$
$$= \det(\lambda I - \sqrt{2}H_{n-1}) \det(\lambda I + \sqrt{2}H_{n-1}).$$

This shows that each eigenvalue μ of H_{n-1} generates two eigenvalues $\pm\sqrt{2}\mu$ of H_n. The assertion then follows by the induction hypothesis, for H_{n-1} has eigenvalues $+2^{(n-1)/2}$ and $-2^{(n-1)/2}$ each of multiplicity 2^{n-2}.

Problem 20. An $n \times n$ matrix A is called *reducible* if there is a permutation matrix P such that

$$P^T A P = \begin{pmatrix} B & C \\ 0 & D \end{pmatrix}$$

where B and D are square matrices of order at least 1. An $n \times n$ matrix A is called *irreducible* if it is not reducible. Show that the $n \times n$ *primary permutation matrix*

$$A := \begin{pmatrix} 0 & 1 & 0 & \cdots & 0 \\ 0 & 0 & 1 & \cdots & 0 \\ \vdots & \vdots & \vdots & \ddots & \vdots \\ 0 & 0 & 0 & \cdots & 1 \\ 1 & 0 & 0 & \cdots & 0 \end{pmatrix}$$

is irreducible.

Solution 20. Suppose the matrix A is reducible. Let

$$P^T A P = J_1 \oplus J_2 \oplus \cdots \oplus J_k, \quad k \geq 2$$

where P is some permutation matrix and the J_j are irreducible matrices of order $< n$. Here $\oplus$ denotes the direct sum. The rank of $A - I$ is $n - 1$ since $\det(A - I) = 0$ and the submatrix of size $n - 1$ by deleting the last row and the last column from $A - I$ is nonsingular. It follows that

$$\operatorname{rank}(P^T A P - I_n) = \operatorname{rank}(P^T (A - I_n)P) = n - 1.$$

By using the above decomposition, we obtain

$$\operatorname{rank}(P^T A P - I_n) = \sum_{j=1}^{k} \operatorname{rank}(J_j - I_n) \leq (n - k) < (n - 1).$$

This is a contradiction. Thus A is irreducible.

Problem 21. Let U be an $n \times n$ unitary matrix. Then U can be written as

$$U = V\text{diag}(\lambda_1, \lambda_2, \ldots, \lambda_n)V^*$$

where $\lambda_1, \lambda_2, \ldots, \lambda_n$ are the eigenvalues of U and V is an $n \times n$ unitary matrix. Let

$$U = \begin{pmatrix} 0 & 1 \\ 1 & 0 \end{pmatrix}.$$

Find the decomposition for U given above.

Solution 21. The eigenvalues of U are $+1$ and -1. Thus we have

$$U = V\text{diag}(1, -1)V^*$$

with

$$V = \frac{1}{\sqrt{2}} \begin{pmatrix} 1 & 1 \\ 1 & -1 \end{pmatrix}.$$

Therefore $V = V^*$. The columns of V are the eigenvectors of U.

Problem 22. An $n \times n$ hermitian matrix A over the complex numbers is called *positive semidefinite* (written as $A \geq 0$), if

$$\mathbf{x}^* A \mathbf{x} \geq 0 \quad \text{for all } \mathbf{x} \in \mathbb{C}^n.$$

Show that for every $A \geq 0$, there exists a unique $B \geq 0$ so that $B^2 = A$.

Solution 22. Let

$$A = U^*\text{diag}(\lambda_1, \ldots, \lambda_n)U$$

where U is unitary. We take

$$B = U^*\text{diag}(\lambda_1^{1/2}, \ldots, \lambda_n^{1/2})U.$$

Then the matrix B is positive semidefinite and $B^2 = A$ since $U^*U = I_n$. To show the uniqueness, suppose that C is an $n \times n$ positive semidefinite matrix satisfying $C^2 = A$. Since the eigenvalues of C are the nonnegative square roots of the eigenvalues of A, we can write

$$C = V\text{diag}(\lambda_1^{1/2}, \ldots, \lambda_n^{1/2})V^*$$

for some unitary matrix V. Then the identity $C^2 = A = B^2$ yields

$$T\text{diag}(\lambda_1, \ldots, \lambda_n) = \text{diag}(\lambda_1, \ldots, \lambda_n)T$$

where $T = UV$. This yields $t_{jk}\lambda_k = \lambda_j t_{jk}$. Thus

$$t_{jk}\lambda_k^{1/2} = \lambda_j^{1/2} t_{jk} \,.$$

Hence

$$T\text{diag}(\lambda_1^{1/2}, \ldots, \lambda_n^{1/2}) = \text{diag}(\lambda_1^{1/2}, \ldots, \lambda_n^{1/2})T \,.$$

Since $T = UV$ it follows that $B = C$.

Problem 23. An $n \times n$ matrix A over the complex numbers is said to be *normal* if it commutes with its conjugate transpose $A^*A = AA^*$. The matrix A can be written

$$A = \sum_{j=1}^{n} \lambda_j E_j$$

where $\lambda_j \in \mathbb{C}$ are the eigenvalues of A and E_j are $n \times n$ matrices satisfying

$$E_j^2 = E_j = E_j^*, \qquad E_j E_k = 0 \text{ if } j \neq k, \qquad \sum_{j=1}^{n} E_j = I_n \,.$$

Consider the *NOT gate*

$$A = \begin{pmatrix} 0 & 1 \\ 1 & 0 \end{pmatrix} \,.$$

Find the decomposition of A given above.

Solution 23. The eigenvalues of A are given by $\lambda_1 = +1$, $\lambda_2 = -1$. The matrices E_j are constructed from the normalized eigenvectors of A. The normalized eigenvectors of A are given by

$$\mathbf{x}_1 = \frac{1}{\sqrt{2}} \begin{pmatrix} 1 \\ 1 \end{pmatrix}, \qquad \mathbf{x}_2 = \frac{1}{\sqrt{2}} \begin{pmatrix} 1 \\ -1 \end{pmatrix} \,.$$

Thus

$$E_1 = \mathbf{x}_1 \mathbf{x}_1^* = \frac{1}{2} \begin{pmatrix} 1 & 1 \\ 1 & 1 \end{pmatrix}, \qquad E_2 = \mathbf{x}_2 \mathbf{x}_2^* = \frac{1}{2} \begin{pmatrix} 1 & -1 \\ -1 & 1 \end{pmatrix} \,.$$

Problem 24. Let X be an $n \times n$ matrix over $\mathbb{C}$ with $X^2 = I_n$, where I_n is the $n \times n$ identity matrix. Let $z \in \mathbb{C}$.
(i) Show that

$$e^{zX} = I_n \cosh(z) + X \sinh(z) \,.$$

(ii) Let $z = i\alpha$, where $\alpha \in \mathbb{R}$. Simplify the result from (i).

Solution 24. (i) We have

$$e^{zX} := \sum_{k=0}^{\infty} \frac{(zX)^k}{k!} = \sum_{k=0}^{\infty} \frac{z^k X^k}{k!} = I_n + zX + \frac{1}{2!}z^2 X^2 + \frac{1}{3!}z^3 X^3 + \cdots.$$

Since $X^2 = I_n$ we have

$$e^{zX} = I_n \left(1 + \frac{1}{2!}z^2 + \frac{1}{4!}z^4 + \cdots \right) + X \left(z + \frac{1}{3!}z^3 + \frac{1}{5!}z^5 + \cdots \right)$$
$$= I_n \cosh(z) + X \sinh(z).$$

(ii) Since $\cosh(i\alpha) = \cos(\alpha)$, $\sinh(i\alpha) = i\sin(\alpha)$ we obtain

$$e^{i\alpha X} = I_n \cos(\alpha) + iX \sin(\alpha).$$

Problem 25. Let

$$V = \exp(i(\pi/4)\sigma_x), \qquad W = \exp(i(\pi/4)\sigma_y)$$

where

$$\sigma_x := \begin{pmatrix} 0 & 1 \\ 1 & 0 \end{pmatrix}, \qquad \sigma_y := \begin{pmatrix} 0 & -i \\ i & 0 \end{pmatrix}.$$

Show that

$$V = \frac{1}{\sqrt{2}} \begin{pmatrix} 1 & i \\ i & 1 \end{pmatrix}, \qquad W = \frac{1}{\sqrt{2}} \begin{pmatrix} 1 & 1 \\ -1 & 1 \end{pmatrix}.$$

Solution 25. Using $\sigma_x^2 = I_2$, $\sigma_y^2 = I_2$ we find

$$V = I_n \cos(\pi/4) + i\sigma_x \sin(\pi/4) = \frac{1}{\sqrt{2}}(I_2 + i\sigma_x) = \frac{1}{\sqrt{2}} \begin{pmatrix} 1 & i \\ i & 1 \end{pmatrix}$$

and

$$W = I_n \cos(\pi/4) + i\sigma_y \sin(\pi/4) = \frac{1}{\sqrt{2}}(I_2 + i\sigma_y) = \frac{1}{\sqrt{2}} \begin{pmatrix} 1 & 1 \\ -1 & 1 \end{pmatrix}.$$

Problem 26. Let

$$V = \exp(i(\pi/4)\sigma_x), \qquad W = \exp(i(\pi/4)\sigma_y)$$

be two unitary 2×2 matrices. Calculate the matrices $V^*\sigma_z V$, $W^*\sigma_z W$.

Solution 26.   We have

$$
V^* \sigma_z V = \left(\frac{1}{\sqrt{2}} I_2 - \frac{i}{\sqrt{2}} \sigma_x \right) \sigma_z \left(\frac{1}{\sqrt{2}} I_2 + \frac{i}{\sqrt{2}} \sigma_x \right)
$$
$$
= \frac{1}{2} \sigma_z + \frac{i}{2} \sigma_z \sigma_x - \frac{i}{2} \sigma_x \sigma_z + \frac{1}{2} \sigma_x \sigma_z \sigma_x
$$
$$
= -\sigma_y
$$

and

$$
W^* \sigma_z W = \left(\frac{1}{\sqrt{2}} I_2 - \frac{i}{\sqrt{2}} \sigma_y \right) \sigma_z \left(\frac{1}{\sqrt{2}} I_2 + \frac{i}{\sqrt{2}} \sigma_y \right)
$$
$$
= \frac{1}{2} \sigma_z + \frac{i}{2} \sigma_z \sigma_y - \frac{i}{2} \sigma_y \sigma_z + \frac{1}{2} \sigma_y \sigma_z \sigma_y
$$
$$
= \sigma_x .
$$

Problem 27.   Let

$$
V = \exp(i(\pi/4)\sigma_x) \otimes \exp(i(\pi/4)\sigma_x), \qquad W = \exp(i(\pi/4)\sigma_y) \otimes \exp(i(\pi/4)\sigma_y).
$$

Calculate $V^*(\sigma_z \otimes \sigma_z)V$, $W^*(\sigma_z \otimes \sigma_z)W$.

Solution 27.   We have

$$
V^*(\sigma_z \otimes \sigma_z)V = (\exp(-i(\pi/4)\sigma_x)\sigma_z \exp(i(\pi/4)\sigma_x))
$$
$$
\otimes (\exp(-i(\pi/4)\sigma_x)\sigma_z \exp(i(\pi/4)\sigma_x))
$$
$$
= \sigma_y \otimes \sigma_y .
$$

Analogously

$$
W^*(\sigma_z \otimes \sigma_z)W = \sigma_x \otimes \sigma_x .
$$

Problem 28.   Let X be an $n \times n$ matrix over $\mathbb{C}$. Assume that $X^2 = I_n$. Let Y be an arbitrary $n \times n$ matrix over $\mathbb{C}$. Let $z \in \mathbb{C}$.
(i) Calculate $\exp(zX)Y \exp(-zX)$ using the *Baker-Campbell-Hausdorff formula*

$$
e^{zX} Y e^{-zX} = Y + z[X,Y] + \frac{z^2}{2!}[X,[X,Y]] + \frac{z^3}{3!}[X,[X,[X,Y]]] + \cdots .
$$

(ii) Calculate $\exp(zX)Y \exp(-zX)$ by first calculating $\exp(zX)$ and $\exp(-zX)$ and then doing the matrix multiplication. Compare the two methods.

Solution 28.   (i) Using $X^2 = I_n$ we find for the first three commutators

$$
[X,[X,Y]] = [X, XY - YX] = 2(Y - XYX)
$$

$$[X,[X,[X,Y]]] = 2^2[X,Y]$$
$$[X,[X,[X,[X,Y]]]] = 2^3(Y - XYX).$$

If the number of X's in the commutator is even (say m, $m \geq 2$) we have

$$[X,[X,\ldots[X,Y]\ldots] = 2^{m-1}(Y - XYX).$$

If the number of X's in the commutator is odd (say m, $m \geq 3$) we have

$$[X,[X,\ldots[X,Y]\ldots] = 2^{m-1}[X,Y].$$

Thus

$$e^{zX}Ye^{-zX} = Y\left(1 + \frac{2^1 z^2}{2!} + \frac{2^3 z^4}{4!} + \cdots\right) + [X,Y]\left(z + \frac{2^2 z^3}{3!} + \cdots\right)$$
$$- XYX\left(\frac{2^1 z^2}{2!} + \frac{2^3 z^4}{4!} + \cdots\right).$$

Consequently

$$e^{zX}Ye^{-zX} = Y\cosh^2(z) + [X,Y]\sinh(z)\cosh(z) - XYX\sinh^2(z).$$

(ii) Using the expansion

$$e^{zX} = \sum_{j=0}^{\infty} \frac{(zX)^j}{j!}$$

and $X^2 = I_n$ we have

$$e^{zX} = I_n\cosh(z) + X\sinh(z), \qquad e^{-zX} = I_n\cosh(z) - X\sinh(z).$$

Matrix multiplication yields

$$e^{zX}Ye^{-zX} = Y\cosh^2(z) + [X,Y]\sinh(z)\cosh(z) - XYX\sinh^2(z).$$

Problem 29. The definition of the Lie group $SU(2)$ is

$$SU(2) := \{\, A \,:\, A \text{ a } 2 \times 2 \text{ complex matrix, } \det(A) = 1,\ AA^* = A^*A = I_2 \,\}.$$

In the name $SU(2)$, the S stands for special and refers to the condition $\det(A) = 1$ and the U stands for unitary and refers to the conditions $AA^* = A^*A = I_2$. Show that $SU(2)$ can also be defined as

$$SU(2) := \{\, x_0 I_2 + i\mathbf{x}^T\boldsymbol{\sigma} \,:\, (x_0, \mathbf{x})^T \in \mathbb{R}^4,\ x_0^2 + \|\mathbf{x}\|^2 = 1 \,\}$$

where $\boldsymbol{\sigma} = (\sigma_1, \sigma_2, \sigma_3)^T$ are the Pauli spin matrices. Here $\mathbf{x}^T\boldsymbol{\sigma} := x_1\sigma_1 + x_2\sigma_2 + x_3\sigma_3$.

Solution 29.   Let A be any 2×2 complex matrix. Then A can be written as

$$A = a_0 I_2 + i \mathbf{a}^T \boldsymbol{\sigma}$$

with $\mathbf{a} = (a_1, a_2, a_3)^T \in \mathbb{C}^3$. Thus we have

$$
\begin{aligned}
AA^* &= (a_0 I_2 + i\mathbf{a}^T \boldsymbol{\sigma})(\overline{a_0} I_2 - i\overline{\mathbf{a}}^T \boldsymbol{\sigma}) \\
&= |a_0|^2 I_2 + i\overline{a_0}\mathbf{a}^T\boldsymbol{\sigma} - ia_0\overline{\mathbf{a}}^T\boldsymbol{\sigma} + \mathbf{a}^T\overline{\mathbf{a}} I_2 + i(\mathbf{a} \times \overline{\mathbf{a}})^T \boldsymbol{\sigma} \\
&= (|a_0|^2 + \|\mathbf{a}\|^2)I_2 + i(\overline{a_0}\mathbf{a} - a_0\overline{\mathbf{a}} + \mathbf{a} \times \overline{\mathbf{a}})^T \boldsymbol{\sigma}
\end{aligned}
$$

where we used that $\sigma_j^* = \sigma_j$ for $j = 1, 2, 3$. It follows that

$$AA^* = I_2 \Leftrightarrow |a_0|^2 + \|\mathbf{a}\|^2 = 1, \quad \overline{a_0}\mathbf{a} - a_0\overline{\mathbf{a}} + \mathbf{a} \times \overline{\mathbf{a}} = \mathbf{0}.$$

First, suppose that $\mathbf{a} \neq \mathbf{0}$. Since $\mathbf{a} \times \overline{\mathbf{a}}$ is orthogonal to both $\mathbf{a}$ and $\overline{\mathbf{a}}$, the equation $\overline{a_0}\mathbf{a} - a_0\overline{\mathbf{a}} + \mathbf{a} \times \overline{\mathbf{a}} = \mathbf{0}$ can only be satisfied if $\mathbf{a} \times \overline{\mathbf{a}} = \mathbf{0}$. That is, only if $\mathbf{a}$ and $\overline{\mathbf{a}}$ are parallel. Since $\mathbf{a}$ and $\overline{\mathbf{a}}$ have the same length, this is the case only if $\overline{\mathbf{a}} = e^{-2i\theta}\mathbf{a}$ for some real number θ. This can be written as

$$\overline{e^{-i\theta}\mathbf{a}} = e^{-i\theta}\mathbf{a}$$

which says that $\mathbf{x} = e^{-i\theta}\mathbf{a}$ is real. Substituting $\mathbf{a} = e^{i\theta}\mathbf{x}$ into $\overline{a_0}\mathbf{a} - a_0\overline{\mathbf{a}} + \mathbf{a} \times \overline{\mathbf{a}} = \mathbf{0}$ gives

$$e^{i\theta}\overline{a_0}\mathbf{x} - e^{-i\theta}a_0\mathbf{x} = \mathbf{0}.$$

This forces $a_0 = e^{i\theta}x_0$ for some real x_0. If $\mathbf{a} = \mathbf{0}$, we may still choose θ so that $a_0 = e^{i\theta}x_0$. We have shown that

$$AA^* = I_2 \Leftrightarrow A = e^{i\theta}(x_0 I_2 + i\mathbf{x}^T \boldsymbol{\sigma})$$

for some $(x_0, \mathbf{x})^T \in \mathbb{R}^4$ with $|x_0|^2 + \|\mathbf{x}\|^2 = 1$ and some $\theta \in \mathbb{R}$. Since

$$
\begin{aligned}
\det A &= \det e^{i\theta}(x_0 I_2 + i\mathbf{x}^T \boldsymbol{\sigma}) \\
&= \det e^{i\theta} \begin{pmatrix} x_0 + ix_3 & ix_1 + x_2 \\ ix_1 - x_2 & x_0 - ix_3 \end{pmatrix} \\
&= e^{2i\theta}(x_0^2 + x_1^2 + x_2^2 + x_3^2) \\
&= e^{2i\theta}
\end{aligned}
$$

we have that $\det A = 1$ if and only if $e^{i\theta} = \pm 1$. If $e^{i\theta} = -1$, we can absorb the -1 into the vector $(x_0, \mathbf{x})^T$.

Problem 30.   Let H be an $n \times n$ matrix which depends on n real parameters $\epsilon_1, \epsilon_2, \dots, \epsilon_n$, where we assume that we can differentiate H with respect to all ϵ's. Let $\beta > 0$ and $Z := \operatorname{tr} \exp(-\beta H)$.

(i) Show that

$$\frac{\partial}{\partial \epsilon_j} e^{-\beta H} \equiv - \int_0^\beta d\tau \, e^{(\tau - \beta)H} \frac{\partial H}{\partial \epsilon_j} e^{-\tau H} . \tag{1}$$

(ii) Show that

$$\frac{\partial}{\partial \epsilon_j} Z = -\beta \mathrm{tr} \left(\frac{\partial}{\partial \epsilon_j} H e^{-\beta H} \right) .$$

Solution 30. (i) We set

$$f(\beta, \epsilon_1, \ldots, \epsilon_n) := e^{\beta H} \frac{\partial}{\partial \epsilon_j} e^{-\beta H} .$$

It follows that

$$\frac{\partial f}{\partial \beta} = H e^{\beta H} \frac{\partial}{\partial \epsilon_j} e^{-\beta H} + e^{\beta H} \frac{\partial}{\partial \beta} \left(\frac{\partial}{\partial \epsilon_j} e^{-\beta H} \right) .$$

Since

$$\frac{\partial}{\partial \beta} \left(\frac{\partial}{\partial \epsilon_j} e^{-\beta H} \right) = \frac{\partial}{\partial \epsilon_j} \left(\frac{\partial}{\partial \beta} e^{-\beta H} \right) = -\frac{\partial}{\partial \epsilon_j} H e^{-\beta H}$$

$$= -\frac{\partial H}{\partial \epsilon_j} e^{-\beta H} - H \frac{\partial}{\partial \epsilon_j} e^{-\beta H}$$

we obtain

$$\frac{\partial f}{\partial \beta} = -e^{\beta H} \frac{\partial H}{\partial \epsilon_j} e^{-\beta H}$$

with the initial value $f(0, \epsilon_1, \ldots, \epsilon_n) = 0$. Integrating provides

$$f(\beta, \epsilon_1, \ldots, \epsilon_n) = - \int_0^\beta e^{\tau H} \frac{\partial H}{\partial \epsilon_j} e^{-\tau H} d\tau .$$

Multiplying by $\exp(-\beta H)$ yields identity (1).
(ii) We have

$$\frac{\partial Z}{\partial \epsilon_j} = \frac{\partial}{\partial \epsilon_j} \mathrm{tr} \exp(-\beta H) = \mathrm{tr} \left(\frac{\partial}{\partial \epsilon_j} \exp(-\beta H) \right)$$

$$= -\mathrm{tr} \left(\int_0^\beta d\tau \, e^{(\tau - \beta)H} \frac{\partial H}{\partial \epsilon_j} e^{-\tau H} \right)$$

$$= -\beta \mathrm{tr} \left(e^{-\beta H} \frac{\partial H}{\partial \epsilon_j} \right)$$

where we used the result from (i) and the cyclic invariance of the trace.

Problem 31. Let A, B be two $n \times n$ matrices over $\mathbb{C}$. We introduce the scalar product

$$\langle A, B \rangle := \frac{\text{tr}(AB^*)}{\text{tr} I_n} = \frac{1}{n}\text{tr}(AB^*).$$

The Lie group $SU(N)$ is defined by the complex $n \times n$ matrices U

$$SU(N) := \{ U : U^*U = UU^* = I_n , \ \det(U) = 1 \}.$$

The dimension is $N^2 - 1$. The semi-simple Lie algebra $su(N)$ is defined by the $n \times n$ matrices X

$$su(N) := \{ X : X^* = -X , \ \text{tr} X = 0 \}.$$

(i) Let $U \in SU(N)$. Calculate the scalar product $\langle U, U \rangle$.
(ii) Let A be an arbitrary complex $n \times n$ matrix. Let $U \in SU(N)$. Calculate $\langle UA, UA \rangle$.
(iii) Consider the Lie algebra $su(2)$. Provide a basis. The elements of the basis should be orthogonal to each other with respect to the scalar product given above. Calculate the commutator of these matrices.

Solution 31. (i) We have

$$\langle U, U \rangle = \frac{1}{n}\text{tr}(UU^*) = 1$$

where we used that $U^* = U^{-1}$.
(ii) We obtain

$$\langle UA, UA \rangle = \frac{1}{n}\text{tr}(UA(UA)^*) = \frac{1}{n}\text{tr}(UAA^*U^*) = \frac{1}{n}\text{tr}(AA^*) = \langle A, A \rangle$$

where we used the *cyclic invariance* of the trace.
(iii) We are looking for three linear independent 2×2 matrices which are traceless and skew-hermitian matrices. A choice is

$$\tau_1 = \begin{pmatrix} i & 0 \\ 0 & -i \end{pmatrix}, \quad \tau_2 = \begin{pmatrix} 0 & -1 \\ 1 & 0 \end{pmatrix}, \quad \tau_3 = \begin{pmatrix} 0 & i \\ i & 0 \end{pmatrix}.$$

The matrices are also orthogonal to each other using the scalar product given above. For the commutators we find

$$[\tau_1, \tau_2] = -2\tau_3, \quad [\tau_3, \tau_1] = -2\tau_2, \quad [\tau_2, \tau_3] = -2\tau_1.$$

Problem 32. To calculate $\exp(A)$ we can also use the Cayley-Hamilton theorem and the *Putzer method*. Using the Cayley-Hamilton theorem we can write

$$f(A) = a_{n-1}A^{n-1} + a_{n-2}A^{n-2} + \cdots + a_2 A^2 + a_1 A + a_0 I_n \qquad (1)$$

where the complex numbers $a_0, a_1, \ldots, a_{n-1}$ are determined as follows:
Let

$$r(\lambda) := a_{n-1}\lambda^{n-1} + a_{n-2}\lambda^{n-2} + \cdots + a_2\lambda^2 + a_1\lambda + a_0$$

which is the right-hand side of (1) with A^j replaced by λ^j ($j = 0, 1, \ldots, n - 1$). For each distinct eigenvalue λ_j of the matrix A, we consider the equation

$$f(\lambda_j) = r(\lambda_j).\tag{2}$$

If λ_j is an eigenvalue of multiplicity k, for $k > 1$, then we consider also the following equations

$$f'(\lambda)|_{\lambda=\lambda_j} = r'(\lambda)|_{\lambda=\lambda_j}, \quad \cdots \quad , f^{(k-1)}(\lambda)\Big|_{\lambda=\lambda_j} = r^{(k-1)}(\lambda)\Big|_{\lambda=\lambda_j}.$$

Any unitary matrix U can be written as $U = \exp(iK)$, where K is hermitian. Apply this method to find K for the *Hadamard gate*

$$U_H = \frac{1}{\sqrt{2}} \begin{pmatrix} 1 & 1 \\ 1 & -1 \end{pmatrix}.$$

Solution 32. The hermitian 2×2 matrix K is given by

$$K = \begin{pmatrix} a & b \\ \bar{b} & c \end{pmatrix}, \qquad a, c \in \mathbb{R}, \quad b \in \mathbb{C}.$$

Then we find the condition on a, b and c such that $e^{iK} = U_H$. The eigenvalues of iK are given by

$$\lambda_{1,2} = \frac{i(a+c)}{2} \pm \frac{1}{2}\sqrt{2ac - a^2 - c^2 - 4b\bar{b}}.$$

We set in the following

$$\Delta := \lambda_1 - \lambda_2 = \sqrt{2ac - a^2 - c^2 - 4b\bar{b}}.$$

To apply the method given above we have

$$r(\lambda) = \alpha_1\lambda + \alpha_0 = f(\lambda) = e^\lambda.$$

Thus we obtain the two equations $e^{\lambda_1} = \alpha_1\lambda_1 + \alpha_0$, $e^{\lambda_2} = \alpha_1\lambda_2 + \alpha_0$. It follows that

$$\alpha_1 = \frac{e^{\lambda_1} - e^{\lambda_2}}{\lambda_1 - \lambda_2}, \qquad \alpha_0 = \frac{e^{\lambda_2}\lambda_1 - e^{\lambda_1}\lambda_2}{\lambda_1 - \lambda_2}.$$

Thus we have the condition

$$e^{iK} = \alpha_1 iK + \alpha_0 I_2 = \begin{pmatrix} i\alpha_1 a + \alpha_0 & i\alpha_1 b \\ i\alpha_1 \bar{b} & i\alpha_1 c + \alpha_0 \end{pmatrix} = \frac{1}{\sqrt{2}} \begin{pmatrix} 1 & 1 \\ 1 & -1 \end{pmatrix}.$$

We obtain the four equations

$$i\alpha_1 a + \alpha_0 = \frac{1}{\sqrt{2}}, \quad i\alpha_1 c + \alpha_0 = -\frac{1}{\sqrt{2}}, \quad i\alpha_1 b = \frac{1}{\sqrt{2}}, \quad i\alpha_1 \bar{b} = \frac{1}{\sqrt{2}}.$$

From the last two equations we find that $\bar{b} = b$, i.e., b is real. From the first two equations we find $\alpha_0 = -i\alpha_1(a+c)/2$ and therefore, using the last two equations, $c = a - 2b$. Thus

$$\begin{pmatrix} i\alpha_1 a + \alpha_0 & i\alpha_1 b \\ i\alpha_1 \bar{b} & i\alpha_1 c + \alpha_0 \end{pmatrix} = \begin{pmatrix} i\alpha_1 b & i\alpha_1 b \\ i\alpha_1 b & -i\alpha_1 b \end{pmatrix}.$$

From the eigenvalues of e^{iK} we find $e^{\lambda_1} - e^{\lambda_2} = 2$ and

$$\Delta = \sqrt{2ac - a^2 - c^2 - 4b^2} = 2\sqrt{2}ib.$$

Furthermore

$$\lambda_1 = i(a - b) + \sqrt{2}ib, \qquad \lambda_2 = i(a - b) - \sqrt{2}ib.$$

Thus we arrive at the equation

$$e^{i(a-b)+\sqrt{2}ib} - e^{i(a-b)-\sqrt{2}ib} = 2.$$

It follows that $ie^{i(a-b)} \sin(\sqrt{2}b) = 1$ and therefore

$$i\cos(a - b)\sin(\sqrt{2}b) - \sin(a - b)\sin(\sqrt{2}b) = 1$$

with a solution

$$b = \frac{\pi}{2\sqrt{2}}, \qquad a = \frac{\pi}{2}\left(3 + \frac{1}{\sqrt{2}}\right), \qquad c = a - 2b = \frac{\pi}{2}\left(3 - \frac{1}{\sqrt{2}}\right).$$

Then the matrix K is given by

$$K = \frac{\pi}{2}\begin{pmatrix} 3 + 1/\sqrt{2} & 1/\sqrt{2} \\ 1/\sqrt{2} & 3 - 1/\sqrt{2} \end{pmatrix} = \frac{3\pi}{2}\begin{pmatrix} 1 & 0 \\ 0 & 1 \end{pmatrix} + \frac{\pi}{2} \cdot \frac{1}{\sqrt{2}}\begin{pmatrix} 1 & 1 \\ 1 & -1 \end{pmatrix}.$$

We note that the second matrix on the right-hand side is the Hadamard gate again.

Problem 33. Let A, B, C_2, ... , C_m, ... be $n \times n$ matrices over $\mathbb{C}$. The *Zassenhaus formula* is given by

$$\exp(A + B) = \exp(A)\exp(B)\exp(C_2)\cdots\exp(C_m)\cdots$$

The left-hand side is called the *disentangled form* and the right hand side is called the *undisentangled form*. Find C_2, C_3, $\ldots$, using the *comparison method.* In the comparison method the disentangled and undisentangled form are expanded in terms of an ordering scalar α and matrix coefficients of equal powers of α are compared. From

$$\exp(\alpha(A+B)) = \exp(\alpha A)\exp(\alpha B)\exp(\alpha^2 C_2)\exp(\alpha^3 C_3)\cdots$$

we obtain

$$\sum_{k=0}^{\infty} \frac{\alpha^k}{k!}(A+B)^k = \sum_{r_0,r_1,r_2,r_3,\ldots=0}^{\infty} \frac{\alpha^{r_0+r_1+2r_2+3r_3+\ldots}}{r_0!r_1!r_2!r_3!\cdots} A^{r_0}B^{r_1}C_2^{r_2}C_3^{r_3}\cdots$$

(i) Find the matrices C_2 and C_3.
(ii) Assume that $[A,[A,B]] = 0$ and $[B,[A,B]] = 0$. What conclusion can we draw for the Zassenhaus formula?

Solution 33. (i) For α^2 we have the decompositions $(r_0, r_1, r_2) = (2,0,0)$, $(1,1,0)$, $(0,2,0)$, $(0,0,1)$. Thus we obtain

$$(A+B)^2 = A^2 + 2AB + B^2 + 2C_2 \,.$$

Thus it follows that

$$C_2 = -\frac{1}{2}[A,B]\,.$$

For α^3 we obtain

$$(A+B)^3 = A^3 + 3A^2B + 3AB^2 + B^3 + 6AC_2 + 6BC_2 + 6C_3\,.$$

Using C_2 given above we obtain

$$C_3 = \frac{1}{3}[B,[A,B]] + \frac{1}{6}[A,[A,B]]\,.$$

(ii) Since $[B,[A,B]] = 0$ and $[A,[A,B]] = 0$ we find that $C_3 = C_4 = \cdots = 0$. Thus

$$\exp(\alpha(A+B)) = \exp(\alpha A)\exp(\alpha B)\exp(-\alpha^2[A,B]/2)\,.$$

Problem 34. Let H be a hermitian $n \times n$ matrix. Show that $\exp(H)$ is a positive definite matrix.

Solution 34. If H is hermitian then H^2, H^3 etc are hermitian and also $\exp(H)$. Let λ_j $(j = 1, 2, \ldots, n)$ be the real eigenvalues of H since H is hermitian. Then e^{λ_j} $(j = 1, 2, \ldots, n)$ are the real eigenvalues of $\exp(H)$ and obviously $e^{\lambda_j} > 0$ for $(j = 1, 2, \ldots, n)$. Thus $\exp(H)$ is a positive definite matrix.

Problem 35.   Let σ_x, σ_y, σ_z be the Pauli spin matrices. Does the set of 4×4 matrices

$$\{\, I_2 \otimes I_2, \quad \sigma_x \otimes \sigma_x, \quad -\sigma_y \otimes \sigma_y, \quad \sigma_z \otimes \sigma_z \,\}$$

form a group under matrix multiplication?

Solution 35.   We have

$$\sigma_x\sigma_y = i\sigma_z, \ \ \sigma_y\sigma_x = -i\sigma_z, \ \ \sigma_y\sigma_z = i\sigma_x, \ \ \sigma_z\sigma_y = i\sigma_x, \ \ \sigma_z\sigma_x = i\sigma_y, \ \ \sigma_x\sigma_z = -i$$

Thus

$$(\sigma_x \otimes \sigma_x)(-\sigma_y \otimes \sigma_y) = -(\sigma_x\sigma_y) \otimes (\sigma_x\sigma_y) = \sigma_z \otimes \sigma_z$$
$$(-\sigma_y \otimes \sigma_y)(\sigma_z \otimes \sigma_z) = -(\sigma_y\sigma_z) \otimes (\sigma_y\sigma_z) = \sigma_x \otimes \sigma_x$$
$$(\sigma_z \otimes \sigma_z)(\sigma_x \otimes \sigma_x) = (\sigma_z\sigma_x) \otimes (\sigma_z\sigma_x) = -\sigma_y \otimes \sigma_y\,.$$

The neutral element is $I_2 \otimes I_2$. Each element is its own inverse. Thus the set forms a group under matrix multiplication.

Problem 36.   Find the matrices

$$e^{(i\pi/4)\sigma_y}\sigma_x e^{-(i\pi/4)\sigma_y}, \qquad e^{(i\pi/4)\sigma_y}\sigma_z e^{-(i\pi/4)\sigma_y}\,.$$

Use the technique of parameter representation

$$f(\epsilon) = e^{\epsilon\sigma_y}\sigma_x e^{-\epsilon\sigma_y}$$

with $f(\epsilon = 0) = \sigma_x$.

Solution 36.   Differentiation of f with respect to ϵ and using $[\sigma_y, \sigma_x] = -2i\sigma_z$ yields

$$\frac{df}{d\epsilon} = -2ie^{\epsilon\sigma_y}\sigma_z e^{-\epsilon\sigma_y}$$

with $df(\epsilon = 0)/d\epsilon = -2i\sigma_z$. The second derivative and using $[\sigma_y, \sigma_z] = 2i\sigma_x$ yields

$$\frac{d^2 f}{d\epsilon^2} = 4f(\epsilon)\,.$$

The solution of this second order linear differential equation with constant coefficients is

$$f(\epsilon) = C_1 \cosh(2\epsilon) + C_2 \sinh(2\epsilon)\,.$$

Inserting the initial values provides

$$f(\epsilon) = \cosh(2\epsilon)\sigma_x - i\sinh(2\epsilon)\sigma_z$$

Since $\epsilon = i\pi/4$ we have $\cosh(i\pi/2) = \cos(\pi/2) = 0$ and $\sinh(i\pi/2) = i\sin(\pi/2) = i$. Thus $f(\epsilon = i\pi/4) = \sigma_z$ or

$$e^{(i\pi/4)\sigma_y}\sigma_x e^{-(i\pi/4)\sigma_y} = \sigma_z \,.$$

Analogously we find

$$e^{(i\pi/4)\sigma_y}\sigma_z e^{-(i\pi/4)\sigma_y} = \sigma_x \,.$$

Problem 37. The spin matrices for spin-2 particles (for example *graviton*) are given by

$$J_x = \frac{1}{2}\begin{pmatrix} 0 & 2 & 0 & 0 & 0 \\ 2 & 0 & \sqrt{6} & 0 & 0 \\ 0 & \sqrt{6} & 0 & \sqrt{6} & 0 \\ 0 & 0 & \sqrt{6} & 0 & 2 \\ 0 & 0 & 0 & 2 & 0 \end{pmatrix},$$

$$J_y = \frac{i}{2}\begin{pmatrix} 0 & -2 & 0 & 0 & 0 \\ 2 & 0 & -\sqrt{6} & 0 & 0 \\ 0 & \sqrt{6} & 0 & -\sqrt{6} & 0 \\ 0 & 0 & \sqrt{6} & 0 & -2 \\ 0 & 0 & 0 & 2 & 0 \end{pmatrix},$$

$$J_z = \begin{pmatrix} 2 & 0 & 0 & 0 & 0 \\ 0 & 1 & 0 & 0 & 0 \\ 0 & 0 & 0 & 0 & 0 \\ 0 & 0 & 0 & -1 & 0 \\ 0 & 0 & 0 & 0 & -2 \end{pmatrix}.$$

(i) Show that the matrices are hermitian.
(ii) Find the eigenvalues and eigenvectors of these matrices.
(iii) Calculate the commutation relations.
(iv) Are the matrices unitary?

Solution 37. (i) Obviously the matrices are hermitian, i.e.

$$J_x^* = J_x, \qquad J_y^* = J_y, \qquad J_z^* = J_z \,.$$

(ii) The eigenvalues of J_x are $-2, 2, -1, 1, 0$ with the corresponding normalized eigenvectors

$$\mathbf{u}_1 = \frac{1}{4}\begin{pmatrix} 1 \\ -2 \\ \sqrt{6} \\ 2 \\ 1 \end{pmatrix}, \quad \mathbf{u}_2 = \frac{1}{4}\begin{pmatrix} 1 \\ 2 \\ \sqrt{6} \\ 2 \\ 1 \end{pmatrix}, \quad \mathbf{u}_3 = \frac{1}{2}\begin{pmatrix} 1 \\ -1 \\ 0 \\ 1 \\ -1 \end{pmatrix},$$

$$\mathbf{u}_4 = \frac{1}{2} \begin{pmatrix} 1 \\ 1 \\ 0 \\ -1 \\ -1 \end{pmatrix}, \quad \mathbf{u}_5 = \frac{\sqrt{3}}{\sqrt{8}} \begin{pmatrix} 1 \\ 0 \\ -\sqrt{2}/\sqrt{3} \\ 0 \\ 1 \end{pmatrix}.$$

The eigenvectors form an orthonormal basis in $\mathbb{C}^5$. The eigenvalues of J_y are 2, 1, 0, −1, −2 with the corresponding normalized eigenvectors

$$\mathbf{u}_1 = \frac{1}{4} \begin{pmatrix} 1 \\ -2i \\ 1 \\ -\sqrt{6} \\ 2i \\ 1 \end{pmatrix}, \quad \mathbf{u}_2 = \frac{1}{4} \begin{pmatrix} 1 \\ 2i \\ -\sqrt{6} \\ -2i \\ 1 \end{pmatrix}, \quad \mathbf{u}_3 = \frac{1}{2} \begin{pmatrix} 1 \\ -i \\ 0 \\ -i \\ -1 \end{pmatrix},$$

$$\mathbf{u}_4 = \frac{1}{2} \begin{pmatrix} 1 \\ i \\ 0 \\ i \\ -1 \end{pmatrix}, \quad \mathbf{u}_5 = \frac{\sqrt{3}}{\sqrt{8}} \begin{pmatrix} 1 \\ 0 \\ \sqrt{2}/\sqrt{3} \\ 0 \\ 1 \end{pmatrix}.$$

The eigenvalues of J_z are 2, 1, 0, −1, −2 with the corresponding eigenvectors (standard basis)

$$\mathbf{u}_1 = \begin{pmatrix} 1 \\ 0 \\ 0 \\ 0 \\ 0 \end{pmatrix}, \quad \mathbf{u}_2 = \begin{pmatrix} 0 \\ 1 \\ 0 \\ 0 \\ 0 \end{pmatrix}, \quad \mathbf{u}_3 = \begin{pmatrix} 0 \\ 0 \\ 1 \\ 0 \\ 0 \end{pmatrix}, \quad \mathbf{u}_4 = \begin{pmatrix} 0 \\ 0 \\ 0 \\ 1 \\ 0 \end{pmatrix}, \quad \mathbf{u}_5 = \begin{pmatrix} 0 \\ 0 \\ 0 \\ 0 \\ 1 \end{pmatrix}.$$

(iii) The commutation relations are

$$[J_x, J_y] = iJ_z, \qquad [J_y, J_z] = iJ_x, \qquad [J_z, J_x] = iJ_y.$$

(iv) No the matrices are not unitary. Note that $\det(J_x) = \det(J_y) = \det(J_z) = 0$ owing to the eigenvalue 0.

Problem 38. Two orthonormal bases in an n-dimensional complex Hilbert space

$$\{ |\mathbf{u}_j\rangle \,:\, j = 1, 2, \ldots, n \}, \qquad \{ |\mathbf{v}_j\rangle \,:\, j = 1, 2, \ldots, n \}$$

are called *mutually unbiased* if the inner products (scalar products) between all possible pairs of vectors taken from distinct bases have the same magnitude $1/\sqrt{n}$, i.e.

$$|\langle \mathbf{u}_j | \mathbf{v}_k \rangle| = \frac{1}{\sqrt{n}} \qquad \text{for all} \quad j, k \in \{ 1, 2, \ldots, n \}.$$

(i) Find such bases for the Hilbert space $\mathbb{C}^2$.
(ii) Find such bases for the Hilbert space $\mathbb{C}^3$.

Solution 38. (i) As the first bases we select the standard basis

$$\left\{ \begin{pmatrix} 1 \\ 0 \end{pmatrix}, \begin{pmatrix} 0 \\ 1 \end{pmatrix} \right\}.$$

For the second basis we could select

$$\left\{ \frac{1}{\sqrt{2}} \begin{pmatrix} 1 \\ 1 \end{pmatrix}, \frac{1}{\sqrt{2}} \begin{pmatrix} 1 \\ -1 \end{pmatrix} \right\}$$

or

$$\left\{ \frac{1}{\sqrt{2}} \begin{pmatrix} 1 \\ i \end{pmatrix}, \frac{1}{\sqrt{2}} \begin{pmatrix} 1 \\ -i \end{pmatrix} \right\}.$$

Applying a unitary matrix to these two sets provide other such sets.
(ii) An example is the standard basis

$$\left\{ \begin{pmatrix} 1 \\ 0 \\ 0 \end{pmatrix}, \begin{pmatrix} 0 \\ 1 \\ 0 \end{pmatrix}, \begin{pmatrix} 0 \\ 0 \\ 1 \end{pmatrix} \right\}$$

and

$$\left\{ \frac{1}{\sqrt{3}} \begin{pmatrix} 1 \\ 1 \\ 1 \end{pmatrix}, \frac{1}{\sqrt{3}} \begin{pmatrix} -1 \\ (1+\sqrt{3})/2 \\ (1-\sqrt{3})/2 \end{pmatrix}, \frac{1}{\sqrt{3}} \begin{pmatrix} (-1+i\sqrt{3})/2 \\ (-1-i\sqrt{3})/2 \\ 1 \end{pmatrix} \right\}.$$

Applying a unitary matrix to these two sets provide other such sets.

Problem 39. Find a 4×4 matrix A such that

$$-A = A^{-1} = A^T = A^*.$$

Solution 39. We find

$$A = \begin{pmatrix} 0_2 & -i\sigma_y \\ -i\sigma_y & 0_2 \end{pmatrix} \equiv \begin{pmatrix} 0 & 0 & 0 & -1 \\ 0 & 0 & 1 & 0 \\ 0 & -1 & 0 & 0 \\ 1 & 0 & 0 & 0 \end{pmatrix}.$$

This matrix plays a role for the charge conjugation in the Dirac equation.

Problem 40. Consider the spin matrices for a spin-1 particle

$$S_1 = \frac{1}{\sqrt{2}} \begin{pmatrix} 0 & 1 & 0 \\ 1 & 0 & 1 \\ 0 & 1 & 0 \end{pmatrix}, \quad S_2 = \frac{1}{\sqrt{2}} \begin{pmatrix} 0 & -i & 0 \\ i & 0 & -i \\ 0 & i & 0 \end{pmatrix}, \quad S_3 = \begin{pmatrix} 1 & 0 & 0 \\ 0 & 0 & 0 \\ 0 & 0 & -1 \end{pmatrix}$$

and the unit vector

$$\mathbf{n} = (\, \sin\theta\cos\phi \quad \sin\theta\sin\phi \quad \cos\theta \,) \, .$$

We define the scalar product

$$\mathbf{n} \cdot \mathbf{S} := n_1 S_1 + n_2 S_2 + n_3 S_3 \, .$$

Consider the Hamilton operator

$$\hat{H} = \hbar\omega(\mathbf{n} \cdot \mathbf{S}) \, .$$

(i) Calculate the Hamilton operator $\hat{H}$. Is the Hamilton operator $\hat{H}$ hermitian?
(ii) Calculate the trace of $\hat{H}$.
(iii) Find the eigenvalues and normalized eigenvectors of $\hat{H}$.
(iv) Do the eigenvectors form a basis in the Hilbert space $\mathbb{C}^3$?

Solution 40. (i) The Hamilton operator is given by

$$\hat{H} = \hbar\omega \begin{pmatrix} \cos\theta & \sin\theta e^{-i\phi}/\sqrt{2} & 0 \\ \sin\theta e^{i\phi}/\sqrt{2} & 0 & \sin\theta e^{-i\phi}/\sqrt{2} \\ 0 & \sin\theta e^{i\phi}/\sqrt{2} & -\cos\theta \end{pmatrix} \, .$$

Since $\overline{\exp(i\phi)} = \exp(-i\phi)$ the Hamilton operator is hermitian. Thus the eigenvalues must be real.
(ii) The trace of $\hat{H}$ is 0. Thus the sum of the three eigenvalues of $\hat{H}$ must be 0.
(iii) The eigenvalues of $\hat{H}$ are

$$\hbar\omega, \qquad 0, \qquad -\hbar\omega \, .$$

The corresponding normalized eigenvectors are

$$\begin{pmatrix} (1+\cos\theta)e^{-i\phi}/2 \\ \sin\theta/\sqrt{2} \\ (1-\cos\theta)e^{i\phi}/2 \end{pmatrix}, \quad \begin{pmatrix} -\sin\theta e^{-i\phi}/\sqrt{2} \\ \cos\theta \\ \sin\theta e^{i\phi}/\sqrt{2} \end{pmatrix}, \quad \begin{pmatrix} (1-\cos\theta)e^{-i\phi}/2 \\ -\sin\theta/\sqrt{2} \\ (1+\cos\theta)e^{i\phi}/2 \end{pmatrix} \, .$$

(iv) The Hamilton operator is hermitian and the three eigenvalues are different. Thus the normalized eigenvectors form an orthonormal basis in the Hilbert space $\mathbb{C}^3$.

Problem 41. Consider a complex Hilbert space $\mathcal{H}$ and $|\phi_1\rangle, |\phi_2\rangle \in \mathcal{H}$. Let $c_1, c_2 \in \mathbb{C}$. An *antilinear operator* K in this Hilbert space $\mathcal{H}$ is characterized by

$$K(c_1|\phi_1\rangle + c_2|\phi_2\rangle) = c_1^* K|\phi_1\rangle + c_2^* K|\phi_2\rangle \, .$$

A *comb* is an antilinear operator K with zero expectation value for all states $|\psi\rangle$ of a certain complex Hilbert space $\mathcal{H}$. This means

$$\langle\psi|K|\psi\rangle = \langle\psi|LC|\psi\rangle = \langle\psi|L|\psi^*\rangle = 0$$

for all states $|\psi\rangle \in \mathcal{H}$, where L is a linear operator and C is the complex conjugation.

(i) Consider the two-dimensional Hilbert space $\mathcal{H} = \mathbb{C}^2$. Find a unitary 2×2 matrix such that

$$\langle\psi|UC|\psi\rangle = 0.$$

(ii) Consider the Pauli spin matrices with $\sigma_0 = I_2$, $\sigma_1 = \sigma_x$, $\sigma_2 = \sigma_y$, $\sigma_3 = \sigma_z$. Find

$$\sum_{\mu=0}^{3}\sum_{\nu=0}^{3}\langle\psi|\sigma_\mu C|\psi\rangle g^{\mu,\nu}\langle\psi|\sigma_\nu C|\psi\rangle$$

where $g^{\mu,\nu} = \text{diag}(-1, 1, 0, 1)$.

Solution 41.   (i) We find $U = \sigma_y$ since

$$\langle\psi|\sigma_y C|\psi\rangle = \langle\psi|\sigma_y|\psi^*\rangle = \begin{pmatrix} \psi_1^* & \psi_2^* \end{pmatrix}\begin{pmatrix} 0 & -i \\ i & 0 \end{pmatrix}\begin{pmatrix} \psi_1^* \\ \psi_2^* \end{pmatrix} = 0.$$

(ii) We have

$$\sum_{\mu=0}^{3}\sum_{\nu=0}^{3}\langle\psi|\sigma_\mu C|\psi\rangle g^{\mu,\nu}\langle\psi|\sigma_\nu C|\psi\rangle = -\langle\psi|\sigma_0|\psi^*\rangle^2 + \langle\psi|\sigma_1|\psi^*\rangle^2 + \langle\psi|\sigma_3|\psi^*\rangle^2 = 0.$$

Problem 42.   Starting from *Maxwell's equations* in vacuum

$$\text{curl}\mathbf{B} = \frac{1}{c^2}\frac{\partial\mathbf{E}}{\partial t}, \qquad \text{curl}\mathbf{E} = -\frac{\partial\mathbf{B}}{\partial t}, \qquad \text{div}\mathbf{E} = 0, \qquad \text{div}\mathbf{B} = 0$$

and *Kramer's vector*

$$\mathbf{F} := \mathbf{E} + ic\mathbf{B}, \qquad \mathbf{F}^* := \mathbf{E} - ic\mathbf{B}$$

show that the *photon* is a spin-1 particle.

Solution 42.   Using Kramer's vector we can write

$$\text{curl}\mathbf{F} = \frac{i}{c}\frac{\partial\mathbf{F}}{\partial t}, \qquad \text{curl}\mathbf{F}^* = -\frac{i}{c}\frac{\partial\mathbf{F}^*}{\partial t}, \qquad \text{div}\mathbf{F} = 0, \qquad \text{div}\mathbf{F}^* = 0.$$

Let

$$\epsilon_{jk\ell} = \begin{cases} +1 & \text{if } jk\ell \text{ are an even permutation of the integers 123} \\ -1 & \text{if } jk\ell \text{ are an odd permutation of the integers 123} \\ 0 & \text{otherwise} \end{cases}.$$

Since

$$(\text{curl}\mathbf{F})_j = \sum_{k=1}^{3}\sum_{\ell=1}^{3} \epsilon_{jk\ell}\frac{\partial}{\partial x_k}F_\ell = -\sum_{k=1}^{3}\sum_{\ell=1}^{3}\frac{\partial}{\partial x_k}\epsilon_{kj\ell}F_\ell$$

we can write

$$\sum_{k=1}^{3}\sum_{\ell=1}^{3}\left(-i\frac{\partial}{\partial x_k}\right)(-i\epsilon_{kj\ell})F_\ell = \frac{i}{c}\frac{\partial F_j}{\partial t}.$$

Introducing the differential operator

$$\hat{p}_k := -i\frac{\partial}{\partial x_k}$$

we find

$$-\sum_{j=1}^{3}\sum_{k=1}^{3}\hat{p}_k i\epsilon_{kj\ell}F_\ell = \frac{i}{c}\frac{\partial F_\ell}{\partial t}.$$

For fixed k, $-i\epsilon_{kj\ell}$ is a 3×3 matrix, $S_{k(j,\ell)}$. The equation for $\mathbf{F}$ then takes the form

$$\left(\sum_{k=1}^{3}\hat{p}_k S_k\right)\mathbf{F} = (\hat{\mathbf{p}}\cdot\mathbf{S})\mathbf{F} = \frac{i}{c}\frac{\partial\mathbf{F}}{\partial t}.$$

Using the definition of $\epsilon_{jk\ell}$, we obtain the representation for the 3×3 matrices

$$S_1 = i\begin{pmatrix} 0 & 0 & 0 \\ 0 & 0 & -1 \\ 0 & 1 & 0 \end{pmatrix}, \quad S_2 = i\begin{pmatrix} 0 & 0 & 1 \\ 0 & 0 & 0 \\ -1 & 0 & 0 \end{pmatrix}, \quad S_3 = i\begin{pmatrix} 0 & -1 & 0 \\ 1 & 0 & 0 \\ 0 & 0 & 0 \end{pmatrix}.$$

The commutators are

$$[S_1, S_2] = iS_3, \quad [S_2, S_3] = iS_1, \quad [S_3, S_1] = iS_2.$$

We have $\mathbf{S} \times \mathbf{S} = i\mathbf{S}$ and

$$S_1^2 + S_2^2 + S_3^2 = 2I_3$$

where I_3 is the 3×3 identity matrix. Thus Maxwell's equations describe a particle of spin-1.

Chapter 4

Density Operators

A *density operator* ρ or *density matrix* is a positive semidefinite operator on a Hilbert space with unit trace. An operator is *positive semidefinite* if it is hermitian and none of its (necessarily real) eigenvalues are less than zero. The state of a quantum-mechanical system is characterized by a density operator ρ with $\mathrm{tr}\rho = 1$. The *expectation value* of an observable $\hat{A}$, determined in an experiment as the average value $\langle \hat{A} \rangle$ is given by

$$\langle \hat{A} \rangle := \mathrm{tr}(\hat{A}\rho).$$

A density matrix or density operator is used in quantum theory to describe the statistical state of a quantum system. If we have a pure state $|\psi\rangle$ in a Hilbert space then

$$\rho = |\psi\rangle\langle\psi|$$

defines a density matrix with ($\langle\psi|\psi\rangle = 1$)

$$\rho^2 = |\psi\rangle\langle\psi|\psi\rangle\langle\psi| = |\psi\rangle\langle\psi| = \rho.$$

For a *mixed state* we have the spectral representation

$$\rho = \sum_{j=1}^{n} p_j |\psi_j\rangle\langle\psi_j|$$

where $p_j \geq 0$ for $j = 1, 2, \ldots, n$, $\sum_{j=1}^{n} p_j = 1$ and $\langle\psi_j|\psi_k\rangle = \delta_{jk}$. Then for the expectation value of a hermitian operator $\hat{A}$ in the Hilbert space we have

$$\mathrm{tr}(\rho\hat{A}) = \sum_{j=1}^{n} p_j \langle\psi_j|\hat{A}|\psi_j\rangle.$$

Problem 1. Is the 2×2 matrix

$$\rho = \begin{pmatrix} 1/2 & -1/2 \\ -1/2 & 1/2 \end{pmatrix}$$

a density matrix? If so find the state in $\mathbb{C}^2$ that provides this density matrix.

Solution 1. We find that $\mathrm{tr}\rho = 1$ and the matrix is hermitian over $\mathbb{C}$. The eigenvalues of ρ are 0 and 1. Thus ρ is a density matrix. The state

$$|\psi\rangle = \frac{1}{\sqrt{2}} \begin{pmatrix} 1 \\ -1 \end{pmatrix}$$

is a state that provides the density matrix $\rho = |\psi\rangle\langle\psi|$. We have a pure state.

Problem 2. Let

$$|\psi\rangle = \begin{pmatrix} \cos\theta \\ e^{i\phi}\sin\theta \end{pmatrix}, \qquad \theta, \phi \in \mathbb{R}.$$

Does $\rho := |\psi\rangle\langle\psi|$ define a density matrix?

Solution 2. We find the 2×2 matrix for ρ

$$\rho = \begin{pmatrix} \cos^2\theta & e^{-i\phi}\cos\theta\sin\theta \\ e^{i\phi}\cos\theta\sin\theta & \sin^2\theta \end{pmatrix}.$$

Thus $\rho = \rho^\dagger$, $\mathrm{tr}\rho = \cos^2\theta + \sin^2\theta = 1$ and

$$
\begin{aligned}
\langle \mathbf{x}|\rho|\mathbf{x}\rangle &= |x_1|^2\cos^2\theta + x_1\overline{x}_2 e^{-i\phi}\cos\theta\sin\theta + \overline{x}_1 x_2 e^{i\phi}\cos\theta\sin\theta + |x_2|^2\sin^2\theta \\
&= |x_1\cos\theta|^2 + x_1\cos\theta\overline{x_2 e^{i\phi}\sin\theta} + \overline{x_1\cos\theta}x_2 e^{i\phi}\sin\theta + |x_2 e^{i\phi}\sin\theta|^2 \\
&= (x_1\cos\theta + x_2 e^{i\phi}\sin\theta)\overline{(x_1\cos\theta + x_2 e^{i\phi}\sin\theta)} \\
&= |x_1\cos\theta + x_2 e^{i\phi}\sin\theta|^2 \\
&\geq 0
\end{aligned}
$$

for all $\mathbf{x} := (x_1, x_2)^T \in \mathbb{C}^2$. Thus ρ defines a density matrix.

Problem 3. Let $r \geq 0$. Is the 2×2 matrix

$$\rho = \frac{1}{2} \begin{pmatrix} 1 + r\cos\theta & r\sin\theta e^{-i\phi} \\ r\sin\theta e^{i\phi} & 1 - r\cos\theta \end{pmatrix}$$

a density matrix? What are the conditions on r, θ, ϕ?

Solution 3. We have that $\mathrm{tr}(\rho) = 1$ and the matrix is hermitian. Thus the eigenvalues are real. The eigenvalues are given by

$$\lambda_{1,2} = \frac{1}{2} \pm \frac{1}{2}r.$$

Thus the condition that ρ is density matrix is $r \leq 1$. There is no condition on ϕ and θ.

Problem 4. Let σ_1, σ_2 and σ_3 be the Pauli spin matrices. Let $\boldsymbol{\sigma} = (\sigma_1, \sigma_2, \sigma_3)$ and $\mathbf{r} \in \mathbb{R}^3$ with $\mathbf{r}^2 \leq 1$. Consider the 2×2 matrix (density matrix)

$$\rho := \frac{1}{2}(I_2 + \mathbf{r} \cdot \boldsymbol{\sigma})$$

where

$$\mathbf{r} \cdot \boldsymbol{\sigma} := r_1\sigma_1 + r_2\sigma_2 + r_3\sigma_3.$$

Let $\mathbf{n}$ be an arbitrary unit length vector in $\mathbb{R}^3$, i.e. $\mathbf{n}^2 \equiv n_1^2 + n_2^2 + n_3^2 = 1$. Calculate

$$\mathrm{tr}((\mathbf{n} \cdot \boldsymbol{\sigma})\rho)$$

i.e., we calculate the expectation value of $\mathbf{n} \cdot \boldsymbol{\sigma}$. Give an interpretation of the result.

Solution 4. We have

$$\mathrm{tr}((\mathbf{n} \cdot \boldsymbol{\sigma})\rho) = \mathrm{tr}\left((\mathbf{n} \cdot \boldsymbol{\sigma})\frac{1}{2}(I_2 + \mathbf{r} \cdot \boldsymbol{\sigma})\right) = \frac{1}{2}\mathrm{tr}(\mathbf{n} \cdot \boldsymbol{\sigma} + (\mathbf{n} \cdot \boldsymbol{\sigma})(\mathbf{r} \cdot \boldsymbol{\sigma}))$$

$$= \frac{1}{2}\mathrm{tr}((\mathbf{n} \cdot \boldsymbol{\sigma})(\mathbf{r} \cdot \boldsymbol{\sigma})) = \frac{1}{2}\mathrm{tr}\left(\sum_{i,j=1}^{3} n_i r_j \sigma_i \sigma_j\right)$$

$$= \frac{1}{2}\mathrm{tr}\left(\sum_{i=1}^{3} n_i r_i I_2\right) = \sum_{i=1}^{3} n_i r_i = \mathbf{n} \cdot \mathbf{r}.$$

Thus the vector $\mathbf{r}$ can be thought of as an expectation value of spin polarization, and it can be obtained by measuring $\mathbf{n} \cdot \boldsymbol{\sigma}$ along each direction $\mathbf{e}_1$, $\mathbf{e}_2$ and $\mathbf{e}_3$.

Problem 5. Let A be a nonzero $n \times n$ matrix over $\mathbb{C}$. Then $\mathrm{tr}(AA^*) > 0$. Consider the map

$$A \to \rho = \frac{AA^*}{\mathrm{tr}(AA^*)}.$$

(i) Show that ρ is a density matrix.
(ii) Show that ρ is invariant under the map $A \to AU$, where U is an $n \times n$ unitary matrix.

(iii) Is $AA^* = A^*A$ in general? A matrix is called a *normal matrix* if $AA^* = A^*A$.

(iv) Consider the map

$$A \to \sigma = \frac{A^*A}{\mathrm{tr}(A^*A)} \,.$$

Is $\sigma = \rho$?

Solution 5. (i) Since A is a nonzero matrix we find that AA^* is nonzero and $\mathrm{tr}(AA^*) \neq 0$. The matrix ρ is positive-semidefinite and $\mathrm{tr}\rho = 1$.

(ii) We set $A' = AU$. Thus we have

$$\begin{aligned}
\rho \to \rho' &= \frac{A'A'^*}{\mathrm{tr}(A'A'^*)} = \frac{(AU)(AU)^*}{\mathrm{tr}((AU)(AU)^*)} \\
&= \frac{AUU^*A^*}{\mathrm{tr}(AUU^*A^*)} = \frac{AA^*}{\mathrm{tr}(AA^*)} \\
&= \rho
\end{aligned}$$

where we used that $UU^* = I$.

(iii) In general we have $AA^* \neq A^*A$. For example, let

$$A = \begin{pmatrix} 0 & 1 \\ 0 & 0 \end{pmatrix}, \qquad A^* = \begin{pmatrix} 0 & 0 \\ 1 & 0 \end{pmatrix}.$$

Then

$$AA^* = \begin{pmatrix} 1 & 0 \\ 0 & 0 \end{pmatrix}, \qquad A^*A = \begin{pmatrix} 0 & 0 \\ 0 & 1 \end{pmatrix}.$$

Thus $AA^* \neq A^*A$. However, we have $\mathrm{tr}(AA^*) = \mathrm{tr}(A^*A)$.

(iv) From (iii) it follows that in general we have $\rho \neq \sigma$.

Problem 6. (i) The *Hilbert-Schmidt distance* between any two density operators ρ_1 and ρ_2 is given by the Frobenius-Hilbert-Schmidt norm of their differences

$$D_{HS}(\rho_1, \rho_2) := \sqrt{\mathrm{tr}((\rho_1 - \rho_2)^2)} \,.$$

Let

$$\rho_1 = \begin{pmatrix} 1 & 0 \\ 0 & 0 \end{pmatrix}, \qquad \rho_2 = \begin{pmatrix} 0 & 0 \\ 0 & 1 \end{pmatrix}.$$

Calculate $D_{HS}(\rho_1, \rho_2)$.

(ii) The *Bures distance* in the space of mixed quantum states described by the density matrices ρ_1 and ρ_2 is defined as

$$D_B(\rho_1, \rho_2) := \sqrt{2(1 - \mathrm{tr}((\rho_1^{1/2}\rho_2\rho_1^{1/2})^{1/2}))} \,.$$

Let

$$\rho_1 = \begin{pmatrix} 1/2 & 0 \\ 0 & 1/2 \end{pmatrix}, \qquad \rho_2 = \begin{pmatrix} 3/4 & 0 \\ 0 & 1/4 \end{pmatrix}.$$

Calculate the Bures distance $D_B(\rho_1, \rho_2)$.

Solution 6. (i) Since

$$\rho_1 - \rho_2 = \begin{pmatrix} -1/4 & 0 \\ 0 & 1/4 \end{pmatrix}$$

we find $D_{HS}(\rho_1, \rho_2) = 1/\sqrt{8}$.
(ii) Since

$$\rho_1^{1/2} \rho_2 \rho_1^{1/2} = \begin{pmatrix} 3/8 & 0 \\ 0 & 1/8 \end{pmatrix}$$

we find

$$D_B(\rho_1, \rho_2) = \sqrt{\frac{2\sqrt{2} - \sqrt{3} - 1}{\sqrt{2}}}.$$

Problem 7. Consider the operator $(4 \times 4$ matrix$)$ in the Hilbert space $\mathbb{C}^4$

$$\rho = \frac{1}{4}(1 - \epsilon)I_4 + \epsilon(|0\rangle \otimes |0\rangle)(\langle 0| \otimes \langle 0|)$$

where ϵ is a real parameter with $\epsilon \in [0, 1]$ and the state

$$|0\rangle = \begin{pmatrix} 1 \\ 0 \end{pmatrix}.$$

Does ρ define a *density matrix*?

Solution 7. We find the diagonal matrix for ρ

$$\rho = \begin{pmatrix} (1 - \epsilon)/4 + \epsilon & 0 & 0 & 0 \\ 0 & (1 - \epsilon)/4 & 0 & 0 \\ 0 & 0 & (1 - \epsilon)/4 & 0 \\ 0 & 0 & 0 & (1 - \epsilon)/4 \end{pmatrix}.$$

Thus $\rho = \rho^\dagger$, $\mathrm{tr}\rho = 1$, and

$$\langle \mathbf{x}|\rho|\mathbf{x}\rangle \geq 0, \qquad \text{for all } \mathbf{x} \in \mathbb{C}^4.$$

The last property follows since all entries on the diagonal are non-negative. Thus ρ defines a density matrix.

Problem 8.   We consider *mixed states*. A mixed state is a statistical mixture of pure states, i.e., the state is described by pairs of probabilities and pure states. Given a mixture $\{ (p_1, |\psi_1\rangle), \ldots, (p_n, |\psi_n\rangle) \}$ we define its *density matrix* to be the positive hermitian matrix

$$\rho = \sum_{j=1}^{n} p_j |\psi_j\rangle \langle \psi_j|$$

where the pure states $|\psi_j\rangle$ are normalized (i.e., $\langle \psi_j | \psi_j \rangle = 1$), and $p_j \geq 0$ for $j = 1, 2, \ldots, n$ with

$$\sum_{j=1}^{n} p_j = 1.$$

(i) Find the probability that measurement in the orthonormal basis

$$\{ |k_1\rangle, \ldots, |k_n\rangle \}$$

will yield $|k_j\rangle$.

(ii) Find the density matrix ρ_U when the mixture is transformed according to the unitary matrix U.

Solution 8.   (i) From the probability distribution of states in the mixture we have for the probability $P(k_j)$ of measuring the state $|k_j\rangle$ $(j = 1, 2, \ldots, n)$

$$P(k_j) = \sum_{l=1}^{n} p_l |\langle k_j | \psi_l \rangle|^2 = \sum_{l=1}^{n} p_l \langle k_j | \psi_l \rangle \langle \psi_l | k_j \rangle = \langle k_j | \rho | k_j \rangle.$$

(ii) After applying the transform U to the states in the mixture we have the new mixture $\{ (p_1, U|\psi_1\rangle), \ldots, (p_n, U|\psi_n\rangle) \}$, with the density matrix

$$\rho_U = \sum_{j=1}^{n} p_j U|\psi_j\rangle \langle \psi_j| U^* = U \left(\sum_{j=1}^{n} p_j |\psi_j\rangle \langle \psi_j| \right) U^* = U \rho U^*.$$

Problem 9.   (i) The Bell state

$$|\psi\rangle = \frac{1}{\sqrt{2}} (|0\rangle \otimes |0\rangle + |1\rangle \otimes |1\rangle)$$

has the density matrix

$$\rho = \frac{1}{2} \begin{pmatrix} 1 & 0 & 0 & 1 \\ 0 & 0 & 0 & 0 \\ 0 & 0 & 0 & 0 \\ 1 & 0 & 0 & 1 \end{pmatrix}.$$

Show that ρ can be written as linear combination of the matrices $\Lambda_{00} = \frac{1}{2}(I_2 \otimes I_2)$, $\Lambda_{11} = \frac{1}{2}(\sigma_x \otimes \sigma_x)$, $\Lambda_{22} = \frac{1}{2}(\sigma_y \otimes \sigma_y)$ and $\Lambda_{33} = \frac{1}{2}(\sigma_z \otimes \sigma_z)$.
(ii) The *Werner state* is described by the density matrix

$$
\rho_W = \begin{pmatrix} (1-x)/4 & 0 & 0 & 0 \\ 0 & (1+x)/4 & -x/2 & 0 \\ 0 & -x/2 & (1+x)/4 & 0 \\ 0 & 0 & 0 & (1-x)/4 \end{pmatrix}
$$

where $x \in [0,1]$. Show that ρ_W can also be written as linear combination of the operators given in (i).

Solution 9. (i) We find the linear combination

$$
\rho = \frac{1}{2}\Lambda_{00} + \frac{1}{2}\Lambda_{11} - \frac{1}{2}\Lambda_{22} + \frac{1}{2}\Lambda_{33} .
$$

(ii) We obtain the linear combination

$$
\rho_W = \frac{1}{2}\Lambda_{00} - \frac{x}{2}\Lambda_{11} - \frac{x}{2}\Lambda_{22} - \frac{x}{2}\Lambda_{33} .
$$

Problem 10. Suppose we expand a density matrix for N qubits in terms of Kronecker products of Pauli spin matrices

$$
\rho = \frac{1}{2^N} \sum_{j_0=0}^{3} \sum_{j_1=0}^{3} \cdots \sum_{j_{N-1}=0}^{3} c_{j_0 j_1 \ldots j_{N-1}} \sigma_{j_0} \otimes \sigma_{j_1} \otimes \cdots \otimes \sigma_{j_{N-1}}
$$

where $\sigma_0 = I_2$.
(i) What is condition on the expansion coefficients if we impose $\rho^\dagger = \rho$?
(ii) What is the condition on the expansion coefficients if we impose $\mathrm{tr}\rho = 1$?
(iii) Calculate

$$
\mathrm{tr}\big(\rho\sigma_{k_0} \otimes \sigma_{k_1} \otimes \cdots \otimes \sigma_{k_{N-1}}\big) .
$$

Solution 10. (i) Since $\sigma_1 = \sigma_1^\dagger$, $\sigma_2 = \sigma_2^\dagger$, $\sigma_3 = \sigma_3^\dagger$ and $I_2 = I_2^\dagger$ we find that the expansion coefficients are real.
(ii) Since $\mathrm{tr}(A \otimes B) = \mathrm{tr}(A)\mathrm{tr}(B)$ for square matrices A and B and

$$
\mathrm{tr}\sigma_1 = \mathrm{tr}\sigma_2 = \mathrm{tr}\sigma_3 = 0, \qquad \mathrm{tr}I_2 = 2
$$

we find $c_{00\ldots0} = 1$.
(iii) Since $\mathrm{tr}(\sigma_1\sigma_2) = 0$, $\mathrm{tr}(\sigma_2\sigma_3) = 0$, $\mathrm{tr}(\sigma_3\sigma_1) = 0$ we find

$$
\mathrm{tr}\big(\rho\sigma_{k_0} \otimes \sigma_{k_1} \otimes \cdots \otimes \sigma_{k_{N-1}}\big) = c_{k_0 k_1 \ldots k_{N-1}} .
$$

Problem 11. Let A and B be a pair of qubits and let the density matrix of the pair be ρ_{AB}, which may be pure or mixed. We define the *spin flipped density matrix* to be

$$\tilde{\rho}_{AB} := (\sigma_y \otimes \sigma_y)\rho_{AB}^*(\sigma_y \otimes \sigma_y)$$

where the asterisk denotes complex conjugation and transpose in the standard basis

$$\{\, |0\rangle \otimes |0\rangle, \quad |0\rangle \otimes |1\rangle, \quad |1\rangle \otimes |0\rangle, \quad |1\rangle \otimes |1\rangle \,\}$$

and

$$\sigma_y = \begin{pmatrix} 0 & -i \\ i & 0 \end{pmatrix}.$$

Since both ρ_{AB} and $\tilde{\rho}_{AB}$ are positive operators, it follows that the product $\rho_{AB}\tilde{\rho}_{AB}$, though non-hermitian, also has only real and non-negative eigenvalues. Consider the Bell state

$$|\psi\rangle := \frac{1}{\sqrt{2}}(|0\rangle \otimes |0\rangle + |1\rangle \otimes |1\rangle)$$

and $\rho := |\psi\rangle\langle\psi|$. Find the eigenvalues of $\rho_{AB}\tilde{\rho}_{AB}$.

Solution 11. Since

$$\rho = |\psi\rangle\langle\psi| = \frac{1}{2}\begin{pmatrix} 1 & 0 & 0 & 1 \\ 0 & 0 & 0 & 0 \\ 0 & 0 & 0 & 0 \\ 1 & 0 & 0 & 1 \end{pmatrix}$$

we have $\rho^* = \rho$. Furthermore

$$\sigma_y \otimes \sigma_y = \begin{pmatrix} 0 & 0 & 0 & -1 \\ 0 & 0 & 1 & 0 \\ 0 & 1 & 0 & 0 \\ -1 & 0 & 0 & 0 \end{pmatrix}.$$

Thus $\tilde{\rho} = \rho$ and $\rho\tilde{\rho} = \rho$ with eigenvalues 1, 0, 0, 0. The *tangle* of the density matrix ρ_{AB} is defined as

$$\tau_{AB} := [\max\{\,\mu_1 - \mu_2 - \mu_3 - \mu_4,\, 0\}]^2$$

where μ_j are the square root of the eigenvalues of $\rho_{AB}\tilde{\rho}_{AB}$ ordered in decreasing order. For the special case in which the state of AB is pure, the matrix $\rho_{AB}\tilde{\rho}_{AB}$ has only one non-zero eigenvalue. One can show that

$$\tau_{AB} = 4\det\rho_A$$

where ρ_A is the density matrix of qubit A, that is, the trace of ρ_{AB} over qubit B.

Problem 12. Consider the density matrix

$$\rho = \frac{1}{2}\left(I_2 + \mathbf{r}\cdot\boldsymbol{\sigma}\right)$$

where $\mathbf{r}\cdot\boldsymbol{\sigma} := r_1\sigma_1 + r_2\sigma_2 + r_3\sigma_3$ and $\mathbf{r}^2 \leq 1$. Consider the four vectors $\mathbf{a}_1$, $\mathbf{a}_2$, $\mathbf{a}_3$, $\mathbf{a}_4$ in $\mathbb{R}^3$

$$\mathbf{a}_1 = \frac{1}{\sqrt{3}}\begin{pmatrix}1\\1\\1\end{pmatrix}, \quad \mathbf{a}_2 = \frac{1}{\sqrt{3}}\begin{pmatrix}1\\-1\\-1\end{pmatrix}, \quad \mathbf{a}_3 = \frac{1}{\sqrt{3}}\begin{pmatrix}-1\\1\\-1\end{pmatrix}, \quad \mathbf{a}_4 = \frac{1}{\sqrt{3}}\begin{pmatrix}-1\\-1\\1\end{pmatrix}$$

such that

$$\mathbf{a}_j^T\mathbf{a}_k = \frac{4}{3}\delta_{jk} - \frac{1}{3} = \begin{cases} 1 & \text{for } j = k \\ -1/3 & \text{for } j \neq k \end{cases}.$$

We have

$$\sum_{j=1}^{4}\mathbf{a}_j = \mathbf{0}, \qquad \frac{3}{4}\sum_{j=1}^{4}\mathbf{a}_j\mathbf{a}_j^T = I_3.$$

Such a quartet of vectors consists of the vectors pointing from the center of a cube to nonadjacent corners. These four vectors can be viewed as the normal vectors for the faces of the *tetrahedron* that is defined by the other four corners of the cube. Owing to the conditions the four vectors are normalized. Each such quartets of $\mathbf{a}_j$'s defines a positive operator-valued measure for minimal four-state tomography owing to

$$\sum_{j=1}^{4}P_j = I_2, \qquad P_j := \frac{1}{4}\left(I_2 + \mathbf{a}_j\cdot\boldsymbol{\sigma}\right).$$

(i) Show that

$$p_j := \langle P_j\rangle = \operatorname{tr}(P_j\rho) = \frac{1}{4}\left(1 + \mathbf{a}_j\cdot\mathbf{r}\right).$$

(ii) Given p_j for $j = 1,2,3,4$ find the density matrix ρ.

Solution 12. (i) Since $\operatorname{tr}\sigma_1 = \operatorname{tr}\sigma_2 = \operatorname{tr}\sigma_3 = 0$ we have

$$\langle P_j\rangle = \operatorname{tr}(P_j\rho) = \frac{1}{8}\operatorname{tr}((I_2 + \mathbf{a}_j\cdot\boldsymbol{\sigma})(I_2 + \mathbf{r}\cdot\boldsymbol{\sigma}))$$

$$= \frac{1}{8}\operatorname{tr}(I_2 + \mathbf{a}_j\cdot\boldsymbol{\sigma} + \mathbf{r}\cdot\boldsymbol{\sigma} + (\mathbf{a}_j\cdot\boldsymbol{\sigma})(\mathbf{r}\cdot\boldsymbol{\sigma})) = \frac{1}{8}\operatorname{tr}(I_2 + (\mathbf{a}_j\cdot\boldsymbol{\sigma})(\mathbf{r}\cdot\boldsymbol{\sigma}))$$

$$= \frac{1}{8}\operatorname{tr}(I_2 + (\mathbf{a}_j\cdot\mathbf{r})I_2)$$

$$= \frac{1}{4}(1 + \mathbf{a}_j\cdot\mathbf{r}).$$

(ii) Since

$$I_3 \mathbf{r} = \left(\frac{3}{4} \sum_{j=1}^{4} \mathbf{a}_j \mathbf{a}_j^T \right) \mathbf{r}$$

the vector $\mathbf{r}$ is obtained as

$$\mathbf{r} = \frac{3}{4} \sum_{j=1}^{4} (\mathbf{a}_j \cdot \mathbf{r}) \mathbf{a}_j = 3 \sum_{j=1}^{4} \frac{1}{4} (1 + \mathbf{a}_j \cdot \mathbf{r}) \, \mathbf{a}_j = 3 \sum_{j=1}^{4} p_j \mathbf{a}_j$$

where we used $\sum_{j=1}^{4} \mathbf{a}_j = \mathbf{0}$. From $\mathbf{a}_j \cdot \boldsymbol{\sigma} = 4 P_j - I_2$ and substituting $\mathbf{r}$ from above yields the density matrix

$$\rho = 6 \sum_{j=1}^{4} p_j P_j - I_2 = \sum_{j=1}^{4} \langle P_j \rangle (6 P_j - I_2) = \frac{1}{2} (I_2 + \mathbf{r} \cdot \boldsymbol{\sigma}).$$

It follows that p_j is restricted to the range $0 \le p_j \le 1/2$ and the probabilities p_j obey the inequalities

$$\frac{1}{4} \le \sum_{j=1}^{4} p_j^2 = \frac{3 + \mathbf{r}^2}{12} \le \frac{1}{3}.$$

The upper bound is reached by all pure states, $\rho = \rho^2$ and $\mathbf{r}^2 = 1$. The lower bound is reached for the completely mixed state, $\rho = 1/2$ and $\mathbf{r}^2 = 0$.

Problem 13. Consider the mixed states

$$\frac{1}{2} \left\{ \frac{3}{5} |0\rangle + \frac{4}{5} |1\rangle \right\} + \frac{1}{2} \left\{ \frac{3}{5} |0\rangle - \frac{4}{5} |1\rangle \right\}$$

and

$$\frac{9}{25} \{ |0\rangle \} + \frac{16}{25} \{ |1\rangle \}.$$

Find the density matrices. Discuss.

Solution 13. In the first case we have

$$\rho_1 = \frac{1}{2} \begin{pmatrix} 9/25 & 12/25 \\ 12/25 & 16/25 \end{pmatrix} + \frac{1}{2} \begin{pmatrix} 9/25 & -12/25 \\ 12/25 & 16/25 \end{pmatrix} = \begin{pmatrix} 9/25 & 0 \\ 0 & 16/25 \end{pmatrix}.$$

In the second case we have

$$\rho_2 = \frac{9}{25} \begin{pmatrix} 1 & 0 \\ 0 & 0 \end{pmatrix} + \frac{16}{25} \begin{pmatrix} 0 & 0 \\ 0 & 1 \end{pmatrix} = \begin{pmatrix} 9/25 & 0 \\ 0 & 16/25 \end{pmatrix}.$$

Thus these two different mixed states correspond to the same density matrix and thus they are indistinguishable.

Problem 14. Let ρ_1 and ρ_2 be $n \times n$ density matrices. Let λ_j denote the eigenvalues of $\rho_1 - \rho_2$ with corresponding orthonormal eigenvectors $|\phi_j\rangle$ where $j = 1, 2, \ldots, n$.
(i) Find the difference $|D_1 - D_2|$ between the probability distributions D_1 and D_2 for the measurement of the mixtures ρ_1 and ρ_2 in the basis $\{\, |\phi_1\rangle, \ldots, |\phi_n\rangle \,\}$.
(ii) Show that measurement in the basis $\{\, |\phi_1\rangle, \ldots, |\phi_n\rangle \,\}$ maximizes the difference $|D_1 - D_2|$.
Hint. Use *Schur's theorem*. For any hermitian matrix A, let

$$a_1 \geq a_2 \geq \cdots \geq a_n$$

be the non increasing diagonal entries of A and

$$\mu_1 \geq \mu_2 \geq \cdots \geq \mu_n$$

the non increasing eigenvalues of A. Then for $1 \leq k \leq n$

$$\sum_{j=1}^{k} \mu_j \geq \sum_{j=1}^{k} a_j$$

where equality holds for $k = n$.

Solution 14. (i) We write ρ_1 and ρ_2 in the basis $\{\, |\phi_1\rangle, \ldots, |\phi_n\rangle \,\}$. In this basis we have

$$|D_1 - D_2| = \sum_{j=1}^{n} |\langle\phi_j|\rho_1|\phi_j\rangle - \langle\phi_j|\rho_2|\phi_j\rangle| = \sum_{j=1}^{n} |\langle\phi_j|(\rho_1 - \rho_2)|\phi_j\rangle| = \sum_{j=1}^{n} |\lambda_j| \, .$$

(ii) Let U be an arbitrary unitary transform (change of basis). We define $P := U\rho_1 U^*$ and $Q := U\rho_2 U^*$. The matrix $P - Q$ is hermitian. Let

$$q_1 \geq q_2 \geq \cdots \geq q_n$$

denote the non decreasing diagonal entries of $P - Q$ in the $\{\, |\phi_1\rangle, \ldots, |\phi_n\rangle \,\}$ basis. Let

$$\nu_1 \geq \nu_2 \geq \cdots \geq \nu_n$$

be the non decreasing eigenvalues (i.e. λ_j) of $P - Q$. Consider the difference $|D_1' - D_2'|$ between the probability distributions D_1' and D_2' for the measurement of the mixtures ρ_1 and ρ_2 in the basis $\{\, U|\phi_1\rangle, \ldots, U|\phi_n\rangle \,\}$.

$$|D_1' - D_2'| = \sum_{j=1}^{n} |\langle\phi_j|U^*\rho_1 U|\phi_j\rangle - \langle\phi_j|U^*\rho_2 U|\phi_j\rangle|$$

$$= \sum_{j=1}^{n} |\langle\phi_j|P|\phi_j\rangle - \langle\phi_j|Q|\phi_j\rangle| = \sum_{j=1}^{n} |q_j| \, .$$

Since $\operatorname{tr}(P - Q) = \operatorname{tr}(P) - \operatorname{tr}(Q) = 1 - 1 = 0$ and

$$\operatorname{tr}(P) - \operatorname{tr}(Q) = \sum_{j=1}^{n}(\langle\phi_j|P|\phi_j\rangle - \langle\phi_j|Q|\phi_j\rangle) = \sum_{j=1}^{n} q_j$$

we have for all $1 \le k \le n$

$$\left|\sum_{j=1}^{k} q_j\right| = \left|\sum_{j=k+1}^{n} q_j\right|.$$

We conclude from the *triangle inequality* that

$$\sum_{j=1}^{n} |q_j| \ge 2\left|\sum_{j=1}^{k} q_j\right|$$

where equality holds for some $1 \le k_0 \le n$. Similarly

$$\sum_{j=1}^{n} |\nu_j| \ge 2\left|\sum_{j=1}^{k} \nu_j\right|.$$

From Schur's theorem we have

$$\sum_{j=1}^{n} |\nu_j| \ge \sum_{j=1}^{k_0} \nu_j \ge \sum_{j=1}^{k_0} q_j = \sum_{j=1}^{n} |q_j|.$$

Thus

$$\sum_{j=1}^{n} |\nu_j| = |D_1 - D_2| \ge |D_1' - D_2'| = \sum_{j=1}^{n} |q_j|.$$

Problem 15.   Let σ_1, σ_2, σ_3 be the Pauli spin matrices. Is the 4×4 matrix

$$\rho = \frac{1}{4}(I_2 \otimes I_2 + \sigma_x \otimes \sigma_x)$$

a density matrix?

Solution 15.   We have $\operatorname{tr}\rho = 1$ and ρ is hermitian. The eigenvalues of ρ are $1/2$, $1/2$, 0, 0. Thus the matrix ρ is a density matrix (mixed state).

Problem 16.   Is the 4×4 matrix

$$\rho = \frac{1}{4}(I_4 - \sigma_1 \otimes \sigma_1 - \sigma_2 \otimes \sigma_2 - \sigma_3 \otimes \sigma_3)$$

a density matrix?

Solution 16. Obviously we have $\rho^* = \rho$, $\text{tr}\rho = 1$. Furthermore the matrix is positive semidefinite. We have $\rho^2 = \rho$. Thus ρ is a density matrix. We have a pure state. The density matrix is given by

$$\rho = \begin{pmatrix} 0 & 0 & 0 & 0 \\ 0 & 1/2 & -1/2 & 0 \\ 0 & -1/2 & 1/2 & 0 \\ 0 & 0 & 0 & 0 \end{pmatrix}.$$

Problem 17. Let ρ, σ be two density operators acting in the same finite dimensional Hilbert space. The *trace distance* between ρ and σ is defined as

$$D(\rho, \sigma) := \frac{1}{2}\text{tr}\sqrt{(\rho^* - \sigma^*)(\rho - \sigma)}.$$

Let

$$\rho = \begin{pmatrix} 1/2 & 0 & 0 & 1/2 \\ 0 & 0 & 0 & 0 \\ 0 & 0 & 0 & 0 \\ 1/2 & 0 & 0 & 1/2 \end{pmatrix}, \qquad \sigma = \begin{pmatrix} 1/2 & 0 & 0 & 0 \\ 0 & 0 & 0 & 0 \\ 0 & 0 & 0 & 0 \\ 0 & 0 & 0 & 1/2 \end{pmatrix}.$$

Find the trace distance.

Solution 17. We have

$$\rho - \sigma = \begin{pmatrix} 0 & 0 & 0 & 1/2 \\ 0 & 0 & 0 & 0 \\ 0 & 0 & 0 & 0 \\ 1/2 & 0 & 0 & 0 \end{pmatrix}.$$

Thus

$$(\rho^* - \sigma^*)(\rho - \sigma) = \begin{pmatrix} 1/4 & 0 & 0 & 0 \\ 0 & 0 & 0 & 0 \\ 0 & 0 & 0 & 0 \\ 0 & 0 & 0 & 1/4 \end{pmatrix}.$$

It follows that $D(\rho, \sigma) = 1/2$.

Problem 18. Let $|0\rangle$, $|1\rangle$ be an orthonormal basis in $\mathbb{C}^2$. The *Kraus operators* are defined by

$$A_0 := (\alpha|0\rangle\langle 0| + \beta|1\rangle\langle 1|) \otimes I_2$$
$$A_1 := (\beta|1\rangle\langle 0| + \alpha|0\rangle\langle 1|) \otimes (|1\rangle\langle 0| + |0\rangle\langle 1|)$$

where $\alpha\alpha^* + \beta\beta^* = 1$.

(i) Show that

$$A_0^\dagger A_0 + A_1^\dagger A_1 = I_2 \otimes I_2 \,.$$

(ii) Let

$$|\phi\rangle = \alpha|0\rangle \otimes |0\rangle + \beta|1\rangle \otimes |1\rangle, \qquad |\psi\rangle = \frac{1}{\sqrt{2}}(|0\rangle \otimes |0\rangle + |1\rangle \otimes |1\rangle)\,.$$

Show that

$$A_0|\psi\rangle\langle\psi|A_0^\dagger + A_1|\psi\rangle\langle\psi|A_1^\dagger = |\phi\rangle\langle\phi|\,.$$

Solution 18.   (i) From

$$A_0^\dagger = (\alpha^*|0\rangle\langle 0| + \beta^*|1\rangle\langle 1|) \otimes I_2$$

we obtain

$$A_0^\dagger A_0 = (\alpha\alpha^*|0\rangle\langle 0| + \beta\beta^*|1\rangle\langle 1|) \otimes I_2\,.$$

Since

$$A_1^\dagger = (\beta^*|0\rangle\langle 1| + \alpha^*|1\rangle\langle 0|) \otimes (|0\rangle\langle 1| + |1\rangle\langle 0|)$$

we have

$$A_1^\dagger A_1 = (\beta\beta^*|0\rangle\langle 0| + \alpha\alpha^*|1\rangle\langle 1|)) \otimes (|0\rangle\langle 0| + |1\rangle\langle 1|)\,.$$

Since $|0\rangle\langle 0| + |1\rangle\langle 1| = I_2$ and $\alpha\alpha^* + \beta\beta^* = 1$ we obtain the result.
(ii) Since

$$A_0|\psi\rangle = \frac{1}{\sqrt{2}}(\alpha|0\rangle \otimes |0\rangle + \beta|1\rangle \otimes |1\rangle)$$

we have

$$A_0|\psi\rangle\langle\psi|A_0^\dagger =$$

$$\frac{1}{2}(\alpha\alpha^*|0\rangle\langle 0|\otimes|0\rangle\langle 0|+\beta\beta^*|1\rangle\langle 1|\otimes|1\rangle\langle 1|+\alpha\beta^*|0\rangle\langle 1|\otimes|0\rangle\langle 1|+\alpha^*\beta|1\rangle\langle 0|\otimes|1\rangle\langle 0|)\,.$$

Since

$$A_1|\psi\rangle = \frac{1}{\sqrt{2}}(\alpha|0\rangle \otimes |0\rangle + \beta|1\rangle \otimes |1\rangle) = A_0|\psi\rangle$$

we have

$$A_1|\psi\rangle\langle\psi|A_1^\dagger = A_0|\psi\rangle\langle\psi|A_0\,.$$

Thus

$$A_0|\psi\rangle\langle\psi|A_0^\dagger + A_1|\psi\rangle\langle\psi|A_1^\dagger =$$

$$\alpha\alpha^*|0\rangle\langle 0| \otimes |0\rangle\langle 0|+\beta\beta^*|1\rangle\langle 1| \otimes |1\rangle\langle 1|+\alpha\beta^*|0\rangle\langle 1| \otimes |0\rangle\langle 1|+\alpha^*\beta|1\rangle\langle 0| \otimes |1\rangle\langle 0|$$

which is the density matrix $|\phi\rangle\langle\phi|$.

Problem 19. Consider the eight *Gell-Mann matrices*

$$G_1 = \begin{pmatrix} 0 & 1 & 0 \\ 1 & 0 & 0 \\ 0 & 0 & 0 \end{pmatrix}, \quad G_2 = \begin{pmatrix} 0 & -i & 0 \\ i & 0 & 0 \\ 0 & 0 & 0 \end{pmatrix}, \quad G_3 = \begin{pmatrix} 1 & 0 & 0 \\ 0 & -1 & 0 \\ 0 & 0 & 0 \end{pmatrix},$$

$$G_4 = \begin{pmatrix} 0 & 0 & 1 \\ 0 & 0 & 0 \\ 1 & 0 & 0 \end{pmatrix}, \quad G_5 = \begin{pmatrix} 0 & 0 & -i \\ 0 & 0 & 0 \\ i & 0 & 0 \end{pmatrix}, \quad G_6 = \begin{pmatrix} 0 & 0 & 0 \\ 0 & 0 & 1 \\ 0 & 1 & 0 \end{pmatrix},$$

$$G_7 = \begin{pmatrix} 0 & 0 & 0 \\ 0 & 0 & -i \\ 0 & i & 0 \end{pmatrix}, \quad G_8 = \frac{1}{\sqrt{3}} \begin{pmatrix} 1 & 0 & 0 \\ 0 & 1 & 0 \\ 0 & 0 & -2 \end{pmatrix}.$$

They all have trace 0.
(i) Find the anticommutation relations for these matrices.
(ii) Consider the matrix

$$\rho = \frac{1}{3} \left(I_3 + \sqrt{3} \sum_{j=1}^{8} n_j G_j \right)$$

where $n_j \in \mathbb{R}$. What is the condition on the vector $\mathbf{n} = (n_1 \ n_2 \ \dots \ n_8)^T$ such that ρ is a density matrix for a pure state?

Solution 19. (i) We obtain for the anticommutators

$$[G_j, G_k]_+ = \frac{4}{3} \delta_{jk} I_3 + 2 \sum_{\ell=1}^{8} d_{jk\ell} G_\ell$$

where the nonzero components of the completely symmetric tensor $d_{jk\ell}$ are

$$d_{118} = d_{228} = d_{338} = -d_{888} = \frac{1}{\sqrt{3}}, \quad d_{448} = d_{558} = d_{668} = d_{778} = -\frac{1}{2\sqrt{3}}$$

$$d_{146} = d_{157} = -d_{247} = d_{256} = d_{344} = d_{355} = -d_{366} = -d_{377} = \frac{1}{2}.$$

(ii) The conditions for a *density matrix* ρ of a *pure state* are

$$\rho^* = \rho, \quad \rho^2 = \rho, \quad \text{tr}\rho = 1.$$

Imposing these conditions we obtain for the vector $\mathbf{n}$ that

$$\bar{\mathbf{n}} = \mathbf{n}, \quad \mathbf{n}^T \mathbf{n} = 1, \quad \mathbf{n} \star \mathbf{n} = \mathbf{n}$$

where

$$(\mathbf{a} \star \mathbf{b})_j := \sqrt{3} \sum_{k,\ell=1}^{8} d_{jk\ell} a_k b_\ell.$$

Problem 20.   Let ρ denote the density matrix

$$\rho := \frac{1}{2}\begin{pmatrix} 1 & 0 \\ 0 & 1 \end{pmatrix}$$

in $\mathbb{C}^2$. Find a pure state $|\Psi\rangle \in \mathbb{C}^2 \otimes \mathbb{C}^2$ such that the reduced density matrix found by taking the partial trace over the second system ($\mathbb{C}^2$) is ρ. In other words *purify the density matrix* ρ to obtain a pure state $|\Psi\rangle$.

Solution 20.   We begin with the *Schmidt decomposition* of $|\Psi\rangle$ over the Hilbert space $\mathbb{C}^2 \otimes \mathbb{C}^2$

$$|\Psi\rangle = \sum_{j=1}^{\mathrm{Sch}(|\Psi\rangle,\mathbb{C}^2,\mathbb{C}^2)} \sqrt{\lambda_j}|\psi_j\rangle \otimes |\phi_j\rangle$$

where λ_1 and λ_2 are the eigenvalues of ρ and $|\psi_1\rangle$ and $|\psi_2\rangle$ are the corresponding orthonormal eigenvectors of ρ. The states $|\phi_1\rangle$ and $|\phi_2\rangle$ in $\mathbb{C}^2$ are also orthonormal. The eigenvalues and eigenvectors of ρ are given by $\lambda_1 = \lambda_2 = 1/2$ and

$$|\psi_1\rangle = \begin{pmatrix} 1 \\ 0 \end{pmatrix}, \qquad |\psi_2\rangle = \begin{pmatrix} 0 \\ 1 \end{pmatrix}.$$

Thus the spectral decomposition of ρ is given by

$$\rho = \frac{1}{2}\begin{pmatrix} 1 \\ 0 \end{pmatrix}\begin{pmatrix} 1 & 0 \end{pmatrix} + \frac{1}{2}\begin{pmatrix} 0 \\ 1 \end{pmatrix}\begin{pmatrix} 0 & 1 \end{pmatrix}.$$

Hence

$$|\Psi\rangle = \frac{1}{\sqrt{2}}\begin{pmatrix} 1 \\ 0 \end{pmatrix} \otimes |\phi_1\rangle + \frac{1}{\sqrt{2}}\begin{pmatrix} 0 \\ 1 \end{pmatrix} \otimes |\phi_2\rangle$$

where $\langle\phi_1|\phi_1\rangle = \langle\phi_2|\phi_2\rangle = 1$ and $\langle\phi_1|\phi_2\rangle = \langle\phi_2|\phi_1\rangle = 0$. Thus we could take $|\Psi\rangle$ as one of the *Bell states*

$$\frac{1}{\sqrt{2}}\begin{pmatrix} 1 \\ 0 \\ 0 \\ 1 \end{pmatrix}, \quad \frac{1}{\sqrt{2}}\begin{pmatrix} 1 \\ 0 \\ 0 \\ -1 \end{pmatrix}, \quad \frac{1}{\sqrt{2}}\begin{pmatrix} 0 \\ 1 \\ 1 \\ 0 \end{pmatrix}, \quad \frac{1}{\sqrt{2}}\begin{pmatrix} 0 \\ 1 \\ -1 \\ 0 \end{pmatrix}$$

but not a product state.

Problem 21.   Consider a spin-1 system. Any pure state can be parameterized, with a suitable choice of its phase as

$$|\psi\rangle = \begin{pmatrix} e^{i\alpha}\sin\theta\cos\phi \\ e^{i\beta}\sin\theta\sin\phi \\ \cos\theta \end{pmatrix}$$

where $0 \le \theta, \phi \le \pi/2$ and $0 \le \alpha, \beta < 2\pi$. Find the density matrix $\rho = |\psi\rangle\langle\psi|$.

Solution 21. We obtain

$$\rho = \begin{pmatrix} \sin^2\theta\cos^2\phi & \sin^2\theta\cos\phi\sin\phi e^{i(\alpha-\beta)} & \sin\theta\cos\theta\cos\phi e^{i\alpha} \\ \sin^2\theta\sin\phi\cos\phi e^{i(\beta-\alpha)} & \sin^2\theta\sin^2\phi & \sin\theta\cos\theta\sin\phi e^{i\beta} \\ \sin\theta\cos\theta\sin\phi e^{-i\alpha} & \sin\theta\cos\theta\sin\phi e^{-i\beta} & \cos^2\theta \end{pmatrix}.$$

Thus the density matrix depends on all four parameters.

Problem 22. Let $r \in [0, 1]$. Consider the density matrix

$$\rho = r|\Phi^+\rangle\langle\Phi^+| + (1-r)|00\rangle\langle00|$$

where $|\Phi^+\rangle$ is the Bell state

$$|\Phi^+\rangle = \frac{1}{\sqrt{2}}(|00\rangle + |11\rangle).$$

Calculate the eigenvalues of ρ.

Solution 22. The matrix representation of ρ is

$$\rho = \begin{pmatrix} 1 - r/2 & 0 & 0 & r/2 \\ 0 & 0 & 0 & 0 \\ 0 & 0 & 0 & 0 \\ r/2 & 0 & 0 & r/2 \end{pmatrix}.$$

The characteristic equation is

$$\left(\left(1 - \frac{r}{2}\right) - \lambda\right)(-\lambda)(-\lambda)\left(\frac{r}{2} - \lambda\right) - \frac{r}{2}(-\lambda)(-\lambda)\frac{r}{2} = 0.$$

Thus two eigenvalues are 0 ($\lambda_2 = 0$, $\lambda_3 = 0$) with the corresponding eigenvectors

$$\mathbf{u}_2 = \begin{pmatrix} 0 \\ 1 \\ 0 \\ 0 \end{pmatrix}, \qquad \mathbf{u}_3 = \begin{pmatrix} 0 \\ 0 \\ 1 \\ 0 \end{pmatrix}.$$

The characteristic equation reduces to

$$\left(\left(1 - \frac{r}{2}\right) - \lambda\right)\left(\frac{r}{2} - \lambda\right) - \frac{r^2}{4} = 0$$

with the eigenvalues

$$\lambda_{1,4} = \frac{1}{2} \pm \frac{1}{2}\sqrt{1 + 2r(r - 1)}.$$

If $r = 0$ the eigenvalues reduce to 1 and 0. If $r = 1$ the eigenvalues also reduce to 1 and 0.

Problem 23. Let $\rho_j^{(1)}$ $(j = 1, 2, \ldots, n)$ be density matrices in a finite-dimensional Hilbert space $\mathcal{H}_1$. Let $\rho_j^{(2)}$ $(j = 1, 2, \ldots, n)$ be density matrices in a finite-dimensional Hilbert space $\mathcal{H}_2$. Show that the *convex combination*

$$\rho = \sum_{j=1}^{n} \lambda_j \rho_j^{(1)} \otimes \rho_j^{(2)}, \quad \lambda_j \geq 0, \quad \sum_{j=1}^{n} \lambda_j = 1$$

is also a density matrix.

Solution 23. We have

$$\text{tr}(\rho) = \text{tr}\left(\sum_{j=1}^{n} \lambda_j \rho_j^{(1)} \otimes \rho_j^{(2)} \right) = \sum_{j=1}^{n} \lambda_j \text{tr}(\rho_j^{(1)} \otimes \rho_j^{(2)})$$

$$= \sum_{j=1}^{n} \lambda_j \text{tr}\rho_j^{(1)} \text{tr}\rho_j^{(2)} = \sum_{j=1}^{n} \lambda_j$$

$$= 1 \,.$$

Obviously $\rho \geq 0$, since $\rho_j^{(1)} \geq 0$, $\rho_j^{(2)} \geq 0$ and $\lambda_j \geq 0$ for $j = 1, 2, \ldots, n$.

Problem 24. Given the *Schrödinger equation*

$$i\hbar \frac{\partial}{\partial t} |\psi\rangle = \hat{H} |\psi\rangle \,.$$

Find the time-evolution of the density matrix

$$\rho(t) := \sum_{j=1}^{n} |\psi^{(j)}(t)\rangle \langle \psi^{(j)}(t)| \,.$$

Solution 24. From the Schrödinger equation we find

$$-i\hbar \frac{\partial}{\partial t} \langle \psi^{(j)}(t)| = \langle \psi^{(j)}(t)| \hat{H} \,.$$

Thus

$$\frac{\partial \rho}{\partial t} = \sum_{j=1}^{n} \left(\left(\frac{\partial}{\partial t} |\psi^{(j)}(t)\rangle \right) \langle \psi^{(j)}(t)| + |\psi^{(j)}(t)\rangle \left(\frac{\partial}{\partial t} \langle \psi^{(j)}(t)| \right) \right)$$

$$= \frac{1}{i\hbar} \sum_{j=1}^{n} \left(\left(\hat{H} | \psi^{(j)}(t) \rangle \right) \langle \psi^{(j)}(t) | - | \psi^{(j)}(t) \rangle \left(\langle \psi^{(j)}(t) | \hat{H} \right) \right)$$

$$= \frac{1}{i\hbar} (\hat{H} \rho(t) - \rho(t) \hat{H})$$

$$= \frac{1}{i\hbar} [\hat{H}, \rho(t)] .$$

Note that the equation of motion for $\rho(t)$ differs from the Heisenberg equation of motion by a minus sign. Since $\rho(t)$ is constructed from state vectors it is not an observable like other hermitian operators, so there is no reason to expect that its time-evolution will be the same. The solution to the equation of motion is given by

$$\rho(t) = e^{-i\hat{H}t/\hbar} \rho(0) e^{i\hat{H}t/\hbar} .$$

Problem 25. Consider a quantum system of spin-1/2 particles. The density matrix describing the spin degree of freedom is a 2×2 matrix which can be written as

$$\rho(\mathbf{n}) = \frac{1}{2}(I_2 + \mathbf{n} \cdot \boldsymbol{\sigma}) \equiv \frac{1}{2}(I_2 + n_1\sigma_1 + n_2\sigma_2 + n_3\sigma_3)$$

where σ_1, σ_2, σ_3 denote the Pauli spin matrices and $|\mathbf{n}| \leq 1$. For $|\mathbf{n}| = 1$ the density matrix describes a pure state, whereas for $|\mathbf{n}| < 1$ one has a mixed state. The density matrix ρ is thus uniquely determined by a point of the unit sphere $|\mathbf{n}| \leq 1$. Consider the Hamilton operator

$$\hat{H}(t) = -\frac{\gamma}{2} \boldsymbol{\sigma} \cdot \mathbf{B}(t) \equiv -\frac{\gamma}{2}(\sigma_1 B_1(t) + \sigma_2 B_2(t) + \sigma_3 B_3(t))$$

where γ denotes the gyromagnetic ratio and $\mathbf{B}(t)$ denotes the time-dependent magnetic induction. The time-evolution of the density matrix $\rho(t)$ obeys the *von Neumann equation*

$$i\hbar \frac{d\rho(t)}{dt} = [\hat{H}(t), \rho(t)]$$

and the time-dependent expectation value of the spin vector is given by

$$\langle \boldsymbol{\sigma}(t) \rangle := \mathrm{tr}(\boldsymbol{\sigma}\rho(t))$$

or, written in components

$$\langle \sigma_1(t) \rangle = \mathrm{tr}(\sigma_1\rho(t)), \quad \langle \sigma_2(t) \rangle = \mathrm{tr}(\sigma_2\rho(t)), \quad \langle \sigma_3(t) \rangle = \mathrm{tr}(\sigma_3\rho(t)) .$$

It follows that the *Bloch vector* $\mathbf{n}(t)$ pertaining to $\rho(t)$ is related to the spin vector as follows

$$\mathbf{n}(t) = \langle \boldsymbol{\sigma}(t) \rangle$$

or, written in components

$$n_1(t) = \langle \sigma_1(t) \rangle, \quad n_2(t) = \langle \sigma_2(t) \rangle, \quad n_3(t) = \langle \sigma_3(t) \rangle.$$

Find the time-evolution of $\mathbf{n}(t)$.

Solution 25. We have

$$\frac{dn_j}{dt} = \left\langle \frac{d\sigma_j(t)}{dt} \right\rangle = \text{tr}\left(\sigma_j \frac{d\rho(t)}{dt} \right)$$

where $j = 1, 2, 3$. Inserting the right-hand side of the von Neumann equation, using the *cyclic invariance* of the trace

$$\text{tr}(XYZ) = \text{tr}(ZXY) = \text{tr}(YZX)$$

and the properties $\sigma_1\sigma_2 = i\sigma_3$, $\sigma_2\sigma_3 = i\sigma_1$, $\sigma_3\sigma_1 = i\sigma_2$, we obtain

$$\frac{d}{dt}\mathbf{n}(t) = \frac{\gamma}{\hbar}\mathbf{n}(t) \times \mathbf{B}(t)$$

where $\times$ denotes the vector product.

Chapter 5

Trace and Partial Trace

Let $\mathcal{H}$ be the finite dimensional Hilbert space $\mathbb{C}^n$ with an orthonormal basis $\{\,|\phi_j\rangle \,:\, j = 1, 2, \ldots, n\,\}$. Let A be a linear operator ($n \times n$ matrix) acting in this Hilbert space. Then the *trace* of A is defined as

$$\operatorname{tr} A := \sum_{j=1}^{n} \langle \phi_j | A | \phi_j \rangle \,.$$

The trace is independent of the chosen orthonormal basis. The calculation of the *partial trace* plays a central role in quantum computing. Suppose a finite dimensional quantum system S^{AB} is a system composed of two subsystems S^A and S^B. The finite dimensional Hilbert space $\mathcal{H}$ of S^{AB} is given by the tensor product of the individual Hilbert spaces $\mathcal{H}_A \otimes \mathcal{H}_B$. Let $N_A := \dim \mathcal{H}_A$ and $N_B := \dim \mathcal{H}_B$. Let ρ^{AB} be the density matrix of S^{AB}. Using the partial trace we can define the density operators ρ^A and ρ^B in the subspaces $\mathcal{H}_A$ and $\mathcal{H}_B$ as follows

$$\rho^A := \operatorname{tr}_B \rho^{AB} \equiv \sum_{j=1}^{N_B} (I_A \otimes \langle \phi_j |) \rho^{AB} (I_A \otimes |\phi_j\rangle)$$

and

$$\rho^B := \operatorname{tr}_A \rho^{AB} \equiv \sum_{j=1}^{N_A} (\langle \psi_j | \otimes I_B) \rho^{AB} (|\psi_j\rangle \otimes I_B)$$

where I_A is the identity operator in $\mathcal{H}_A$, I_B is the identity operator in $\mathcal{H}_B$, $|\phi_j\rangle$ ($j = 1, 2, \ldots, N_B$) is an orthonormal basis in $\mathcal{H}_B$ and $|\psi_j\rangle$ ($j = 1, 2, \ldots, N_A$) is an orthonormal basis in $\mathcal{H}_A$.

Problem 1. Consider the Pauli spin matrix σ_x. Calculate the trace of σ_x with respect to the standard basis. Calculate the trace of σ_x with respect to the *Hadamard basis*

$$\left\{ \frac{1}{\sqrt{2}} \begin{pmatrix} 1 \\ 1 \end{pmatrix}, \quad \frac{1}{\sqrt{2}} \begin{pmatrix} 1 \\ -1 \end{pmatrix} \right\}.$$

Solution 1. Obviously for the standard basis we find $\mathrm{tr}\,\sigma_x = 0$. Since

$$\frac{1}{\sqrt{2}}(1 \ 1) \begin{pmatrix} 0 & 1 \\ 1 & 0 \end{pmatrix} \frac{1}{\sqrt{2}} \begin{pmatrix} 1 \\ 1 \end{pmatrix} = 1, \quad \frac{1}{\sqrt{2}}(1 \ -1) \begin{pmatrix} 0 & 1 \\ 1 & 0 \end{pmatrix} \frac{1}{\sqrt{2}} \begin{pmatrix} 1 \\ -1 \end{pmatrix} = -1$$

we also obtain (as expected since the trace is independent of the chosen orthonormal basis) for the Hadamard basis the result 0. In the Hadamard basis we have the Pauli matrix σ_z.

Problem 2. Assume that the density matrix ρ^{AB} is *separable*, i.e.,

$$\rho^{AB} = \rho_A \otimes \rho_B.$$

Calculate ρ^A and ρ^B.

Solution 2. Obviously we find $\rho^A = \rho_A$ and $\rho^B = \rho_B$. This can be seen as follows

$$\rho^A = \sum_{j=1}^{N_B} (I_A \otimes \langle \phi_j |)(\rho_A \otimes \rho_B)(I_A \otimes |\phi_j\rangle) = \sum_{j=1}^{N_B} (I_A \otimes \langle \phi_j |)(\rho_A \otimes \rho_B |\phi_j\rangle).$$

Thus

$$\rho^A = \sum_{j=1}^{N_B} \rho_A \otimes \langle \phi_j | \rho_B | \phi_j \rangle = \rho_A \otimes \sum_{j=1}^{N_B} \langle \phi_j | \rho_B | \phi_j \rangle.$$

Since

$$\sum_{j=1}^{N_B} \langle \phi_j | \rho_B | \phi_j \rangle = 1$$

we find $\rho^A = \rho_A$. Analogously we can show that $\rho^B = \rho_B$.

Problem 3. Consider the 4×4 matrix (*density matrix*)

$$|\mathbf{u}\rangle\langle\mathbf{u}| = (u_j \bar{u}_k), \qquad j, k = 1, \ldots, 4$$

in the product Hilbert space $\mathcal{H}_A \otimes \mathcal{H}_B \equiv \mathbb{C}^4$, where $\mathcal{H}_A = \mathcal{H}_B = \mathbb{C}^2$.

(i) Calculate

$$\mathrm{tr}_A(|\mathbf{u}\rangle\langle\mathbf{u}|)$$

where the basis is given by

$$\begin{pmatrix} 1 \\ 0 \end{pmatrix} \otimes I_2, \qquad \begin{pmatrix} 0 \\ 1 \end{pmatrix} \otimes I_2$$

and I_2 denotes the 2×2 unit matrix.

(ii) Find the partial trace

$$\mathrm{tr}_B(|\mathbf{u}\rangle\langle\mathbf{u}|)$$

where the basis is given by

$$I_2 \otimes \begin{pmatrix} 1 \\ 0 \end{pmatrix}, \qquad I_2 \otimes \begin{pmatrix} 0 \\ 1 \end{pmatrix}.$$

Solution 3. (i) Since

$$\begin{pmatrix} 1 \\ 0 \end{pmatrix} \otimes I_2 = \begin{pmatrix} 1 & 0 \\ 0 & 1 \\ 0 & 0 \\ 0 & 0 \end{pmatrix}, \qquad \begin{pmatrix} 0 \\ 1 \end{pmatrix} \otimes I_2 = \begin{pmatrix} 0 & 0 \\ 0 & 0 \\ 1 & 0 \\ 0 & 1 \end{pmatrix}$$

we find, using the transpose of these matrices on the left-hand side of $|\mathbf{u}\rangle\langle\mathbf{u}|$, that

$$\mathrm{tr}_A(|\mathbf{u}\rangle\langle\mathbf{u}|) = \begin{pmatrix} 1 & 0 & 0 & 0 \\ 0 & 1 & 0 & 0 \end{pmatrix} \begin{pmatrix} u_1\bar{u}_1 & u_1\bar{u}_2 & u_1\bar{u}_3 & u_1\bar{u}_4 \\ u_2\bar{u}_1 & u_2\bar{u}_2 & u_2\bar{u}_3 & u_2\bar{u}_4 \\ u_3\bar{u}_1 & u_3\bar{u}_2 & u_3\bar{u}_3 & u_3\bar{u}_4 \\ u_4\bar{u}_1 & u_4\bar{u}_2 & u_4\bar{u}_3 & u_4\bar{u}_4 \end{pmatrix} \begin{pmatrix} 1 & 0 \\ 0 & 1 \\ 0 & 0 \\ 0 & 0 \end{pmatrix}$$

$$+ \begin{pmatrix} 0 & 0 & 1 & 0 \\ 0 & 0 & 0 & 1 \end{pmatrix} \begin{pmatrix} u_1\bar{u}_1 & u_1\bar{u}_2 & u_1\bar{u}_3 & u_1\bar{u}_4 \\ u_2\bar{u}_1 & u_2\bar{u}_2 & u_2\bar{u}_3 & u_2\bar{u}_4 \\ u_3\bar{u}_1 & u_3\bar{u}_2 & u_3\bar{u}_3 & u_3\bar{u}_4 \\ u_4\bar{u}_1 & u_4\bar{u}_2 & u_4\bar{u}_3 & u_4\bar{u}_4 \end{pmatrix} \begin{pmatrix} 0 & 0 \\ 0 & 0 \\ 1 & 0 \\ 0 & 1 \end{pmatrix}.$$

Using matrix multiplication and matrix addition we obtain

$$\mathrm{tr}_A(|\mathbf{u}\rangle\langle\mathbf{u}|) = \begin{pmatrix} u_1\bar{u}_1 + u_3\bar{u}_3 & u_1\bar{u}_2 + u_3\bar{u}_4 \\ u_2\bar{u}_1 + u_4\bar{u}_3 & u_2\bar{u}_2 + u_4\bar{u}_4 \end{pmatrix}.$$

(ii) Since

$$I_2 \otimes \begin{pmatrix} 1 \\ 0 \end{pmatrix} = \begin{pmatrix} 1 & 0 \\ 0 & 0 \\ 0 & 1 \\ 0 & 0 \end{pmatrix}, \qquad I_2 \otimes \begin{pmatrix} 0 \\ 1 \end{pmatrix} = \begin{pmatrix} 0 & 0 \\ 1 & 0 \\ 0 & 0 \\ 0 & 1 \end{pmatrix}$$

we find

$$
\mathrm{tr}_B(|\mathbf{u}\rangle\langle\mathbf{u}|) = \begin{pmatrix} 1 & 0 & 0 & 0 \\ 0 & 0 & 1 & 0 \end{pmatrix} \begin{pmatrix} u_1\bar{u}_1 & u_1\bar{u}_2 & u_1\bar{u}_3 & u_1\bar{u}_4 \\ u_2\bar{u}_1 & u_2\bar{u}_2 & u_2\bar{u}_3 & u_2\bar{u}_4 \\ u_3\bar{u}_1 & u_3\bar{u}_2 & u_3\bar{u}_3 & u_3\bar{u}_4 \\ u_4\bar{u}_1 & u_4\bar{u}_2 & u_4\bar{u}_3 & u_4\bar{u}_4 \end{pmatrix} \begin{pmatrix} 1 & 0 \\ 0 & 0 \\ 0 & 1 \\ 0 & 0 \end{pmatrix}
$$

$$
+ \begin{pmatrix} 0 & 1 & 0 & 0 \\ 0 & 0 & 0 & 1 \end{pmatrix} \begin{pmatrix} u_1\bar{u}_1 & u_1\bar{u}_2 & u_1\bar{u}_3 & u_1\bar{u}_4 \\ u_2\bar{u}_1 & u_2\bar{u}_2 & u_2\bar{u}_3 & u_2\bar{u}_4 \\ u_3\bar{u}_1 & u_3\bar{u}_2 & u_3\bar{u}_3 & u_3\bar{u}_4 \\ u_4\bar{u}_1 & u_4\bar{u}_2 & u_4\bar{u}_3 & u_4\bar{u}_4 \end{pmatrix} \begin{pmatrix} 0 & 0 \\ 1 & 0 \\ 0 & 0 \\ 0 & 1 \end{pmatrix}.
$$

Using matrix multiplication and matrix addition yields

$$
\mathrm{tr}_B(|\mathbf{u}\rangle\langle\mathbf{u}|) = \begin{pmatrix} u_1\bar{u}_1 + u_2\bar{u}_2 & u_1\bar{u}_3 + u_2\bar{u}_4 \\ u_3\bar{u}_1 + u_4\bar{u}_2 & u_3\bar{u}_3 + u_4\bar{u}_4 \end{pmatrix}.
$$

We see that

$$
\mathrm{tr}_A(|\mathbf{u}\rangle\langle\mathbf{u}|) \neq \mathrm{tr}_B(|\mathbf{u}\rangle\langle\mathbf{u}|).
$$

However

$$
\mathrm{tr}\,(\mathrm{tr}_A(|\mathbf{u}\rangle\langle\mathbf{u}|)) = \mathrm{tr}\,(\mathrm{tr}_B(|\mathbf{u}\rangle\langle\mathbf{u}|)), \quad \det\,(\mathrm{tr}_A(|\mathbf{u}\rangle\langle\mathbf{u}|)) = \det\,(\mathrm{tr}_B(|\mathbf{u}\rangle\langle\mathbf{u}|)).
$$

Problem 4.   Consider the 9×9 matrix (density matrix)

$$
|\mathbf{u}\rangle\langle\mathbf{u}| = (u_j\bar{u}_k), \qquad j, k = 1, \ldots, 9.
$$

Find the partial trace

$$
\mathrm{tr}_{\mathbb{C}^3}(|\mathbf{u}\rangle\langle\mathbf{u}|)
$$

where the basis is given by

$$
\begin{pmatrix} 1 \\ 0 \\ 0 \end{pmatrix} \otimes I_3, \qquad \begin{pmatrix} 0 \\ 1 \\ 0 \end{pmatrix} \otimes I_3, \qquad \begin{pmatrix} 0 \\ 0 \\ 1 \end{pmatrix} \otimes I_3
$$

and I_3 denotes the 3×3 unit matrix.

Solution 4.   We have

$$
\begin{pmatrix} 1 \\ 0 \\ 0 \end{pmatrix} \otimes I_3 = \begin{pmatrix} 1 & 0 & 0 \\ 0 & 1 & 0 \\ 0 & 0 & 1 \\ 0 & 0 & 0 \\ 0 & 0 & 0 \\ 0 & 0 & 0 \\ 0 & 0 & 0 \\ 0 & 0 & 0 \\ 0 & 0 & 0 \end{pmatrix}, \qquad \begin{pmatrix} 0 \\ 1 \\ 0 \end{pmatrix} \otimes I_3 = \begin{pmatrix} 0 & 0 & 0 \\ 0 & 0 & 0 \\ 0 & 0 & 0 \\ 1 & 0 & 0 \\ 0 & 1 & 0 \\ 0 & 0 & 1 \\ 0 & 0 & 0 \\ 0 & 0 & 0 \\ 0 & 0 & 0 \end{pmatrix}
$$

and

$$
\begin{pmatrix} 0 \\ 0 \\ 1 \end{pmatrix} \otimes I_3 = \begin{pmatrix} 0 & 0 & 0 \\ 0 & 0 & 0 \\ 0 & 0 & 0 \\ 0 & 0 & 0 \\ 0 & 0 & 0 \\ 0 & 0 & 0 \\ 1 & 0 & 0 \\ 0 & 1 & 0 \\ 0 & 0 & 1 \end{pmatrix}.
$$

The respective transposes of the above matrices are given by

$$
(1\ 0\ 0) \otimes I_3 = \begin{pmatrix} 1 & 0 & 0 & 0 & 0 & 0 & 0 & 0 & 0 \\ 0 & 1 & 0 & 0 & 0 & 0 & 0 & 0 & 0 \\ 0 & 0 & 1 & 0 & 0 & 0 & 0 & 0 & 0 \end{pmatrix}
$$

$$
(0\ 1\ 0) \otimes I_3 = \begin{pmatrix} 0 & 0 & 0 & 1 & 0 & 0 & 0 & 0 & 0 \\ 0 & 0 & 0 & 0 & 1 & 0 & 0 & 0 & 0 \\ 0 & 0 & 0 & 0 & 0 & 1 & 0 & 0 & 0 \end{pmatrix}
$$

$$
(0\ 0\ 1) \otimes I_3 = \begin{pmatrix} 0 & 0 & 0 & 0 & 0 & 0 & 1 & 0 & 0 \\ 0 & 0 & 0 & 0 & 0 & 0 & 0 & 1 & 0 \\ 0 & 0 & 0 & 0 & 0 & 0 & 0 & 0 & 1 \end{pmatrix}.
$$

Taking this basis we find

$$
\mathrm{tr}_A(|\mathbf{u}\rangle\langle\mathbf{u}|) =
$$

$$
\begin{pmatrix} u_1\bar{u}_1 + u_4\bar{u}_4 + u_7\bar{u}_7 & u_1\bar{u}_2 + u_4\bar{u}_5 + u_7\bar{u}_8 & u_1\bar{u}_3 + u_4\bar{u}_6 + u_7\bar{u}_9 \\ u_2\bar{u}_1 + u_5\bar{u}_4 + u_8\bar{u}_7 & u_2\bar{u}_2 + u_5\bar{u}_5 + u_8\bar{u}_8 & u_2\bar{u}_3 + u_5\bar{u}_6 + u_8\bar{u}_9 \\ u_3\bar{u}_1 + u_6\bar{u}_4 + u_9\bar{u}_7 & u_3\bar{u}_2 + u_6\bar{u}_5 + u_9\bar{u}_8 & u_3\bar{u}_3 + u_6\bar{u}_6 + u_9\bar{u}_9 \end{pmatrix}.
$$

Problem 5. The partial trace can also be calculated as follows. Consider a bipartite state

$$
|\psi\rangle = \sum_{j=0}^{n-1}\sum_{k=0}^{n-1} c_{jk}|jk\rangle \equiv \sum_{j=0}^{n-1}\sum_{k=0}^{n-1} c_{jk}|j\rangle \otimes |k\rangle, \qquad \sum_{j=0}^{n-1}\sum_{k=0}^{n-1} c_{jk}c_{jk}^* = 1
$$

in the finite-dimensional Hilbert space $\mathcal{H} = \mathbb{C}^n \otimes \mathbb{C}^n$. We can define the $n \times n$ matrix

$$
\Lambda_{jk} := c_{jk}, \qquad j,k = 0,1,\dots,n-1.
$$

Then we have (prove it)

$$
\rho_A = \mathrm{tr}_B\rho = \mathrm{tr}_B|\psi\rangle\langle\psi| = \Lambda\Lambda^\dagger.
$$

(i) Consider the Bell states. Find ρ_A.

(ii) Under a unitary transformation U, V (U and V are $n \times n$ unitary matrices) the matrix Λ is changed to

$$\Lambda \to U^T \Lambda V$$

where ^{T} denotes the transpose. Apply the transformation to $\Lambda\Lambda^\dagger$. Calculate the Wigner function $\text{tr}(\Lambda\Lambda^\dagger)^2$.

Solution 5. (i) Consider the *Bell state*

$$|\psi\rangle = \frac{1}{\sqrt{2}}(|00\rangle + |11\rangle).$$

Since $c_{00} = c_{11} = 1/\sqrt{2}$ and $c_{01} = c_{10} = 0$, we find the matrix

$$\Lambda = \begin{pmatrix} \frac{1}{\sqrt{2}} & 0 \\ 0 & \frac{1}{\sqrt{2}} \end{pmatrix}.$$

Thus

$$\rho_A = \Lambda\Lambda^\dagger = \begin{pmatrix} \frac{1}{2} & 0 \\ 0 & \frac{1}{2} \end{pmatrix}.$$

For the other three Bell states we find the same result.

(ii) We have

$$\Lambda\Lambda^\dagger \to (U^T \Lambda V)(V^\dagger \Lambda^\dagger U^{T\dagger}) = U^T \Lambda\Lambda^\dagger U^{T\dagger}$$

since $V^\dagger V = VV^\dagger = I_n$. Furthermore, $\text{tr}(\Lambda\Lambda^\dagger)^2$ stays invariant under the transformation since $U^T U^{T\dagger} = (U^\dagger U)^T = I_n$.

Problem 6. Let $\{\,|0\rangle, |1\rangle, \ldots, |N-1\rangle\,\}$ be an orthonormal basis in the Hilbert space $\mathbf{C}^N$. The discrete *Wigner operator* is defined as

$$\hat{A}(q,p) := \sum_{r=0}^{N-1}\sum_{s=0}^{N-1} \delta_{2q,r+s} \exp\left(i\frac{2\pi}{N}p(r-s)\right) |r\rangle\langle s|$$

where q and p take integer values from 0 to $N-1$ and $\delta_{m,u}$ denotes the Kronecker delta. The arithmetic in the subscript is modulo N arithmetic, i.e., $2q \bmod N$ and $(r+s) \bmod N$. The (p,q) pairs constitute the discrete phase space. For a state described by the density matrix ρ the discrete *Wigner function* is defined as

$$W(p,q) := \frac{1}{N}\text{tr}(\rho\hat{A}).$$

Let $\rho = |0\rangle\langle 0|$. Calculate $W(p,q)$.

Solution 6. Since $\langle 0|r \rangle = \delta_{0r}$ we obtain

$$W(p,q) = \frac{1}{N}\mathrm{tr}\left(|0\rangle \sum_{s=0}^{N-1} \delta_{2q,s}\exp\left(-i\frac{2\pi}{N}ps\right)\langle s|\right).$$

To calculate the trace we have

$$W(p,q) = \frac{1}{N}\sum_{k=0}^{N-1}\left(\langle k|0\rangle \sum_{s=0}^{N-1} \delta_{2q,s}\exp\left(-i\frac{2\pi}{N}ps\right)\langle s|k\rangle\right).$$

Using $\langle k|0\rangle = \delta_{k0}$ and $\langle s|k\rangle = \delta_{sk}$ we arrive at

$$W(p,q) = \frac{1}{N}\delta_{2q,0}.$$

Problem 7. For a bipartite state with subsystems 1 and 2 described by the joint density matrix the joint Wigner function is given by

$$W(q_1,q_2,p_1,p_2) := \frac{1}{N^2}\mathrm{tr}(\rho^{(12)}(\hat{A}_1(q_1,p_1)\otimes\hat{A}_2(q_2,p_2)))$$

where the Wigner operators are given by

$$\hat{A}_1(q_1,p_1) := \sum_{r=0}^{N-1}\sum_{s=0}^{N-1}\delta_{2q_1,r+s}\exp\left(i\frac{2\pi}{N}p_1(r-s)\right)|r\rangle\langle s|$$

and

$$\hat{A}_2(q_2,p_2) := \sum_{r=0}^{N-1}\sum_{s=0}^{N-1}\delta_{2q_2,r+s}\exp\left(i\frac{2\pi}{N}p_2(r-s)\right)|r\rangle\langle s|.$$

Wigner functions describing a subsystem are obtained by summing the joint Wigner functions in the corresponding set of the respective variables, e.g.,

$$W(q_1,p_1) = \sum_{q_2=0}^{N-1}\sum_{p_2=0}^{N-1}W(q_1,p_1,q_2,p_2)$$

$$W(q_2,p_2) = \sum_{q_1=0}^{N-1}\sum_{p_1=0}^{N-1}W(q_1,p_1,q_2,p_2).$$

Consider the *EPR state*

$$|\psi\rangle = \frac{1}{\sqrt{N}}\sum_{k=0}^{N-1}|k\rangle\otimes|k\rangle.$$

Let $\rho = |\psi\rangle\langle\psi|$. Find $W(q_1, q_2, p_1, p_2)$. Discuss.

Solution 7. Straightforward calculation yields the Wigner function

$$W(q_1, q_2, p_1, p_2) = \frac{1}{N^2} \delta_{q_1, q_2} \delta_{p_1, -p_2}.$$

The Wigner function given above shows the connection with the EPR state for continuous-variable teleportation

$$\delta(q_1 - q_2) \otimes \delta(p_1 + p_2)$$

where δ denotes the Dirac delta function.

Problem 8. Consider the state in $\mathbb{C}^4$

$$|\psi\rangle = \frac{1}{\sqrt{2}} \left(\begin{pmatrix} 1 \\ 0 \end{pmatrix} \otimes \begin{pmatrix} 0 \\ 1 \end{pmatrix} - \begin{pmatrix} 0 \\ 1 \end{pmatrix} \otimes \begin{pmatrix} 1 \\ 0 \end{pmatrix} \right) = \frac{1}{\sqrt{2}} \begin{pmatrix} 0 \\ 1 \\ -1 \\ 0 \end{pmatrix}$$

and the density matrix $\rho = |\psi\rangle\langle\psi|$. Find

$$\rho_1 = \text{tr}_2(\rho), \qquad \rho_2 = \text{tr}_1(\rho)$$

i.e. calculate the partial trace.

Solution 8. We have

$$\langle\psi| = \frac{1}{\sqrt{2}} \begin{pmatrix} 0 & 1 & -1 & 0 \end{pmatrix}.$$

Thus

$$\rho = \frac{1}{2} \begin{pmatrix} 0 & 0 & 0 & 0 \\ 0 & 1 & -1 & 0 \\ 0 & -1 & 1 & 0 \\ 0 & 0 & 0 & 0 \end{pmatrix}.$$

Using the basis

$$\begin{pmatrix} 1 \\ 0 \end{pmatrix} \otimes \begin{pmatrix} 1 & 0 \\ 0 & 1 \end{pmatrix}, \qquad \begin{pmatrix} 0 \\ 1 \end{pmatrix} \otimes \begin{pmatrix} 1 & 0 \\ 0 & 1 \end{pmatrix}$$

and

$$\begin{pmatrix} 1 & 0 \\ 0 & 1 \end{pmatrix} \otimes \begin{pmatrix} 1 \\ 0 \end{pmatrix}, \qquad \begin{pmatrix} 1 & 0 \\ 0 & 1 \end{pmatrix} \otimes \begin{pmatrix} 0 \\ 1 \end{pmatrix}$$

we find

$$\rho_1 = \rho_2 = \frac{1}{2} \begin{pmatrix} 1 & 0 \\ 0 & 1 \end{pmatrix}.$$

Problem 9. The trace of an $m \times m$ matrix A over $\mathbb{C}$ is defined as

$$\mathrm{tr}_m(A) = \sum_{j=1}^{m} \mathbf{b}_{m,j}^* \, A \, \mathbf{b}_{m,j}$$

where $\{\, \mathbf{b}_{m,1}, \mathbf{b}_{m,2}, \ldots, \mathbf{b}_{m,m} \,\}$ is an orthonormal basis for $\mathbb{C}^m$. The partial traces of an $mn \times mn$ matrix C over $\mathbb{C}$ are defined as

$$\mathrm{tr}_{m,n}^1(C) = \sum_{j=1}^{m} \left(\mathbf{b}_{m,j}^* \otimes I_n \right) C \left(\mathbf{b}_{m,j} \otimes I_n \right)$$

$$\mathrm{tr}_{m,n}^2(C) = \sum_{k=1}^{n} \left(I_m \otimes \mathbf{b}_{n,k}^* \right) C \left(I_m \otimes \mathbf{b}_{n,k} \right)$$

where $\{\, \mathbf{b}_{n,1}, \mathbf{b}_{n,2}, \ldots, \mathbf{b}_{n,n} \,\}$ is an orthonormal basis for $\mathbb{C}^n$ and I_m and I_n denote the $m \times m$ and $n \times n$ identity matrices respectively. Let B be an $n \times n$ matrix over $\mathbb{C}$.

(i) Show that the above definition of $\mathrm{tr}_m(A)$ is independent of the choice of orthonormal basis

$$\{\, \mathbf{b}_{m,1}, \mathbf{b}_{m,2}, \ldots, \mathbf{b}_{m,m} \,\}.$$

(ii) Show that

$$\mathrm{tr}_{mn}(C) = \mathrm{tr}_n \left(\mathrm{tr}_{m,n}^1(C) \right)$$

and

$$\mathrm{tr}_{mn}(C) = \mathrm{tr}_m \left(\mathrm{tr}_{m,n}^2(C) \right).$$

(iv) Use your favourite orthonormal basis for $\mathbb{C}^2$ to calculate

$$\mathrm{tr}_{2,2}^1 \left(\begin{pmatrix} 0 & 1 \\ 1 & 0 \end{pmatrix} \otimes \begin{pmatrix} 1 & 2 \\ 3 & 4 \end{pmatrix} \right) \qquad \text{and} \qquad \mathrm{tr}_{2,2}^2 \begin{pmatrix} 0 & 1 & 1 & 0 \\ 1 & 0 & 0 & 1 \\ 1 & 0 & 0 & -1 \\ 0 & 1 & -1 & 0 \end{pmatrix}.$$

Solution 9. (i) Let $\{\, \phi_1, \phi_2, \ldots, \phi_m \,\}$ be an orthonormal basis for $\mathbb{C}^n$. Expanding each of the $\mathbf{b}_{m,k}$ in terms of the ϕ_j and vice versa yields

$$\mathbf{b}_{m,j} = \sum_{k=1}^{m} \phi_k^* \mathbf{b}_{m,j} \phi_k, \qquad \phi_j = \sum_{k=1}^{m} \mathbf{b}_{m,k}^* \phi_j \mathbf{b}_{m,k}$$

where we require

$$\mathbf{b}_{m,j}^* \mathbf{b}_{m,k} = \sum_{u,v=1}^{m} \mathbf{b}_{m,j}^* \phi_u \phi_u^* \phi_v^* \mathbf{b}_{m,k} \phi_v = \sum_{u,v=1}^{m} \mathbf{b}_{m,j}^* \phi_u \phi_v^* \mathbf{b}_{m,k} (\phi_u^* \phi_v)$$

$$= \sum_{u=1}^{m} \phi_u^* \mathbf{b}_{m,k} \mathbf{b}_{m,j}^* \phi_u = \delta_{j,k}$$

so that orthonormality holds. It follows that

$$\sum_{k=1}^{m} \phi_k^* A \phi_k = \sum_{k=1}^{m} \sum_{j=1}^{m} \sum_{l=1}^{m} \phi_k^* \mathbf{b}_{m,j} \mathbf{b}_{m,j}^* A \mathbf{b}_{m,l}^* \phi_k \mathbf{b}_{m,l}$$

$$= \sum_{j=1}^{m} \sum_{l=1}^{m} \left(\sum_{k=1}^{m} \phi_k^* \mathbf{b}_{m,j} \mathbf{b}_{m,l}^* \phi_k \right) \mathbf{b}_{m,j}^* A \mathbf{b}_{m,l}$$

$$= \sum_{j=1}^{m} \sum_{l=1}^{m} \delta_{j,l} \mathbf{b}_{m,j}^* A \mathbf{b}_{m,l}$$

$$= \sum_{j=1}^{m} \mathbf{b}_{m,j}^* A \mathbf{b}_{m,j} = \operatorname{tr}_m(A).$$

(ii) We have

$$\operatorname{tr}_n \left(\operatorname{tr}_{m,n}^1(C) \right) = \sum_{j=1}^{n} \mathbf{b}_{n,j}^* \left(\sum_{k=1}^{m} (\mathbf{b}_{m,k}^* \otimes I_n) C (\mathbf{b}_{m,k} \otimes I_n) \right) \mathbf{b}_{n,j}$$

$$= \sum_{j=1}^{n} \sum_{k=1}^{m} (I_1 \otimes \mathbf{b}_{n,j}^*)(\mathbf{b}_{m,k}^* \otimes I_n) C (\mathbf{b}_{m,k} \otimes I_n)(I_1 \otimes \mathbf{b}_{n,j})$$

$$= \sum_{j=1}^{n} \sum_{k=1}^{m} (\mathbf{b}_{m,k}^* \otimes \mathbf{b}_{n,j}^*) C (\mathbf{b}_{m,k} \otimes \mathbf{b}_{n,j})$$

$$= \operatorname{tr}_{mn}(C)$$

where the last equivalence follows from taking the trace of C using the basis

$$\{ \mathbf{b}_{m,j} \otimes \mathbf{b}_{n,k} \,|\, j \in \{1, 2, \ldots, m\} \; k \in \{1, 2, \ldots, n\} \}.$$

The result

$$\operatorname{tr}_{mn}(C) = \operatorname{tr}_m \left(\operatorname{tr}_{m,n}^2(C) \right)$$

follows similarly.

(iii) Using any basis, for example

$$\left\{ \begin{pmatrix} 1 \\ 0 \end{pmatrix}, \begin{pmatrix} 0 \\ 1 \end{pmatrix} \right\}$$

we find

$$\operatorname{tr}_{2,2}^1 \left(\begin{pmatrix} 0 & 1 \\ 1 & 0 \end{pmatrix} \otimes \begin{pmatrix} 1 & 2 \\ 3 & 4 \end{pmatrix} \right) = \begin{pmatrix} 0 & 0 \\ 0 & 0 \end{pmatrix}$$

and

$$\operatorname{tr}_{2,2}^2 \begin{pmatrix} 0 & 1 & 1 & 0 \\ 1 & 0 & 0 & 1 \\ 1 & 0 & 0 & -1 \\ 0 & 1 & -1 & 0 \end{pmatrix} = \begin{pmatrix} 0 & 2 \\ 2 & 0 \end{pmatrix}.$$

Chapter 6

Reversible Logic Circuits

Reversible gates are gates that function in both directions. CMOS implementations of such gates have been designed. A special pass transistor logic family has been applied: reversible MOS. Many different reversible logic gates are candidates as universal building blocks. The controlled NOT gate, the Fredkin gate can be implemented. They dissipate very little energy. Owing to their use of reversible truth tables, they are even candidates for zero-power computing. Circuit synthesis takes advantage of mathematical group theory. Algorithms have been developed for the synthesis of arbitrary reversible circuits. A reversible logic gate has a corresponding quantum version, whose properties are completely defined by the truth table for the classical version.

Reversible circuits are applicable to nanotechnolgy, quantum and optical computing as well as reducing power in CMOS implementations. In adiabatic circuits, current is restricted to flow across devices with low voltage drop and the energy stored on their capacitors is recycled. One uses reversible energy recovery gates capable of realizing the functions $\{AND, OR\}$ or $\{NAND, NOR\}$.

A truth table is a tabular description of a logic operation listing all possible combinations of the truth values of the operands together with the truth value of the outcome for each of the possible combinations.

The AND operation is denoted by $\cdot$, the OR operation is denoted by $+$ and the XOR operation is denoted by $\oplus$.

Problem 1.   For reversible gates the following boolean expression plays an important role

$$(a_{11} \cdot a_{22}) \oplus (a_{12} \cdot a_{21})$$

where $a_{11}, a_{12}, a_{21}, a_{22} \in \{0, 1\}$. It could be considered as the *determinant* of the 2×2 binary matrix

$$\begin{pmatrix} a_{11} & a_{12} \\ a_{21} & a_{22} \end{pmatrix}.$$

Discuss. Find the inverse of the matrix when it exists.

Solution 1.   The inverse exists iff $(a_{11} \cdot a_{22}) \oplus (a_{12} \cdot a_{21}) = 1$. The inverse is given by

$$\begin{pmatrix} a_{22} & a_{12} \\ a_{21} & a_{11} \end{pmatrix}$$

since

$$\begin{pmatrix} a_{11} & a_{12} \\ a_{21} & a_{22} \end{pmatrix} \begin{pmatrix} a_{22} & a_{12} \\ a_{21} & a_{11} \end{pmatrix} =$$

$$\begin{pmatrix} (a_{11} \cdot a_{22}) \oplus (a_{12} \cdot a_{21}) & (a_{11} \cdot a_{12}) \oplus (a_{12} \cdot a_{11}) \\ (a_{21} \cdot a_{22}) \oplus (a_{22} \cdot a_{21}) & (a_{21} \cdot a_{12}) \oplus (a_{11} \cdot a_{22}) \end{pmatrix}.$$

Problem 2.   Consider a two input gate (x, y) / two output gate (x', y') given by

$$x' = a \cdot x \oplus b \cdot y \oplus c$$
$$y' = a' \cdot x \oplus b' \cdot y \oplus c'$$

where $a, b, a', b', c, c' \in \{0, 1\}$.
(i) Let $a = 0, b = 1, a' = 1, b' = 0$ and $c = c' = 0$. Find the output (x', y') for all possible inputs (x, y). Is the transformation invertible?
(ii) Let $a = 1, b = 1, a' = 1, b' = 1$ and $c = c' = 0$. Find the output (x', y') for all possible inputs (x, y). Is the transformation invertible?

Solution 2.   (i) We have

$$x' = 0 \cdot x \oplus 1 \cdot y \oplus 0$$
$$y' = 1 \cdot x \oplus 0 \cdot y \oplus 0.$$

Thus

$$x' = 0 \oplus y \oplus 0 = y$$
$$y' = x \oplus 0 \oplus 0 = x.$$

The truth table follows as

x	y	x'	y'
0	0	0	0
0	1	1	0
1	0	0	1
1	1	1	1

Therefore the transformation is invertible.
(ii) We have

$$x' = 1 \cdot x \oplus 1 \cdot y \oplus 0$$
$$y' = 1 \cdot x \oplus 1 \cdot y \oplus 0 \,.$$

Thus

$$x' = x \oplus y \oplus 0 = x \oplus y$$
$$y' = x \oplus y \oplus 0 = x \oplus y \,.$$

The truth table follows as

x	y	x'	y'
0	0	0	0
0	1	1	1
1	0	1	1
1	1	0	0

Therefore the transformation is not invertible.

Problem 3. Consider the *Toffoli gate*

$$T : \{0,1\}^3 \to \{0,1\}^3, \qquad T(a,b,c) := (a,b,(a \cdot b) \oplus c)$$

and the *Fredkin gate*

$$F : \{0,1\}^3 \to \{0,1\}^3, \qquad F(a,b,c) := (a, a \cdot b + \bar{a} \cdot c, a \cdot c + \bar{a} \cdot b)$$

where $\bar{a}$ is the NOT operation, $+$ is the OR operation, $\cdot$ is the AND operation and $\oplus$ is the XOR operation.

1. Express $NOT(a)$ exclusively in terms of the TOFFOLI gate.

2. Express $NOT(a)$ exclusively in terms of the FREDKIN gate.

3. Express $AND(a,b)$ exclusively in terms of the TOFFOLI gate.

4. Express $AND(a,b)$ exclusively in terms of the FREDKIN gate.

5. Express $OR(a,b)$ exclusively in terms of the TOFFOLI gate.

6. Express $OR(a, b)$ exclusively in terms of the FREDKIN gate.

7. Show that the TOFFOLI gate is invertible.

8. Show that the FREDKIN gate is invertible.

Thus the TOFFOLI and FREDKIN gates are each universal and reversible (invertible).

Solution 3.   (1) We have $NOT(a) = BIT3(T(a, a, 1))$, where $BIT3(a, b, c) = c$. This follows from $T(a, a, 1) = (a, a, a \oplus 1)$.

(2) $NOT(a) = BIT3(F(a, 1, 0))$.

(3) $AND(a, b) = BIT3(T(a, b, 0))$.

(4) $AND(a, b) = BIT3(F(a, 0, b))$.

(5) We can use

$$OR(a, b) = NOT(AND(NOT(a), NOT(b)))$$

and simply expand NOT and AND in terms of TOFFOLI gate. Noting that

$$T(\bar{a}, \bar{b}, 1) = (\bar{a}, \bar{b}, (\bar{a} \cdot \bar{b}) \oplus 1) = (\bar{a}, \bar{b}, \overline{\bar{a} \cdot \bar{b}} = (\bar{a}, \bar{b}, a + b)$$

we implement the OR operation as

$$\begin{aligned} OR(a, b) &= BIT3(T(NOT(a), NOT(b), 1)) \\ &= BIT3(T(BIT3(T(a, a, 1)), BIT3(T(b, b, 1)), 1)) . \end{aligned}$$

(6) We can use

$$OR(a, b) = NOT(AND(NOT(a), NOT(b)))$$

and simply expand NOT and AND in terms of the FREDKIN gate. Noting that

$$F(a, b, 1) = (a, a \cdot b + \bar{a} \cdot 1, a \cdot 1 + \bar{a} \cdot b) = (a, b + \bar{a}, a + b)$$

we implement the OR operation as

$$OR(a, b) = BIT3(F(a, b, 1)).$$

(7) We have $T^{-1}(a, b, c) = T(a, b, c)$ since the XOR ($\oplus$) is invertible if we remember at least one of the arguments. In other words

$$\begin{aligned} T(T(a, b, c)) &= T(a, b, (a \cdot b) \oplus c) \\ &= (a, b, (a \cdot b) \oplus ((a \cdot b) \oplus c)) \\ &= (a, b, ((a \cdot b) \oplus (a \cdot b)) \oplus c) \\ &= (a, b, c). \end{aligned}$$

(8) We have $F^{-1}(a, b, c) = F(a, b, c)$ since the swap operation on two bits is its own inverse, and the FREDKIN gate swaps the last two bits whenever the first argument is 0. In other words

$$
\begin{aligned}
F(F(a, b, c)) &= F(a, a \cdot b + \bar{a} \cdot c, a \cdot c + \bar{a} \cdot b) \\
&= (a, a \cdot (a \cdot b + \bar{a} \cdot c) + \bar{a} \cdot (a \cdot c + \bar{a} \cdot b), a \cdot (a \cdot c + \bar{a} \cdot b) \\
&\quad + \bar{a} \cdot (a \cdot b + \bar{a} \cdot c)) \\
&= (a, a \cdot b + \bar{a} \cdot b, a \cdot c + \bar{a} \cdot c) \\
&= (a, b, c).
\end{aligned}
$$

Problem 4. The *Toffoli gate* $T(x_1, x_2, x_3)$ has 3 inputs (x_1, x_2, x_3) and three outputs (y_1, y_2, y_3) and is given by

$$
(x_1, x_2, x_3) \rightarrow (x_1, x_2, x_3 \oplus (x_1 \cdot x_2))
$$

where $x_1, x_2, x_3 \in \{0, 1\}$, $\oplus$ is the XOR operation and $\cdot$ the AND operation. Give the truth table.

Solution 4. We have

x_1	x_2	x_3	y_1	y_2	y_3
0	0	0	0	0	0
0	0	1	0	0	1
0	1	0	0	1	0
0	1	1	0	1	1
1	0	0	1	0	0
1	0	1	1	0	1
1	1	0	1	1	1
1	1	1	1	1	0

Problem 5. A *generalized Toffoli gate* $T(x_1, x_2, \ldots, x_n, x_{n+1})$ is a gate that maps a boolean pattern $(x_1, x_2, \ldots, x_n, x_{n+1})$ to

$$
(x_1, x_2, \ldots, x_n, x_{n+1} \oplus (x_1 \cdot x_2 \cdot \ldots \cdot x_n))
$$

where $\oplus$ is the XOR operation and $\cdot$ the AND operation. Show that the generalized Toffoli gate includes the NOT gate, CNOT gate and the original Toffoli gate.

Solution 5. The NOT gate is given by $T(1, a)$, where $a \in \{0, 1\}$. The CNOT-gate is given by $T(a, b)$, where $a, b \in \{0, 1\}$. The original Toffoli gate is given by $T(a, b, c)$.

Problem 6. The *Fredkin gate* $F(x_1, x_2, x_3)$ has 3 inputs (x_1, x_2, x_3) and three outputs (y_1, y_2, y_3). It maps boolean patterns

$$(x_1, x_2, x_3) \to (x_1, x_3, x_2)$$

if and only if $x_1 = 1$, otherwise it passes the boolean pattern unchanged. Give the truth table.

Solution 6. We have

x_1	x_2	x_3	y_1	y_2	y_3
0	0	0	0	0	0
0	0	1	0	0	1
0	1	0	0	1	0
0	1	1	0	1	1
1	0	0	1	0	0
1	0	1	1	1	0
1	1	0	1	0	1
1	1	1	1	1	1

Problem 7. The *generalized Fredkin gate* $F(x_1, x_2, \ldots, x_n, x_{n+1}, x_{n+2})$ is the mapping of the boolean pattern

$$(x_1, x_2, \ldots, x_n, x_{n+1}, x_{n+2}) \to (x_1, x_2, \ldots, x_n, x_{n+2}, x_{n+1})$$

if and only if the boolean product $x_1 \cdot x_2 \cdot \ldots \cdot x_n = 1$ ($\cdot$ is the bitwise AND operation), otherwise the boolean pattern passes unchanged. Let $n = 2$ and $(x_1, x_2, x_3, x_4) = (1, 1, 0, 1)$. Find the output.

Solution 7. Since $x_1 \cdot x_2 = 1$, we find

x_1	x_2	x_3	x_4	y_1	y_2	y_3	y_4
1	1	0	1	1	1	1	0

Problem 8. Is the gate ($a, b, c \in \{0, 1\}$)

$$(a, b, c) \to (a, a \cdot b \oplus c, \bar{a} \cdot \bar{c} \oplus \bar{b})$$

reversible?

Solution 8. We set $x = a$, $y = a \cdot b \oplus c$ and $z = \bar{a} \cdot \bar{c} \oplus \bar{b}$. We find the truth table

a	b	c	x	y	z
0	0	0	0	0	0
0	0	1	0	1	1
0	1	0	0	0	1
0	1	1	0	1	0
1	0	0	1	0	1
1	0	1	1	1	1
1	1	0	1	1	0
1	1	1	1	0	0

Thus from the truth table we see that the gate is reversible.

Problem 9. Show that one Fredkin gate

$$(a, b, c) \to (a, \overline{a} \cdot b + a \cdot c, \overline{a} \cdot c + a \cdot b)$$

is sufficient to implement the XOR gate. Assume that either $\overline{b}$ or $\overline{c}$ are available.

Solution 9. Choosing $b = \overline{c}$ (equivalently $c = \overline{b}$) we find that the Fredkin gate yields

$$(a, b, c) \to (a, \overline{a} \cdot b + a \cdot \overline{b}, \overline{a} \cdot c + a \cdot \overline{c}) \equiv (a, a \oplus b, a \oplus c).$$

Thus we can apply the Fredkin gate to $(a, b, \overline{b})$ and use the second bit to obtain $a \oplus b$ or equivalently apply the Fredkin gate to $(a, \overline{c}, c)$ and use the third bit to obtain $a \oplus c$.

Problem 10. Consider the 2×2 identity matrix and the Pauli spin matrices σ_x, σ_y, σ_z using the following notation

$$\sigma_{00} = \tau_{00} = I_2 = \begin{pmatrix} 1 & 0 \\ 0 & 1 \end{pmatrix},$$

$$\sigma_{01} = \tau_{01} = \sigma_x = \begin{pmatrix} 0 & 1 \\ 1 & 0 \end{pmatrix},$$

$$\sigma_{10} = \tau_{10} = \sigma_z = \begin{pmatrix} 1 & 0 \\ 0 & -1 \end{pmatrix},$$

$$\sigma_{11} = \sigma_y = \begin{pmatrix} 0 & -i \\ i & 0 \end{pmatrix},$$

$$\tau_{11} = i\sigma_y = \begin{pmatrix} 0 & 1 \\ -1 & 0 \end{pmatrix}.$$

Let n be a positive integer. If $\mathbf{v}, \mathbf{w} \in \mathbb{Z}_2^n$ with

$$\mathbf{v} = \begin{pmatrix} v_1 \\ \vdots \\ v_n \end{pmatrix}, \qquad \mathbf{w} = \begin{pmatrix} w_1 \\ \vdots \\ w_n \end{pmatrix}$$

and

$$\mathbf{b} := \begin{pmatrix} \mathbf{v} \\ \mathbf{w} \end{pmatrix} \in \mathbb{Z}_2^{2n}$$

we define

$$\sigma_{\mathbf{b}} := \sigma_{v_1 w_1} \otimes \sigma_{v_2 w_2} \otimes \cdots \otimes \sigma_{v_n w_n}$$

and

$$\tau_{\mathbf{b}} := \tau_{v_1 w_1} \otimes \tau_{v_2 w_2} \otimes \cdots \otimes \tau_{v_n w_n}.$$

Thus we can associate a bit string $\mathbf{b}$ with each $\sigma_{\mathbf{b}}$ and vice versa.
(i) Let $n = 3$ and

$$\sigma_{\mathbf{b}_1} = \sigma_x \otimes \sigma_z \otimes \sigma_x, \qquad \sigma_{\mathbf{b}_2} = I_2 \otimes \sigma_z \otimes \sigma_x.$$

Find the corresponding bit strings $\mathbf{b}_1$ and $\mathbf{b}_2$ for the given $\sigma_{\mathbf{b}_1}$ and $\sigma_{\mathbf{b}_2}$. Then XOR the two bit strings and find the corresponding $\sigma_{\mathbf{b}_3}$. Calculate the matrix product $\sigma_{\mathbf{b}_1} \sigma_{\mathbf{b}_2}$. Discuss.

Solution 10.   (i) We have

$$\mathbf{b}_1 = \begin{pmatrix} 0 \\ 1 \\ 0 \\ 1 \\ 0 \\ 1 \end{pmatrix}, \qquad \mathbf{b}_2 = \begin{pmatrix} 0 \\ 1 \\ 0 \\ 0 \\ 0 \\ 1 \end{pmatrix}.$$

The XOR operation provides the bit string

$$\mathbf{b}_3 = \begin{pmatrix} 0 \\ 0 \\ 0 \\ 1 \\ 0 \\ 0 \end{pmatrix}$$

with the corresponding matrix

$$\sigma_{\mathbf{b}_3} = \sigma_x \otimes I_2 \otimes I_2$$

which is $\sigma_{\mathbf{b}_3} = \sigma_{\mathbf{b}_1} \sigma_{\mathbf{b}_2}$.

Chapter 7

Unitary Transforms and Quantum Gates

Quantum gates are realised as *unitary operators*. Let $\mathcal{H}$ denote a Hilbert space. A linear operator U in a Hilbert space is unitary if

$$U^* = U^{-1}.$$

In other words, U is unitary if and only if U is invertible and $U^{-1} = U^*$. An equivalent form of definition is: U is unitary iff

$$\langle U\mathbf{x}, U\mathbf{y} \rangle = \langle \mathbf{x}, \mathbf{y} \rangle, \qquad \text{for all} \quad \mathbf{x}, \mathbf{y} \in \mathcal{H}.$$

Thus we have

$$U^*U = UU^* = I$$

where $*$ denotes the adjoint and I is the identity operator. The composite of two unitary operators is again unitary.

Here we consider $n \times n$ unitary matrices. All the eigenvalues of a unitary matrix have absolute value equal to 1, i.e., $|\lambda| = 1$. Thus $|\det U| = 1$. All $n \times n$ unitary matrices form a group under matrix multiplication. An important subgroup are all $n \times n$ matrices with $\det U = 1$. If $\det U_1 = 1$ and $\det U_2 = 1$, then $\det(U_1 U_2) = 1$. If K is a hermitian matrix, then $\exp(iK)$ is a unitary matrix. Thus if $\hat{H}$ is a hermitian matrix describing the Hamilton operator we find that the exponential function $\exp(-i\hat{H}t/\hbar)$ is a unitary matrix. If U and V are unitary, then $U \otimes V$ and $U \oplus V$ are unitary, where $\otimes$ is the Kronecker product and $\oplus$ the direct sum.

Problem 1.   (i) Let $A := |0\rangle\langle 0| - |1\rangle\langle 1|$ in the Hilbert space $\mathbb{C}^2$. Calculate

$$U_H A U_H |0\rangle, \qquad U_H A U_H |1\rangle$$

where U_H is the *Walsh-Hadamard transform*. The unitary transform U_H is defined by

$$U_H |k\rangle = \frac{1}{\sqrt{2}}(|0\rangle + (-1)^k |1\rangle), \qquad k \in \{0, 1\}.$$

(ii) Calculate

$$(U_H \otimes U_H) U_{CNOT} (U_H \otimes U_H) |j, k\rangle$$

where $|j, k\rangle \equiv |j\rangle \otimes |k\rangle$ with $j, k \in \{0, 1\}$, and the answer is in the form of a ket $|m, n\rangle$ with $m, n \in \{0, 1\}$. The unitary transform

$$U_{CNOT} := |0\rangle\langle 0| \otimes I_2 + |1\rangle\langle 1| \otimes U_{NOT}$$

is the *controlled NOT* operation and the unitary transform

$$U_{NOT} = |0\rangle\langle 1| + |1\rangle\langle 0|$$

is the *NOT operation*.

Solution 1.   (i) Let $j \in \{0, 1\}$. Then

$$U_H A U_H |j\rangle = \frac{1}{\sqrt{2}} U_H A(|0\rangle + (-1)^j |1\rangle) = \frac{1}{\sqrt{2}} U_H(|0\rangle + (-1)^{j+1}|1\rangle)$$

$$= \frac{1}{\sqrt{2}} U_H(|0\rangle + (-1)^{\bar{j}}|1\rangle) = |\bar{j}\rangle$$

where $\bar{j} := 1 - j$. In other words $U_H A U_H = U_{NOT}$.
(ii) Straightforward calculation yields

$$(U_H \otimes U_H) U_{CNOT} (U_H \otimes U_H) |j, k\rangle$$

$$= \frac{1}{2}(U_H \otimes U_H) U_{CNOT} \left((|0\rangle + (-1)^j |1\rangle) \otimes (|0\rangle + (-1)^k |1\rangle)\right)$$

$$= \frac{1}{2}(U_H \otimes U_H)(|00\rangle + (-1)^k |01\rangle + (-1)^j |11\rangle + (-1)^{j+k}|10\rangle)$$

$$= \frac{1}{2}(U_H \otimes U_H)(|0\rangle \otimes (|0\rangle + (-1)^k |1\rangle) + (-1)^j |1\rangle \otimes (|1\rangle + (-1)^k |0\rangle))$$

$$= \frac{1}{2}(U_H \otimes U_H)(|0\rangle \otimes (|0\rangle + (-1)^k |1\rangle) + (-1)^{j+k} |1\rangle \otimes (|0\rangle + (-1)^k |1\rangle))$$

$$= \frac{1}{2}(U_H \otimes U_H)(|0\rangle + (-1)^{j+k}|1\rangle) \otimes (|0\rangle + (-1)^k |1\rangle)$$

$$= |j \oplus k, k\rangle$$

where $\oplus$ is the XOR operation. In other words we have the *controlled NOT* operation, where the control qubit is the second qubit and the target qubit is the first qubit.

Problem 2. Consider the linear operator (Hamilton operator)

$$H := i\hbar\omega(|0\rangle\langle 1| - |1\rangle\langle 0|)$$

operating in the Hilbert space $\mathbb{C}^2$, where $\{\,|0\rangle,\,|1\rangle\,\}$ is an orthonormal basis in $\mathbb{C}^2$ and ω is a real parameter (frequency).
(i) Is H self-adjoint?
(ii) Find the eigenvalues and corresponding normalized eigenvectors of H.
(iii) Find the unitary matrix

$$U(t) := \exp(-iHt/\hbar)\,.$$

Find the values of t such that $U(t)$ performs the *NOT operation*

$$U(t)|0\rangle \to |1\rangle, \qquad U(t)|1\rangle \to |0\rangle\,.$$

(iv) Calculate $U(t = \pi/4\omega)$ and $(U(t = \pi/4\omega))^2$.

Solution 2. (i) The adjoint of an operator can be obtained by simply swapping the labels of the corresponding bra and ket vectors in the sum, and taking the complex conjugate of all complex coefficients. Thus

$$H^* = \overline{i}\hbar\omega(|1\rangle\langle 0| - |0\rangle\langle 1|) = -i\hbar\omega(|1\rangle\langle 0| - |0\rangle\langle 1|) = H$$

i.e., H is self-adjoint. We can determine H^* as follows. Let

$$H^* = a_{00}|0\rangle\langle 0| + a_{01}|0\rangle\langle 1| + a_{10}|1\rangle\langle 0| + a_{11}|1\rangle\langle 1|, \qquad a_{00}, a_{01}, a_{10}, a_{11} \in \mathbb{C}.$$

The bra vector corresponding to the ket $H|y\rangle$ is $\langle y|H^*$. We require that $\langle H^*y|x\rangle = \langle y|Hx\rangle$ for all $|x\rangle = x_0|0\rangle + x_1|1\rangle$ and $|y\rangle = y_0|0\rangle + y_1|1\rangle$. We find

$$H|x\rangle = i\hbar\omega(x_1|0\rangle - x_0|1\rangle)$$
$$H^*|y\rangle = (y_0 a_{00} + y_1 a_{01})|0\rangle + (y_0 a_{10} + y_1 a_{11})|1\rangle$$
$$\langle y|Hx\rangle = i\hbar\omega(x_1\overline{y_0} - x_0\overline{y_1})$$
$$\langle H^*y|x\rangle = x_0(\overline{y_0 a_{00}} + \overline{y_1 a_{01}}) + x_1(\overline{y_0 a_{10}} + \overline{y_1 a_{11}}).$$

Since $\langle H^*y|x\rangle = \langle y|Hx\rangle$ for all $|x\rangle$ and $|y\rangle$, we obtain

$$i\hbar\omega\overline{y_0} = (\overline{y_0 a_{10}} + \overline{y_1 a_{11}}), \qquad -i\hbar\omega\overline{y_1} = (\overline{y_0 a_{00}} + \overline{y_1 a_{01}})\,.$$

Consequently $a_{00} = 0$, $a_{01} = i\hbar\omega$, $a_{10} = -i\hbar\omega$, $a_{11} = 0$.

(ii) The eigenvalue equation for H is

$$H(a|0\rangle + b|1\rangle) = \lambda(a|0\rangle + b|1\rangle).$$

Thus we have the two equations

$$-i\hbar\omega a = \lambda b, \qquad i\hbar\omega b = \lambda a.$$

If $\lambda = 0$ we have $a = 0$ and $b = 0$. Therefore we only have to consider $\lambda \neq 0$. Obviously we may assume $b \neq 0$ (thus $a \neq 0$). We obtain

$$\lambda = -\frac{i\hbar\omega a}{b}.$$

Hence $ib^2 = -ia^2$ so that $b = \pm ia$. Using $|a|^2 + |b|^2 = 1$ we find $|a| = \frac{1}{\sqrt{2}}$. We obtain the eigenvalues and corresponding orthonormal eigenvectors

$$\lambda_1 = -\hbar\omega, \quad \frac{1}{\sqrt{2}}(|0\rangle + i|1\rangle), \qquad \lambda_2 = \hbar\omega, \quad \frac{1}{\sqrt{2}}(|0\rangle - i|1\rangle).$$

(iii) We find H^n ($n \in \mathbb{N}$) by observing that

$$H^2 = (\hbar\omega)^2(|0\rangle\langle 0| + |1\rangle\langle 1|) = (\hbar\omega)^2 I_2, \qquad H^3 = (\hbar\omega)^2 H, \qquad H^4 = (\hbar\omega)^4 I_2.$$

Thus

$$H^n = \begin{cases} (\hbar\omega)^{n-1}H & n \text{ odd} \\ (\hbar\omega)^n I_2 & n \text{ even} \end{cases}.$$

Since $U(t) := \exp(-iHt/\hbar)$ we have

$$\begin{aligned}
U(t) &= \sum_{j=0}^{\infty} \frac{(-\frac{it}{\hbar})^j H^j}{j!} = \sum_{j=0}^{\infty} \frac{(-i\omega t)^{2j}}{(2j)!} I_2 + \frac{1}{\hbar\omega} \sum_{j=0}^{\infty} \frac{(-i\omega t)^{2j+1}}{(2j+1)!} H \\
&= \sum_{j=0}^{\infty} \frac{(-1)^j (\omega t)^{2j}}{(2j)!} I_2 - i\frac{1}{\hbar\omega} \sum_{j=0}^{\infty} \frac{(-1)^j (\omega t)^{2j+1}}{(2j+1)!} H \\
&= \cos(\omega t) I_2 - \frac{i}{\hbar\omega} \sin(\omega t) H \\
&= \cos(\omega t)(|0\rangle\langle 0| + |1\rangle\langle 1|) + \sin(\omega t)(|0\rangle\langle 1| - |1\rangle\langle 0|).
\end{aligned}$$

For the NOT operation we use

$$U(t = \pi/2\omega) = |0\rangle\langle 1| - |1\rangle\langle 0|.$$

The unitary transforms $U((2k+1)\pi/2\omega)$, $k \in \mathbb{N}_0$ implement the NOT operation.

(iv) We have

$$U(t = \pi/4\omega) = \frac{1}{\sqrt{2}}(|0\rangle - |1\rangle)\langle 0| + \frac{1}{\sqrt{2}}(|0\rangle + |1\rangle)\langle 1|$$

$$U(t = \pi/4\omega)^2 = U(t = \pi/2\omega) = |0\rangle\langle 1| - |1\rangle\langle 0|.$$

Thus we find $(U(t = \pi/4\omega))^2 = U(t = \pi/2\omega)$, i.e. $U(t = \pi/4\omega)$ acts as the *square root* of the *NOT operation*. Traditionally in quantum computation we use

$$U_{NOT} = |0\rangle\langle1| + |1\rangle\langle0|.$$

In this case for the $\sqrt{\text{NOT}}$ operation we use

$$U_{\sqrt{NOT}} = \frac{1}{2}(1 + i)(|0\rangle\langle0| + |1\rangle\langle1|) + \frac{1}{2}(1 - i)(|0\rangle\langle1| + |1\rangle\langle0|).$$

Problem 3. (i) Let σ_x, σ_y, σ_z be the *Pauli spin matrices*. Find

$$R_{1x}(\alpha) := \exp(-i\alpha(\sigma_x \otimes I_2)), \qquad R_{1y}(\alpha) := \exp(-i\alpha(\sigma_y \otimes I_2))$$

where $\alpha \in \mathbb{R}$ and I_2 denotes the 2×2 unit matrix.
(ii) Consider the special case $R_{1x}(\alpha = \pi/2)$ and $R_{1y}(\alpha = \pi/4)$. Calculate the matrix $R_{1x}(\pi/2)R_{1y}(\pi/4)$. Discuss.

Solution 3. (i) We have

$$\exp(-i\alpha(\sigma_x \otimes I_2)) := \sum_{k=0}^{\infty} \frac{(-i\alpha(\sigma_x \otimes I_2))^k}{k!}.$$

Since $\sigma_x^2 = I_2$ we have $(\sigma_x \otimes I_2)^2 = I_2 \otimes I_2$. Thus we find

$$\exp(-i\alpha(\sigma_x \otimes I_2)) = (I_2 \otimes I_2)\cos\alpha + e^{-i\pi/2}(\sigma_x \otimes I_2)\sin\alpha$$

where we used $\exp(-i\pi/2) = -i$. Analogously, we find

$$\exp(-i\alpha(\sigma_y \otimes I_2)) = (I_2 \otimes I_2)\cos\alpha + e^{-i\pi/2}(\sigma_y \otimes I_2)\sin\alpha$$

since

$$(\sigma_y \otimes I_2)^2 = I_2 \otimes I_2.$$

(ii) Since $\sin(\pi/2) = 1$, $\cos(\pi/2) = 0$ we arrive at

$$R_{1x}(\pi/2) = e^{-i\pi/2}(\sigma_x \otimes I_2).$$

From $\sin(\pi/4) = \sqrt{2}/2$, $\cos(\pi/4) = \sqrt{2}/2$ it follows that

$$R_{1y}(\pi/4) = \frac{1}{\sqrt{2}}(I_2 \otimes I_2) + \frac{1}{\sqrt{2}}e^{-i\pi/2}(\sigma_y \otimes I_2).$$

Thus

$$R_{1x}(\pi/2)R_{1y}(\pi/4) = \frac{e^{-i\pi/2}}{\sqrt{2}}(\sigma_x \otimes I_2) + \frac{e^{-i\pi/2}}{\sqrt{2}}(\sigma_z \otimes I_2)$$

where we used that $\sigma_x \sigma_y = i\sigma_z$. Therefore

$$R_{1x}(\pi/2)R_{1y}(\pi/4) = \frac{e^{-i\pi/2}}{\sqrt{2}}(\sigma_x + \sigma_z) \otimes I_2$$

where

$$\frac{1}{\sqrt{2}}(\sigma_x + \sigma_z) = \frac{1}{\sqrt{2}}\begin{pmatrix} 1 & 1 \\ 1 & -1 \end{pmatrix}$$

is the *Walsh-Hadamard gate*. All the single operations are in the Lie group $SU(2)$ whose determinant is $+1$, while the determinant of the Walsh-Hadamard gate is -1. Thus the overall phase is unavoidable.

Problem 4.   Consider the state in the Hilbert space $\mathcal{H} = \mathbb{C}^{16}$

$$|\psi_0\rangle = |0101\rangle \equiv |0\rangle \otimes |1\rangle \otimes |0\rangle \otimes |1\rangle$$

where $\{\,|0\rangle, |1\rangle\,\}$ is the standard basis in $\mathbb{C}^2$. Let

$$|\psi_1\rangle = B|\psi_0\rangle = \frac{1}{\sqrt{2}}(|0101\rangle + |0110\rangle)$$

$$|\psi_2\rangle = U|\psi_1\rangle = \frac{1}{\sqrt{2}}(|0101\rangle + |1010\rangle)$$

$$|\psi_3\rangle = S|\psi_2\rangle = \frac{1}{\sqrt{2}}(|0101\rangle - |1010\rangle)$$

$$|\psi_4\rangle = U^*|\psi_3\rangle = \frac{1}{\sqrt{2}}(|0101\rangle - |0110\rangle)$$

$$|\psi_5\rangle = B^*|\psi_4\rangle = -|0110\rangle.$$

Find the 16×16 unitary matrices B, U, S which perform these transformations.

Solution 4.   From the above equations we find the following

$$B|0101\rangle = \frac{1}{\sqrt{2}}(|0101\rangle + |0110\rangle)$$

$$U\frac{1}{\sqrt{2}}(|0101\rangle + |0110\rangle) = \frac{1}{\sqrt{2}}(|0101\rangle + |1010\rangle)$$

$$S\frac{1}{\sqrt{2}}(|0101\rangle + |1010\rangle) = \frac{1}{\sqrt{2}}(|0101\rangle - |1010\rangle)$$

$$U^*\frac{1}{\sqrt{2}}(|0101\rangle - |1010\rangle) = \frac{1}{\sqrt{2}}(|0101\rangle - |0110\rangle)$$

$$B^*\frac{1}{\sqrt{2}}(|0101\rangle - |0110\rangle) = -|0110\rangle.$$

A unitary transform maps an orthonormal basis to an orthonormal basis. The above equations do not determine B, U and S uniquely. For simplicity let B, U and S act as the identity on subspaces for which the unitary transformations are not constrained by the above equations. For B we have

$$B|0101\rangle = |01\rangle \otimes \frac{1}{\sqrt{2}}(|01\rangle + |10\rangle), \qquad B|0110\rangle = |01\rangle \otimes \frac{1}{\sqrt{2}}(|10\rangle - |01\rangle).$$

One solution is

$$B = \frac{1}{\sqrt{2}} I_4 \otimes (|\gamma\rangle\langle 01| + |\delta\rangle\langle 10| + |\alpha\rangle\langle 00| + |\beta\rangle\langle 11|)$$

where

$$|\alpha\rangle = |00\rangle + |11\rangle, \quad |\beta\rangle = |00\rangle - |11\rangle, \quad |\gamma\rangle = |01\rangle + |10\rangle, \quad |\delta\rangle = |10\rangle - |01\rangle.$$

This means that B maps from the computational basis to the Bell basis in the second two qubits. For U we have

$$U\left(|01\rangle \otimes \frac{1}{\sqrt{2}}(|01\rangle + |10\rangle)\right) = \frac{1}{\sqrt{2}}(|0101\rangle + |1010\rangle)$$

and

$$U\left(|01\rangle \otimes \frac{1}{\sqrt{2}}(|01\rangle - |10\rangle)\right) = \frac{1}{\sqrt{2}}(|0101\rangle - |1010\rangle).$$

We rewrite these equations in the simpler form

$$U|0101\rangle = |0101\rangle, \qquad U|0110\rangle = |1010\rangle.$$

A solution for U is then

$$U = I_{16} + (|1010\rangle - |0110\rangle)(\langle 0110| - \langle 1010|)$$

i.e., U is the identity except on the subspace spanned by $|0110\rangle$ and $|1010\rangle$, where U swaps $|0110\rangle$ and $|1010\rangle$. For S we have

$$S\frac{1}{\sqrt{2}}(|0101\rangle + |1010\rangle) = \frac{1}{\sqrt{2}}(|0101\rangle - |1010\rangle).$$

A solution for S is

$$S = I_{16} - 2|1010\rangle\langle 1010|$$

i.e., S is the identity except for changing the sign of $|1010\rangle$.

Problem 5. Consider the Hilbert space $\mathbb{C}^{2^n}$. Let

$$\{\, |0\rangle, \ |1\rangle, \ \ldots, |2^n - 1\rangle \,\}$$

be an orthonormal basis in this Hilbert space. We define the linear operator

$$U_{QFT} := \frac{1}{\sqrt{2^n}} \sum_{j=0}^{2^n-1} \sum_{k=0}^{2^n-1} e^{-i2\pi kj/2^n} |k\rangle\langle j| . \tag{1}$$

This transform is called the *quantum Fourier transform* . Show that U_{QFT} is unitary. In other words show that $U_{QFT}U_{QFT}^* = I_{2^n}$, where we use the *completeness relation*

$$I_{2^n} = \sum_{j=0}^{2^n-1} |j\rangle\langle j| .$$

Thus I_{2^n} is the $2^n \times 2^n$ unit matrix.

Solution 5.   From the definition (1) we find

$$U_{QFT}^* = \frac{1}{\sqrt{2^n}} \sum_{j=0}^{2^n-1} \sum_{k=0}^{2^n-1} e^{i2\pi kj/2^n} |j\rangle\langle k|$$

where * denotes the adjoint. Therefore

$$U_{QFT}U_{QFT}^* = \frac{1}{2^n} \sum_{j=0}^{2^n-1} \sum_{k=0}^{2^n-1} \sum_{l=0}^{2^n-1} \sum_{m=0}^{2^n-1} e^{i2\pi(kj-lm)/2^n} |j\rangle\langle k|l\rangle\langle m|$$

$$= \frac{1}{2^n} \sum_{j=0}^{2^n-1} \sum_{k=0}^{2^n-1} \sum_{m=0}^{2^n-1} e^{i2\pi(kj-km)/2^n} |j\rangle\langle m| .$$

We have for $j = m$, $e^{i2\pi(kj-km)/2^n} = 1$. Thus for $j, m = 0, 1, \ldots, 2^n - 1$

$$\sum_{k=0}^{2^n-1} (e^{i2\pi(j-m)/2^n})^k = 2^n, \quad j = m$$

$$\sum_{k=0}^{2^n-1} (e^{i2\pi(j-m)/2^n})^k = \frac{1 - e^{i2\pi(j-m)}}{1 - e^{i2\pi(j-m)/2^n}} = 0, \quad j \neq m .$$

Thus

$$U_{QFT}U_{QFT}^* = \sum_{j=0}^{2^n-1} |j\rangle\langle j| = I_{2^n} .$$

Problem 6.   Apply the quantum Fourier transform to the state in the Hilbert space $\mathbb{C}^8$

$$\frac{1}{2} \sum_{j=0}^{7} \cos(2\pi j/8)|j\rangle$$

where the quantum Fourier transform is given by

$$U_{QFT} = \frac{1}{2\sqrt{2}} \sum_{j=0}^{7} \sum_{k=0}^{7} e^{-i2\pi kj/8} |k\rangle\langle j|.$$

We use

$$\{ |j\rangle \ : \ j = 0,1,\dots,7 \}$$

as an orthonormal basis in the Hilbert space $\mathbb{C}^8$, where $|7\rangle = |111\rangle \equiv |1\rangle \otimes |1\rangle \otimes |1\rangle$.

Solution 6. We use *Euler's identity* $e^{i\theta} \equiv \cos\theta + i\sin\theta$ and

$$\sum_{k=0}^{N-1} e^{i2\pi k(n-m)/N} = N\delta_{nm}.$$

Thus we have

$$\hat{x}(k) = \sum_{j=0}^{7} e^{-i2\pi kj/8} \cos(2\pi j/8) = \frac{1}{2} \sum_{j=0}^{7} \left(e^{i2\pi(1-k)j/8} + e^{-i2\pi(1+k)j/8} \right)$$

$$= 4(\delta_{k1} + \delta_{k7}).$$

and

$$U_{QFT}\frac{1}{2} \sum_{j=0}^{7} \cos(2\pi j/8)|j\rangle = \frac{1}{2\sqrt{8}} \sum_{k=0}^{7} \hat{x}(k)|k\rangle = \frac{1}{\sqrt{2}}(|1\rangle + |7\rangle).$$

Problem 7. Let

$$U_{IA} := \sum_{j=0}^{2^n-1} \sum_{k=0}^{2^n-1} \left(\frac{2}{2^n} - \delta_{jk} \right) |k\rangle\langle j|. \tag{1}$$

U_{IA} is called the *inversion about average operator*. Show that U_{IA} is unitary. In other words show that $U_{IA}U_{IA}^* = I_{2^n}$.
Hint: Use the *completeness relation*

$$\sum_{j=0}^{2^n-1} |j\rangle\langle j| = I_{2^n}.$$

Solution 7. From (1) we find

$$U_{IA}^* = \sum_{j=0}^{2^n-1} \sum_{k=0}^{2^n-1} \left(\frac{2}{2^n} - \delta_{jk} \right) |k\rangle\langle j| = U_{IA}.$$

Thus

$$
U_{IA}U_{IA}^* = U_{IA}^2 = \sum_{j,k,l,m=0}^{2^n-1} \left(\frac{2}{2^n} - \delta_{jk} \right) \left(\frac{2}{2^n} - \delta_{lm} \right) |k\rangle\langle j|m\rangle\langle l|
$$

$$
= \sum_{j,k,l=0}^{2^n-1} \left(\frac{2}{2^n} - \delta_{jk} \right) \left(\frac{2}{2^n} - \delta_{lj} \right) |k\rangle\langle l|
$$

where we used that $\langle j|m\rangle = \delta_{jm}$. Furthermore, we find

$$
\sum_{j=0}^{2^n-1} \left(\frac{2}{2^n} - \delta_{jk} \right) \left(\frac{2}{2^n} - \delta_{lj} \right) = \sum_{j=0}^{2^n-1} \left(\frac{4}{2^{2n}} - \delta_{jk}\frac{2}{2^n} - \delta_{lj}\frac{2}{2^n} + \delta_{jk}\delta_{lj} \right)
$$

$$
= \frac{4}{2^n} - \frac{2}{2^n} - \frac{2}{2^n} + \sum_{j=0}^{2^n-1} \delta_{jk}\delta_{lk}
$$

$$
= \delta_{lk} \sum_{j=0}^{2^n-1} \delta_{jk}
$$

$$
= \delta_{lk}.
$$

Therefore

$$
U_{IA}U_{IA}^* = \sum_{j=0}^{2^n-1} |j\rangle\langle j| = I_{2^n}.
$$

Problem 8.   Let $\{\, |0\rangle, |1\rangle \,\}$ be an orthonormal basis in the two-dimensional Hilbert space $\mathbb{C}^2$ and

$$
U_H|k\rangle := \frac{1}{\sqrt{2}} \left(|0\rangle + (-1)^k|1\rangle \right), \quad k \in \{0,1\}
$$

$$
U_{PS(\theta)} := |00\rangle\langle00| + |01\rangle\langle01| + |10\rangle\langle10| + e^{i\theta}|11\rangle\langle11|
$$

$$
U_{CNOT} := |00\rangle\langle00| + |01\rangle\langle01| + |10\rangle\langle11| + |11\rangle\langle10|.
$$

(i) From these definitions show that $U_H U_H = I_2$.
(ii) Calculate

$$
(I_2 \otimes U_H)U_{PS(\pi)}(I_2 \otimes U_H)|ab\rangle
$$

and

$$
(I_2 \otimes U_H)U_{CNOT}(I_2 \otimes U_H)|ab\rangle
$$

where $a, b \in \{0,1\}$. What is the use of these transforms?

Solution 8.   (i) An arbitrary state in the Hilbert space $\mathbb{C}^2$ can be written as

$$
|\psi\rangle := a|0\rangle + b|1\rangle
$$

where $a, b \in \mathbb{C}$ and $|a|^2 + |b|^2 = 1$. We find

$$U_H U_H |\psi\rangle = U_H \frac{1}{\sqrt{2}} (a|0\rangle + a|1\rangle + b|0\rangle - b|1\rangle) = \frac{1}{2} (2a|0\rangle + 2b|1\rangle)$$
$$= a|0\rangle + b|1\rangle \,.$$

Thus, $U_H U_H = I_2$.
(ii) We find

$$(I_2 \otimes U_H) U_{PS(\pi)} (I_2 \otimes U_H) |ab\rangle = (I_2 \otimes U_H) U_{PS(\pi)} \frac{1}{\sqrt{2}} |a\rangle \otimes (|0\rangle + (-1)^b |1\rangle)$$
$$= (I_2 \otimes U_H) \frac{1}{\sqrt{2}} |a\rangle \otimes (|0\rangle + (-1)^{a+b} |1\rangle)$$
$$= \frac{1}{2} |a, a \oplus b\rangle$$

where $a \oplus b = a + b$ (modulo 2) is the XOR operation. We obtain

$$(I_2 \otimes U_H) U_{CNOT} (I_2 \otimes U_H) |ab\rangle = (I_2 \otimes U_H) U_{CNOT} \frac{1}{\sqrt{2}} |a\rangle \otimes (|0\rangle + (-1)^b |1\rangle)$$
$$= \begin{cases} (I_2 \otimes U_H) \frac{1}{\sqrt{2}} |a\rangle \otimes (|0\rangle + (-1)^b |1\rangle) & a = 0 \\ (I_2 \otimes U_H) \frac{1}{\sqrt{2}} |a\rangle \otimes (|1\rangle + (-1)^b |0\rangle) & a = 1 \end{cases}$$
$$= (I_2 \otimes U_H) \frac{1}{\sqrt{2}} |a\rangle \otimes (-1)^{ab} (|0\rangle + (-1)^b |1\rangle)$$
$$= (-1)^{ab} |ab\rangle \,.$$

The first computation is U_{CNOT}, the second is $U_{PS(\pi)}$.

Problem 9. The *XOR gate* is given by

$$U_{XOR} |m\rangle \otimes |n\rangle = |m\rangle \otimes |m \oplus n\rangle$$

where $m, n \in \{0, 1\}$ and $\oplus$ denotes addition modulo 2. The transformation has the following properties: (a) it is unitary and thus reversible, (b) it is hermitian, (c) $m \oplus n = 0$ if and only if $m = n$. The first index denotes the state of the control qubit and the second index denotes the state of the target qubit.

A generalized quantum XOR gate (*GXOR gate*) acts on two d-dimensional quantum systems ($d > 2$). In analogy with qubits one calls these two systems *qudits*. The basis states $|m\rangle$ of each qudit are labeled by elements in the ring $\mathbb{Z}_d$ which we denote by the numbers, $m = 0, 1, \ldots, d - 1$, with the usual rules for addition and multiplication modulo d. We define two operators

$$U_{GXOR1} |m\rangle \otimes |n\rangle := |m\rangle \otimes |m \oplus n\rangle$$

and

$$U_{GXOR2}|m\rangle \otimes |n\rangle := |m\rangle \otimes |m \ominus n\rangle$$

where $m \ominus n := (m - n)$ modulo d. Discuss the properties of these two operators.

Solution 9.   For U_{GXOR1} we find that the operator is unitary but not hermitian for $d > 2$. Therefore it is no longer its own inverse. We have to obtain the inverse of the U_{GXOR1} gate by iteration, i.e.,

$$U_{GXOR1}^{-1} = U_{GXOR1}^{d-1} = U_{GXOR1}^{\dagger} \neq U_{GXOR1}.$$

For the operator U_{GXOR2} we find that in the special case for $d = 2$ it reduces to the XOR-gate. Furthermore, the operator is unitary, hermitian and

$$m \ominus n = 0 \text{ modulo } d$$

if and only if $m = n$.

Problem 10.   Given an orthonormal basis in $\mathbb{C}^N$ denoted by

$$|\phi_0\rangle, |\phi_1\rangle, \ldots, |\phi_{N-1}\rangle.$$

(i) Show that

$$U := \sum_{k=0}^{N-2} |\phi_k\rangle\langle\phi_{k+1}| + |\phi_{N-1}\rangle\langle\phi_0|$$

is a unitary matrix.
(ii) Find $\operatorname{tr}(U)$.
(iii) Find U^N.
(iv) Does U depend on the chosen basis? Prove or disprove.
Hint. Consider $N = 2$, the standard basis $(1,0)^T$, $(0,1)^T$ and the basis $\frac{1}{\sqrt{2}}(1,1)^T$, $\frac{1}{\sqrt{2}}(1,-1)^T$.
(v) Show that the set

$$\{U, U^2, \ldots, U^N\}$$

forms a *commutative group* (*abelian group*) under matrix multiplication. The set is a subgroup of the group of all permutation matrices.
(vi) Assume that the set given above is the standard basis. Show that the unitary matrix U is given by the permutation matrix

$$U = \begin{pmatrix} 0 & 1 & 0 & \ldots & 0 \\ 0 & 0 & 1 & \ldots & 0 \\ \vdots & \vdots & \vdots & \vdots & \vdots \\ 0 & 0 & 0 & \ldots & 1 \\ 1 & 0 & 0 & \ldots & 0 \end{pmatrix}.$$

Solution 10. (i) Since $\langle \phi_j | \phi_k \rangle = \delta_{jk}$ we have

$$UU^* = \left(\sum_{k=0}^{N-2} |\phi_k\rangle\langle\phi_{k+1}| + |\phi_{N-1}\rangle\langle\phi_0| \right) \left(\sum_{k=0}^{N-2} |\phi_{k+1}\rangle\langle\phi_k| + |\phi_0\rangle\langle\phi_{N-1}| \right)$$

$$= \sum_{k=0}^{N-1} |\phi_k\rangle\langle\phi_k| = I_N .$$

(ii) Obviously we have

$$\mathrm{tr}(U) = 0$$

since the terms $|\phi_k\rangle\langle\phi_k|$ do not appear in the sum (i.e. we calculate the trace in the basis $\{ |\phi_0\rangle, \ldots, |\phi_{N-1}\rangle \}$).

(iii) We notice that U maps $|\phi_k\rangle$ to $|\phi_{k-1}\rangle$. Applying this N times and using modulo N arithmetic we obtain (i.e., U^N maps $|\phi_k\rangle$ to $|\phi_{k-N}\rangle$)

$$U^N = I_N .$$

(iv) For the standard basis in $\mathbb{C}^2$ $\{ (1,0)^T, (0,1)^T \}$ we obtain

$$U_{std} = \begin{pmatrix} 0 & 1 \\ 1 & 0 \end{pmatrix} .$$

For the basis in $\mathbb{C}^2$ $\{ \frac{1}{\sqrt{2}}(1,1)^T, \frac{1}{\sqrt{2}}(1,-1)^T \}$ we obtain

$$\sigma_z = \begin{pmatrix} 1 & 0 \\ 0 & -1 \end{pmatrix} .$$

Obviously the two unitary matrices are different. Of course there is a unitary matrix V such that $U_{std} = V^{-1}\sigma_z V$.

(v) Since $U_N = I_N = U^0$ we have that

$$U^s U^t = U^{s+t} = U^{s+t \bmod N}.$$

Thus the set of matrices $\{U, U^2, \ldots, U^N\}$ forms an abelian group under matrix multiplication, because $\{0, 1, \ldots, N-1\}$ forms a group under addition modulo N. The two groups are isomorphic.

(vi) Let $\mathbf{e}_j$ denote the element of the standard basis in $\mathbb{C}^n$ with a 1 in the jth position (numbered from 0) and 0 in all other positions. Then U is given by

$$U = \sum_{k=0}^{N-2} \mathbf{e}_k \mathbf{e}_{k+1}^T + \mathbf{e}_{N-1} \mathbf{e}_0^T .$$

In the product $\mathbf{e}_k \mathbf{e}_{k+1}^T$, $\mathbf{e}_k$ denotes the row and $\mathbf{e}_{k+1}^T$ denotes the column in the matrix U. Thus we obtain the matrix described above.

Problem 11.   (i) Let σ_x, σ_y and σ_z be the Pauli spin matrices and I_2 be the 2×2 unit matrix. Find

$$(\sigma_z \otimes \sigma_z \otimes \sigma_z \otimes \sigma_z)(\sigma_x \otimes \sigma_x \otimes I_2 \otimes I_2)(\sigma_z \otimes \sigma_z \otimes \sigma_z \otimes \sigma_z),$$

$$(\sigma_z \otimes \sigma_z \otimes \sigma_z \otimes \sigma_z)(I_2 \otimes \sigma_x \otimes \sigma_x \otimes I_2)(\sigma_z \otimes \sigma_z \otimes \sigma_z \otimes \sigma_z),$$

$$(\sigma_z \otimes \sigma_z \otimes \sigma_z \otimes \sigma_z)(I_2 \otimes I_2 \otimes \sigma_x \otimes \sigma_x)(\sigma_z \otimes \sigma_z \otimes \sigma_z \otimes \sigma_z).$$

(ii) Replace σ_x by σ_y in the expressions given above and calculate the expressions.

(iii) Given the one-dimensional XY-*model* with open boundary conditions

$$\hat{H}_{XY} = - \sum_{j=-N/2+1}^{N/2-1} \left(\frac{1+\gamma}{2} \sigma_{x,j} \sigma_{x,j+1} + \frac{1-\gamma}{2} \sigma_{y,j} \sigma_{y,j+1} \right) - \lambda \sum_{j=-N/2+1}^{N/2} \sigma_{z,j}$$

where the parameter λ is the intensity of the magnetic field applied in the z-direction and the parameter γ determines the degree of anisotropy of the spin-spin interaction, which is restricted to the xy-plane in spin space. Find

$$\left(\prod_{j=-N/2+1}^{N/2} \sigma_{z,j} \right) \hat{H}_{XY} \left(\prod_{j=-N/2+1}^{N/2} \sigma_{z,j} \right).$$

Solution 11.   (i) Since $\sigma_z^2 = I_2$ and $\sigma_z \sigma_x \sigma_z = -\sigma_x$ we find for the first expression

$$(\sigma_z \otimes \sigma_z \otimes \sigma_z \otimes \sigma_z)(\sigma_x \otimes \sigma_x \otimes I_2 \otimes I_2)(\sigma_z \otimes \sigma_z \otimes \sigma_z \otimes \sigma_z) = \sigma_x \otimes \sigma_x \otimes I_2 \otimes I_2 .$$

Analogously, we find

$$(\sigma_z \otimes \sigma_z \otimes \sigma_z \otimes \sigma_z)(I_2 \otimes \sigma_x \otimes \sigma_x \otimes I_2)(\sigma_z \otimes \sigma_z \otimes \sigma_z \otimes \sigma_z) = I_2 \otimes \sigma_x \otimes \sigma_x \otimes I_2$$

$$(\sigma_z \otimes \sigma_z \otimes \sigma_z \otimes \sigma_z)(I_2 \otimes I_2 \otimes \sigma_x \otimes \sigma_x)(\sigma_z \otimes \sigma_z \otimes \sigma_z \otimes \sigma_z) = I_2 \otimes I_2 \otimes \sigma_x \otimes \sigma_x .$$

(ii) Replacing σ_x by σ_y and using $\sigma_z \sigma_y \sigma_z = -\sigma_y$ yields

$$(\sigma_z \otimes \sigma_z \otimes \sigma_z \otimes \sigma_z)(\sigma_y \otimes \sigma_y \otimes I_2 \otimes I_2)(\sigma_z \otimes \sigma_z \otimes \sigma_z \otimes \sigma_z) = \sigma_y \otimes \sigma_y \otimes I_2 \otimes I_2$$

$$(\sigma_z \otimes \sigma_z \otimes \sigma_z \otimes \sigma_z)(I_2 \otimes \sigma_y \otimes \sigma_y \otimes I_2)(\sigma_z \otimes \sigma_z \otimes \sigma_z \otimes \sigma_z) = I_2 \otimes \sigma_y \otimes \sigma_y \otimes I_2$$

$$(\sigma_z \otimes \sigma_z \otimes \sigma_z \otimes \sigma_z)(I_2 \otimes I_2 \otimes \sigma_y \otimes \sigma_y)(\sigma_z \otimes \sigma_z \otimes \sigma_z \otimes \sigma_z) = I_2 \otimes I_2 \otimes \sigma_y \otimes \sigma_y .$$

(iii) Using the results from (i) and (ii) and extending from $N = 4$ to arbitrary N, we find

$$\left(\prod_{j=-N/2+1}^{N/2} \sigma_{z,j} \right) \hat{H}_{XY} \left(\prod_{j=-N/2+1}^{N/2} \sigma_{z,j} \right) = \hat{H}_{XY} .$$

From (ii) and (iii) we find that the Hamilton operator $\hat{H}_{XY}$ is invariant under this transformation.

Problem 12. (i) Consider the product state

$$|D\rangle \otimes |P\rangle$$

where $|D\rangle$ is a state to describe a m-qubit data register and $|P\rangle$ is a state to describe an n-qubit program register. Let G be a unitary operator acting on this product state

$$|D\rangle \otimes |P\rangle \rightarrow G(|D\rangle \otimes |P\rangle).$$

The unitary operator is implemented as follows. A unitary operator U acting on the m-qubits of the data register, is said to be implemented by this gate array if there exists a state $|P_U\rangle$ of the program register such that

$$G(|D\rangle \otimes |P_U\rangle) = (U|D\rangle) \otimes |P'_U\rangle$$

for all states $|D\rangle$ of the data register and some state $|P'_U\rangle$ of the program register. Show that $|P'_U\rangle$ does not depend on $|D\rangle$.
(ii) Suppose distinct (up to a global phase) unitary operators $U_1, \ldots, U_N$ are implemented by some programmable quantum gate array. Show that the corresponding programs $|P_1\rangle, \ldots, |P_N\rangle$ are mutually orthogonal.

Solution 12. (i) Consider

$$G(|D_1\rangle \otimes |P\rangle) = (U|D_1\rangle) \otimes |P'_1\rangle, \qquad G(|D_2\rangle \otimes |P\rangle) = (U|D_2\rangle) \otimes |P'_2\rangle.$$

Taking the scalar product of these two equations and using $G^\dagger G = I$, $U^\dagger U = I$ and $\langle P|P\rangle = 1$ we find

$$\langle D_1|D_2\rangle = \langle D_1|D_2\rangle\langle P'_1|P'_2\rangle.$$

If $\langle D_1|D_2\rangle \neq 0$ we find $\langle P'_1|P'_2\rangle = 1$. Thus

$$|P'_1\rangle = |P'_2\rangle.$$

Consequently, there is no $|D\rangle$ dependence of $|P'_U\rangle$. What happens for $\langle D_1|D_2\rangle = 0$?

(ii) Suppose that $|P\rangle$ and $|Q\rangle$ are programs which implement unitary operators U_p and U_q which are distinct up to global phase changes. Then for an arbitrary data state $|D\rangle$ we have

$$G(|D\rangle \otimes |P\rangle) = (U_p|D\rangle) \otimes |P'\rangle$$

$$G(|D\rangle \otimes |Q\rangle) = (U_q|D\rangle) \otimes |Q'\rangle$$

where $|P'\rangle$ and $|Q'\rangle$ are states of the program register. Taking the scalar product of these two equations and using $G^\dagger G = I$, $\langle D|D\rangle = 1$ we obtain

$$\langle Q|P\rangle = \langle Q'|P'\rangle \langle D|U_q^\dagger U_p|D\rangle.$$

Suppose that $\langle Q'|P'\rangle \neq 0$. Then we have

$$\frac{\langle Q|P\rangle}{\langle Q'|P'\rangle} = \langle D|U_q^\dagger U_p|D\rangle.$$

The left-hand side of this equation has no $|D\rangle$ dependence. Thus we have $U_q^\dagger U_p = cI$ for some complex number c. It follows that we can only have $\langle P'|Q'\rangle \neq 0$ if U_p and U_q are the same up to a global phase. However we assumed that this is not the case and therefore $\langle Q'|P'\rangle = 0$. Hence

$$\langle Q|P\rangle = 0.$$

This means the states $|Q\rangle$ and $|P\rangle$ are orthogonal.

Problem 13.   (i) Let

$$M := \frac{1}{\sqrt{2}} \begin{pmatrix} 1 & i & 0 & 0 \\ 0 & 0 & i & 1 \\ 0 & 0 & i & -1 \\ 1 & -i & 0 & 0 \end{pmatrix}.$$

Is the matrix M unitary? In quantum computing M is called the *magic gate*.

(ii) Let

$$U_H := \frac{1}{\sqrt{2}} \begin{pmatrix} 1 & 1 \\ 1 & -1 \end{pmatrix}, \qquad U_S := \begin{pmatrix} 1 & 0 \\ 0 & i \end{pmatrix}$$

and

$$U_{CNOT2} = \begin{pmatrix} 1 & 0 & 0 & 0 \\ 0 & 0 & 0 & 1 \\ 0 & 0 & 1 & 0 \\ 0 & 1 & 0 & 0 \end{pmatrix}.$$

Show that the matrix M can be written as

$$M = U_{CNOT2}(I_2 \otimes U_H)(U_S \otimes U_S).$$

(iii) Let $SO(4)$ be the special orthogonal Lie group. Let $SU(2)$ be the special unitary Lie group. Show that for every real orthogonal matrix $U \in SO(4)$, the matrix MUM^{-1} is the Kronecker product of two 2-dimensional special unitary matrices, i.e.,

$$MUM^{-1} \in SU(2) \otimes SU(2).$$

Solution 13.   (i) Since $MM^* = I_4$ we find that M is unitary.
(ii) We obtain

$$
M = \begin{pmatrix} 1 & 0 & 0 & 0 \\ 0 & 0 & 0 & 1 \\ 0 & 0 & 1 & 0 \\ 0 & 1 & 0 & 0 \end{pmatrix}
\begin{pmatrix} \frac{1}{\sqrt{2}} & \frac{1}{\sqrt{2}} & 0 & 0 \\ \frac{1}{\sqrt{2}} & -\frac{1}{\sqrt{2}} & 0 & 0 \\ 0 & 0 & \frac{1}{\sqrt{2}} & \frac{1}{\sqrt{2}} \\ 0 & 0 & \frac{1}{\sqrt{2}} & -\frac{1}{\sqrt{2}} \end{pmatrix}
\begin{pmatrix} 1 & 0 & 0 & 0 \\ 0 & i & 0 & 0 \\ 0 & 0 & i & 0 \\ 0 & 0 & 0 & -1 \end{pmatrix}.
$$

(iii) We show that for every $A \otimes B \in SU(2) \otimes SU(2)$, we have

$$
M^{-1}(A \otimes B)M \in SO(4).
$$

Now every matrix $A \in SU(2)$ can be written as

$$
R_z(\alpha)R_y(\theta)R_z(\beta)
$$

for some $\alpha, \beta, \theta \in \mathbb{R}$, where

$$
R_y(\theta) = \begin{pmatrix} \cos(\theta/2) & \sin(\theta/2) \\ -\sin(\theta/2) & \cos(\theta/2) \end{pmatrix}, \qquad R_z(\alpha) = \begin{pmatrix} e^{-i\alpha/2} & 0 \\ 0 & e^{i\alpha/2} \end{pmatrix}.
$$

Therefore any matrix $A \otimes B \in SU(2) \otimes SU(2)$ can be written as a product of the matrices of the form $V \otimes I_2$ or $I_2 \otimes V$, where V is either $R_y(\theta)$ or $R_z(\alpha)$. Next we have to show that $M^{-1}(V \otimes I_2)M$ and $M^{-1}(I_2 \otimes V)M$ are in $SO(4)$. We have

$$
M^{-1}(R_y(\theta) \otimes I_2)M = \begin{pmatrix} \cos(\theta/2) & 0 & 0 & -\sin(\theta/2) \\ 0 & \cos(\theta/2) & \sin(\theta/2) & 0 \\ 0 & -\sin(\theta/2) & \cos(\theta/2) & 0 \\ \sin(\theta/2) & 0 & 0 & \cos(\theta/2) \end{pmatrix}
$$

$$
M^{-1}(R_z(\alpha) \otimes I_2)M = \begin{pmatrix} \cos(\alpha/2) & \sin(\alpha/2) & 0 & 0 \\ -\sin(\alpha/2) & \cos(\alpha/2) & 0 & 0 \\ 0 & 0 & \cos(\alpha/2) & -\sin(\alpha/2) \\ 0 & 0 & \sin(\alpha/2) & \cos(\alpha/2) \end{pmatrix}.
$$

We have similar equations for the cases of $I_2 \otimes R_y(\theta)$ and $I_2 \otimes R_z(\alpha)$. Since the mapping

$$
A \otimes B \to M^{-1}(A \otimes B)M
$$

is one-to-one (invertible) and the Lie groups $SU(2) \otimes SU(2)$ and $SO(4)$ have the same topological dimension, we conclude that the mapping is an isomorphism between these two Lie groups.

Problem 14.   Consider three two-dimensional Hilbert spaces $\mathcal{H}_1$, $\mathcal{H}_2$ and $\mathcal{H}_3$. Consider the normalized product state

$$
|\psi\rangle = \sum_{j=0}^{1}\sum_{k=0}^{1}\sum_{\ell=0}^{1} c_{jk\ell}|j\rangle \otimes |k\rangle \otimes |\ell\rangle
$$

in the product Hilbert space $\mathcal{H}_1 \otimes \mathcal{H}_2 \otimes \mathcal{H}_3$. Let U_1, U_2, U_3 be unitary operators acting in these Hilbert spaces. By the *First Fundamental Theorem* of invariant theory applied to U_1, U_2, U_3, any polynomial in $c_{jk\ell}$ which is invariant under the action on $|\psi\rangle$ of the local unitary transformation $U_1 \otimes U_2 \otimes U_3$ is a sum of homogeneous polynomials of even degree (say $2r$). For $r = 1$ we have

$$P_{\sigma_1\sigma_2}(\mathbf{c}) = \sum_{j_1=0}^{1}\sum_{k_1=0}^{1}\sum_{\ell_1=0}^{1} c_{j_1 k_1 \ell_1} c^*_{j_1 k_{\sigma_1(1)} \ell_{\sigma_2(1)}}$$

where σ_1 and σ_2 are permutations of 1. We denote by e the identity permutation. For $r = 2$ we have

$$P_{\sigma_1\sigma_2}(\mathbf{c}) = \sum_{j_1=0}^{1}\sum_{k_1=0}^{1}\sum_{\ell_1=0}^{1}\sum_{j_2=0}^{1}\sum_{k_2=0}^{1}\sum_{\ell_2=0}^{1} c_{j_1 k_1 \ell_1} c_{j_2 k_2 \ell_2} c^*_{j_1 k_{\sigma_1(1)} \ell_{\sigma_2(1)}} c^*_{j_2 k_{\sigma_1(2)} \ell_{\sigma_2(2)}}.$$

(i) Calculate the invariants.
(ii) Describe the connection with the partial traces

$$\rho_1 := \mathrm{tr}_{23}(|\psi\rangle\langle\psi|), \quad \rho_2 := \mathrm{tr}_{31}(|\psi\rangle\langle\psi|), \quad \rho_3 := \mathrm{tr}_{12}(|\psi\rangle\langle\psi|)$$

of the density operator $\rho := |\psi\rangle\langle\psi|$.

Solution 14.   (i) Obviously for the case $r = 1$ (degree 2) we only have the identity permutation, i.e., $\sigma_1 = \sigma_2 = e$ with $e(1) = 1$, $e(2) = 2$. Thus we find only one invariant, namely

$$I_0 = \sum_{j=0}^{1}\sum_{k=0}^{1}\sum_{\ell=0}^{1} c_{jk\ell} c^*_{jk\ell} = \langle\psi|\psi\rangle = 1$$

which is the normalization condition. For the case $r = 2$ (degree 4) we find four linearly independent quartic invariants since

$$e(1) = 1, \quad e(2) = 2, \quad \sigma(1) = 2, \quad \sigma(2) = 1.$$

Thus

$$I_1 = P_{ee}(\mathbf{c}) = \sum_{j_1=0}^{1}\sum_{k_1=0}^{1}\sum_{\ell_1=0}^{1}\sum_{j_2=0}^{1}\sum_{k_2=0}^{1}\sum_{\ell_2=0}^{1} c_{j_1 k_1 \ell_1} c^*_{j_1 k_1 \ell_1} c_{j_2 k_2 \ell_2} c^*_{j_2 k_2 \ell_2} = \langle\psi|\psi\rangle^2$$

$$I_2 = P_{e\sigma}(\mathbf{c}) = \sum_{j_1=0}^{1}\sum_{k_1=0}^{1}\sum_{\ell_1=0}^{1}\sum_{j_2=0}^{1}\sum_{k_2=0}^{1}\sum_{\ell_2=0}^{1} c_{j_1 k_1 \ell_1} c^*_{j_1 k_1 \ell_2} c_{j_2 k_2 \ell_2} c^*_{j_2 k_2 \ell_1}$$

$$I_3 = P_{\sigma e}(\mathbf{c}) = \sum_{j_1=0}^{1}\sum_{k_1=0}^{1}\sum_{\ell_1=0}^{1}\sum_{j_2=0}^{1}\sum_{k_2=0}^{1}\sum_{\ell_2=0}^{1} c_{j_1 k_1 \ell_1} c^*_{j_1 k_2 \ell_1} c_{j_2 k_2 \ell_2} c^*_{j_2 k_1 \ell_2}$$

$$I_4 = P_{\sigma\sigma}(\mathbf{c}) = \sum_{j_1=0}^{1}\sum_{k_1=0}^{1}\sum_{\ell_1=0}^{1}\sum_{j_2=0}^{1}\sum_{k_2=0}^{1}\sum_{\ell_2=0}^{1} c_{j_1 k_1 \ell_1} c^*_{j_1 k_2 \ell_2} c_{j_2 k_2 \ell_2} c^*_{j_2 k_1 \ell_1}.$$

(ii) We have

$$I_2 = \mathrm{tr}(\rho_3^2), \qquad I_3 = \mathrm{tr}(\rho_2^2), \qquad I_4 = \mathrm{tr}(\rho_1^2).$$

Problem 15. Consider two Hilbert spaces $\mathcal{H}_{reg}$ and $\mathcal{H}_{sys}$ and the product state

$$|\psi\rangle = (\alpha|0^{reg}\rangle + \beta|1^{reg}\rangle) \otimes |0^{sys}\rangle$$

in the Hilbert space $\mathcal{H}_{reg} \otimes \mathcal{H}_{sys}$, where *reg* stands for register and *sys* for system. Consider the swap operation (swap gate)

$$U_{swap}((\alpha|0^{reg}\rangle + \beta|1^{reg}\rangle) \otimes |0^{sys}\rangle) = |0^{reg}\rangle \otimes (\alpha|0^{sys}\rangle + \beta|1^{sys}\rangle).$$

Discuss the operation on physical grounds.

Solution 15. Creating such a superposition could violate conservation laws (for example charge) and in this case is forbidden by superselection rules.

Problem 16. The *Toffoli gate* is the unitary operator acting as

$$U_T|a,b,c\rangle = |a,b,ab+c\rangle$$

in the Hilbert space $\mathbb{C}^8$, where $a,b,c \in \{0,1\}$ and ab denotes the AND operation of a and b. The addition $+$ is modulo 2.
(i) Find the truth table.
(ii) Find the matrix representation for the standard basis.

The Toffoli gate is an extension of the CNOT gate.

Solution 16. (i) We have the truth table

a	b	c	a	b	$ab+c$
0	0	0	0	0	0
0	0	1	0	0	1
0	1	0	0	1	0
0	1	1	0	1	1
1	0	0	1	0	0
1	0	1	1	0	1
1	1	0	1	1	1
1	1	1	1	1	0

(ii) The matrix representation of the Toffoli gate is given by the permutation matrix

$$\begin{pmatrix} 1 & 0 & 0 & 0 & 0 & 0 & 0 & 0 \\ 0 & 1 & 0 & 0 & 0 & 0 & 0 & 0 \\ 0 & 0 & 1 & 0 & 0 & 0 & 0 & 0 \\ 0 & 0 & 0 & 1 & 0 & 0 & 0 & 0 \\ 0 & 0 & 0 & 0 & 1 & 0 & 0 & 0 \\ 0 & 0 & 0 & 0 & 0 & 1 & 0 & 0 \\ 0 & 0 & 0 & 0 & 0 & 0 & 0 & 1 \\ 0 & 0 & 0 & 0 & 0 & 0 & 1 & 0 \end{pmatrix}.$$

Problem 17. The *Fredkin gate* is the unitary operator acting as

$$U_F|c, x, y\rangle = |c, cx + \bar{c}y, \bar{c}x + cy\rangle$$

in the Hilbert space $\mathbb{C}^8$, where $c, x, y \in \{0, 1\}$.
(i) Consider the cases $c = 0$ and $c = 1$.
(ii) Find the matrix representation for the standard basis.

Solution 17.   (i) For $c = 0$ we have $\bar{c} = 1$. Therefore

$$cx = 0, \quad \bar{c}x = x, \quad cy = 0, \quad \bar{c}y = y.$$

Thus

$$U_F|0, x, y\rangle = |0, y, x\rangle.$$

For $c = 1$ we have $\bar{c} = 0$. Therefore $cx = x$, $\bar{c}x = 0$, $cy = y$, $\bar{c}y = 0$. Thus

$$U_F|1, x, y\rangle = |1, x, y\rangle.$$

Consequently c is a control bit. If $c = 0$ then x and y swap around. If $c = 1$ then x and y stay the same.
(ii) The matrix representation of the Fredkin gate is given by the permutation matrix

$$\begin{pmatrix} 1 & 0 & 0 & 0 & 0 & 0 & 0 & 0 \\ 0 & 0 & 1 & 0 & 0 & 0 & 0 & 0 \\ 0 & 1 & 0 & 0 & 0 & 0 & 0 & 0 \\ 0 & 0 & 0 & 1 & 0 & 0 & 0 & 0 \\ 0 & 0 & 0 & 0 & 1 & 0 & 0 & 0 \\ 0 & 0 & 0 & 0 & 0 & 1 & 0 & 0 \\ 0 & 0 & 0 & 0 & 0 & 0 & 1 & 0 \\ 0 & 0 & 0 & 0 & 0 & 0 & 0 & 1 \end{pmatrix} = (1) \oplus U_{NOT} \oplus I_6.$$

Problem 18. Consider the 8×8 matrix

$$U(\alpha) = \frac{e^{i\alpha}}{\sqrt{2}}(I_2 \otimes I_2 \otimes I_2 + i\sigma_x \otimes \sigma_x \otimes \sigma_x)$$

where $\alpha \in \mathbb{R}$.
(i) Show that $U(\alpha)$ is unitary.
(ii) Let

$$|\uparrow\rangle = \begin{pmatrix} 1 \\ 0 \end{pmatrix}, \qquad |\downarrow\rangle = \begin{pmatrix} 0 \\ 1 \end{pmatrix}.$$

Consider the product state

$$|\psi\rangle = |\downarrow\rangle \otimes |\downarrow\rangle \otimes |\downarrow\rangle \equiv \begin{pmatrix} 0 \\ 1 \end{pmatrix} \otimes \begin{pmatrix} 0 \\ 1 \end{pmatrix} \otimes \begin{pmatrix} 0 \\ 1 \end{pmatrix}.$$

Calculate the state $U|\psi\rangle$.
(iii) Consider $U(\alpha = 0)$ and the unitary 8×8 diagonal matrix

$$V = \mathrm{diag}(e^{i3\phi/2},\ 1,\ 1,\ 1,\ 1,\ 1,\ 1,\ e^{-i3\phi/2}).$$

Calculate the state $VU(\alpha = 0)|\psi\rangle$.
(iv) Calculate the state $U(\alpha = 0)VU(\alpha = 0)|\psi\rangle$.
(v) Let

$$|\xi_1\rangle = |\downarrow\rangle \otimes |\downarrow\rangle \otimes |\downarrow\rangle, \qquad |\xi_2\rangle = |\uparrow\rangle \otimes |\uparrow\rangle \otimes |\uparrow\rangle.$$

Calculate the probabilities

$$|\langle\xi_1|U(\alpha = 0)VU(\alpha = 0)|\psi\rangle|^2, \qquad |\langle\xi_2|U(\alpha = 0)VU(\alpha = 0)|\psi\rangle|^2.$$

Draw $|\langle\xi_1|U(\alpha = 0)VU(\alpha = 0)|\psi\rangle|^2$ and $|\langle\xi_2|U(\alpha = 0)VU(\alpha = 0)|\psi\rangle|^2$ as functions of ϕ.

Solution 18. (i) Since $\sigma_x^* = \sigma_x$ we have

$$U^*(\alpha) = \frac{e^{-i\alpha}}{\sqrt{2}}(I_2 \otimes I_2 \otimes I_2 - i\sigma_x \otimes \sigma_x \otimes \sigma_x).$$

Since $\sigma_x^2 = I_2$ we obtain

$$U^*(\alpha)U(\alpha) = \frac{1}{2}(I_2 \otimes I_2 \otimes I_2 + I_2 \otimes I_2 \otimes I_2) = I_2 \otimes I_2 \otimes I_2 = I_8.$$

Thus U is a unitary matrix.
(ii) We find

$$U(\alpha)|\psi\rangle = U(\alpha)|\downarrow\rangle \otimes |\downarrow\rangle \otimes |\downarrow\rangle$$

$$= \frac{e^{i\alpha}}{\sqrt{2}}(I_2 \otimes I_2 \otimes I_2 + i\sigma_x \otimes \sigma_x \otimes \sigma_x)(|\downarrow\rangle \otimes |\downarrow\rangle \otimes |\downarrow\rangle)$$

$$= \frac{e^{i\alpha}}{\sqrt{2}}(|\downarrow\rangle \otimes |\downarrow\rangle \otimes |\downarrow\rangle + i|\uparrow\rangle \otimes |\uparrow\rangle \otimes |\uparrow\rangle)$$

$$= \frac{e^{i\alpha}}{\sqrt{2}}(|\downarrow\rangle \otimes |\downarrow\rangle \otimes |\downarrow\rangle + e^{i\pi/2}|\uparrow\rangle \otimes |\uparrow\rangle \otimes |\uparrow\rangle).$$

This is the *Greenberger-Horne-Zeilinger state* (GHZ-state).
(iii) We find

$$VU(\alpha = 0)|\psi\rangle = \frac{1}{\sqrt{2}}(e^{-i3\phi/2}|\downarrow\rangle \otimes |\downarrow\rangle \otimes |\downarrow\rangle + ie^{i3\phi/2}|\uparrow\rangle \otimes |\uparrow\rangle \otimes |\uparrow\rangle)$$

$$= \frac{1}{\sqrt{2}}(\cos(3\phi/2)(|\downarrow\rangle \otimes |\downarrow\rangle \otimes |\downarrow\rangle + i|\uparrow\rangle \otimes |\uparrow\rangle \otimes |\uparrow\rangle)$$
$$-i\sin(3\phi/2)(|\downarrow\rangle \otimes |\downarrow\rangle \otimes |\downarrow\rangle - i|\uparrow\rangle \otimes |\uparrow\rangle \otimes |\uparrow\rangle)).$$

(iv) Using the result from (iii) we find

$$U(\alpha = 0)VU(\alpha = 0)|\psi\rangle = i\cos(\frac{3}{2}\phi)|\uparrow\rangle \otimes |\uparrow\rangle \otimes |\uparrow\rangle - i\sin(\frac{3}{2}\phi)|\downarrow\rangle \otimes |\downarrow\rangle \otimes |\downarrow\rangle.$$

(v) We obtain for the probabilities

$$|\langle\xi_1|U(\alpha = 0)VU(\alpha = 0)|\psi\rangle|^2 = \sin^2(3\phi/2) \equiv \frac{1}{2} - \frac{1}{2}\cos(3\phi)$$

and

$$|\langle\xi_2|U(\alpha = 0)VU(\alpha = 0)|\psi\rangle|^2 = \cos^2(3\phi/2) \equiv \frac{1}{2} + \frac{1}{2}\cos(3\phi).$$

Problem 19. A *quantum 2-torus* is based on a C^*-algebra A generated by the elements U_1 and U_2 with the relations

$$U_1^{-1} = U_1^*, \qquad U_2^* = U_2^{-1}, \qquad U_1U_2 = zU_2U_1$$

where z is a fixed complex unit. The algebra A will be commutative if and only if $z = 1$, and in this case it describes the classical 2-torus. If z is different from 1 then A describes a purely quantum object. This space is called a quantum torus. A non commutative d-torus T_θ^d is a C^*-algebra generated by d-unitaries $U_1, U_2, \ldots, U_d$ subject to the relations

$$U_\alpha U_\beta = e^{2\pi i\theta_{\alpha\beta}}U_\beta U_\alpha, \qquad \alpha, \beta = 1, 2, \ldots, d$$

where $\theta = (\theta_{\alpha\beta})$ is a skew-symmetric matrix with real entries. Consider the C^*-algebra given by the 2×2 matrices.
(i) Let

$$U_1 = \begin{pmatrix} 0 & 1 \\ 1 & 0 \end{pmatrix}, \qquad U_2 = \begin{pmatrix} 0 & 1 \\ -1 & 0 \end{pmatrix}.$$

Can we find $z \in \mathbb{C}$ such that $U_1U_2 = zU_2U_1$?
(ii) Let

$$U_1 = \frac{1}{\sqrt{2}}\begin{pmatrix} 1 & 1 \\ 1 & -1 \end{pmatrix}, \qquad U_2 = \begin{pmatrix} 0 & 1 \\ 1 & 0 \end{pmatrix}.$$

Can we find $z \in \mathbb{C}$ such that $U_1 U_2 = z U_2 U_1$?

Solution 19. (i) Since

$$U_1 U_2 = \begin{pmatrix} -1 & 0 \\ 0 & 1 \end{pmatrix}, \qquad U_2 U_1 = \begin{pmatrix} 1 & 0 \\ 0 & -1 \end{pmatrix}$$

we have $U_1 U_2 = e^{i\pi} U_2 U_1$.
(ii) Since

$$U_1 U_2 = \frac{1}{\sqrt{2}} \begin{pmatrix} 1 & 1 \\ -1 & 1 \end{pmatrix}, \qquad U_2 U_1 = \frac{1}{\sqrt{2}} \begin{pmatrix} 1 & -1 \\ 1 & 1 \end{pmatrix}$$

we cannot find $z \in \mathbb{C}$ such that $U_1 U_2 = z U_2 U_1$, but we find the unitary matrix

$$Z = \begin{pmatrix} 0 & 1 \\ -1 & 0 \end{pmatrix} = i\sigma_y$$

such that $U_1 U_2 = Z U_2 U_1$.

Problem 20. Let $\alpha \in \mathbb{R}$. Find the unitary matrix

$$U(\alpha) = \exp(-i\alpha(\sigma_x \otimes \sigma_x + \sigma_y \otimes \sigma_y + \sigma_z \otimes \sigma_z)).$$

Solution 20. Since

$$[\sigma_x \otimes \sigma_x, \sigma_y \otimes \sigma_y] = 0, \quad [\sigma_x \otimes \sigma_x, \sigma_z \otimes \sigma_z] = 0, \quad [\sigma_y \otimes \sigma_y, \sigma_z \otimes \sigma_z] = 0$$

we can write

$$U(\alpha) = \exp(-i\alpha(\sigma_x \otimes \sigma_x)) \exp(-i\alpha(\sigma_y \otimes \sigma_y)) \exp(-i\alpha(\sigma_z \otimes \sigma_z)).$$

It follows that

$$U(\alpha) = (\cos^3 \alpha - i \sin^3 \alpha) I_4 - i \sin \alpha \cos \alpha e^{i\alpha} (\sigma_x \otimes \sigma_x + \sigma_y \otimes \sigma_y + \sigma_z \otimes \sigma_z).$$

Problem 21. The *Hadamard gate* is defined by

$$U_H |0\rangle = \frac{1}{\sqrt{2}} (|0\rangle + |1\rangle), \qquad U_H |1\rangle = \frac{1}{\sqrt{2}} (|0\rangle - |1\rangle)$$

and the *CNOT gate* is defined as

$$U_{CNOT} |a\rangle \otimes |b\rangle = |a\rangle \otimes |a \oplus b\rangle$$

where $a, b \in \{0, 1\}$ and $\oplus$ is the XOR operation. Calculate the state

$$U_{CNOT}(U_H \otimes I)|0\rangle \otimes |0\rangle, \quad U_{CNOT}(U_H \otimes I)|0\rangle \otimes |1\rangle$$

$$U_{CNOT}(U_H \otimes I)|1\rangle \otimes |0\rangle, \quad U_{CNOT}(U_H \otimes I)|1\rangle \otimes |1\rangle$$

and discuss.

Solution 21. We obtain

$$U_{CNOT}(U_H \otimes I)|0\rangle \otimes |0\rangle = \frac{1}{\sqrt{2}}(|0\rangle \otimes |0\rangle + |1\rangle \otimes |1\rangle)$$

$$U_{CNOT}(U_H \otimes I)|0\rangle \otimes |1\rangle = \frac{1}{\sqrt{2}}(|0\rangle \otimes |1\rangle + |1\rangle \otimes |0\rangle)$$

$$U_{CNOT}(U_H \otimes I)|1\rangle \otimes |0\rangle = \frac{1}{\sqrt{2}}(|0\rangle \otimes |0\rangle - |1\rangle \otimes |1\rangle)$$

$$U_{CNOT}(U_H \otimes I)|1\rangle \otimes |1\rangle = \frac{1}{\sqrt{2}}(|0\rangle \otimes |1\rangle - |1\rangle \otimes |0\rangle).$$

These are the *Bell states*. Thus we generated the Bell states (which are maximally entangled) from non-entangled states.

Problem 22. Let U be a 2×2 unitary matrix and I_2 be the 2×2 identity matrix. Is the 4×4 matrix

$$V(\alpha) = \begin{pmatrix} 0 & 0 \\ 0 & 1 \end{pmatrix} \otimes U + \begin{pmatrix} e^{i\alpha} & 0 \\ 0 & 0 \end{pmatrix} \otimes I_2, \quad \alpha \in \mathbb{R}$$

unitary?

Solution 22. Yes, $V(\alpha)$ is unitary. Since

$$\begin{pmatrix} 0 & 0 \\ 0 & 1 \end{pmatrix} \begin{pmatrix} e^{i\alpha} & 0 \\ 0 & 0 \end{pmatrix} = \begin{pmatrix} e^{i\alpha} & 0 \\ 0 & 0 \end{pmatrix} \begin{pmatrix} 0 & 0 \\ 0 & 1 \end{pmatrix} = \begin{pmatrix} 0 & 0 \\ 0 & 0 \end{pmatrix}$$

and $UU^* = I_2$ we obtain

$$V(\alpha)V^*(\alpha) = \begin{pmatrix} 0 & 0 \\ 0 & 1 \end{pmatrix} \otimes I_2 + \begin{pmatrix} 1 & 0 \\ 0 & 0 \end{pmatrix} \otimes I_2 = I_2 \otimes I_2.$$

Problem 23. Consider the unitary 2×2 matrix

$$U = \frac{1}{\sqrt{2}} \begin{pmatrix} 1 & -1 \\ 1 & 1 \end{pmatrix}.$$

Calculate the *logarithm* of U, i.e., $\log U$ using

$$\log U = \int_0^1 (U - I_2)(t(U - I_2) + I_2)^{-1} dt$$

to find B given by $U = \exp(B)$. This equation can be applied if the matrix U has no eigenvalues on $\mathbb{R}^-$ (the closed negative real axis). Set $B = iK$. Find K.

Solution 23. The eigenvalues of U are given by

$$\lambda_1 = \frac{1}{\sqrt{2}}(1 + i), \qquad \lambda_2 = \frac{1}{\sqrt{2}}(1 - i).$$

Thus the condition to apply the equation is satisfied. We consider first the general case $U = (u_{jk})$ and then simplify to $u_{11} = u_{22} = 1/\sqrt{2}$ and $u_{21} = -u_{12} = 1/\sqrt{2}$. We obtain

$$t(U - I_2) + I_2 = \begin{pmatrix} 1 + t(u_{11} - 1) & tu_{12} \\ tu_{21} & 1 + t(u_{22} - 1) \end{pmatrix}$$

and

$$\det(t(U - I_2) + I_2) = d(t) = 1 + t(-2 + \text{tr}U) + t^2(1 - \text{tr}U + \det U).$$

Let $X \equiv \det U - \text{tr}U + 1$. Then

$$(U - I_2)(t(U - I_2) + I_2)^{-1} = \frac{1}{d(t)} \begin{pmatrix} tX + u_{11} - 1 & u_{12} \\ u_{21} & tX + u_{22} - 1 \end{pmatrix}.$$

With $u_{11} = u_{22} = 1/\sqrt{2}$, $u_{21} = -u_{12} = 1/\sqrt{2}$ we obtain

$$d(t) = 1 + t(-2 + \sqrt{2}) + t^2(2 - \sqrt{2})$$

and $X \equiv \det U - \text{tr}U + 1 = 2 - \sqrt{2}$. Thus the matrix takes the form

$$\frac{1}{d(t)} \begin{pmatrix} t(2 - \sqrt{2}) + 1/\sqrt{2} - 1 & -1/\sqrt{2} \\ 1/\sqrt{2} & t(2 - \sqrt{2}) + 1/\sqrt{2} - 1 \end{pmatrix}.$$

Since

$$\int_0^1 \frac{1}{d(t)} dt = \frac{2}{\sqrt{2}} \left| \arctan\left(\frac{2(2 - \sqrt{2})t + \sqrt{2} - 2}{\sqrt{2}}\right) \right|_0^1 = \sqrt{2}\frac{\pi}{4}$$

and

$$\int_0^1 \frac{t}{d(t)} dt = \frac{1}{\sqrt{2}}\frac{\pi}{4}$$

we obtain

$$B = \begin{pmatrix} 0 & -\pi/4 \\ \pi/4 & 0 \end{pmatrix}, \qquad K = -iB = \begin{pmatrix} 0 & i\pi/4 \\ -i\pi/4 & 0 \end{pmatrix}.$$

Problem 24. Let $\{ |0\rangle, |1\rangle, \ldots, |n-1\rangle \}$ be an orthonormal basis in the Hilbert space $\mathbb{C}^n$.
(i) Is the linear operator

$$Z_n := \sum_{j=0}^{n-1} \exp(2\pi i j/n) |j\rangle\langle j|$$

unitary?
(ii) Can the operator Z_n be expressed as the exponent of a hermitian operator?

Solution 24. (i) Since

$$Z_n^* = \sum_{j=0}^{n-1} \exp(-2\pi i j/n) |j\rangle\langle j|$$

we find $Z_n Z_n^* = I$. Therefore Z_n is unitary.
(ii) Since Z_n is unitary we can find a hermitian operator, say $\hat{K}$, such that $Z_n = \exp(i\hat{K})$. We obviously find

$$Z_n = \exp(2\pi i \hat{\theta}/n), \qquad \hat{\theta} := \sum_{j=0}^{n-1} j |j\rangle\langle j|.$$

The operator $\hat{\theta}$ is the $SU(2)$ *phase operator*.

Problem 25. The four *Bell states* are given by

$$|\phi^+\rangle = \frac{1}{\sqrt{2}}(|0\rangle \otimes |0\rangle + |1\rangle \otimes |1\rangle), \qquad |\phi^-\rangle = \frac{1}{\sqrt{2}}(|0\rangle \otimes |0\rangle - |1\rangle \otimes |1\rangle)$$

$$|\psi^+\rangle = \frac{1}{\sqrt{2}}(|0\rangle \otimes |1\rangle + |1\rangle \otimes |0\rangle), \qquad |\psi^-\rangle = \frac{1}{\sqrt{2}}(|0\rangle \otimes |1\rangle - |1\rangle \otimes |0\rangle).$$

Show that the Bell states can be transformed to each other under local unitary transformations (i.e. the Kronecker product $U \otimes V$ of two unitary 2×2 matrices U and V). Hint. Consider the Pauli spin matrices and the 2×2 identity matrix.

Solution 25. We find

$$|\phi^-\rangle = (I_2 \otimes \sigma_3)|\phi^+\rangle = (\sigma_3 \otimes I_2)|\phi^+\rangle$$
$$|\psi^+\rangle = (I_2 \otimes \sigma_1)|\phi^+\rangle = (\sigma_1 \otimes I_2)|\phi^+\rangle$$
$$|\psi^-\rangle = (I_2 \otimes (-i\sigma_2))|\phi^+\rangle = (i\sigma_2 \otimes I_2)|\phi^+\rangle .$$

Problem 26. Consider the unitary matrix

$$V = \begin{pmatrix} 0 & 0 & 1 \\ 0 & 1 & 0 \\ 1 & 0 & 0 \end{pmatrix} .$$

Find the hermitian matrix K such that $V = \exp(iK)$.

Solution 26. The matrix V is also hermitian. We calculate the eigenvalues and normalized eigenvectors of V. From the normalized eigenvectors we construct a unitary matrix W such that W^*VW is a diagonal matrix. Then

$$W^*VW = W^*e^{iK}W = e^{iW^*KW} = e^{iL}$$

where $L = W^*KW$. Since W^*VW is a diagonal matrix L is also a diagonal matrix. Thus $(W^*VW)_{jj} = e^{i\ell_{jj}}$. Finally we find K from $K = WLW^*$. Now the eigenvalues of V are given by $\lambda_1 = 1$, $\lambda_2 = 1$, $\lambda_3 = -1$ with the corresponding normalized eigenvectors

$$\begin{pmatrix} 0 \\ 1 \\ 0 \end{pmatrix}, \qquad \frac{1}{\sqrt{2}}\begin{pmatrix} 1 \\ 0 \\ 1 \end{pmatrix}, \qquad \frac{1}{\sqrt{2}}\begin{pmatrix} 1 \\ 0 \\ -1 \end{pmatrix} .$$

This leads to the unitary (orthogonal) matrix

$$W = \begin{pmatrix} 0 & 1/\sqrt{2} & 1/\sqrt{2} \\ 1 & 0 & 0 \\ 0 & 1/\sqrt{2} & -1/\sqrt{2} \end{pmatrix} .$$

Thus we obtain the diagonal matrix

$$W^*VW = \begin{pmatrix} 1 & 0 & 0 \\ 0 & 1 & 0 \\ 0 & 0 & -1 \end{pmatrix} = e^{iL} .$$

From $W^*VW = \exp(iL)$ we find the three equations

$$1 = e^{i\ell_{11}}, \qquad 1 = e^{i\ell_{22}}, \qquad -1 = e^{i\ell_{33}} .$$

The solution is $\ell_{11} = \ell_{22} = 0$, $\ell_{33} = \pi$. Hence we obtain the hermitian matrix

$$K = WLW^* = \frac{\pi}{2} \begin{pmatrix} 1 & 0 & -1 \\ 0 & 0 & 0 \\ -1 & 0 & 1 \end{pmatrix} .$$

Problem 27. Let U be an $n \times n$ unitary matrix. Show that if the bipartite states $|\psi\rangle, |\phi\rangle \in \mathbb{C}^n \otimes \mathbb{C}^m$ satisfy

$$|\phi\rangle = (U \otimes I_m)|\psi\rangle$$

then the ranks of the corresponding reduced density matrices satisfy

$$r(\rho_1^\psi) \geq r(\rho_1^\phi), \qquad r(\rho_2^\psi) \geq r(\rho_2^\phi) .$$

Solution 27. We consider the *Schmidt decomposition* of the state $|\psi\rangle$

$$|\psi\rangle = \sum_{j=1}^s \sqrt{\lambda_j^\psi} |j\rangle \otimes |j\rangle, \qquad \lambda_j^\psi > 0, \quad s \leq \min(n, m)$$

where s is the number of non vanishing terms in the Schmidt decomposition. We write the unitary operator as

$$U = \sum_{j=1}^n |\mu_j\rangle\langle j|$$

where $|\mu_j\rangle \in \mathbb{C}^n$. Then we find that

$$\rho_1^\psi = \sum_{j=1}^s |j\rangle\langle j|, \qquad \rho_1^\phi = U\rho_1^\psi U^* = \sum_{j=1}^s |\mu_j\rangle\langle \mu_j|.$$

Thus $r(\rho_1^\phi) \leq s$. The second inequality follows from the fact that for any bipartite state $r(\rho_1) = r(\rho_2)$.

Problem 28. Consider the hermitian matrix

$$R := \sigma_x \otimes \sigma_x + \sigma_y \otimes \sigma_y + \sigma_z \otimes \sigma_z$$

and the Hamilton operator

$$\hat{H}(t) = \frac{1}{2}\hbar\omega(t)R .$$

Let

$$\phi = \frac{1}{2} \int_0^T \omega(t) dt.$$

and

$$U(T) = \exp\left(-i \int_0^T \hat{H}(t) dt/\hbar\right) = \exp\left(-\frac{1}{2} \left(\int_0^T \omega(t) dt\right) R\right) = \exp(-i\phi R).$$

Calculate $U(T)$ and express it using the *swap gate*

$$U_{sw} = \begin{pmatrix} 1 & 0 & 0 & 0 \\ 0 & 0 & 1 & 0 \\ 0 & 1 & 0 & 0 \\ 0 & 0 & 0 & 1 \end{pmatrix}.$$

Solution 28. Note that $U_{sw} = I_4 + R$. We have

$$U(T) = e^{-i\phi R}$$
$$= \frac{1}{4} e^{-i\phi}(3I_4 + R) + \frac{1}{4} e^{-3i\phi}(I_4 - R)$$
$$= e^{i\phi}\left(\cos(2\phi)I_4 - \frac{i}{2}\sin(2\phi)(I_4 + R)\right)$$
$$= e^{i\phi}(\cos(2\phi)I_4 - i\sin(2\phi)U_{sw}).$$

Problem 29. Consider the *Bell basis* in the form

$$\frac{1}{\sqrt{2}} \begin{pmatrix} 1 \\ 0 \\ 0 \\ -i \end{pmatrix}, \quad \frac{1}{\sqrt{2}} \begin{pmatrix} 0 \\ 1 \\ -i \\ 0 \end{pmatrix}, \quad \frac{1}{\sqrt{2}} \begin{pmatrix} 0 \\ -i \\ 1 \\ 0 \end{pmatrix}, \quad \frac{1}{\sqrt{2}} \begin{pmatrix} -i \\ 0 \\ 0 \\ 1 \end{pmatrix}.$$

From the Bell states we form the matrix

$$B = \frac{1}{\sqrt{2}} \begin{pmatrix} 1 & 0 & 0 & -i \\ 0 & 1 & -i & 0 \\ 0 & -i & 1 & 0 \\ -i & 0 & 0 & 1 \end{pmatrix}.$$

(i) Show that B is unitary. Calculate $B^2, B^3, B^4, \ldots, B^8$ and show that B^k ($k = 1, 2 \ldots, 8$) form a group under matrix multiplication.

(ii) Let γ_1 and γ_4 the Dirac gamma matrices

$$\gamma_1 = \begin{pmatrix} 0 & 0 & 0 & -i \\ 0 & 0 & -i & 0 \\ 0 & i & 0 & 0 \\ i & 0 & 0 & 0 \end{pmatrix}, \quad \gamma_4 = \begin{pmatrix} 1 & 0 & 0 & 0 \\ 0 & 1 & 0 & 0 \\ 0 & 0 & -1 & 0 \\ 0 & 0 & 0 & -1 \end{pmatrix}.$$

Calculate $\exp(\pi\gamma_4\gamma_1/4)$ and show that $\exp(\pi\gamma_4\gamma_1/4) = B$.
(iii) Let $T_2 = B$ and

$$T_1 = \exp(-i\pi I_2 \otimes \sigma_z/4), \qquad T_3 = \exp(-i\pi\sigma_z \otimes I_2/4).$$

Calculate T_1 and T_3. Show that

$$T_1 T_2 T_1 = T_2 T_1 T_2, \qquad T_3 T_2 T_3 = T_2 T_3 T_2$$

and $T_1 T_3 = T_3 T_1$, i.e we have a *braid like relation*.

Solution 29. (i) We have $B^* B = I_4$. Thus $B^* = B^{-1}$ and B is unitary.
We define

$$N := \begin{pmatrix} 0 & 0 & 0 & 1 \\ 0 & 0 & 1 & 0 \\ 0 & 1 & 0 & 0 \\ 1 & 0 & 0 & 0 \end{pmatrix}.$$

Thus $N^2 = I_4$ and we can write $B = \frac{1}{\sqrt{2}}(I_4 - iN)$. We find

$$B^2 = \frac{1}{2}(I_4 - iN)(I_4 - iN) = -iN$$

$$B^3 = \frac{1}{\sqrt{2}}(I_4 - iN)(-iN) = -\frac{1}{\sqrt{2}}(I_4 + iN)$$

$$B^4 = B^2 B^2 = (-iN)(-iN) = -I_4$$

$$B^5 = -B$$

$$B^6 = -B^2 = iN$$

$$B^7 = -B^3 = \frac{1}{\sqrt{2}}(I_4 + iN)$$

$$B^8 = B^4 B^4 = I_4$$

(ii) We have

$$\gamma_4\gamma_1 = \begin{pmatrix} 0 & 0 & 0 & -i \\ 0 & 0 & -i & 0 \\ 0 & -i & 0 & 0 \\ -i & 0 & 0 & 0 \end{pmatrix} = -iN = B^2.$$

Thus $\gamma_4\gamma_1 = -iN$ and $(\gamma_4\gamma_1)^2 = -I_4$, $(\gamma_4\gamma_1)^3 = iN$, $(\gamma_4\gamma_1)^4 = I_4$. Using
this result we find

$$\exp(\pi\gamma_4\gamma_1/4) = I_4 \cos(\pi/4) - iN \sin(\pi/4) = \frac{1}{\sqrt{2}}(I_4 - iN) = B.$$

(iii) Using that $\exp(-i\pi/2) = -i$ we find

$$T_3 T_2 T_3 = \frac{1}{\sqrt{2}} \begin{pmatrix} -i & 0 & 0 & -i \\ 0 & -i & -i & 0 \\ 0 & -i & i & 0 \\ -i & 0 & 0 & i \end{pmatrix}.$$

Using that $e^{-i\pi/4} - e^{i\pi/4} = -i\sqrt{2}$ and $e^{-i\pi/4} + e^{i\pi/4} = \sqrt{2}$ we find

$$
T_2 T_3 T_2 = \frac{1}{\sqrt{2}} \begin{pmatrix} -i & 0 & 0 & -i \\ 0 & -i & -i & 0 \\ 0 & -i & i & 0 \\ -i & 0 & 0 & i \end{pmatrix}.
$$

Thus $T_2 T_3 T_2 = T_3 T_2 T_3$. Analogously we find $T_1 T_2 T_1 = T_2 T_1 T_2$. Obviously $T_1 T_3 = T_3 T_1$ since T_1 and T_3 are diagonal matrices.

Problem 30. Consider the *swap gate*

$$
U_{swap} = \begin{pmatrix} 1 & 0 & 0 & 0 \\ 0 & 0 & 1 & 0 \\ 0 & 1 & 0 & 0 \\ 0 & 0 & 0 & 1 \end{pmatrix}.
$$

Can the swap gate be written as

$$
U_{swap} = R \begin{pmatrix} 1 & 0 & 0 & 0 \\ 0 & 1 & 0 & 0 \\ 0 & 0 & -1 & 0 \\ 0 & 0 & 0 & 1 \end{pmatrix} R^{-1}
$$

where R is an orthogonal matrix?

Solution 30. The matrix U_{swap} is not only unitary but also hermitian with eigenvalues $+1, +1, +1, -1$. To construct the matrix R we just find the normalized eigenvectors of U. This yields

$$
R = R^{-1} = R^* = \begin{pmatrix} 1 & 0 & 0 & 0 \\ 0 & 1/\sqrt{2} & 1/\sqrt{2} & 0 \\ 0 & 1/\sqrt{2} & -1/\sqrt{2} & 0 \\ 0 & 0 & 0 & 1 \end{pmatrix}.
$$

Problem 31. Let σ_x, σ_y, σ_z be the Pauli spin matrices and

$$
\sigma_{j,x} = I_2 \otimes \cdots \otimes I_2 \otimes \sigma_x \otimes I_2 \otimes \cdots \otimes I_2
$$

where σ_x is a the j position (counting from left to right) and $j = 1, 2, \ldots, N$. Analogously we have $\sigma_{j,y}$ and $\sigma_{j,z}$. Let

$$
\sigma_{j,\pm} := \sigma_{j,x} \pm i\sigma_{j,y}.
$$

Consider the unitary matrix

$$
U := \exp\left(i \sum_{j=1}^{n} \chi_j \sigma_{j,+} \sigma_{j,-} \right).
$$

Find
$$U\sigma_{j,x}U^*, \qquad U\sigma_{j,-}U^*, \qquad U\sigma_{j,z}U^*.$$

Solution 31. We obtain
$$U\sigma_{j,x}U^* = e^{i\chi_j}\sigma_{j,+}, \qquad U\sigma_{j,-}U^* = e^{-i\chi_j}\sigma_{j,-}, \qquad U\sigma_{j,z}U^* = \sigma_{j,z}.$$

Problem 32. Let σ_x, σ_z be the Pauli spin matrices. Let
$$\Pi = \frac{1}{2}(I_2 + \sigma_z) \otimes I_2.$$
Show that Π is a projection matrix. Show that the 4×4 matrix
$$U = \frac{1}{2}(I_2 + \sigma_z) \otimes I_2 + \frac{1}{2}(I_2 - \sigma_z) \otimes \sigma_x$$
is unitary.

Solution 32. We have $\Pi = \Pi^*$ and $\Pi^2 = \Pi$. Straightforward calculation yields
$$U = \begin{pmatrix} 1 & 0 & 0 & 0 \\ 0 & 1 & 0 & 0 \\ 0 & 0 & 0 & 1 \\ 0 & 0 & 1 & 0 \end{pmatrix}.$$
Thus $U = U^*$ and $U^2 = I_4$. Thus U is unitary.

Chapter 8

Entropy

For any density operator (where $\langle \psi_j | \psi_k \rangle = \delta_{jk}$)

$$\rho = \sum_{j=1}^{n} \lambda_j |\psi_j\rangle\langle\psi_j|, \qquad \lambda_j \geq 0$$

the *von Neumann entropy* is defined as

$$S(\rho) := -\mathrm{tr}(\rho \log \rho)$$

or equivalently

$$S(\rho) := -\sum_{j=1}^{n} \lambda_j \log \lambda_j$$

with $0 \log 0 = 0$. Thus the von Neumann entropy is equal to the Shannon entropy of the eigenvalues. We have $S(\rho) \geq 0$ with equality iff ρ is a pure state. We also have $S(\rho) \leq \log n$, where n is the dimension of ρ. We find equality iff $\rho = (1/n)I$, where I is the identity operator. The entropy is unchanged under unitary transformation $S(U\rho U^*) = S(\rho)$. For a joint system AB we have

$$S(\rho^{AB}) \leq S(\rho^A) + S(\rho^B).$$

For density operators ρ and σ we have the *quantum relative entropy*

$$S(\rho||\sigma) = \mathrm{tr}(\rho \log \rho) - \mathrm{tr}(\rho \log \sigma).$$

We have *Klein's inequality* $S(\rho||\sigma) \geq 0$ with equality iff $\rho = \sigma$.

Problem 1.   Consider the density matrix

$$\rho = \begin{pmatrix} 1/2 & 0 \\ 0 & 1/2 \end{pmatrix}.$$

Calculate the von Neumann entropy

$$S(\rho) = -\mathrm{tr}(\rho \log_2 \rho)$$

Solution 1.   Since $\log_2(1/2) = -1$ we obtain $S(\rho) = 1$. Note that ρ describes a mixed state.

Problem 2.   Let W be a positive semidefinite $n \times n$ matrix over $\mathbb{C}$ with $\mathrm{tr}W = 1$. Let λ_1, λ_2, ..., λ_n be the eigenvalues (which obviously are real and nonnegative) and $\mathbf{w}_1$, $\mathbf{w}_2$, ..., $\mathbf{w}_n$ be the corresponding normalized eigenvectors of W. We can assume that the normalized eigenvectors form an orthonormal basis in $\mathbb{C}^n$. If eigenvalues are degenerate and the corresponding normalized eigenvectors are not orthogonal we can apply the Gram-Schmidt algorithm. Calculate $\mathrm{tr}(W \ln W)$.

Solution 2.   Let U be a unitary matrix such that $U^{-1}WU$ is a diagonal matrix. Note that $U^{-1} = U^*$. Obviously the unitary matrix U is constructed from the normalized eigenvectors of W. Then we have

$$\mathrm{tr}(W \ln W) = \mathrm{tr}(U^{-1}(W \ln W)U) = \mathrm{tr}(U^{-1}WUU^{-1}(\ln W)U)$$
$$= \mathrm{tr}(U^{-1}WU \ln(U^{-1}WU)).$$

The diagonal elements of $U^{-1}WU$ are the nonnegative real eigenvalues of W. Consequently

$$\mathrm{tr}(W \ln W) = \sum_{j=1}^{n} \lambda_j \ln \lambda_j.$$

Note that $0 \cdot \ln 0 = 0$ if an eigenvalue of W is 0.

Problem 3.   Let W be a positive semidefinite $n \times n$ matrix over $\mathbb{C}$ with $\mathrm{tr}W = 1$. Let λ_1, λ_2, ..., λ_n be the eigenvalues (which obviously are real and nonnegative) and $\mathbf{w}_1$, $\mathbf{w}_2$, ..., $\mathbf{w}_n$ be the corresponding normalized eigenvectors (column vectors) of W. We can assume that the normalized eigenvectors form an orthonormal basis in $\mathbb{C}^n$. If eigenvalues are degenerate and the corresponding normalized eigenvectors are not orthogonal we can apply the Gram-Schmidt algorithm. Let $\widetilde{W}$ be a positive definite $n \times n$ matrix over $\mathbb{C}$ with $\mathrm{tr}\widetilde{W} = 1$. Let $\widetilde{\lambda}_1$, $\widetilde{\lambda}_2$, ..., $\widetilde{\lambda}_n$ be the eigenvalues (which obviously are real and positive) and $\widetilde{\mathbf{w}}_1$, $\widetilde{\mathbf{w}}_2$, ..., $\widetilde{\mathbf{w}}_n$ be the corresponding normalized eigenvectors of $\widetilde{W}$. We can assume that the normalized eigenvectors form an orthonormal basis in $\mathbb{C}^n$. If eigenvalues are degenerate and the corresponding eigenvectors are not orthogonal we can apply the Gram-Schmidt algorithm. Calculate

$$\text{tr}(W \ln \widetilde{W}).$$

What happens if we allow the matrix $\widetilde{W}$ to be positive semidefinite?

Solution 3.   Let $\widetilde{U}$ be a unitary matrix such that $\widetilde{U}^{-1}\widetilde{W}\widetilde{U}$ is a diagonal matrix. Note that $\widetilde{U}^{-1} = \widetilde{U}^*$. Obviously the unitary matrix $\widetilde{U}$ is constructed from the normalized eigenvectors of $\widetilde{W}$. Then we have

$$\text{tr}(W \ln \widetilde{W}) = \text{tr}(\widetilde{U}^{-1}(W \ln \widetilde{W})\widetilde{U}) = \text{tr}(\widetilde{U}^{-1}W\widetilde{U}\widetilde{U}^{-1}(\ln \widetilde{W})\widetilde{U})$$
$$= \text{tr}(\widetilde{U}^{-1}W\widetilde{U} \ln(\widetilde{U}^{-1}\widetilde{W}\widetilde{U})) = \text{tr}(\widetilde{U}^{-1}W\widetilde{U} \ln(\widetilde{W}_D))$$

where $\widetilde{W}_D = \widetilde{U}^{-1}\widetilde{W}\widetilde{U}$ is a diagonal matrix. Using the spectral representation of W

$$W = \sum_{j=1}^{n} \lambda_j \mathbf{w}_j \mathbf{w}_j^*$$

we obtain

$$\text{tr}(W \ln \widetilde{W}) = \sum_{j=1}^{n} \lambda_j \text{tr}(\widetilde{U}^{-1}(\mathbf{w}_j \mathbf{w}_j^*)\widetilde{U} \ln(\widetilde{W}_D)).$$

Since

$$\widetilde{U} = \sum_{\ell=1}^{n} \widetilde{\mathbf{w}}_\ell \mathbf{e}_\ell^*, \qquad \widetilde{U}^* = \sum_{k=1}^{n} \mathbf{e}_k \widetilde{\mathbf{w}}_k^*$$

where $\{\, \mathbf{e}_k \; : \; k = 1, 2, \ldots, n \,\}$ is the standard basis in $\mathbb{C}^n$ we obtain

$$\text{tr}(W \ln \widetilde{W}) = \sum_{j=1}^{n} \lambda_j \text{tr}(\widetilde{U}^{-1}(\mathbf{w}_j \mathbf{w}_j^*)\widetilde{U} \ln(\widetilde{W}_D))$$
$$= \sum_{j=1}^{n} \sum_{k=1}^{n} \sum_{\ell=1}^{n} \lambda_j \text{tr}(\mathbf{e}_k \widetilde{\mathbf{w}}_k^*(\mathbf{w}_j \mathbf{w}_j^*)\widetilde{\mathbf{w}}_\ell \mathbf{e}_\ell^* \ln(\widetilde{W}_D)).$$

To calculate the trace we use the standard basis $\{\, \mathbf{e}_j \; : \; j = 1, 2, \ldots, n \,\}$. Thus

$$\text{tr}(W \ln \widetilde{W}) = \sum_{j=1}^{n} \sum_{k=1}^{n} \sum_{\ell=1}^{n} \sum_{r=1}^{n} \lambda_j (\mathbf{e}_r^* \mathbf{e}_k \widetilde{\mathbf{w}}_k^*(\mathbf{w}_j \mathbf{w}_j^*)\widetilde{\mathbf{w}}_\ell \mathbf{e}_\ell^* \ln(\widetilde{W}_D)\mathbf{e}_r)$$
$$= \sum_{j=1}^{n} \sum_{k=1}^{n} \sum_{\ell=1}^{n} \sum_{r=1}^{n} \lambda_j (\delta_{rk} \widetilde{\mathbf{w}}_k^*(\mathbf{w}_j \mathbf{w}_j^*)\widetilde{\mathbf{w}}_\ell \mathbf{e}_\ell^* \ln(\widetilde{W}_D)\mathbf{e}_r)$$

$$= \sum_{j=1}^{n}\sum_{k=1}^{n}\sum_{\ell=1}^{n} \lambda_j(\widetilde{\mathbf{w}}_k^*(\mathbf{w}_j\mathbf{w}_j^*)\widetilde{\mathbf{w}}_\ell \mathbf{e}_\ell^* \ln(\widetilde{W}_D)\mathbf{e}_k)$$

$$= \sum_{j=1}^{n}\sum_{k=1}^{n}\sum_{\ell=1}^{n} \lambda_j(\widetilde{\mathbf{w}}_k^*(\mathbf{w}_j\mathbf{w}_j^*)\widetilde{\mathbf{w}}_\ell \delta_{\ell k} \ln(\widetilde{\lambda}_k))$$

$$= \sum_{j=1}^{n}\sum_{k=1}^{n} \lambda_j(\widetilde{\mathbf{w}}_k^*(\mathbf{w}_j\mathbf{w}_j^*)\widetilde{\mathbf{w}}_k \ln(\widetilde{\lambda}_k)) .$$

Consequently

$$\text{tr}(W \ln \widetilde{W}) = \sum_{j=1}^{n}\sum_{k=1}^{n} \lambda_j(\widetilde{\mathbf{w}}_k^*\mathbf{w}_j)(\mathbf{w}_j^*\widetilde{\mathbf{w}}_k) \ln(\widetilde{\lambda}_k) .$$

Note that $\widetilde{\mathbf{w}}_k^*\mathbf{w}_j = (\mathbf{w}_j^*\widetilde{\mathbf{w}}_k)^*$. Can the condition on $\widetilde{W}$ be extended to positive semidefinite, i.e., some of the eigenvalues of $\widetilde{W}$ could be zero? Assume that one eigenvalue of $\widetilde{W}$ is 0, say $\widetilde{\lambda}_p = 0$. Thus we should have

$$\sum_{j=1}^{n} \lambda_j\widetilde{\mathbf{w}}_p^*(\mathbf{w}_j\mathbf{w}_j^*)\widetilde{\mathbf{w}}_p = \widetilde{\mathbf{w}}_p^*W\widetilde{\mathbf{w}}_p = 0$$

in order to apply $0 \cdot \ln 0 = 0$. This is in general not true. Consider, for example the density matrices

$$W = \begin{pmatrix} 1/2 & 0 \\ 0 & 1/2 \end{pmatrix}, \qquad \widetilde{W} = \begin{pmatrix} 1 & 0 \\ 0 & 0 \end{pmatrix} .$$

Problem 4. The *relative entropy of entanglement* for bipartite states (say, A and B), where the quantum state is described by the density matrix W, is defined as

$$E_r(W) := \min_{\widetilde{W} \in D} S(W\|\widetilde{W})$$

where D is the set of all convex combinations of separable density matrices (i.e., $\widetilde{W} = W_A \otimes W_B$) and

$$S(W\|\widetilde{W}) := \text{tr}(W(\log_2 W - \log_2 \widetilde{W})) \equiv \text{tr}(W \log_2 W - W \log_2 \widetilde{W})$$

is the *quantum relative entropy*. Let $\widetilde{W}_{min}$ denote the separable state that minimizes the relative entropy. Thus to calculate $E_r(W)$ is to find the state $\widetilde{W}_{min}$. Consider the Bell state

$$|\Phi^+\rangle = \frac{1}{\sqrt{2}}(|0\rangle \otimes |0\rangle + |1\rangle \otimes |1\rangle)$$

and thus $W = |\Phi^+\rangle\langle\Phi^+|$. Let

$$\widetilde{W} = \frac{1}{2}(|0\rangle\langle 0| + |1\rangle\langle 1|) \otimes \frac{1}{2}(|0\rangle\langle 0| + |1\rangle\langle 1|).$$

Calculate $S(W\|\widetilde{W})$.

Solution 4. Using the standard basis in $\mathbb{C}^4$ we obtain the density matrices

$$W = |\Phi^+\rangle\langle\Phi^+| = \frac{1}{2}\begin{pmatrix} 1 & 0 & 0 & 1 \\ 0 & 0 & 0 & 0 \\ 0 & 0 & 0 & 0 \\ 1 & 0 & 0 & 1 \end{pmatrix}$$

and

$$\widetilde{W} = \frac{1}{4}I_4$$

where I_4 is the 4×4 identity matrix. To calculate $\text{tr}(W \log_2 W)$ and $\text{tr}(W \log_2 \widetilde{W})$ we need the eigenvalues and normalized eigenvectors of W and $\widetilde{W}$. For the density matrix W we find the eigenvalues $\lambda_1 = 1$, $\lambda_2 = 0$, $\lambda_3 = 0$, $\lambda_4 = 0$ with the corresponding normalized eigenvectors

$$\mathbf{w}_1 = \frac{1}{\sqrt{2}}\begin{pmatrix} 1 \\ 0 \\ 0 \\ 1 \end{pmatrix}, \quad \mathbf{w}_2 = \frac{1}{\sqrt{2}}\begin{pmatrix} 1 \\ 0 \\ 0 \\ -1 \end{pmatrix}, \quad \mathbf{w}_3 = \begin{pmatrix} 0 \\ 1 \\ 0 \\ 0 \end{pmatrix}, \quad \mathbf{w}_4 = \begin{pmatrix} 0 \\ 0 \\ 1 \\ 0 \end{pmatrix}.$$

Thus we find

$$\text{tr}(W \log_2 W) = \sum_{j=1}^{4} \lambda_j \log_2 \lambda_j = 0$$

where we used $0 \log_2 0 = 0$ and $1 \log_2 1 = 0$. For $\widetilde{W}$ we obtain the eigenvalues $\widetilde{\lambda}_1 = 1/4$, $\widetilde{\lambda}_2 = 1/4$, $\widetilde{\lambda}_3 = 1/4$, $\widetilde{\lambda}_4 = 1/4$ and the corresponding normalized eigenvectors (standard basis)

$$\widetilde{\mathbf{w}}_1 = \begin{pmatrix} 1 \\ 0 \\ 0 \\ 0 \end{pmatrix}, \quad \widetilde{\mathbf{w}}_2 = \begin{pmatrix} 0 \\ 1 \\ 0 \\ 0 \end{pmatrix}, \quad \widetilde{\mathbf{w}}_3 = \begin{pmatrix} 0 \\ 0 \\ 1 \\ 0 \end{pmatrix}, \quad \widetilde{\mathbf{w}}_4 = \begin{pmatrix} 0 \\ 0 \\ 0 \\ 1 \end{pmatrix}.$$

Since

$$\text{tr}(W \log_2 \widetilde{W}) = \sum_{j=1}^{4}\sum_{k=1}^{4} \lambda_j (\widetilde{\mathbf{w}}_k^* \mathbf{w}_j)(\mathbf{w}_j^* \widetilde{\mathbf{w}}_k) \log_2(\widetilde{\lambda}_k)$$

and

$$\mathbf{w}_1^* \widetilde{\mathbf{w}}_1 = \frac{1}{\sqrt{2}}, \quad \mathbf{w}_1^* \widetilde{\mathbf{w}}_2 = 0, \quad \mathbf{w}_1^* \widetilde{\mathbf{w}}_3 = 0, \quad \mathbf{w}_1^* \widetilde{\mathbf{w}}_4 = \frac{1}{\sqrt{2}}$$

$$\mathbf{w}_2^*\widetilde{\mathbf{w}}_1 = \frac{1}{\sqrt{2}}, \quad \mathbf{w}_2^*\widetilde{\mathbf{w}}_2 = 0, \quad \mathbf{w}_2^*\widetilde{\mathbf{w}}_3 = 0, \quad \mathbf{w}_2^*\widetilde{\mathbf{w}}_4 = -\frac{1}{\sqrt{2}}$$

$$\mathbf{w}_3^*\widetilde{\mathbf{w}}_1 = 0, \quad \mathbf{w}_3^*\widetilde{\mathbf{w}}_2 = 1, \quad \mathbf{w}_3^*\widetilde{\mathbf{w}}_3 = 0, \quad \mathbf{w}_3^*\widetilde{\mathbf{w}}_4 = 0$$

$$\mathbf{w}_4^*\widetilde{\mathbf{w}}_1 = 0, \quad \mathbf{w}_4^*\widetilde{\mathbf{w}}_2 = 0, \quad \mathbf{w}_4^*\widetilde{\mathbf{w}}_3 = 1, \quad \mathbf{w}_4^*\widetilde{\mathbf{w}}_4 = 0$$

we obtain
$$\text{tr}(W \log_2 \widetilde{W}) = -2 \,.$$

Consequently, $S(W||\widetilde{W}) = 2$.

Problem 5.   Let ρ^{AB} be a density matrix defined on a $(N \times N)$-dimensional Hilbert space $\mathcal{H} \otimes \mathcal{H}$. The *classical information capacity* is defined as

$$C(\rho) := \log_2 N + S(\rho^B) - S(\rho^{AB})$$

where ρ^B is the reduced quantum state obtained by $\rho^B = \text{tr}_A \rho^{AB}$ and $S(\rho)$ is the von Neumann entropy of a quantum state (density matrix) $S(\rho) = -\text{tr}(\rho \log_2 \rho)$. Consider the Bell state

$$|\psi\rangle = \frac{1}{\sqrt{2}}(|0\rangle \otimes |0\rangle + |1\rangle \otimes |1\rangle) \,.$$

Calculate the density matrix ρ^{AB}, the reduced density matrix ρ^B and then the classical information capacity.

Solution 5.   We find

$$\rho^{AB} = \frac{1}{2}\begin{pmatrix} 1 & 0 & 0 & 1 \\ 0 & 0 & 0 & 0 \\ 0 & 0 & 0 & 0 \\ 1 & 0 & 0 & 1 \end{pmatrix} \,.$$

Thus

$$\rho^B = \begin{pmatrix} 1/2 & 0 \\ 0 & 1/2 \end{pmatrix} \,.$$

We obtain with $N = 2$ and $S(\rho^{AB}) = 0$

$$C(\rho) = \log_2 N + S(\rho^B) - S(\rho^{AB}) = 2 \,.$$

Problem 6.   The *quantum relative entropy* between two density operators ρ and σ is defined by

$$S_b(\rho||\sigma) := \text{tr}(\rho \log_b \rho - \rho \log_b \sigma) = -S_b(\rho) - \text{tr}\rho \log_b \sigma.$$

Here S_b is the Von Neumann entropy where the log is taken with the base b. Show that $S_b(\rho||\sigma) \geq 0$. This inequality is known as *Klein's inequality*.

Solution 6. Consider the term $\rho \log_b \sigma$ where ρ and σ are density operators on a finite-dimensional Hilbert space of dimension n. Let λ_j and $|\phi_j\rangle$ be the corresponding eigenvalues and (orthonormal) eigenstates of ρ. Similarly, let μ_k and $|\psi_k\rangle$ be the corresponding eigenvalues and (orthonormal) eigenstates of σ. Thus we have

$$
\rho \log_2 \sigma = \left(\sum_{j=1}^{n} \lambda_j |\phi_j\rangle\langle\phi_j| \right) \left(\sum_{k=1}^{n} \log_b \mu_k |\psi_k\rangle\langle\psi_k| \right)
$$

$$
= \sum_{j,k=1}^{n} \lambda_j \log_b \mu_k \langle\phi_j|\psi_k\rangle |\phi_j\rangle\langle\psi_k|.
$$

Taking the trace using the basis $\{|\psi_k\rangle,\ k = 1, 2, \ldots, n\}$ yields

$$
\mathrm{tr}(\rho \log_b \sigma) = \sum_{j,k,l=1}^{n} \lambda_j (\log_b \mu_k)\langle\phi_j|\psi_k\rangle\langle\psi_l|\phi_j\rangle\langle\psi_k|\psi_l\rangle
$$

$$
= \sum_{j,k=1}^{n} \lambda_j (\log_b \mu_k)\langle\phi_j|\psi_k\rangle\langle\psi_k|\phi_j\rangle
$$

$$
= \sum_{j,k=1}^{n} \lambda_j (\log_b \mu_k)|\langle\phi_j|\psi_k\rangle|^2.
$$

Thus we obtain

$$
S_b(\rho||\sigma) = \sum_{j=1}^{n} \lambda_j \log_b \lambda_j - \sum_{j,k=1}^{n} \lambda_j (\log_b \mu_k)|\langle\phi_j|\psi_k\rangle|^2
$$

$$
= \sum_{j=1}^{n} \lambda_j \log_b \lambda_j - \sum_{j=1}^{n} \lambda_j \log_b \left(\prod_{k=1}^{n} \mu_k^{|\langle\phi_j|\psi_k\rangle|^2} \right)
$$

$$
= \sum_{j=1}^{n} \lambda_j \log_b \lambda_j - \sum_{j=1}^{n} \lambda_j \log_b \nu_j
$$

where

$$
\nu_j := \left(\prod_{k=1}^{n} \mu_k^{|\langle\phi_j|\psi_k\rangle|^2} \right).
$$

If $\nu_j = 0$ for some j where $\lambda_j \neq 0$ then $S_B(\rho||\sigma) \geq 0$ trivially. In the following we assume $\nu_j = 0$ if and only if $\lambda_j = 0$. Since

$$
\sum_{k=1}^{n} \mu_k = \sum_{k=1}^{n} |\langle\phi_j|\psi_k\rangle|^2 = \sum_{j=1}^{n} |\langle\phi_j|\psi_k\rangle|^2 = 1
$$

we find that

$$
0 \leq \nu_j \leq 1, \qquad \alpha := \sum_{j=1}^{n} \nu_j \leq 1.
$$

In the following we assume that the λ_j are ordered in non-decreasing order and $m \leq n$ is chosen such that $\lambda_j = 0$ iff $j > m$. We determine the maximum value of

$$\sum_{j=1}^{m} \lambda_j \log_b x_j, \qquad \sum_{j=1}^{m} x_j = \alpha$$

which can be formulated as a *Lagrange multiplier problem*. Thus we find the critical points of

$$f(x_1, \ldots, x_m) := \sum_{j=1}^{m} \lambda_j \log_b x_j - \theta \left(\sum_{j=1}^{m} x_j - \alpha \right)$$

where θ is the Lagrange multiplier. We obtain

$$\frac{\partial f}{\partial x_j} = \frac{\lambda_j}{x_j \ln b} - \theta.$$

Since $\partial f / \partial x_j = \partial f / \partial x_k = 0$ for the critical points we have, since $\lambda_j \neq 0$ and $\lambda_k \neq 0$,

$$x_k = \frac{\lambda_k}{\lambda_j} x_j.$$

Inserting $x_k \neq 0$ into the constraint yields

$$\sum_{j=1}^{n} x_j = x_k + \sum_{\substack{j=1 \\ j \neq k}}^{n} \frac{\lambda_j}{\lambda_k} x_k = \alpha.$$

Thus $x_k = \alpha \lambda_k$, since $\sum_{j=1}^{n} \lambda_j = 1$. Since $\sum_{j=1}^{m} \lambda_j \log_b x_j$ is unbounded from below we have a maximum. Consequently

$$S_b(\rho||\sigma) = \sum_{j=1}^{n} \lambda_j \log_b \lambda_j - \sum_{j=1}^{n} \lambda_j \log_b \nu_j \geq \sum_{j=1}^{n} \lambda_j \log_b \lambda_j - \sum_{j=1}^{n} \lambda_j \log_b(\alpha \lambda_j)$$

$$= -\log_b \alpha \geq 0$$

since $0 \leq \alpha \leq 1$.

Problem 7. Show that for the quantum relative entropy

$$S_b(\rho||\sigma) := \operatorname{tr}(\rho \log_b \rho - \rho \log_b \sigma) = -S_b(\rho) - \operatorname{tr}\rho \log_b \sigma$$

the equality

$$S_b(\rho||\rho_A \otimes \rho_B) = S_b(\rho_A) + S_b(\rho_B) - S_b(\rho)$$

holds.

Solution 7. From $\sigma = \rho_A \otimes \rho_B$ we find that the eigenvalues of σ are the products of eigenvalues $\lambda_{A,j}$ of ρ_A and $\lambda_{B,k}$ of ρ_B. The corresponding (orthonormal) eigenstates are $|\phi_{A,j}\rangle \otimes |\phi_{B,k}\rangle$ composed of the orthonormal eigenstates of ρ_A and ρ_B corresponding to $\lambda_{A,j}$ and $\lambda_{B,k}$. Consequently

$$
\begin{aligned}
S_b(\rho\|\rho_A \otimes \rho_B) &= -S_b(\rho) - \mathrm{tr}\rho \log_b(\rho_A \otimes \rho_B) \\
&= -S_b(\rho) - \mathrm{tr}_A \left(\mathrm{tr}_B \left(\rho \log_b(\rho_A \otimes \rho_B) \right) \right) \\
&= -S_b(\rho) - \mathrm{tr}_A \left(\mathrm{tr}_B \left(\rho \sum_{j=1}^{m} \log_b \lambda_{A,j} |\phi_{A,j}\rangle\langle\phi_{A,j}| \otimes I_n \right) \right) \\
&\quad -\mathrm{tr}_B \left(\mathrm{tr}_A \left(\rho \sum_{k=1}^{n} \log_b \lambda_{B,k} I_m \otimes |\phi_{B,k}\rangle\langle\phi_{B,k}| \right) \right) \\
&= -S_b(\rho) - \mathrm{tr}_A \left(\rho_A \sum_{j=1}^{m} \log_b \lambda_{A,j} |\phi_{A,j}\rangle\langle\phi_{A,j}| \right) \\
&\quad -\mathrm{tr}_B \left(\rho_B \sum_{k=1}^{n} \log_b \lambda_{B,k} |\phi_{B,k}\rangle\langle\phi_{B,k}| \right) \\
&= S_b(\rho_A) + S_b(\rho_B) - S_b(\rho).
\end{aligned}
$$

From Klein's inequality, $S(\rho\|\rho_A \otimes \rho_B) \geq 0$, we obtain the property of *subadditivity* for the von Neumann entropy

$$S_b(\rho) \leq S_b(\rho_A) + S_b(\rho_B).$$

Problem 8. An $n \times n$ density matrix ρ is a positive semidefinite matrix such that $\mathrm{tr}(\rho) = 1$. The nonnegative eigenvalues of ρ are the probabilities of the physical states described by the corresponding eigenvectors. The entropy of the statistical state described by the density matrix ρ is defined by

$$S(\rho) := -\mathrm{tr}(\rho \ln \rho)$$

with the convention $0 \ln 0 = 0$. For the $n \times n$ hermitian matrix H (Hamilton operator) the statistical average of the energy E is defined by

$$E := \mathrm{tr}(H\rho).$$

Let

$$\psi(\rho) := \mathrm{tr}(H\rho) - \mathrm{tr}(\rho \ln \rho).$$

(i) Show that

$$\ln \mathrm{tr}(e^H) = \max \{ \, \mathrm{tr}(H\rho) + S(\rho) \, \}.$$

(ii) Show that
$$-S(\rho) = \max\{\operatorname{tr}(H\rho) - \ln\operatorname{tr}(e^H)\}.$$

Solution 8.   (i) For every $n \times n$ hermitian matrix Q, and for each $\epsilon \in \mathbb{R}$ in a neighbourhood of 0, consider the differentiable function

$$f(\epsilon) = \psi(e^{-i\epsilon Q}\rho e^{i\epsilon Q}).$$

Since $e^{i\epsilon Q}$ is a unitary matrix and the trace is invariant under unitary transformation we have

$$f(\epsilon) = \operatorname{tr}(He^{-i\epsilon Q}\rho e^{i\epsilon Q}) - \operatorname{tr}(\rho\ln\rho).$$

By the extremum condition it follows that

$$\frac{df(0)}{d\epsilon} = i\operatorname{tr}(Q[H,\rho]) = 0$$

where $[\,,\,]$ denotes the commutator. Since Q is arbitrary, we conclude that $[H,\rho] = 0_n$ and therefore e^H and ρ also commute. Using $\operatorname{tr}\rho = 1$, we obtain

$$\operatorname{tr}(H\rho) + S(\rho) = \operatorname{tr}(\rho(\ln e^H - \ln\rho))$$
$$= \ln\operatorname{tr} e^H - \operatorname{tr}(\rho(\ln\operatorname{tr}(e^H) + \ln\rho - \ln e^H)).$$

Since e^H and ρ commute, it follows that

$$\operatorname{tr}(H\rho) + S(\rho) = \ln\operatorname{tr} e^H - \operatorname{tr}(e^H\rho e^{-H}\ln(\operatorname{tr}(e^H)\rho e^{-H})).$$

Setting $C := \operatorname{tr}(e^H)\rho e^{-H}$ we obtain

$$\ln\operatorname{tr} e^H - \operatorname{tr}(e^H\rho e^{-H}\ln(\operatorname{tr}(e^H)\rho e^{-H})) =$$

$$\ln\operatorname{tr} e^H - (\operatorname{tr} e^H)^{-1}\operatorname{tr}(e^H(C\ln C - C + I_n)).$$

Since $x\ln x - x + 1 \geq 0$ for $x \geq 0$, we conclude that the last expression is less than or equal to $\ln\operatorname{tr} e^H$ and equality occurs only if $C = I_n$. This means

$$\rho = \frac{e^H}{\operatorname{tr} e^H}.$$

If $\rho = e^H/\operatorname{tr} e^H$ we find that $\psi(e^H/\operatorname{tr}(e^H)) = \ln\operatorname{tr}(e^H)$.
(ii) Since

$$\operatorname{tr}((H + kI_n)\rho) - \ln\operatorname{tr}(e^{H+kI_n}) = \operatorname{tr}(H\rho) - \ln\operatorname{tr} e^H, \quad k \in \mathbb{R}$$

we may assume that $\operatorname{tr}(e^H) = 1$. Following an argument similar to the one in (i) we can show that the maximum of $\operatorname{tr}(H\rho) - \ln\operatorname{tr}(e^H)$ for hermitian matrices H occurs when $[H, \rho] = 0_n$. Thus $[\rho, e^H] = 0_n$. Since $\text{tr}(\rho) - \text{tr}(e^H) = 0_n$, we have

$$-\text{tr}(H\rho) + \text{tr}(\rho \ln \rho) = \text{tr}(e^H \rho e^{-H} \ln(\rho e^{-H})) - \text{tr}(e^H \rho e^{-H}) + \text{tr} e^H$$
$$= \text{tr}(e^H(Z \ln Z - Z + I_n))$$
$$\geq 0$$

where $Z := \rho e^{-H}$. Hence the maximum occurs when $H = \ln \rho$.

Problem 9. Consider the normalized states $|\psi_k\rangle$, $k = 0, 1, \ldots, N - 1$ in the Hilbert space $\mathbb{C}^N$. A positive operator valued measure is specified by a decomposition of the identity matrix I_N into M positive semidefinite matrices P_m, i.e.

$$I_N = \sum_{m=0}^{M-1} P_m.$$

The mutual information is defined by

$$I = \sum_{n=0}^{N-1} \sum_{m=0}^{M-1} p_{nm} \log_N \left(\frac{p_{nm}}{p_{n\cdot} p_{\cdot m}} \right)$$

where

$$p_{nm} := \langle \psi_n | P_m | \psi_n \rangle$$

are the joint probabilities and

$$p_{n\cdot} := \sum_{m=0}^{M-1} p_{nm}, \qquad p_{\cdot m} := \sum_{n=0}^{N-1} p_{nm}$$

are their marginals. Let $M = N = 2$ and

$$P_0 = \frac{1}{2} \begin{pmatrix} 1 & 1 \\ 1 & 1 \end{pmatrix}, \qquad P_1 = \frac{1}{2} \begin{pmatrix} 1 & -1 \\ -1 & 1 \end{pmatrix}$$

$$|\psi_0\rangle = \frac{1}{\sqrt{2}} \begin{pmatrix} 1 \\ -1 \end{pmatrix}, \qquad |\psi_1\rangle = \begin{pmatrix} 0 \\ 1 \end{pmatrix}.$$

Find p_{nm}, $p_{n\cdot}$, $p_{\cdot m}$ and then I.

Solution 9. Straightforward calculation yields

$$p_{00} = \langle \psi_0 | P_0 | \psi_0 \rangle = 0, \qquad p_{10} = \langle \psi_1 | P_0 | \psi_1 \rangle = \frac{1}{2}$$

$$p_{01} = \langle \psi_0 | P_1 | \psi_0 \rangle = 1, \qquad p_{11} = \langle \psi_1 | P_1 | \psi_1 \rangle = \frac{3}{2}.$$

Thus

$$p_{0\cdot} = 1, \qquad p_{1\cdot} = 1, \qquad p_{\cdot 0} = \frac{1}{2}, \qquad p_{\cdot 1} = \frac{3}{2}$$

and $I = 0$.

Problem 10. The *Kullback-Leibler distance* between two probability mass functions

$$\mathbf{w} = (w_1, \dots, w_n), \qquad \widetilde{\mathbf{w}} = (\widetilde{w}_1, \dots, \widetilde{w}_n)$$

is defined by

$$D(\mathbf{w} \,\|\, \widetilde{\mathbf{w}}) := \sum_{j=1}^{n} w_j \log \frac{w_j}{\widetilde{w}_j}.$$

Show that the Kullback-Leibler distance between two mixtures densities

$$\sum_{j=1}^{n} w_j f_j, \qquad \sum_{j=1}^{n} \widetilde{w}_j \widetilde{f}_j$$

has the upper bound

$$D\left(\sum_{j=1}^{n} w_j f_j \,\Big\|\, \sum_{j=1}^{n} \widetilde{w}_j \widetilde{f}_j\right) \le D(\mathbf{w} \,\|\, \widetilde{\mathbf{w}}) + \sum_{j=1}^{n} w_j D(f_j \,\|\, \widetilde{f}_j)$$

with equality if and only if

$$\frac{w_j f_j}{\sum_{j=1}^{n} w_j f_j} = \frac{\widetilde{w}_j \widetilde{f}_j}{\sum_{j=1}^{n} \widetilde{w}_j \widetilde{f}_j}$$

for all j.

Solution 10. Utilizing the log-sum inequality we have

$$D\left(\sum_{j=1}^{n} w_j f_j \,\Big\|\, \sum_{j=1}^{n} \widetilde{w}_j \widetilde{f}_j\right) = \int \left(\sum_{j=1}^{n} w_j f_j\right) \log \left(\frac{\sum_{j=1}^{n} w_j f_j}{\sum_{j=1}^{n} \widetilde{w}_j \widetilde{f}_j}\right)$$

$$\le \int \sum_{j=1}^{n} w_j f_j \log \frac{w_j f_j}{\widetilde{w}_j \widetilde{f}_j}$$

$$= \sum_{j=1}^{n} w_j \log \frac{w_j}{\widetilde{w}_j} + \sum_{j=1}^{n} w_j \int f_j \log \frac{f_j}{\widetilde{f}_j}$$

$$= D(\mathbf{w} \,\|\, \widetilde{\mathbf{w}}) + \sum_{j=1}^{n} w_j D(f_j \,\|\, \widetilde{f}_j).$$

Chapter 9

Measurement

In quantum measurement models we consider what kind of measurements can be made on quantum systems as well as how to determine the probability that a measurement yields a given result. The effect that measurement has on the state of a quantum system is also important.

The pure states of a quantum system, S, are described by normalized vectors ψ which are elements of a Hilbert space $\mathcal{H}$ that describes S. The pure states of a quantum mechanical system are rays in a Hilbert space $\mathcal{H}$ (i.e., unit vectors with an arbitrary phase). The concept of a state as a ray in a Hilbert space leads to the probability interpretation in quantum mechanics. Given a physical system in the state ψ, the probability that it is in the state χ is

$$|\langle \psi | \chi \rangle|^2 \,.$$

Obviously

$$0 \leq |\langle \psi | \chi \rangle|^2 \leq 1 \,.$$

While the phase of a vector ψ has no physical significance, the relative phase of two vectors does. Consider the Schrödinger equation with time independent $\hat{H}$

$$i\hbar \frac{d}{dt} |\psi(t)\rangle = \hat{H} |\psi(t)\rangle$$

and the initial state $|\psi(t = 0)\rangle$. Then $|\psi(t)\rangle = \exp(-i\hat{H}t/\hbar)|\psi(t = 0)\rangle$ and

$$|\langle \psi(0) | e^{-i\hat{H}t/\hbar} | \psi(0) \rangle|^2$$

is the probability to find the state $|\psi(t)\rangle$ in the initial state $|\psi(0)\rangle$.

Problem 1.   Consider the state

$$|\psi\rangle = \frac{1}{\sqrt{3}}|00\rangle + \sqrt{\frac{2}{3}}|11\rangle \equiv \frac{1}{\sqrt{3}}|0\rangle \otimes |0\rangle + \sqrt{\frac{2}{3}}|1\rangle \otimes |1\rangle$$

and the product state

$$|\phi\rangle = |11\rangle \equiv |1\rangle \otimes |1\rangle .$$

Find $p := |\langle \phi | \psi \rangle|^2$, i.e., the probability of finding $|\psi\rangle$ in the state $|\phi\rangle$.

Solution 1.   Since $\langle 11|00\rangle = 0$ and $\langle 11|11\rangle = 1$ we obtain $p = 2/3$.

Problem 2.   Let $\{\,|0\rangle, |1\rangle\,\}$ denote an orthonormal basis in $\mathbb{C}^2$. In other words

$$\langle 0|1\rangle = \langle 1|0\rangle = 0, \qquad \langle 0|0\rangle = \langle 1|1\rangle = 1.$$

(i) Show that for $\alpha, \beta \in \mathbb{R}$,

$$A := \alpha|0\rangle\langle 0| + \beta|1\rangle\langle 1|$$

is an observable. Describe the measurement outcomes and associated probabilities when measuring the first qubit of the two qubit system described by

$$|\psi\rangle = \frac{1}{\sqrt{2}}(|0\rangle \otimes |0\rangle + |1\rangle \otimes |1\rangle)$$

where the measurement is described by A.
(ii) Let $|\phi\rangle$ be the state of the system after the measurement in (i). Describe the measurement outcomes and associated probabilities when measuring the second qubit of A.

Solution 2.   (i) To show that A is an observable it suffices to show that $A^* = A$. Using

$$(|0\rangle\langle 0|)^* = \langle 0|^*|0\rangle^* = |0\rangle\langle 0|, \quad (|0\rangle\langle 1|)^* = \langle 1|^*|0\rangle^* = |1\rangle\langle 0|,$$

$$(|1\rangle\langle 0|)^* = \langle 0|^*|1\rangle^* = |0\rangle\langle 1|, \quad (|1\rangle\langle 1|)^* = \langle 1|^*|1\rangle^* = |1\rangle\langle 1|$$

we find that

$$A^* = (\alpha|0\rangle\langle 0| + \beta|1\rangle\langle 1|)^* = (\alpha|0\rangle\langle 0|)^* + (\beta|1\rangle\langle 1|)^* = \overline{\alpha}|0\rangle\langle 0| + \overline{\beta}|1\rangle\langle 1|$$
$$= \alpha|0\rangle\langle 0| + \beta|1\rangle\langle 1| = A$$

since $\overline{\alpha} = \alpha$ for $\alpha \in \mathbb{R}$. Since we work in $\mathbb{C}^2$, A has two eigenvalues. From

$$A|0\rangle = (\alpha|0\rangle\langle 0| + \beta|1\rangle\langle 1|)|0\rangle = \alpha|0\rangle\langle 0|0\rangle + \beta|1\rangle\langle 1|0\rangle = \alpha|0\rangle$$

and

$$A|1\rangle = (\alpha|0\rangle\langle 0| + \beta|1\rangle\langle 1|)|1\rangle = \alpha|0\rangle\langle 0|1\rangle + \beta|1\rangle\langle 1|1\rangle = \beta|1\rangle$$

the two eigenvalues (i.e. measurement outcomes) are α and β with corresponding orthonormal eigenstates $|0\rangle$ and $|1\rangle$. Thus for measuring the first qubit we consider the observable $A \otimes I_2$ with eigenvalues α (eigenstates $|0\rangle \otimes |0\rangle$ and $|0\rangle \otimes |1\rangle$) and β (eigenstates $|1\rangle \otimes |0\rangle$ and $|1\rangle \otimes |1\rangle$). This is not the only choice for the corresponding eigenstates, but is a convenient one. It will be useful to calculate some scalar products in advance:

$$(|0\rangle \otimes |0\rangle)^* |\psi\rangle = (\langle 0| \otimes \langle 0|)\frac{1}{\sqrt{2}}(|0\rangle \otimes |1\rangle - |1\rangle \otimes |0\rangle)$$

$$= \frac{1}{\sqrt{2}}(\langle 0|0\rangle \otimes \langle 0|1\rangle - \langle 0|1\rangle \otimes \langle 0|0\rangle) = 0,$$

$$(|0\rangle \otimes |1\rangle)^* |\psi\rangle = \frac{1}{\sqrt{2}}, \quad (|1\rangle \otimes |0\rangle)^* |\psi\rangle = -\frac{1}{\sqrt{2}}, \quad (|1\rangle \otimes |1\rangle)^* |\psi\rangle = 0.$$

We need to consider two possibilities, namely $\alpha = \beta$ and $\alpha \neq \beta$.

For $\alpha = \beta$ there is only one measurement outcome: α. The corresponding projection operator onto the eigenspace is determined from the eigenstates

$$\Pi_\alpha := (|0\rangle \otimes |0\rangle)(|0\rangle \otimes |0\rangle)^* + (|0\rangle \otimes |1\rangle)(|0\rangle \otimes |1\rangle)^*$$
$$+ (|1\rangle \otimes |0\rangle)(|1\rangle \otimes |0\rangle)^* + (|1\rangle \otimes |1\rangle)(|1\rangle \otimes |1\rangle)^*$$
$$= |0\rangle\langle 0| \otimes |0\rangle\langle 0| + |0\rangle\langle 0| \otimes |1\rangle\langle 1| + |1\rangle\langle 1| \otimes |0\rangle\langle 0| + |1\rangle\langle 1| \otimes |1\rangle\langle 1|$$
$$= I_2 \otimes I_2 = I_4$$

i.e. the identity operator. The probability of obtaining the measurement outcome α is

$$p_\alpha = \|\Pi_\alpha|\psi\rangle\|^2 = \||\psi\rangle\|^2 = \langle\psi|\psi\rangle = \frac{1}{\sqrt{2}}(\langle 0| \otimes \langle 0| - \langle 1| \otimes \langle 1|)|\psi\rangle$$

$$= \frac{1}{2} - \left(-\frac{1}{2}\right) = 1.$$

The state after measurement is the projected state $\Pi_\alpha|\psi\rangle = I_4|\psi\rangle = |\psi\rangle$ which when normalized yields $|\phi\rangle = |\psi\rangle$ since $\langle\psi|\psi\rangle = 1$.

$\alpha \neq \beta$: Measurement outcome α: The corresponding projection operator onto the eigenspace is determined from the eigenstates

$$\Pi_\alpha := (|0\rangle \otimes |0\rangle)(|0\rangle \otimes |0\rangle)^* + (|0\rangle \otimes |1\rangle)(|0\rangle \otimes |1\rangle)^*$$
$$= |0\rangle\langle 0| \otimes |0\rangle\langle 0| + |0\rangle\langle 0| \otimes |1\rangle\langle 1|$$
$$= |0\rangle\langle 0| \otimes I_2.$$

Note that

$$\Pi_\alpha |\psi\rangle = \frac{1}{\sqrt{2}}(|0\rangle\langle 0| \otimes I_2)(|0\rangle \otimes |1\rangle - |1\rangle \otimes |0\rangle)$$

$$= \frac{1}{\sqrt{2}}(|0\rangle\langle 0|0\rangle) \otimes |1\rangle - (|0\rangle\langle 0|1\rangle) \otimes |0\rangle = \frac{1}{\sqrt{2}}|0\rangle \otimes |1\rangle.$$

The probability of obtaining the measurement outcome α is

$$p_\alpha = \|\Pi_\alpha|\psi\rangle\|^2 = (\Pi_\alpha|\psi\rangle)^*\Pi_\alpha|\psi\rangle = \frac{1}{\sqrt{2}}(\langle 0| \otimes \langle 1|)\frac{1}{\sqrt{2}}(|0\rangle \otimes |1\rangle) = \frac{1}{2}.$$

The state after measurement is the projected state $\Pi_\alpha|\psi\rangle = \frac{1}{\sqrt{2}}|0\rangle \otimes |1\rangle$ which when normalized yields

$$|\phi\rangle = \frac{\Pi_\alpha|\psi\rangle}{\|\Pi_\alpha|\psi\rangle\|} = \frac{\Pi_\alpha|\psi\rangle}{\sqrt{p_\alpha}} = |0\rangle \otimes |1\rangle.$$

Measurement outcome β: The corresponding projection onto the eigenspace is determined from the eigenstates

$$\Pi_\beta := (|1\rangle \otimes |0\rangle)(|1\rangle \otimes |0\rangle)^* + (|1\rangle \otimes |1\rangle)(|1\rangle \otimes |1\rangle)^*$$

$$= |1\rangle\langle 1| \otimes I_2.$$

Note that $\Pi_\beta|\psi\rangle = -\frac{1}{\sqrt{2}}|1\rangle \otimes |0\rangle$. The probability of obtaining the measurement outcome β is

$$p_\beta = \|\Pi_\beta|\psi\rangle\|^2 = \frac{1}{2}.$$

The state after measurement is the projected state

$$\Pi_\beta|\psi\rangle = \frac{1}{\sqrt{2}}|1\rangle \otimes |0\rangle$$

which when normalized yields

$$|\phi\rangle = \frac{\Pi_\beta|\psi\rangle}{\|\Pi_\beta|\psi\rangle\|} = \frac{\Pi_\alpha|\psi\rangle}{\sqrt{p_\beta}} = -|1\rangle \otimes |0\rangle.$$

(ii) For measuring the second qubit we consider the observable $I_2 \otimes A$ with eigenvalues α (eigenstates $|0\rangle \otimes |0\rangle$ and $|1\rangle \otimes |0\rangle$) and β (eigenstates $|0\rangle \otimes |1\rangle$ and $|1\rangle \otimes |1\rangle$). This is not the only choice for the corresponding eigenstates, but is a convenient one. The measurement of the second qubit depends on the results of the first measurement. Thus we need to consider three cases $\alpha = \beta$, and the two outcomes α and β when $\alpha \neq \beta$.

$\alpha = \beta$: We have

$$|\phi\rangle = |\psi\rangle = \frac{1}{\sqrt{2}}(|0\rangle \otimes |1\rangle - |1\rangle \otimes |0\rangle).$$

There is only one measurement outcome: α. The corresponding projection onto the eigenspace is determined from the eigenstates

$$\Pi_\alpha := (|0\rangle \otimes |0\rangle)(|0\rangle \otimes |0\rangle)^* + (|1\rangle \otimes |0\rangle)(|1\rangle \otimes |0\rangle)^*$$
$$+ (|0\rangle \otimes |1\rangle)(|0\rangle \otimes |1\rangle)^* + (|1\rangle \otimes |1\rangle)(|1\rangle \otimes |1\rangle)^*$$
$$= |0\rangle\langle 0| \otimes |0\rangle\langle 0| + |1\rangle\langle 1| \otimes |0\rangle\langle 0| + |0\rangle\langle 0| \otimes |1\rangle\langle 1| + |1\rangle\langle 1| \otimes |1\rangle\langle 1|$$
$$= I_2 \otimes I_2 = I_4$$

i.e. the identity operator. As above, the probability of the measurement outcome α is $p_{\alpha,\alpha} = 1$.

$\alpha \neq \beta$: First measurement outcome was α: We have $|\phi\rangle = |0\rangle \otimes |1\rangle$.
Measurement outcome α: The corresponding projection onto the eigenspace is determined from the eigenstates

$$\Pi_{\alpha,\alpha} := (|0\rangle \otimes |0\rangle)(|0\rangle \otimes |0\rangle)^* + (|1\rangle \otimes |0\rangle)(|1\rangle \otimes |0\rangle)^* = I_2 \otimes |0\rangle\langle 0|.$$

Note that $\Pi_\alpha|\phi\rangle = 0$. The probability of obtaining the measurement outcome α is $p_{\alpha,\alpha} = 0$.

Measurement outcome β: The corresponding projection onto the eigenspace is determined from the eigenstates

$$\Pi_{\alpha,\beta} = I_2 \otimes |1\rangle\langle 1|.$$

Note that $\Pi_{\alpha,\beta}|\phi\rangle = |0\rangle \otimes |1\rangle = |\phi\rangle$. The probability of obtaining the measurement outcome β is $p_{\alpha,\beta} = 1$.

First measurement outcome was β: We have

$$|\phi\rangle = -|1\rangle \otimes |0\rangle.$$

Measurement outcome α: The corresponding projection onto the eigenspace is determined from the eigenstates:

$$\Pi_{\beta,\alpha} = I_2 \otimes |0\rangle\langle 0|.$$

Note that $\Pi_\alpha|\phi\rangle = -|1\rangle \otimes |0\rangle = |\phi\rangle$. The probability of obtaining the measurement outcome α is $p_{\beta,\alpha} = 1$.

Measurement outcome β: The corresponding projection operator onto the eigenspace is determined from the eigenstates

$$\Pi_{\beta,\beta} = I_2 \otimes |1\rangle\langle 1|.$$

Note that $\Pi_{\beta,\beta}|\phi\rangle = 0$. The probability of obtaining the measurement outcome β is $p_{\beta,\beta} = 0$.

Tabulating the probabilities for $\alpha \neq \beta$ we find

Outcomes	α, α	α, β	β, α	β, β
Probability	$p_\alpha p_{\alpha,\alpha} = 0$	$p_\alpha p_{\alpha,\beta} = \frac{1}{2}$	$p_\beta p_{\beta,\alpha} = \frac{1}{2}$	$p_\beta p_{\beta,\beta} = 0$

Consequently, for $\alpha \neq \beta$, the probability that the two measurement outcomes are the same is 0 (impossible) and the probability that the two measurement outcomes are different is 1 (certain).

Problem 3. Consider the states

$$|0\rangle := \begin{pmatrix} 1 \\ 0 \end{pmatrix}, \qquad |1\rangle := \begin{pmatrix} 0 \\ 1 \end{pmatrix}$$

in the Hilbert space $\mathbb{C}^2$ and the Bell state

$$|\psi\rangle = \frac{1}{\sqrt{2}}(|0\rangle \otimes |1\rangle - |1\rangle \otimes |0\rangle)$$

in the Hilbert space $\mathbb{C}^4$. Let $(\alpha, \beta \in \mathbb{R})$

$$|\alpha\rangle := \cos\alpha|0\rangle + \sin\alpha|1\rangle, \qquad |\beta\rangle := \cos\beta|0\rangle + \sin\beta|1\rangle$$

be states in $\mathbb{C}^2$. Find the probability

$$p(\alpha, \beta) := |(\langle\alpha| \otimes \langle\beta|)|\psi\rangle|^2.$$

Discuss p as a function of α and β.

Solution 3. Since $\langle 0|0\rangle = \langle 1|1\rangle = 1$, $\langle 0|1\rangle = \langle 1|0\rangle = 0$ it follows that

$$(\langle 0| \otimes \langle 1|)(|0\rangle \otimes |1\rangle) = 1, \quad (\langle 1| \otimes \langle 0|)(|1\rangle \otimes |0\rangle) = 1.$$

We find

$$p(\alpha, \beta) = \frac{1}{2}(\cos\alpha\sin\beta - \sin\alpha\cos\beta)^2.$$

Using a trigonometric identity we arrive at

$$p(\alpha, \beta) = \frac{1}{2}\sin^2(\alpha - \beta).$$

Thus $p(\alpha, \beta) \leq 1/2$ for all α, β since $\sin^2\phi \leq 1$ for all $\phi \in \mathbb{R}$. For example, if $\alpha = \beta$ we have $p = 0$. If $\alpha - \beta = \pi/2$ we have $p = 1/2$.

Problem 4. Assume that Alice operates a device that prepares a quantum system and Bob does subsequent measurement on the system and records the results. The preparation device indicates the state the system is prepared in. A preparation readout event j, where $j = 1, 2, \ldots, m$ of the preparation device is associated with a linear non-negative definite operator

Λ_j acting on the state space of the system. The operators Λ_j need not be orthogonal to each other. The measurement device has a readout event k, where $k = 1, 2, \ldots, n$ that shows the result of the measurement. A measurement device is associated with a measurement device operator Γ_k which is also linear and non-negative definite. For a *von Neumann measurement* this operator would be a pure state projector. Let

$$\Lambda := \sum_{j=1}^{m} \Lambda_j, \qquad \Gamma := \sum_{k=1}^{n} \Gamma_k.$$

Give an interpretation of the following probabilities

$$p(j,k) = \frac{\mathrm{tr}(\Lambda_j \Gamma_k)}{\mathrm{tr}(\Lambda \Gamma)} \tag{1}$$

$$p(j) = \frac{\mathrm{tr}(\Lambda_j \Gamma)}{\mathrm{tr}(\Lambda \Gamma)} \tag{2}$$

$$p(k) = \frac{\mathrm{tr}(\Lambda \Gamma_k)}{\mathrm{tr}(\Lambda \Gamma)} \tag{3}$$

$$p(k|j) = \frac{\mathrm{tr}(\Lambda_j \Gamma_k)}{\mathrm{tr}(\Lambda_j \Gamma)} \tag{4}$$

$$p(j|k) = \frac{\mathrm{tr}(\Lambda_j \Gamma_k)}{\mathrm{tr}(\Lambda \Gamma_k)}. \tag{5}$$

Solution 4. Expression (1) is the probability associated with a particular point (j, k) in the sample space. Expression (2) is the probability that, if an experiment chosen at random has a recorded combined event, this event includes preparation event j. Expression (3) is the probability that the recorded combined event includes the measurement event k. Expression (4) is the probability that, if the recorded combined event includes event j, it also includes event k. Thus it is the probability that the event recorded by Bob is the detection of the state corresponding to Γ_k if the state prepared by Alice in the experiment corresponds to Λ_j. This expression can be used for prediction. To calculate the required probability from the operator Λ_j associated with the preparation event j, every possible operator Γ_k must be known, that is, the mathematical description of the operation of the measuring device must be known. Analogously, (5) is the probability that the state prepared by Alice corresponds to Λ_j if the event recorded by Bob is the detection of the state corresponding to Γ_k. This expression can be used for retrodiction if Γ_k and all the Λ_j operators of the preparation device are known.

Problem 5.   Let A be an $n \times n$ hermitian matrix. Then the eigenvalues λ_j, $j = 1, 2, \ldots, n$ are real. Assume that all eigenvalues are distinct. The matrix A can be written as (*spectral representation*)

$$A := \sum_{j=1}^{n} \lambda_j P_j, \qquad P_j := |u_j\rangle\langle u_j| \tag{1}$$

where $|u_j\rangle$ are the normalized eigenvectors of A with eigenvalue λ_j. For the projectors P_j we have $P_j P_k = \delta_{jk} P_j$. Every observable A defines a *projective measurement*. A state $|\psi\rangle$ in $\mathbb{C}^n$ subject to projective measurement by observable (1) goes into the state

$$\frac{P_j|\psi\rangle}{\sqrt{\langle\psi|P_j|\psi\rangle}}$$

with *probability*

$$p(j) = \langle\psi|P_j|\psi\rangle \equiv \langle\psi|u_j\rangle\langle u_j|\psi\rangle = |\langle\psi|u_j\rangle|^2 \,.$$

The eigenvalues λ_j are registered as the measured value. If the system is subjected to the same measurement immediately after a projective measurement, the same outcome occurs with certainty. The *expectation* of the measured value is

$$\langle A \rangle = \sum_{j=1}^{n} \lambda_j p(j) = \langle\psi|A|\psi\rangle \,.$$

(i) Let

$$A = \begin{pmatrix} 0 & -i \\ i & 0 \end{pmatrix} \,.$$

Find the spectral representation of A.
(ii) Let

$$|\psi\rangle = \frac{1}{\sqrt{2}} \begin{pmatrix} 1 \\ 1 \end{pmatrix} \,.$$

Calculate the probabilities corresponding to the eigenvalues λ_1 and λ_2 of A

$$p(\lambda_1) = \langle\psi|P_{\lambda_1}|\psi\rangle, \qquad p(\lambda_2) = \langle\psi|P_{\lambda_2}|\psi\rangle \,.$$

Solution 5.   (i) The eigenvalues of A are $\lambda_1 = 1$ and $\lambda_2 = -1$. The corresponding eigenvectors are

$$|u_1\rangle = \frac{1}{\sqrt{2}} \begin{pmatrix} 1 \\ i \end{pmatrix}, \qquad |u_2\rangle = \frac{1}{\sqrt{2}} \begin{pmatrix} 1 \\ -i \end{pmatrix} \,.$$

Thus

$$P_{\lambda_1} = |u_1\rangle\langle u_1| = \frac{1}{2}\begin{pmatrix} 1 \\ i \end{pmatrix}(1 \ -i) = \frac{1}{2}\begin{pmatrix} 1 & -i \\ i & 1 \end{pmatrix}$$

$$P_{\lambda_2} = |u_2\rangle\langle u_2| = \frac{1}{2}\begin{pmatrix} 1 \\ -i \end{pmatrix}(1 \ i) = \frac{1}{2}\begin{pmatrix} 1 & i \\ -i & 1 \end{pmatrix}$$

with $I_2 = P_{\lambda_1} + P_{\lambda_2}$ and $A = P_{\lambda_1} - P_{\lambda_2}$.
(ii) We have

$$P_{\lambda_1}|\psi\rangle = \frac{1}{2}\begin{pmatrix} 1 & -i \\ i & 1 \end{pmatrix}\frac{1}{\sqrt{2}}\begin{pmatrix} 1 \\ 1 \end{pmatrix} = \frac{1}{2\sqrt{2}}\begin{pmatrix} 1-i \\ 1+i \end{pmatrix}$$

$$P_{\lambda_2}|\psi\rangle = \frac{1}{2}\begin{pmatrix} 1 & i \\ -i & 1 \end{pmatrix}\frac{1}{\sqrt{2}}\begin{pmatrix} 1 \\ 1 \end{pmatrix} = \frac{1}{2\sqrt{2}}\begin{pmatrix} 1+i \\ 1-i \end{pmatrix}.$$

Thus

$$p(\lambda_1) = \langle\psi|P_{\lambda_1}|\psi\rangle = \frac{1}{2}, \qquad p(\lambda_2) = \langle\psi|P_{\lambda_2}|\psi\rangle = \frac{1}{2}.$$

Problem 6. A *positive operator-valued measure (POVM)* is a collection

$$\{\, E_j \,:\, j = 1, 2, \ldots, n \,\}$$

of nonnegative (positive semi-definite) operators, satisfying

$$\sum_{j=1}^{n} E_j = I$$

where I is the identity operator. In other words a partition of unity (identity operator) by nonnegative operators is called a positive operator-valued measure (POVM). When a state $|\psi\rangle$ is subjected to such a POVM, outcome j occurs with probability

$$p(j) = \langle\psi|E_j|\psi\rangle.$$

Consider a qubit system. Let

$$E_1 = |0\rangle\langle 0|, \quad E_2 = |1\rangle\langle 1|, \quad |\psi\rangle = \frac{1}{\sqrt{2}}(|0\rangle + |1\rangle).$$

Find $p(1)$ and $p(2)$.

Solution 6. Since $\langle 0|0\rangle = \langle 1|1\rangle = 1$ and $\langle 0|1\rangle = \langle 1|0\rangle = 0$ we find

$$p(1) = \langle\psi|E_1|\psi\rangle = \frac{1}{2}, \qquad p(2) = \langle\psi|E_2|\psi\rangle = \frac{1}{2}.$$

Problem 7.   Consider the state

$$|\psi\rangle = \frac{1}{\sqrt{2}}(|01\rangle - |10\rangle) \equiv \frac{1}{\sqrt{2}}(|0\rangle \otimes |1\rangle - |1\rangle \otimes |0\rangle)$$

and $\langle 0| \otimes I_2$, where I_2 is the 2×2 unit matrix. Find $(\langle 0| \otimes I_2)|\psi\rangle$. Discuss.

Solution 7.   Since $\langle 0|0\rangle = 1$, $\langle 0|1\rangle = 0$ and $I_2|1\rangle = |1\rangle$, we obtain

$$(\langle 0| \otimes I_2)|\psi\rangle = \frac{1}{\sqrt{2}}|1\rangle.$$

The first system is measured with probability $1/2$ and the system collapses to the state $|1\rangle$ (*partial measurement*).

Problem 8.   Let $(\theta \in \mathbb{R})$

$$P(\theta) := e^{i\theta}|0\rangle\langle 0| + e^{-i\theta}|1\rangle\langle 1| \equiv e^{i\theta}(|0\rangle\langle 0| + e^{-i2\theta}|1\rangle\langle 1|)$$

denote the *phase change transform* on a single qubit.
(i) Calculate $(\phi \in \mathbb{R})$

$$|s(\theta, \phi)\rangle := P\left(\frac{\pi}{4} - \frac{\phi}{2}\right) U_H P\left(\frac{\theta}{2}\right) U_H |0\rangle.$$

(ii) Determine the probability that the state $|s(\theta, \phi)\rangle$ is in the state

$$(a)\ |0\rangle \qquad (b)\ |1\rangle \qquad (c)\ |s(\theta', \phi')\rangle.$$

The real parameters θ and ϕ can be interpreted as spherical co-ordinates which define any qubit on the unit sphere called the *Bloch sphere*.

Solution 8.   (i) We have

$$
\begin{aligned}
|s(\theta, \phi)\rangle &= P\left(\frac{\pi}{4} - \frac{\phi}{2}\right) U_H P\left(\frac{\theta}{2}\right) U_H |0\rangle \\
&= P\left(\frac{\pi}{4} - \frac{\phi}{2}\right) U_H P\left(\frac{\theta}{2}\right) \frac{1}{\sqrt{2}}(|0\rangle + |1\rangle) \\
&= P\left(\frac{\pi}{4} - \frac{\phi}{2}\right) U_H \frac{1}{\sqrt{2}}(e^{i\theta/2}|0\rangle + e^{-i\theta/2}|1\rangle) \\
&= P\left(\frac{\pi}{4} - \frac{\phi}{2}\right) \frac{1}{2}\left(e^{i\theta/2}(|0\rangle + |1\rangle) + e^{-i\theta/2}(|0\rangle - |1\rangle)\right) \\
&= P\left(\frac{\pi}{4} - \frac{\phi}{2}\right)\left(\cos\frac{\theta}{2}|0\rangle + i\sin\frac{\theta}{2}|1\rangle\right) \\
&= e^{i\frac{\pi}{4}}\left(e^{-i\frac{\phi}{2}}\cos\frac{\theta}{2}|0\rangle + e^{i\frac{\phi}{2}}\sin\frac{\theta}{2}|1\rangle\right) \\
&= e^{i(\frac{\pi}{4} - \frac{\phi}{2})}\left(\cos\frac{\theta}{2}|0\rangle + e^{i\phi}\sin\frac{\theta}{2}|1\rangle\right).
\end{aligned}
$$

The most general state of a single qubit is described by three real parameters $\theta, \phi, \sigma \in \mathbb{R}$

$$e^{i\sigma} \left(\cos\frac{\theta}{2}|0\rangle + e^{i\phi}\sin\frac{\theta}{2}|1\rangle \right).$$

The parameter σ represents the *global phase*, and can be ignored since it cannot be detected in the measurement model. The same applies to the global phase $\exp(i(\pi/4 - \phi/2))$ in the derivation. Thus θ and ϕ can be used to define any single qubit $|s(\theta, \phi)\rangle$.

(ii) For the probabilities (a) we have

$$|\langle 0|s(\theta, \phi)\rangle|^2 = \cos^2\frac{\theta}{2}.$$

For the probability (b) we have

$$|\langle 1|s(\theta, \phi)\rangle|^2 = \sin^2\frac{\theta}{2}.$$

For the probability $|\langle s(\theta', \phi')|s(\theta, \phi)\rangle|^2$ we find

$$\left| e^{-i\frac{\pi}{4}} \left(e^{i\frac{\phi'}{2}}\cos\frac{\theta'}{2}\langle 0| + e^{-i\frac{\phi'}{2}}\sin\frac{\theta'}{2}\langle 1| \right) e^{i\frac{\pi}{4}} \left(e^{-i\frac{\phi}{2}}\cos\frac{\theta}{2}|0\rangle + e^{i\frac{\phi}{2}}\sin\frac{\theta}{2}|1\rangle \right) \right|^2.$$

Thus

$$|\langle s(\theta', \phi')|s(\theta, \phi)\rangle|^2 = \left| \cos\frac{\theta}{2}\cos\frac{\theta'}{2}e^{\frac{i}{2}(\phi'-\phi)} + \sin\frac{\theta}{2}\sin\frac{\theta'}{2}e^{\frac{i}{2}(\phi-\phi')} \right|^2$$

where we used $\langle 0|0\rangle = \langle 1|1\rangle = 1$ and $\langle 0|1\rangle = \langle 1|0\rangle = 0$. It follows that

$$|\langle s(\theta', \phi')|s(\theta, \phi)\rangle|^2 = \cos^2\frac{1}{2}(\phi'-\phi)\cos^2\frac{1}{2}(\theta'-\theta) + \sin^2\frac{1}{2}(\phi'-\phi)\cos^2\frac{1}{2}(\theta'+\theta).$$

If $\theta' = \theta$ and $\phi' = \phi$ we find 1 for the probability.

Problem 9. Consider the finite-dimensional Hilbert space $\mathbb{C}^n$ with $n > 2$. Consider an orthonormal basis

$$\{ |0\rangle, |1\rangle, \dots, |n-1\rangle \}.$$

Let E be any projector in it, and $E_j := |j\rangle\langle j|$, where $j = 0, 1, \dots, n-1$. Let the probability of obtaining 1 when measuring E be $P(E)$. Then

$$P(I) = 1, \quad 0 \le P(E) \le 1, \quad P(0) = 0, \quad E_j E_k = \delta_{jk} E_j.$$

$$P(E_0 + E_1 + \cdots + E_{n-1}) = P(E_0) + P(E_1) + \cdots + P(E_{n-1}). \quad (1)$$

A state s is determined by the function $P(E)$ which satisfies (1). *Gleason's theorem* states that for any $P(E)$ which satisfies (1) there exists a density matrix ρ such that

$$P(E) = \text{tr}(\rho E).$$

In other words, s is described by the density matrix ρ. Show that Gleason's theorem does not hold in two-dimensional Hilbert spaces.

Solution 9. In the two-dimensional Hilbert space consider the eigenvalue equation

$$(\boldsymbol{\sigma} \cdot \mathbf{n})|\mathbf{m}\rangle = |\mathbf{m}\rangle$$

where $\boldsymbol{\sigma} \cdot \mathbf{n} := \sigma_1 n_1 + \sigma_2 n_2 + \sigma_3 n_3$, $\mathbf{n}$ is the unit vector ($\|\mathbf{n}\| = 1$) with the parameter representation

$$\mathbf{n} := (\sin\theta\cos\phi, \sin\theta\sin\phi, \cos\theta)$$

with $0 \leq \theta \leq \pi$, $0 \leq \phi < 2\pi$ and

$$|\mathbf{m}\rangle := \begin{pmatrix} \cos\frac{\theta}{2} \\ e^{i\phi}\sin\frac{\theta}{2} \end{pmatrix}.$$

The projector onto $|\mathbf{m}\rangle$ is given by

$$E_{\mathbf{m}} \equiv |\mathbf{m}\rangle\langle\mathbf{m}| = \begin{pmatrix} \cos^2\frac{\theta}{2} & \frac{1}{2}e^{-i\phi}\sin\theta \\ \frac{1}{2}e^{i\phi}\sin\theta & \sin^2\frac{\theta}{2} \end{pmatrix} = E(\theta,\phi)$$

since $\cos\frac{\theta}{2}\sin\frac{\theta}{2} \equiv \frac{1}{2}\sin\theta$. Equation (1) holds with

$$P(E_{\mathbf{m}} + E_{-\mathbf{m}}) = P(E_{\mathbf{m}}) + P(E_{-\mathbf{m}}) = P(I) = 1, \quad E_{\mathbf{m}}E_{-\mathbf{m}} = 0.$$

It is not difficult to find probability distribution functions $P_{\mathbf{m}} = P(\theta,\phi)$ such that no density matrix ρ exists. An example is

$$P(\theta,\phi) = \frac{1}{2} + \frac{\cos^3\theta}{2}.$$

Problem 10. Consider the two qubits in the Hilbert space $\mathbb{C}^2$

$$|\psi_1\rangle := \cos(\theta_1/2)|0\rangle + \sin(\theta_1/2)e^{i\phi_1}|1\rangle$$

$$|\psi_2\rangle := \cos(\theta_2/2)|0\rangle + \sin(\theta_2/2)e^{i\phi_2}|1\rangle.$$

(i) Find the product state $|\psi\rangle = |\psi_1\rangle \otimes |\psi_2\rangle$ in $\mathbb{C}^4$.
(ii) Consider the *qutrit state* in the Hilbert space $\mathbb{C}^3$

$$|\phi\rangle = \frac{1}{\sqrt{3}}(|0\rangle + |1\rangle + |2\rangle).$$

To encode the state $|\psi_1\rangle \otimes |\psi_2\rangle$ we use the state $|\phi\rangle$ and perform projective measurements on the state $|\phi\rangle \otimes (|\psi_1\rangle \otimes |\psi_2\rangle)$ given by the projection operators P_0, P_1, P_2, P_3 acting in the Hilbert space $\mathbb{C}^3 \otimes \mathbb{C}^4$

$$
\begin{aligned}
P_0 &:= |0\rangle\langle 0| \otimes (|1\rangle \otimes |0\rangle\langle 1| \otimes \langle 0|) \\
&\quad + |1\rangle\langle 1| \otimes (|0\rangle \otimes |1\rangle\langle 0| \otimes \langle 1|) + |2\rangle\langle 2| \otimes (|1\rangle \otimes |1\rangle\langle 1| \otimes \langle 1|) \\
P_1 &:= |0\rangle\langle 0| \otimes (|0\rangle \otimes |1\rangle\langle 0| \otimes \langle 1|) \\
&\quad + |1\rangle\langle 1| \otimes (|1\rangle \otimes |1\rangle\langle 1| \otimes \langle 1|) + |2\rangle\langle 2| \otimes (|0\rangle \otimes |0\rangle\langle 0| \otimes \langle 0|) \\
P_2 &:= |0\rangle\langle 0| \otimes (|1\rangle \otimes |1\rangle\langle 1| \otimes \langle 1|) \\
&\quad + |1\rangle\langle 1| \otimes (|0\rangle \otimes |0\rangle\langle 0| \otimes \langle 0|) + |2\rangle\langle 2| \otimes (|1\rangle \otimes |0\rangle\langle 1| \otimes \langle 0|) \\
P_3 &:= |0\rangle\langle 0| \otimes (|0\rangle \otimes |0\rangle\langle 0| \otimes \langle 0|) \\
&\quad + |1\rangle\langle 1| \otimes (|1\rangle \otimes |0\rangle\langle 1| \otimes \langle 0|) + |2\rangle\langle 2| \otimes (|0\rangle \otimes |1\rangle\langle 0| \otimes \langle 1|).
\end{aligned}
$$

Find the probability

$$
p_0 := (\langle \phi| \otimes \langle \psi|)P_0(|\phi\rangle \otimes |\psi\rangle).
$$

Solution 10. (i) We have

$$
\begin{aligned}
|\psi\rangle &= |\psi_1\rangle \otimes |\psi_2\rangle \\
&= \cos\frac{\theta_1}{2} \cos\frac{\theta_2}{2}|0\rangle \otimes |0\rangle + \sin\frac{\theta_1}{2}e^{i\phi_1}\cos\frac{\theta_2}{2}|1\rangle \otimes |0\rangle \\
&\quad + \cos\frac{\theta_1}{2}\sin\frac{\theta_2}{2}e^{i\phi_2}|0\rangle \otimes |1\rangle + \sin\frac{\theta_1}{2}e^{i\phi_1}\sin\frac{\theta_2}{2}e^{i\phi_2}|1\rangle \otimes |1\rangle.
\end{aligned}
$$

(ii) Using (i) we have

$$
\begin{aligned}
&P_0(|\phi\rangle \otimes |\psi\rangle) \\
&= \frac{1}{\sqrt{3}}|0\rangle \otimes \sin\frac{\theta_1}{2}e^{i\phi_1}\cos\frac{\theta_2}{2}|1\rangle \otimes |0\rangle + \frac{1}{\sqrt{3}}|1\rangle \otimes \cos\frac{\theta_1}{2}\sin\frac{\theta_2}{2}e^{i\phi_2}|0\rangle \otimes |1\rangle \\
&\quad + \frac{1}{\sqrt{3}}|2\rangle \otimes \sin\frac{\theta_1}{2}e^{i\phi_1}\sin\frac{\theta_2}{2}e^{i\phi_2}|1\rangle \otimes |1\rangle.
\end{aligned}
$$

Then

$$
\begin{aligned}
&(\langle \phi| \otimes \langle \psi|)P_0(|\phi\rangle \otimes |\psi\rangle) \\
&= \frac{1}{3}\sin\frac{\theta_1}{2}e^{i\phi_1}\cos\frac{\theta_2}{2}\langle\psi|(|1\rangle \otimes |0\rangle) + \frac{1}{3}\cos\frac{\theta_1}{2}\sin\frac{\theta_2}{2}e^{i\phi_2}\langle\psi|(|0\rangle \otimes |1\rangle) \\
&\quad + \frac{1}{3}\sin\frac{\theta_1}{2}e^{i\phi_1}\sin\frac{\theta_2}{2}e^{i\phi_2}\langle\psi|(|1\rangle \otimes |1\rangle).
\end{aligned}
$$

Since

$$
\langle\psi|(|1\rangle \otimes |0\rangle) = e^{-i\phi_1}\sin\frac{\theta_1}{2}\cos\frac{\theta_2}{2}
$$

$$\langle\psi|(|0\rangle \otimes |1\rangle) = e^{-i\phi_2} \cos \frac{\theta_1}{2} \sin \frac{\theta_2}{2}$$

$$\langle\psi|(|1\rangle \otimes |1\rangle) = e^{-i\phi_1} e^{-i\phi_2} \sin \frac{\theta_1}{2} \sin \frac{\theta_2}{2}$$

we obtain the probability

$$(\langle\phi| \otimes \langle\psi|)P_0(|\phi\rangle \otimes |\psi\rangle) = \frac{1}{3}\left(1 - \cos^2 \frac{\theta_1}{2} \cos^2 \frac{\theta_2}{2}\right)$$

where we used $\sin^2 \alpha + \cos^2 \alpha \equiv 1$.

Problem 11. Let B be an observable with k possible measurement outcomes (b_j)

$$B = \sum_{j=1}^{k} b_j \Pi_j, \qquad \sum_{j=1}^{k} \Pi_j = I.$$

where Π_j denotes mutually orthogonal projection operators. The measurement of a system described by the density operator ρ yields the following *orthogonal measurement*:

1. The outcome (eigenvalue) b_j is obtained with probability

$$p(b_j) = \text{tr}(\Pi_j \rho) = \text{tr}(\Pi_j \rho \Pi_j)$$

where we used the cyclic invariance of the trace and $\Pi_j^2 = \Pi_j$.

2. The expectation value (average measurement value) is given by $\text{tr}(\rho B)$.

3. The state of the system after measurement is in the measured state of the system (i.e. the system is projected onto the state corresponding to the measurement outcome)

$$\rho_{b_j} = \frac{\Pi_j \rho \Pi_j}{p(b_j)}.$$

Discuss orthogonal measurement of the W state

$$|W\rangle\langle W| = \frac{1}{3}\begin{pmatrix} 0 & 0 & 0 & 0 & 0 & 0 & 0 & 0 \\ 0 & 1 & 1 & 0 & 1 & 0 & 0 & 0 \\ 0 & 1 & 1 & 0 & 1 & 0 & 0 & 0 \\ 0 & 0 & 0 & 0 & 0 & 0 & 0 & 0 \\ 0 & 1 & 1 & 0 & 1 & 0 & 0 & 0 \\ 0 & 0 & 0 & 0 & 0 & 0 & 0 & 0 \\ 0 & 0 & 0 & 0 & 0 & 0 & 0 & 0 \\ 0 & 0 & 0 & 0 & 0 & 0 & 0 & 0 \end{pmatrix}$$

with respect to the observable B given by the diagonal matrix

$$\mathrm{diag}(0,1,0,0,0,0,0,0)+2\mathrm{diag}(0,0,1,0,0,0,0,0)+3\mathrm{diag}(0,0,0,0,1,0,0,0)\,.$$

Solution 11. The measurement outcomes are $b_1 = 0$, $b_2 = 1$, $b_3 = 2$, $b_4 = 3$ (eigenvalues) with corresponding projection operators

$$\Pi_1 = \mathrm{diag}(1,0,0,1,0,1,1,1), \quad \Pi_2 = \mathrm{diag}(0,1,0,0,0,0,0,0),$$
$$\Pi_3 = \mathrm{diag}(0,0,1,0,0,0,0,0), \quad \Pi_4 = \mathrm{diag}(0,0,0,0,1,0,0,0).$$

The probability of the outcome $b_2 = 1$ is given by

$$p_2 := \mathrm{tr}(\Pi_2 W \Pi_2) = \frac{1}{3}.$$

Similarly $p_1 = 0$ and $p_3 = p_4 = \frac{1}{3}$. However, the density operators

$$\rho_1 := \frac{1}{3}\mathrm{diag}(0,1,1,0,1,0,0,0)$$

$$\rho_2 := \frac{1}{6}\begin{pmatrix} 0 & 0 & 0 & 0 & 0 & 0 & 0 & 0 \\ 0 & 2 & 1 & 0 & 1 & 0 & 0 & 0 \\ 0 & 1 & 2 & 0 & 1 & 0 & 0 & 0 \\ 0 & 0 & 0 & 0 & 0 & 0 & 0 & 0 \\ 0 & 1 & 1 & 0 & 2 & 0 & 0 & 0 \\ 0 & 0 & 0 & 0 & 0 & 0 & 0 & 0 \\ 0 & 0 & 0 & 0 & 0 & 0 & 0 & 0 \\ 0 & 0 & 0 & 0 & 0 & 0 & 0 & 0 \end{pmatrix}$$

yield the same probabilities for the measurement outcomes with respect to the measurement B. For the measurement outcome b_2 the state of the system becomes $\mathrm{diag}(0,1,0,0,0,0,0,0)$, for b_3 $\mathrm{diag}(0,0,1,0,0,0,0,0)$ and for b_4 $\mathrm{diag}(0,0,0,0,1,0,0,0)$. It is not possible to obtain the measurement outcome b_1.

Problem 12. Measurement can be generalized in the sense that an ancilla system (in a well defined state), identified by the Hilbert space $\mathcal{H}_A$, is introduced and allowed to interact with the quantum system identified by the Hilbert space $\mathcal{H}$. The ancilla system is subsequently measured, which may disturb the original system. Let ρ be a density operator on $\mathcal{H}$ and B an observable on $\mathcal{H}_A$ with k possible measurement outcomes

$$B = \sum_{j=1}^{k} b_j \Pi_j, \qquad \sum_{j=1}^{k} \Pi_j = I$$

where Π_j denotes mutually orthogonal projection operators. Thus $I \otimes B$ is an observable on the product Hilbert space $\mathcal{H} \otimes \mathcal{H}_A$. Let $|b\rangle$ be a normalized eigenvector of B, and consequently an eigenvector of only one of the Π_j. The generalized measurement of a system described by the density operator ρ yields the following

1. The system ρ is extended with the ancilla in the normalized pure state $|b\rangle$ which gives the density operator $\rho \otimes |b\rangle\langle b|$.

2. The two systems interact via a unitary operator U, i.e. the system is transformed according to $U(\rho \otimes |b\rangle\langle b|)U^*$.

3. The outcome b_j is obtained with probability

$$p(b_j) = \mathrm{tr}((I \otimes \Pi_j)U(\rho \otimes |b\rangle\langle b|)U^*(I \otimes \Pi_j)).$$

4. The expectation value (average measurement value) is given by

$$\mathrm{tr}(U(\rho \otimes |b\rangle\langle b|)U^*(I \otimes B)).$$

5. The state of the system after measurement is in the measured state of the system (i.e. the system is projected onto the state corresponding to the measurement outcome)

$$\sigma_{b_j} = \frac{(I \otimes \Pi_j)U(\rho \otimes |b\rangle\langle b|)U^*(I \otimes \Pi_j)}{p(b_j)}.$$

6. The state of the original system after discarding the ancilla system is given by

$$\rho_{b_j} = \mathrm{tr}_{\mathcal{H}_A}(\sigma_{b_j}).$$

Let ϕ_B denote an orthonormal basis for $\mathcal{H}_A$ of normalized eigenvectors of B, and so $|b\rangle \in \phi_B$. Thus we write the unitary operator U as

$$U = \sum_{|j\rangle,|k\rangle \in \phi_B} U_{jk} \otimes |j\rangle\langle k|.$$

Describe generalized measurement in terms of the operators U_{jk} where $|j\rangle, |k\rangle \in \phi_B$.

Solution 12.   We find that the constraint $UU^* = U^*U = I$ yields

$$U^*U = \sum_{|j\rangle,|k\rangle,|l\rangle,|m\rangle \in \phi_B} U_{jk}^* U_{lm} \otimes |k\rangle\langle j|l\rangle\langle m| = \sum_{|j\rangle,|k\rangle,|m\rangle \in \phi_B} U_{jk}^* U_{jm} \otimes |k\rangle\langle m|$$
$$= I$$

and

$$UU^* = \sum_{|j\rangle,|k\rangle,|l\rangle,|m\rangle \in \phi_B} U_{jk}U_{lm}^* \otimes |j\rangle\langle k|m\rangle\langle l| = \sum_{|j\rangle,|k\rangle,|l\rangle \in \phi_B} U_{jk}U_{lk}^* \otimes |j\rangle\langle l|$$
$$= I.$$

Equating to I we find

$$\sum_{|j\rangle \in \phi_B} U_{jk}^* U_{jm} = \delta_{km} I, \qquad \sum_{|k\rangle \in \phi_B} U_{jk}U_{lk}^* = \delta_{jl} I.$$

Consequently we find

$$(I \otimes \Pi_j)U(\rho \otimes |b\rangle\langle b|)U^*(I \otimes \Pi_j)$$
$$= (I \otimes \Pi_j)\left(\sum_{|m\rangle,|n\rangle \in \phi_B} U_{mb}\rho U_{nb}^* \otimes |m\rangle\langle n| \right)(I \otimes \Pi_j)$$
$$= \sum_{|m\rangle,|n\rangle \in \phi_B} (U_{mb}\rho U_{nb}^*) \otimes (\Pi_j|m\rangle\langle n|\Pi_j).$$

Applying the definition of ρ_{b_j} yields

$$\rho_{b_j} = \sum_{|m\rangle \in \phi_B, \Pi_j|m\rangle = |m\rangle} \frac{1}{p(b_j)} U_{mb}\rho U_{mb}^*$$

where

$$p(b_j) = \mathrm{tr}\left(\sum_{|m\rangle \in \phi_B, \Pi_j|m\rangle = |m\rangle} U_{mb}\rho U_{mb}^* \right).$$

When the b_j are non-degenerate we find

$$p(b_j) = \mathrm{tr}\left(U_{mb}\rho U_{mb}^* \right).$$

Thus for a given $|b\rangle$ the measurement is described by the operators U_{mb} for $|m\rangle \in \phi_B$.

Problem 13. Assuming that the b_j are non-degenerate (i.e. that measurement yields maximal information) and that $|j\rangle$ is the eigenvector of Π_j (for the eigenvalue 1) we have k operators U_m where $m = 1, 2, \dots, k$ and the *generalized measurement*

1. $\displaystyle\sum_{m=1}^{k} U_m^* U_m = I.$

2. The outcome b_j is obtained with probability

$$p(b_j) = \text{tr}\left(U_j \rho U_j^*\right).$$

3. The state of the system ρ after the measurement outcome b_j is given by

$$\rho_{b_j} = \frac{1}{p(b_j)} U_j \rho U_j^*.$$

Consider the Hilbert space $\mathcal{H} = \mathbb{C}^2$ and the ancilla Hilbert space $\mathcal{H}_A = \mathbb{C}^2$. We construct a generalized measurement described by $\alpha, \beta \in \mathbb{C}$

$$U_0 = \alpha|0\rangle\langle 0| + \beta|1\rangle\langle 1|, \qquad U_1 = \alpha|1\rangle\langle 1| + \beta|0\rangle\langle 0|$$

where $\{|0\rangle, |1\rangle\}$ forms an orthonormal basis in $\mathbb{C}^2$ and $|\alpha|^2 + |\beta|^2 = 1$. Construct the generalized measurement for U_0 and U_1.

Solution 13. We have

$$
\begin{aligned}
U_0^* U_0 + U_1^* U_1 &= (\overline{\alpha}|0\rangle\langle 0| + \overline{\beta}|1\rangle\langle 1|)(\alpha|0\rangle\langle 0| + \beta|1\rangle\langle 1|) \\
&\quad + (\overline{\alpha}|1\rangle\langle 1| + \overline{\beta}|0\rangle\langle 0|)(\alpha|1\rangle\langle 1| + \beta|0\rangle\langle 0|) \\
&= |0\rangle\langle 0| + |1\rangle\langle 1| = I_2.
\end{aligned}
$$

For $\alpha = 0$ or $\beta = 0$, U_0 and U_1 describe an orthogonal measurement. To construct the unitary operator

$$U := U_{00} \otimes |0\rangle\langle 0| + U_{01} \otimes |0\rangle\langle 1| + U_{10} \otimes |1\rangle\langle 0| + U_{11} \otimes |1\rangle\langle 1|$$

we use the the commutators given by

$$[U_0, U_0^*] = [U_1, U_1^*] = [U_0, U_1] = [U_0, U_1^*] = 0$$

and set

$$U_{00} := U_0, \qquad U_{01} := U_1^*, \qquad U_{10} := U_1, \qquad U_{11} := -U_0^*$$

to satisfy that U is unitary. One orthogonal measurement that implements the generalized measurement is described by the projection operators

$$\Pi_1 = |0\rangle\langle 0|, \qquad \Pi_2 = |1\rangle\langle 1|$$

and the corresponding observable

$$B := -1\Pi_1 + 1\Pi_2.$$

Thus the generalized measurement of ρ, a density operator on $\mathcal{H}$, may proceed as follows

1. Perform U on the density matrix $\rho \otimes |0\rangle\langle 0|$.

2. Measure the ancillary system with respect to the observable B.

Problem 14. Consider the hermitian matrix (Hamilton operator)

$$\hat{S}_x = \frac{1}{\sqrt{2}} \hbar \omega A$$

where A is the 3×3 matrix

$$A = \begin{pmatrix} 0 & 1 & 0 \\ 1 & 0 & 1 \\ 0 & 1 & 0 \end{pmatrix}.$$

(i) Calculate $\exp(-i\hat{S}_x t/\hbar)$.
(ii) Consider the normalized vector

$$|\psi(0)\rangle = \frac{1}{\sqrt{3}} \begin{pmatrix} 1 \\ 1 \\ 1 \end{pmatrix}.$$

Calculate $|\psi(t)\rangle = \exp(-i\hat{S}_x t/\hbar)|\psi(0)\rangle$.
(iii) Find the probability of finding $\psi(t)$ in the initial state $\psi(0)$, i.e.

$$|\langle \psi(t)|\psi(0)\rangle|^2.$$

Solution 14. (i) We have

$$A^2 = \begin{pmatrix} 1 & 0 & 1 \\ 0 & 2 & 0 \\ 1 & 0 & 1 \end{pmatrix}, \qquad A^3 = \begin{pmatrix} 0 & 2 & 0 \\ 2 & 0 & 2 \\ 0 & 2 & 0 \end{pmatrix}$$

and in general

$$A^{2n} = 2^{n-1}A^2, \qquad A^{2n+1} = 2^n A.$$

Using this result we find

$$\exp(-i\hat{S}_x t/\hbar) = I_3 - \frac{i}{\sqrt{2}}\sin(\omega t)A + \frac{1}{2}(\cos(\omega t) - 1)A^2$$

$$= \begin{pmatrix} \frac{1}{2}\cos(\omega t) + \frac{1}{2} & -\frac{i}{\sqrt{2}}\sin(\omega t) & \frac{1}{2}\cos(\omega t) - \frac{1}{2} \\ -\frac{i}{\sqrt{2}}\sin(\omega t) & \cos(\omega t) & -\frac{i}{\sqrt{2}}\sin(\omega t) \\ \frac{1}{2}\cos(\omega t) - \frac{1}{2} & -\frac{i}{\sqrt{2}}\sin(\omega t) & \frac{1}{2}\cos(\omega t) + \frac{1}{2} \end{pmatrix}.$$

(ii) We obtain

$$|\psi(t)\rangle = \exp(-i\hat{S}_x t/\hbar)|\psi(0)\rangle = \frac{1}{\sqrt{3}} \begin{pmatrix} \cos(\omega t) - \frac{i}{\sqrt{2}}\sin(\omega t) \\ \cos(\omega t) - i\sqrt{2}\sin(\omega t) \\ \cos(\omega t) - \frac{i}{\sqrt{2}}\sin(\omega t) \end{pmatrix}.$$

(iii) We find the probability

$$|\langle\psi(t)|\psi(0)\rangle|^2 = \frac{1}{9}|3\cos(\omega t) - i2\sqrt{2}\sin(\omega t)|^2 = 1 - \frac{1}{9}\sin^2(\omega t).$$

Chapter 10

Entanglement

Entanglement is the characteristic trait of quantum mechanics which enforces its entire departure from classical lines of thought. Let $\mathcal{H}_1$ and $\mathcal{H}_2$ be two finite-dimensional Hilbert spaces and let $|\psi\rangle \in \mathcal{H}_1 \otimes \mathcal{H}_2$. Then $|\psi\rangle$ is said to be *disentangled, separable* or a *product state* if there exist states $|\psi_1\rangle \in \mathcal{H}_1$ and $|\psi_2\rangle \in \mathcal{H}_2$ such that $|\psi\rangle = |\psi_1\rangle \otimes |\psi_2\rangle$ otherwise $|\psi\rangle$ is said to be *entangled*. For example, a polarization entangled state is

$$\frac{1}{\sqrt{2}}(|H\rangle \otimes |V\rangle + e^{i\phi}|V\rangle \otimes |H\rangle)$$

where H denotes horizontal polarization and V vertical polarization. A density operator is said to be separable if there exists $m \in \mathbb{N}$ and density operators $\rho_{1,j}$, $\rho_{2,j}$ $(j = 1, \ldots, m)$ such that

$$\rho = \sum_{j=1}^{m} p_j \rho_{1,j} \otimes \rho_{2,j}, \quad \sum_{j=1}^{m} p_j = 1, \quad p_j \in [0,1].$$

There are several measures of entanglement, for example the von Neumann entropy, the tangle and the Schmidt number. An entanglement measure E has to satisfy several requirements. For example, if the density matrix ρ is separable then $E(\rho) = 0$. The entanglement of a maximally entangled state of two n-dimensional systems should be given by $\log n$. There should be no increase in entanglement under $LOCC$, i.e. local operations to the density matrix ρ and classically communicating cannot increase the entanglement of ρ. The entanglement measure should be a convex function.

Problem 1.   Can the *Bell state*

$$\frac{1}{\sqrt{2}}(|01\rangle - |10\rangle) \equiv \frac{1}{\sqrt{2}}(|0\rangle \otimes |1\rangle - |1\rangle \otimes |0\rangle)$$

in the Hilbert space $\mathbb{C}^4$ be written as a product state?

Solution 1.   This state cannot be written as product state. Assume that

$$(c_0|0\rangle + c_1|1\rangle) \otimes (d_0|0\rangle + d_1|1\rangle) = \frac{1}{\sqrt{2}}(|0\rangle \otimes |1\rangle - |1\rangle \otimes |0\rangle), \qquad c_0, c_1, d_0, d_1 \in \mathbb{C}$$

where $|c_0|^2 + |c_1|^2 = 1$ and $|d_0|^2 + |d_1|^2 = 1$. Then we obtain the system of equations

$$c_0 d_0 = 0, \quad c_0 d_1 = \frac{1}{\sqrt{2}}, \quad c_1 d_0 = -\frac{1}{\sqrt{2}}, \quad c_1 d_1 = 0.$$

This set of equations admits no solution. Thus the Bell state cannot be written as a product state. The Bell state is *entangled*.

Problem 2.   Consider the Hilbert space $\mathbb{C}^2 \otimes \mathbb{C}^2$ and the unitary 2×2 matrix

$$U(\theta, \phi) := \begin{pmatrix} \cos\frac{\theta}{2} & e^{-i\phi}\sin\frac{\theta}{2} \\ -e^{i\phi}\sin\frac{\theta}{2} & \cos\frac{\theta}{2} \end{pmatrix}.$$

Which of the following states are entangled?

$$\text{(i)} \quad \Big(U(\theta_1, \phi_1) \otimes U(\theta_2, \phi_2) \Big)(1, 0, 0, 0)^T$$

$$\text{(ii)} \quad \Big(U(\theta_1, \phi_1) \otimes U(\theta_2, \phi_2) \Big)\frac{1}{\sqrt{2}}(1, 0, 0, 1)^T$$

where $\otimes$ denotes the Kronecker product and T denotes the transpose.

Solution 2.   We use the fact that the vector $(x_1, x_2, x_3, x_4)^T \in \mathbb{C}^4$ is separable if and only if $x_1 x_4 = x_2 x_3$. Thus

$$U(\theta_1, \phi_1)(1,0)^T \otimes U(\theta_2, \phi_2)(1,0)^T = \begin{pmatrix} \cos\frac{\theta_1}{2} \\ -e^{i\phi_1}\sin\frac{\theta_1}{2} \end{pmatrix} \otimes \begin{pmatrix} \cos\frac{\theta_2}{2} \\ -e^{i\phi_2}\sin\frac{\theta_2}{2} \end{pmatrix}$$

$$U(\theta_1, \phi_1)(0,1)^T \otimes U(\theta_2, \phi_2)(0,1)^T = \begin{pmatrix} e^{-i\phi_1}\sin\frac{\theta_1}{2} \\ \cos\frac{\theta_1}{2} \end{pmatrix} \otimes \begin{pmatrix} e^{-i\phi_2}\sin\frac{\theta_2}{2} \\ \cos\frac{\theta_2}{2} \end{pmatrix}.$$

Now we apply the separability criteria.
For (i) we have

$$\cos\frac{\theta_1}{2}\cos\frac{\theta_2}{2}e^{i\phi_1}\sin\frac{\theta_1}{2}e^{i\phi_2}\sin\frac{\theta_2}{2} = e^{i\phi_1}\sin\frac{\theta_1}{2}\cos\frac{\theta_2}{2}e^{i\phi_2}\sin\frac{\theta_2}{2}\cos\frac{\theta_1}{2}.$$

Thus the state is not entangled.
For (ii) we obtain

$$U(\theta_1,\phi_1)(1,0)^T \otimes U(\theta_2,\phi_2)(1,0)^T + U(\theta_1,\phi_1)(0,1)^T \otimes U(\theta_2,\phi_2)(0,1)^T$$

$$= \begin{pmatrix} \cos\frac{\theta_1}{2}\cos\frac{\theta_2}{2} + e^{-i(\phi_1+\phi_2)}\sin\frac{\theta_1}{2}\sin\frac{\theta_2}{2} \\ \cos\frac{\theta_2}{2}e^{-i\phi_1}\sin\frac{\theta_1}{2} - \cos\frac{\theta_1}{2}e^{i\phi_2}\sin\frac{\theta_2}{2} \\ \cos\frac{\theta_1}{2}e^{-i\phi_2}\sin\frac{\theta_2}{2} - \cos\frac{\theta_2}{2}e^{i\phi_1}\sin\frac{\theta_1}{2} \\ \cos\frac{\theta_1}{2}\cos\frac{\theta_2}{2} + e^{i(\phi_1+\phi_2)}\sin\frac{\theta_1}{2}\sin\frac{\theta_2}{2} \end{pmatrix}.$$

Thus

$$\cos^2\frac{\theta_1}{2}\cos^2\frac{\theta_2}{2} + \sin^2\frac{\theta_1}{2}\sin^2\frac{\theta_2}{2} \neq -\cos^2\frac{\theta_1}{2}\sin^2\frac{\theta_2}{2} - \sin^2\frac{\theta_1}{2}\cos^2\frac{\theta_2}{2}$$

$$= \cos^2\frac{\theta_1}{2}\cos^2\frac{\theta_2}{2} + \sin^2\frac{\theta_1}{2}\sin^2\frac{\theta_2}{2} - 1.$$

Thus the state is entangled.

Problem 3. Let $|0\rangle$, $|1\rangle$ be an arbitrary orthonormal basis. Can the state in $\mathbb{C}^4$

$$|\psi\rangle = \frac{1}{\sqrt{2}}|0\rangle \otimes |0\rangle + \frac{1}{\sqrt{8}}|0\rangle \otimes |1\rangle + \frac{1}{\sqrt{8}}|1\rangle \otimes |0\rangle + \frac{1}{\sqrt{4}}|1\rangle \otimes |1\rangle$$

be written as a product state?

Solution 3. From

$$(c_1|0\rangle + c_2|1\rangle) \otimes (d_1|0\rangle + d_2|1\rangle) = |\psi\rangle$$

we obtain the two conditions

$$c_1 d_1 c_2 d_2 = \frac{1}{\sqrt{8}}, \qquad c_1 d_1 c_2 d_2 = \frac{1}{8}.$$

Consequently we have a contradiction and the state $|\psi\rangle$ cannot be written as a product state.

Problem 4. (i) Consider the Hilbert space $\mathcal{H}_A \otimes \mathcal{H}_B$, where $\mathcal{H}_A = \mathcal{H}_B = \mathbb{C}^2$ and the state

$$|\psi\rangle := \frac{1}{2} \begin{pmatrix} 1 \\ -1 \\ -1 \\ 1 \end{pmatrix}.$$

Calculate the density matrices using the partial trace

$$\rho_A := \mathrm{tr}_{\mathcal{H}_B}(|\psi\rangle\langle\psi|), \qquad \rho_B := \mathrm{tr}_{\mathcal{H}_A}(|\psi\rangle\langle\psi|)$$

and

$$-\text{tr}(\rho_A \log_2 \rho_A), \qquad -\text{tr}(\rho_B \log_2 \rho_B)$$

where $-\text{tr}(\rho_A \log_2 \rho_A)$ denotes the *von Neumann entropy*.
(ii) Consider the *Bell state*

$$|\psi\rangle := \frac{1}{\sqrt{2}} \begin{pmatrix} 1 \\ 0 \\ 0 \\ -1 \end{pmatrix}.$$

Calculate

$$\rho_A := \text{tr}_{\mathcal{H}_B}(|\psi\rangle\langle\psi|), \qquad -\text{tr}(\rho_A \log_2 \rho_A).$$

(iii) Consider the state

$$|\psi\rangle := \frac{1}{2}(U_1 \otimes U_2) \begin{pmatrix} 1 \\ -1 \\ -1 \\ 1 \end{pmatrix}$$

where U_1 and U_2 are unitary matrices acting on $\mathbb{C}^2$. Calculate

$$\rho_A := \text{tr}_{\mathcal{H}_B}(|\psi\rangle\langle\psi|), \qquad -\text{tr}(\rho_A \log_2 \rho_A).$$

(iv) Consider the state

$$|\psi\rangle := \frac{1}{\sqrt{2}}(U_1 \otimes U_2) \begin{pmatrix} 1 \\ 0 \\ 0 \\ -1 \end{pmatrix}$$

where U_1 and U_2 are unitary matrices acting on $\mathbb{C}^2$. Calculate

$$\rho_A := \text{tr}_{\mathcal{H}_B}(|\psi\rangle\langle\psi|), \qquad -\text{tr}(\rho_A \log_2 \rho_A).$$

Solution 4.   (i) We choose the standard basis in $\mathbb{C}^2$ to calculate the trace. For the density matrix ρ we find

$$\rho \equiv |\psi\rangle\langle\psi| = \frac{1}{4} \begin{pmatrix} 1 & -1 & -1 & 1 \\ -1 & 1 & 1 & -1 \\ -1 & 1 & 1 & -1 \\ 1 & -1 & -1 & 1 \end{pmatrix}.$$

Therefore

$$\rho_A = \begin{pmatrix} 1 & 0 \\ 0 & 1 \end{pmatrix} \otimes (1 \quad 0) \, |\psi\rangle\langle\psi| \begin{pmatrix} 1 & 0 \\ 0 & 1 \end{pmatrix} \otimes \begin{pmatrix} 1 \\ 0 \end{pmatrix}$$

$$+ \begin{pmatrix} 1 & 0 \\ 0 & 1 \end{pmatrix} \otimes (0 \ \ 1) |\psi\rangle\langle\psi| \begin{pmatrix} 1 & 0 \\ 0 & 1 \end{pmatrix} \otimes \begin{pmatrix} 0 \\ 1 \end{pmatrix}$$

$$= \begin{pmatrix} 1 & 0 & 0 & 0 \\ 0 & 0 & 1 & 0 \end{pmatrix} |\psi\rangle\langle\psi| \begin{pmatrix} 1 & 0 \\ 0 & 0 \\ 0 & 1 \\ 0 & 0 \end{pmatrix} + \begin{pmatrix} 0 & 1 & 0 & 0 \\ 0 & 0 & 0 & 1 \end{pmatrix} |\psi\rangle\langle\psi| \begin{pmatrix} 0 & 0 \\ 1 & 0 \\ 0 & 0 \\ 0 & 1 \end{pmatrix}$$

$$= \frac{1}{4} \begin{pmatrix} 1 & -1 \\ -1 & 1 \end{pmatrix} + \frac{1}{4} \begin{pmatrix} 1 & -1 \\ -1 & 1 \end{pmatrix}$$

$$= \frac{1}{2} \begin{pmatrix} 1 & -1 \\ -1 & 1 \end{pmatrix}.$$

Analogously

$$\rho_B = \frac{1}{4} \begin{pmatrix} 1 & -1 \\ -1 & 1 \end{pmatrix} + \frac{1}{4} \begin{pmatrix} 1 & -1 \\ -1 & 1 \end{pmatrix} = \frac{1}{2} \begin{pmatrix} 1 & -1 \\ -1 & 1 \end{pmatrix}.$$

In this case, $\rho_A = \rho_B$. We diagonalise ρ_A. The eigenvalues are 0 and 1 with corresponding orthonormal eigenvectors $\frac{1}{\sqrt{2}}(1,1)^T$ and $\frac{1}{\sqrt{2}}(1,-1)^T$, respectively. Thus

$$-\text{tr}(\rho_A \log_2 \rho_A) = -\text{tr}\left(\frac{1}{2} \begin{pmatrix} 1 & 1 \\ 1 & -1 \end{pmatrix} \begin{pmatrix} 0 & 0 \\ 0 & 1 \end{pmatrix} \begin{pmatrix} 1 & 1 \\ 1 & -1 \end{pmatrix} \right.$$

$$\left. \times \log_2\left(\frac{1}{2} \begin{pmatrix} 1 & 1 \\ 1 & -1 \end{pmatrix} \begin{pmatrix} 0 & 0 \\ 0 & 1 \end{pmatrix} \begin{pmatrix} 1 & 1 \\ 1 & -1 \end{pmatrix} \right) \right)$$

$$= -\text{tr}\left(\frac{1}{2} \begin{pmatrix} 1 & 1 \\ 1 & -1 \end{pmatrix} \begin{pmatrix} 0 & 0 \\ 0 & 1 \end{pmatrix} \begin{pmatrix} 1 & 1 \\ 1 & -1 \end{pmatrix} \right.$$

$$\left. \times \frac{1}{2} \begin{pmatrix} 1 & 1 \\ 1 & -1 \end{pmatrix} \log_2 \begin{pmatrix} 0 & 0 \\ 0 & 1 \end{pmatrix} \begin{pmatrix} 1 & -1 \\ 1 & 1 \end{pmatrix} \right)$$

$$= -\text{tr}\left(\frac{1}{2} \begin{pmatrix} 1 & 1 \\ 1 & -1 \end{pmatrix} \begin{pmatrix} 0\log_2 0 & 0 \\ 0 & 1\log_2 1 \end{pmatrix} \begin{pmatrix} 1 & -1 \\ 1 & 1 \end{pmatrix} \right)$$

$$= -\text{tr} \begin{pmatrix} 0 & 0 \\ 0 & 0 \end{pmatrix} = 0.$$

where $0\log_2 0 = 0$ and $1\log_2 1 = 0$.

(ii) We choose the standard basis in $\mathbb{C}^2$ to calculate the trace. We have

$$\rho = |\psi\rangle\langle\psi| = \frac{1}{2} \begin{pmatrix} 1 & 0 & 0 & -1 \\ 0 & 0 & 0 & 0 \\ 0 & 0 & 0 & 0 \\ -1 & 0 & 0 & 1 \end{pmatrix}.$$

Thus

$$\rho_A = \begin{pmatrix} 1 & 0 \\ 0 & 1 \end{pmatrix} \otimes (1 \ \ 0) |\psi\rangle\langle\psi| \begin{pmatrix} 1 & 0 \\ 0 & 1 \end{pmatrix} \otimes \begin{pmatrix} 1 \\ 0 \end{pmatrix}$$

$$+ \begin{pmatrix} 1 & 0 \\ 0 & 1 \end{pmatrix} \otimes (0 \quad 1) |\psi\rangle\langle\psi| \begin{pmatrix} 1 & 0 \\ 0 & 1 \end{pmatrix} \otimes \begin{pmatrix} 0 \\ 1 \end{pmatrix}$$

$$= \frac{1}{2} \begin{pmatrix} 0 & 0 \\ 0 & 1 \end{pmatrix} + \frac{1}{2} \begin{pmatrix} 1 & 0 \\ 0 & 0 \end{pmatrix} = \frac{1}{2} \begin{pmatrix} 1 & 0 \\ 0 & 1 \end{pmatrix}.$$

Therefore

$$-\text{tr}(\rho_A \log_2 \rho_A) = -\text{tr}\left(\frac{1}{2} \begin{pmatrix} 1 & 0 \\ 0 & 1 \end{pmatrix} \log_2 \frac{1}{2} \begin{pmatrix} 1 & 0 \\ 0 & 1 \end{pmatrix} \right) = 1$$

where $\log_2 \frac{1}{2} = -1$.

(iii) We choose the basis

$$\left\{ U_2 \begin{pmatrix} 1 \\ 0 \end{pmatrix}, U_2 \begin{pmatrix} 0 \\ 1 \end{pmatrix} \right\}$$

to calculate the partial trace. We have

$$|\psi\rangle\langle\psi| = \frac{1}{4}(U_1 \otimes U_2) \begin{pmatrix} 1 & -1 & -1 & 1 \\ -1 & 1 & 1 & -1 \\ -1 & 1 & 1 & -1 \\ 1 & -1 & -1 & 1 \end{pmatrix} (U_1^* \otimes U_2^*)$$

$$= \frac{1}{4}(U_1 \otimes U_2) \left(\begin{pmatrix} 1 & -1 \\ -1 & 1 \end{pmatrix} \otimes \begin{pmatrix} 1 & -1 \\ -1 & 1 \end{pmatrix} \right) (U_1^* \otimes U_2^*).$$

Therefore

$$\rho_A = \begin{pmatrix} 1 & 0 \\ 0 & 1 \end{pmatrix} \otimes ((1 \quad 0) U_2^*) |\psi\rangle\langle\psi| \begin{pmatrix} 1 & 0 \\ 0 & 1 \end{pmatrix} \otimes U_2 \begin{pmatrix} 1 \\ 0 \end{pmatrix}$$

$$+ \begin{pmatrix} 1 & 0 \\ 0 & 1 \end{pmatrix} \otimes ((0 \quad 1) U_2^*) |\psi\rangle\langle\psi| \begin{pmatrix} 1 & 0 \\ 0 & 1 \end{pmatrix} \otimes U_2 \begin{pmatrix} 0 \\ 1 \end{pmatrix}$$

$$= \frac{1}{4} U_1 \begin{pmatrix} 1 & -1 \\ -1 & 1 \end{pmatrix} U_1^* \otimes \left((1 \quad 0) \begin{pmatrix} 1 & -1 \\ -1 & 1 \end{pmatrix} \begin{pmatrix} 1 \\ 0 \end{pmatrix} \right)$$

$$+ \frac{1}{4} U_1 \begin{pmatrix} 1 & -1 \\ -1 & 1 \end{pmatrix} U_1^* \otimes \left((0 \quad 1) \begin{pmatrix} 1 & -1 \\ -1 & 1 \end{pmatrix} \begin{pmatrix} 0 \\ 1 \end{pmatrix} \right)$$

$$= \frac{1}{4} U_1 \begin{pmatrix} 1 & -1 \\ -1 & 1 \end{pmatrix} U_1^* + \frac{1}{4} U_1 \begin{pmatrix} 1 & -1 \\ -1 & 1 \end{pmatrix} U_1^*$$

$$= \frac{1}{2} U_1 \begin{pmatrix} 1 & -1 \\ -1 & 1 \end{pmatrix} U_1^*.$$

We diagonalise ρ_A. The eigenvalues are 0 and 1 with corresponding orthonormal eigenvectors $\frac{1}{\sqrt{2}} U_1 (1 \quad 1)^T$ and $\frac{1}{\sqrt{2}} U_1 (1 \quad -1)^T$, respectively. Thus

$$-\text{tr}(\rho_A \log_2 \rho_A) = -\text{tr}\left(\frac{1}{2} U_1 \begin{pmatrix} 1 & 1 \\ 1 & -1 \end{pmatrix} U_1^* \begin{pmatrix} 0 & 0 \\ 0 & 1 \end{pmatrix} U_1 \begin{pmatrix} 1 & 1 \\ 1 & -1 \end{pmatrix} U_1^* \right.$$

$$\times \log_2\left(\frac{1}{2}U_1\begin{pmatrix}1 & 1\\1 & -1\end{pmatrix}U_1^*\begin{pmatrix}0 & 0\\0 & 1\end{pmatrix}U_1\begin{pmatrix}1 & 1\\1 & -1\end{pmatrix}U_1^*\right)\Bigg)$$

$$= -\operatorname{tr}\left(\frac{1}{2}U_1\begin{pmatrix}1 & 1\\1 & -1\end{pmatrix}U_1^*\begin{pmatrix}0 & 0\\0 & 1\end{pmatrix}U_1\begin{pmatrix}1 & 1\\1 & -1\end{pmatrix}U_1^*\right.$$

$$\left.\times \frac{1}{2}U_1\begin{pmatrix}1 & 1\\1 & -1\end{pmatrix}U_1^*\log_2\begin{pmatrix}0 & 0\\0 & 1\end{pmatrix}U_1^*\begin{pmatrix}1 & -1\\1 & 1\end{pmatrix}U_1^*\right)$$

$$= -\operatorname{tr}\begin{pmatrix}0 & 0\\0 & 0\end{pmatrix} = 0$$

where we used

$$\log_2\begin{pmatrix}0 & 0\\0 & 1\end{pmatrix} = \begin{pmatrix}0\log_2 0 & 0\\0 & 1\log_2 1\end{pmatrix} = \begin{pmatrix}0 & 0\\0 & 0\end{pmatrix}.$$

(iv) We choose the basis

$$\left\{U_2\begin{pmatrix}1\\0\end{pmatrix}, U_2\begin{pmatrix}0\\1\end{pmatrix}\right\}$$

to calculate the partial trace. We have

$$|\psi\rangle\langle\psi| = \frac{1}{4}(U_1\otimes U_2)\begin{pmatrix}1 & 0 & 0 & -1\\0 & 0 & 0 & 0\\0 & 0 & 0 & 0\\-1 & 0 & 0 & 1\end{pmatrix}(U_1^*\otimes U_2^*)$$

and therefore

$$U_1^*\rho_A U_1 = \begin{pmatrix}1 & 0\\0 & 1\end{pmatrix}\otimes(1\ \ 0)\begin{pmatrix}1 & 0 & 0 & -1\\0 & 0 & 0 & 0\\0 & 0 & 0 & 0\\-1 & 0 & 0 & -1\end{pmatrix}\begin{pmatrix}1 & 0\\0 & 1\end{pmatrix}\otimes\begin{pmatrix}1\\0\end{pmatrix}$$

$$+\begin{pmatrix}1 & 0\\0 & 1\end{pmatrix}\otimes(0\ \ 1)\begin{pmatrix}1 & 0 & 0 & -1\\0 & 0 & 0 & 0\\0 & 0 & 0 & 0\\-1 & 0 & 0 & -1\end{pmatrix}\begin{pmatrix}1 & 0\\0 & 1\end{pmatrix}\otimes\begin{pmatrix}0\\1\end{pmatrix}$$

$$= \frac{1}{2}\begin{pmatrix}0 & 0\\0 & 1\end{pmatrix} + \frac{1}{2}\begin{pmatrix}1 & 0\\0 & 0\end{pmatrix} = \frac{1}{2}\begin{pmatrix}1 & 0\\0 & 1\end{pmatrix}.$$

Thus

$$\rho_A = U_1(U_1^*\rho_A U_1)U_1^* = \frac{1}{2}U_1\begin{pmatrix}1 & 0\\0 & 1\end{pmatrix}U_1^*.$$

We choose the basis

$$\left\{U_1\begin{pmatrix}1\\0\end{pmatrix}, U_1\begin{pmatrix}0\\1\end{pmatrix}\right\}$$

to calculate the trace. Thus

$$-\mathrm{tr}(\rho_A \log_2 \rho_A) = -\mathrm{tr}\left(\frac{1}{2}U_1 \begin{pmatrix} 1 & 0 \\ 0 & 1 \end{pmatrix} U_1^* \log_2 \frac{1}{2} U_1 \begin{pmatrix} 1 & 0 \\ 0 & 1 \end{pmatrix} U_1^*\right)$$

$$= -\mathrm{tr}\left(\frac{1}{2}U_1 \begin{pmatrix} \log_2 \frac{1}{2} & 0 \\ 0 & \log_2 \frac{1}{2} \end{pmatrix} U_1^*\right) = 1$$

where we used the cyclic invariance of the trace, $\log_2 \frac{1}{2} = -1$ and that U_1 is a unitary matrix, i.e., $U_1 U_1^* = I_2$.

Problem 5.   Let $\mathcal{H}_A$ and $\mathcal{H}_B$ be two finite-dimensional Hilbert spaces over $\mathbb{C}$. Let $|\psi\rangle$ denote a pure state in the Hilbert space $\mathcal{H}_A \otimes \mathcal{H}_B$. Let $\{|0\rangle, |1\rangle\}$ denote an orthonormal basis in $\mathbb{C}^2$. The *Schmidt number* (also called the *Schmidt rank*) of $|\psi\rangle \in \mathcal{H}_A \otimes \mathcal{H}_B$ over $\mathcal{H}_A \otimes \mathcal{H}_B$ is the smallest non-negative integer $\mathrm{Sch}(|\psi\rangle, \mathcal{H}_A, \mathcal{H}_B)$ such that $|\psi\rangle$ can be written as

$$|\psi\rangle = \sum_{j=1}^{\mathrm{Sch}(|\psi\rangle, \mathcal{H}_A, \mathcal{H}_B)} |\psi_j\rangle_A \otimes |\psi_j\rangle_B$$

where $|\psi_j\rangle_A \in \mathcal{H}_A$ and $|\psi_j\rangle_B \in \mathcal{H}_B$.

Let

$$|\psi\rangle = \sum_{j=1}^{\min(d_1, d_2)} \lambda_j |j\rangle_A \otimes |j\rangle_B$$

be the *Schmidt decomposition* of $|\psi\rangle$ over $\mathcal{H}_A \otimes \mathcal{H}_B$, where d_1 and d_2 are the dimensions of the subsystems. Then the Schmidt number is the number of non-zero λ_j. The λ_j^2 are the eigenvalues of the matrix $\mathrm{tr}_B(|\psi\rangle\langle\psi|)$. A separable state has Schmidt number 1 and an entangled state has Schmidt number greater than 1.

Let $f : \{0,1\}^2 \to \{0,1\}$ be a boolean function. We define the state

$$|\psi_f\rangle := \frac{1}{2} \sum_{a,b \in \{0,1\}} (-1)^{f(a,b)} |a\rangle \otimes |b\rangle. \tag{1}$$

For f we select the AND, OR and XOR operations. The AND, OR and XOR operations are given by

a	b	AND(a,b)	OR(a,b)	XOR(a,b)
0	0	0	0	0
0	1	0	1	1
1	0	0	1	1
1	1	1	1	0

Find the Schmidt numbers of $|\psi_{AND}\rangle$, $|\psi_{OR}\rangle$ and $|\psi_{XOR}\rangle$ over $\mathbb{C}^2 \otimes \mathbb{C}^2$.

Solution 5. From (1) we obtain

$$
\begin{aligned}
|\psi_{AND}\rangle &= \frac{1}{2}((-1)^{0 \cdot 0}|00\rangle + (-1)^{0 \cdot 1}|01\rangle + (-1)^{1 \cdot 0}|10\rangle + (-1)^{1 \cdot 1}|11\rangle) \\
&= \frac{1}{2}(|00\rangle + |01\rangle + |10\rangle - |11\rangle)
\end{aligned}
$$

where $\cdot$ denotes the AND operation. Analogously we find for the OR and XOR operations

$$
|\psi_{OR}\rangle = \frac{1}{2}(|00\rangle - |01\rangle - |10\rangle - |11\rangle)
$$

$$
|\psi_{XOR}\rangle = \frac{1}{2}(|00\rangle - |01\rangle - |10\rangle + |11\rangle).
$$

Next we take the partial trace of $|\psi_{AND}\rangle\langle\psi_{AND}|$. We obtain

$$
\begin{aligned}
\mathrm{tr}_B(|\psi_{AND}\rangle\langle\psi_{AND}|) &= (I_2 \otimes \langle 0|)|\psi_{AND}\rangle\langle\psi_{AND}|(I_2 \otimes |0\rangle) \\
&\quad + (I_2 \otimes \langle 1|)|\psi_{AND}\rangle\langle\psi_{AND}|(I_2 \otimes |1\rangle) \\
&= \frac{1}{4}(2|0\rangle\langle 0| + 2|1\rangle\langle 1|) \\
&= \frac{1}{2}I_2.
\end{aligned}
$$

In the above calculation we used the fact that

$$
(I_2 \otimes \langle 0|)|ab\rangle\langle cd|(I_2 \otimes |0\rangle) + (I_2 \otimes \langle 1|)|ab\rangle\langle cd|(I_2 \otimes |1\rangle) = \delta_{bd}|a\rangle\langle c|
$$

where δ_{bd} denotes the Kronecker delta and $|ab\rangle \equiv |a\rangle \otimes |b\rangle$. Similarly we find

$$
\mathrm{tr}_B(|\psi_{OR}\rangle\langle\psi_{OR}|) = \frac{1}{2}I_2
$$

$$
\begin{aligned}
\mathrm{tr}_B(|\psi_{XOR}\rangle\langle\psi_{XOR}|) &= \frac{1}{2}(|0\rangle\langle 0| - |0\rangle\langle 1| - |1\rangle\langle 0| + |1\rangle\langle 1|) \\
&= \frac{1}{2}(|0\rangle - |1\rangle)(\langle 0| - \langle 1|).
\end{aligned}
$$

Clearly the eigenvalues of $\mathrm{tr}_B(|\psi_{AND}\rangle\langle\psi_{AND}|)$ and $\mathrm{tr}_B(|\psi_{OR}\rangle\langle\psi_{OR}|)$ are $\frac{1}{2}$. Thus

$$
\mathrm{Sch}(|\psi_{AND}\rangle, \mathbb{C}^2, \mathbb{C}^2) = 2, \qquad \mathrm{Sch}(|\psi_{OR}\rangle, \mathbb{C}^2, \mathbb{C}^2) = 2.
$$

The eigenvalues of $\mathrm{tr}_B(|\psi_{XOR}\rangle\langle\psi_{XOR}|)$ are 0 and 1. Thus

$$
\mathrm{Sch}(|\psi_{XOR}\rangle, \mathbb{C}^2, \mathbb{C}^2) = 1.
$$

We note that

$$|\psi_{XOR}\rangle = \frac{1}{2}(|0\rangle - |1\rangle) \otimes (|0\rangle - |1\rangle).$$

Problem 6.   (i) Let $\{\, |0\rangle, |1\rangle, \ldots, |n-1\rangle \,\}$ be an orthonormal basis in $\mathbb{C}^n$. Is

$$|\psi\rangle := \frac{1}{\sqrt{n}} \sum_{j=0}^{n-1} |j\rangle \otimes |j\rangle$$

independent of the chosen orthonormal basis?
(ii) Find the density matrix $|\psi\rangle\langle\psi|$.
(iii) Consider the linear operator

$$P := \frac{1}{4} \sum_{j=0}^{n-1} \sum_{\substack{k=0 \\ k \neq j}}^{n-1} (|jk\rangle - |kj\rangle)(\langle jk| - \langle kj|)$$

where we used the short-cut notation $|jk\rangle \equiv |j\rangle \otimes |k\rangle$. Calculate P^* and P^2. What is the use of this operator?

Solution 6.   (i) Let

$$\{\, |\phi_0\rangle, |\phi_1\rangle, \ldots, |\phi_{n-1}\rangle \,\}$$

be an orthonormal basis in $\mathbb{C}^n$. Then we have the expansion for the state $|j\rangle$

$$|j\rangle = \sum_{k=0}^{n-1} \langle j|\phi_k\rangle |\phi_k\rangle.$$

Thus $|\psi\rangle$ can be written as

$$|\psi\rangle = \frac{1}{\sqrt{n}} \sum_{j=0}^{n-1} \left(\left(\sum_{k=0}^{n-1} \langle j|\phi_k\rangle |\phi_k\rangle \right) \otimes \left(\sum_{l=0}^{n-1} \langle j|\phi_l\rangle |\phi_l\rangle \right) \right)$$

$$= \frac{1}{\sqrt{n}} \sum_{j=0}^{n-1} \sum_{k=0}^{n-1} \sum_{l=0}^{n-1} \overline{\langle \phi_k|j\rangle} \langle j|\phi_l\rangle |\phi_k\rangle \otimes |\phi_l\rangle$$

$$= \frac{1}{\sqrt{n}} \sum_{k=0}^{n-1} \sum_{l=0}^{n-1} \left(\sum_{j=0}^{n-1} \overline{\langle \phi_k|j\rangle} \langle j|\phi_l\rangle \right) |\phi_k\rangle \otimes |\phi_l\rangle$$

where we used $\langle j|\phi_k\rangle = \overline{\langle \phi_k|j\rangle}$. Note that for the sum

$$\sum_{j=0}^{n-1} \overline{\langle \phi_k|j\rangle} \langle j|\phi_l\rangle$$

we cannot apply *Parseval's relation*. Parseval's relation would apply to

$$\sum_{j=0}^{n-1} \overline{\langle\phi_k|j\rangle}\langle\phi_l|j\rangle = \langle\phi_k|\phi_l\rangle = \delta_{kl} \,.$$

Thus the Bell state $|\psi\rangle$ is dependent on the chosen basis. However, if all scalar products $\langle j|\phi_k\rangle$ are real numbers then $|\psi\rangle$ is independent of the chosen basis.

(ii) We have

$$|\psi\rangle\langle\psi| = \frac{1}{n}\sum_{j=0}^{n-1}\sum_{k=0}^{n-1}(|j\rangle\otimes|j\rangle)(\langle k|\otimes\langle k|) = \frac{1}{n}\sum_{j=0}^{n-1}\sum_{k=0}^{n-1}|j\rangle\langle k|\otimes|j\rangle\langle k| \,.$$

(iii) Clearly, $P^* = P$. Furthermore

$$\begin{aligned}
P^2 &= \frac{1}{16}\sum_{j=0}^{n-1}\sum_{j\neq k}\sum_{l=0}^{n-1}\sum_{l\neq m}(|jk\rangle - |kj\rangle)(\langle jk| - \langle kj|)(|lm\rangle - |ml\rangle)(\langle lm| - \langle ml|) \\
&= \frac{1}{16}\sum_{j=0}^{n-1}\sum_{j\neq k}\sum_{l=0}^{n-1}\sum_{l\neq m}(|jk\rangle - |kj\rangle)(2\delta_{jl}\delta_{km} - 2\delta_{jm}\delta_{lk})(\langle lm| - \langle ml|) \\
&= P.
\end{aligned}$$

Thus P is a projection matrix. It projects onto the space spanned by

$$\left\{ \frac{1}{\sqrt{2}}(|jk\rangle - |kj\rangle) \,:\, j,k \in \{0,1,\ldots,n-1\},\ k > j \right\}.$$

Problem 7. One particularly interesting state in quantum computing is the *Greenberger-Horne-Zeilinger state* (GHZ state). This state of three qubits acts in the Hilbert space $\mathbb{C}^8$ and is given by

$$|\psi\rangle = \frac{1}{\sqrt{2}}\left[\begin{pmatrix}1\\0\end{pmatrix}\otimes\begin{pmatrix}1\\0\end{pmatrix}\otimes\begin{pmatrix}1\\0\end{pmatrix} + \begin{pmatrix}0\\1\end{pmatrix}\otimes\begin{pmatrix}0\\1\end{pmatrix}\otimes\begin{pmatrix}0\\1\end{pmatrix}\right].$$

(i) Find the density matrix $\rho = |\psi\rangle\langle\psi|$.

(ii) Let $\sigma_0 \equiv I_2$, σ_1, σ_2 and σ_3 be the Pauli spin matrices, where I_2 is the 2×2 unit matrix. Show that ρ can be written as a linear combination in terms of Kronecker products of Pauli matrices (including σ_0), i.e.,

$$\rho = \frac{1}{2^3}\sum_{j_1=0}^{3}\sum_{j_2=0}^{3}\sum_{j_3=0}^{3} c_{j_1,j_2,j_3}\sigma_{j_1}\otimes\sigma_{j_2}\otimes\sigma_{j_3} \,.$$

Solution 7.   (i) We find the dual state

$$\langle \psi | = \frac{1}{\sqrt{2}}(1\, 0\, 0\, 0\, 0\, 0\, 0\, 1).$$

Thus

$$\rho = \frac{1}{2}\begin{pmatrix} 1 & 0 & 0 & 0 & 0 & 0 & 0 & 1 \\ 0 & 0 & 0 & 0 & 0 & 0 & 0 & 0 \\ 0 & 0 & 0 & 0 & 0 & 0 & 0 & 0 \\ 0 & 0 & 0 & 0 & 0 & 0 & 0 & 0 \\ 0 & 0 & 0 & 0 & 0 & 0 & 0 & 0 \\ 0 & 0 & 0 & 0 & 0 & 0 & 0 & 0 \\ 0 & 0 & 0 & 0 & 0 & 0 & 0 & 0 \\ 1 & 0 & 0 & 0 & 0 & 0 & 0 & 1 \end{pmatrix}.$$

(ii) We find

$$\rho = \frac{1}{8}(I_2 \otimes I_2 \otimes I_2 + I_2 \otimes \sigma_3 \otimes \sigma_3 + \sigma_3 \otimes I_2 \otimes \sigma_3 + \sigma_3 \otimes \sigma_3 \otimes I_2$$
$$+ \sigma_1 \otimes \sigma_1 \otimes \sigma_1 - \sigma_1 \otimes \sigma_2 \otimes \sigma_2 - \sigma_2 \otimes \sigma_1 \otimes \sigma_2 - \sigma_2 \otimes \sigma_2 \otimes \sigma_1)$$

with $I_8 = I_2 \otimes I_2 \otimes I_2$.

Problem 8.   Consider a symmetric matrix A over $\mathbb{R}$

$$A = \begin{pmatrix} a_{11} & a_{12} & a_{13} & a_{14} \\ a_{12} & a_{22} & a_{23} & a_{24} \\ a_{13} & a_{23} & a_{33} & a_{34} \\ a_{14} & a_{24} & a_{34} & a_{44} \end{pmatrix}$$

and the *Bell basis*

$$|\Phi^+\rangle = \frac{1}{\sqrt{2}}\begin{pmatrix} 1 \\ 0 \\ 0 \\ 1 \end{pmatrix}, \qquad |\Phi^-\rangle = \frac{1}{\sqrt{2}}\begin{pmatrix} 1 \\ 0 \\ 0 \\ -1 \end{pmatrix}$$

$$|\Psi^+\rangle = \frac{1}{\sqrt{2}}\begin{pmatrix} 0 \\ 1 \\ 1 \\ 0 \end{pmatrix}, \qquad |\Psi^-\rangle = \frac{1}{\sqrt{2}}\begin{pmatrix} 0 \\ 1 \\ -1 \\ 0 \end{pmatrix}.$$

The Bell basis forms an orthonormal basis in $\mathbb{R}^4$. Let $\widetilde{A}$ denote the matrix A in the Bell basis. What is the condition on the entries a_{ij} such that the matrix A is diagonal in the Bell basis?

Solution 8. Obviously we have $\tilde{a}_{ij} = \tilde{a}_{ji}$ i.e., the matrix $\tilde{A}$ is also symmetric. Straightforward calculation yields

$$\tilde{a}_{11} = (\Phi^+)^T A \Phi^+ = \frac{1}{2}(a_{11} + 2a_{14} + a_{44})$$

$$\tilde{a}_{12} = (\Phi^+)^T A \Phi^- = \frac{1}{2}(a_{11} - a_{44})$$

$$\tilde{a}_{13} = (\Phi^+)^T A \Psi^+ = \frac{1}{2}(a_{12} + a_{13} + a_{24} + a_{34})$$

$$\tilde{a}_{14} = (\Phi^+)^T A \Psi^- = \frac{1}{2}(a_{12} - a_{13} + a_{24} - a_{34})$$

$$\tilde{a}_{22} = (\Phi^-)^T A \Phi^- = \frac{1}{2}(a_{11} - 2a_{14} + a_{44})$$

$$\tilde{a}_{23} = (\Phi^-)^T A \Psi^+ = \frac{1}{2}(a_{12} + a_{13} - a_{24} - a_{23})$$

$$\tilde{a}_{24} = (\Phi^-)^T A \Psi^- = \frac{1}{2}(a_{12} - a_{13} - a_{24} + a_{34})$$

$$\tilde{a}_{33} = (\Psi^+)^T A \Psi^+ = \frac{1}{2}(a_{22} + 2a_{23} + a_{33})$$

$$\tilde{a}_{34} = (\Psi^+)^T A \Psi^- = \frac{1}{2}(a_{22} - a_{33})$$

$$\tilde{a}_{44} = (\Psi^-)^T A \Psi^- = \frac{1}{2}(a_{22} - 2a_{23} + a_{33}).$$

The condition that the matrix $\tilde{A}$ should be diagonal leads to

$$a_{11} - a_{44} = 0, \qquad a_{22} - a_{33} = 0$$

and $a_{12} = a_{13} = a_{24} = a_{34} = 0$ with the entries a_{14} and a_{23} arbitrary. Thus the matrix A has the form

$$A = \begin{pmatrix} a_{11} & 0 & 0 & a_{14} \\ 0 & a_{22} & a_{23} & 0 \\ 0 & a_{23} & a_{22} & 0 \\ a_{14} & 0 & 0 & a_{11} \end{pmatrix}.$$

Problem 9. Let $|\psi\rangle$ be a given state in the Hilbert space $\mathbb{C}^n$. Let X and Y be two $n \times n$ hermitian matrices. We define the *correlation* for a given state $|\psi\rangle$ as

$$C_{XY}(|\psi\rangle) := \langle \psi | XY | \psi \rangle - \langle \psi | X | \psi \rangle \langle \psi | Y | \psi \rangle.$$

Let $n = 4$,

$$X = \begin{pmatrix} 0 & 0 & 0 & 1 \\ 0 & 0 & 1 & 0 \\ 0 & 1 & 0 & 0 \\ 1 & 0 & 0 & 0 \end{pmatrix}, \qquad Y = \begin{pmatrix} 1 & 0 & 0 & 0 \\ 0 & 0 & 0 & 1 \\ 0 & 0 & 1 & 0 \\ 0 & 1 & 0 & 0 \end{pmatrix}$$

and consider the Bell state

$$|\psi\rangle = \frac{1}{\sqrt{2}} \begin{pmatrix} 1 \\ 0 \\ 0 \\ 1 \end{pmatrix}.$$

Find the correlation.

Solution 9.   Since $X|\psi\rangle = |\psi\rangle$ we have $\langle\psi|X|\psi\rangle = 1$ and $\langle\psi|X = \langle\psi|$. Thus

$$\langle\psi|XY|\psi\rangle - \langle\psi|X|\psi\rangle\langle\psi|Y|\psi\rangle = \langle\psi|Y|\psi\rangle - \langle\psi|Y|\psi\rangle = 0.$$

Problem 10.   Consider a bipartite qutrit system $\mathcal{H}_A = \mathcal{H}_B = \mathbb{C}^3$ with an arbitrary orthonormal basis $\{\,|0\rangle, |1\rangle, |2\rangle\,\}$ in $\mathcal{H}_A$ and $\mathcal{H}_B$, respectively.
(i) Find the *antisymmetric subspace* $\mathcal{H}_-$ on $\mathcal{H} \otimes \mathcal{H}_B$.
(ii) Find an arbitrary antisymmetric state on $\mathcal{H}^{\otimes n}$.

Solution 10.   (i) The antisymmetric subspace $\mathcal{H}_-$ on $\mathcal{H}_A \otimes \mathcal{H}_B$ is defined as

$$\mathcal{H}_- := \text{span}_{\mathbb{C}}\{\,|01\rangle - |10\rangle,\, |12\rangle - |21\rangle,\, |20\rangle - |02\rangle\,\} \subset \mathcal{H}_A \otimes \mathcal{H}_B.$$

(ii) An antisymmetric state on $\mathcal{H}^{\otimes n}$ is given by

$$|\psi\rangle = \sum_{j_1,j_2,\ldots,j_n=0}^{2} \sum_{k_1,k_2,\ldots,k_n=0}^{2} a_{j_1,j_2,\ldots,j_n;k_1,k_2,\ldots,k_n}|j_1,\ldots,j_n;k_1,\ldots,k_n\rangle$$

where

$$a_{j_1,j_2,\ldots,j_n;k_1,k_2,\ldots,k_n} := \left(\frac{1}{\sqrt{2}}\right)^n \sum_{i_1,i_2,\ldots,i_n=0}^{2} b_{i_1,i_2,\ldots,i_n} \prod_{m=1}^{n} \epsilon_{i_m j_m k_m}$$

and ϵ is the *Levi-Civita symbol*, i.e. $\epsilon_{ijk} = 1$ for $(i,j,k) = (1,2,3)$ and its even permutations, and -1 for odd permutations and 0 otherwise.

Problem 11.   Consider the density matrix (*Werner state*) in $\mathbb{C}^4$

$$\rho_w := r|\phi^+\rangle\langle\phi^+| + \frac{1-r}{4}I_4$$

where $|\phi^+\rangle = \frac{1}{\sqrt{2}}(1,0,0,1)^T$ is the Bell state, and $0 \le r \le 1$.
(i) Find $\text{tr}(\rho_w)$ and the eigenvalues of ρ_w.
(ii) Determine the *concurrence*

$$C(\rho_w) := \max\{\,\lambda_1 - \lambda_2 - \lambda_3 - \lambda_4,\, 0\,\}$$

where $\lambda_1 \geq \lambda_2 \geq \lambda_3 \geq \lambda_4$ are the eigenvalues of ρ_w.

Solution 11. (i) We have

$$\rho_w = \begin{pmatrix} (1+r)/4 & 0 & 0 & r/2 \\ 0 & (1-r)/4 & 0 & 0 \\ 0 & 0 & (1-r)/4 & 0 \\ r/2 & 0 & 0 & (1+r)/4 \end{pmatrix}.$$

Thus $\mathrm{tr}(\rho_w) = 1$. The eigenvalues of ρ_w are $(1+r)/4 + r/2 = (1+3r)/4$ and $(1-r)/4$ with multiplicity 3.

(ii) From (i) it follows that

$$\lambda_1 - \lambda_2 - \lambda_3 - \lambda_4 = (1+3r)/4 - 3(1-r)/4 = (3r-1)/2.$$

The concurrence is

$$C(\rho_w) = \max\{(3r-1)/2, 0\}.$$

If $r = 0$ we have $C(\rho_w) = 0$ and if $r = 1$ then $C(\rho_w) = 1$. For $r = \frac{1}{2}$ we find $C(\rho_w) = \frac{1}{4}$.

Problem 12. Let ρ be a density matrix over $\mathbb{C}^2 \otimes \mathbb{C}^2 = \mathbb{C}^4$. We define the *entanglement of formation* as

$$E_f(\rho) := \min_{\{p_k, |\psi_k\rangle\}} \sum_{j=0}^{|\{p_k, |\psi_k\rangle\}|} p_j S(\mathrm{tr}_{\mathbb{C}^2}(|\psi_j\rangle\langle\psi_j|))$$

where $\{p_k, |\psi_k\rangle\}$ indicates that the minimum should be taken over all mixtures which realize ρ. $|\{p_k, |\psi_k\rangle\}|$ is the number of pure states comprising the mixture and

$$S(\sigma) := -\mathrm{tr}(\sigma \log_2 \sigma)$$

is the *von Neumann entropy*. The minimum is taken over all mixtures

$$\{ (p_0, |\psi_0\rangle), (p_1, |\psi_1\rangle), \ldots \}$$

which realize ρ where the cardinality of the set is obviously determined by the mixture and is finite. We can calculate $E_f(\rho)$ from

$$E_f(\rho) = h\left(\frac{1 + \sqrt{1 - C(\rho)^2}}{2}\right)$$

where

$$C(\rho) := \max\left\{ \sqrt{\lambda_1} - \sqrt{\lambda_2} - \sqrt{\lambda_3} - \sqrt{\lambda_4}, 0 \right\}$$

is the *concurrence*, $\lambda_1 \geq \lambda_2 \geq \lambda_3 \geq \lambda_4$ are the eigenvalues of

$$\rho(\sigma_y \otimes \sigma_y)\rho^*(\sigma_y \otimes \sigma_y)$$

and

$$h(p) := -p\log_2 p - (1-p)\log_2(1-p)$$

is the *Shannon entropy*. Find $E_f(\rho)$ for the *Werner state*

$$\rho_w := \frac{5}{8}|\phi^+\rangle\langle\phi^+| + \frac{1}{8}\left(|\phi^-\rangle\langle\phi^-| + |\psi^+\rangle\langle\psi^+| + |\psi^-\rangle\langle\psi^-|\right) = \frac{1}{2}|\phi^+\rangle\langle\phi^+| + \frac{1}{8}I_4$$

where $|\phi^+\rangle = \frac{1}{\sqrt{2}}(1,0,0,1)^T$ is a Bell state.

Solution 12.   We have

$$\rho_w = \frac{1}{8}\begin{pmatrix} 3 & 0 & 0 & 2 \\ 0 & 1 & 0 & 0 \\ 0 & 0 & 1 & 0 \\ 2 & 0 & 0 & 3 \end{pmatrix}.$$

Hence

$$\tilde{\rho}_w := (\sigma_y \otimes \sigma_y)\rho_w^*(\sigma_y \otimes \sigma_y) = \rho_w$$

where

$$\sigma_y \otimes \sigma_y = \begin{pmatrix} 0 & 0 & 0 & -1 \\ 0 & 0 & 1 & 0 \\ 0 & 1 & 0 & 0 \\ -1 & 0 & 0 & 0 \end{pmatrix}.$$

Thus

$$\rho_w(\sigma_y \otimes \sigma_y)\rho_w^*(\sigma_y \otimes \sigma_y) = \rho_w^2 = \frac{1}{64}\begin{pmatrix} 13 & 0 & 0 & 12 \\ 0 & 1 & 0 & 0 \\ 0 & 0 & 1 & 0 \\ 12 & 0 & 0 & 13 \end{pmatrix}.$$

The eigenvalues are $\frac{25}{64}$, $\frac{1}{64}$, $\frac{1}{64}$ and $\frac{1}{64}$. The concurrence is

$$C(\rho) = \max\left\{\frac{5}{8} - \frac{1}{8} - \frac{1}{8} - \frac{1}{8}, 0\right\} = \frac{1}{4}.$$

This result is consistent with solution 11 when $r = \frac{1}{2}$. Thus $E_f(\rho) = 0.1176$.

Problem 13.   Let $\mathcal{H}_A$ and $\mathcal{H}_B$ denote two finite-dimensional Hilbert spaces. Consider the Hamilton operator

$$\hat{H} = X_A \otimes X_B$$

where the linear operator $X_A = X_A^{-1}$ acts on $\mathcal{H}_A$ and the linear operator $X_B = X_B^{-1}$ acts on $\mathcal{H}_B$. Consequently $\hat{H} = \hat{H}^{-1}$. Let $|\psi\rangle \in \mathcal{H}_A \otimes \mathcal{H}_B$. The *von Neumann entropy* is given by

$$E(|\psi\rangle) := -\text{tr}_A(\rho_A \log_2 \rho_A)$$

where $\rho_A = \text{tr}_B(|\psi\rangle\langle\psi|)$. The *entanglement capability* of $\hat{H}$ is defined as

$$E(\hat{H}) := \max_{|\psi\rangle \in \mathcal{H}_A \otimes \mathcal{H}_B} \Gamma(t)|_{t \to 0}$$

where

$$\Gamma(t) := \frac{dE(\exp(-i\hat{H}t)|\psi(0)\rangle)}{dt}$$

is the *state entanglement rate*.

(i) Show that

$$\Gamma(t) = i\text{tr}_A\left(\text{tr}_B([\hat{H}, |\psi\rangle\langle\psi|] \log_2 \rho_A\right)$$

where $[\,,\,]$ denotes the commutator.

(ii) Show that an upper bound on $\Gamma(t)|_{t \to 0}$ is given by

$$\Gamma(t)|_{t \to 0} \leq 1.9123.$$

Solution 13. Let $\rho_{AB}(t) := |\psi(t)\rangle\langle\psi(t)|$ and $\rho_A(t) := \text{tr}_B(\rho_{AB}(t))$. We have

$$\rho_{AB}(t) = \exp(-i\hat{H}t)\rho_{AB}(0)\exp(i\hat{H}t)$$

and the time evolution of $\rho_{AB}(t)$ (*von Neumann equation*) is given by

$$i\frac{d\rho_{AB}(t)}{dt} = [\hat{H}, \rho_{AB}(t)].$$

Thus

$$i\frac{d\rho_A(t)}{dt} = \text{tr}_B[\hat{H}, \rho_{AB}(t)].$$

It follows that

$$\begin{aligned}
\Gamma(t) &= -\frac{d}{dt}\text{tr}_A(\rho_A \log_2 \rho_A) = -\text{tr}_A(\frac{d}{dt}\rho_A \log_2 \rho_A) \\
&= -\text{tr}_A(\frac{d\rho_A}{dt}\log_2 \rho_A + \rho_A\frac{d}{dt}\log_2 \rho_A) = -\text{tr}_A(\frac{d\rho_A}{dt}\log_2 \rho_A) \\
&= i\text{tr}_A(\text{tr}_B[\hat{H}, \rho_{AB}]\log_2 \rho_A)
\end{aligned}$$

since

$$\text{tr}_A(\rho_A\frac{d}{dt}\log_2 \rho_A) = 0.$$

Let

$$|\psi(0)\rangle = \sum_{j=1}^{\mathrm{Sch}(|\psi(0)\rangle)} \sqrt{\lambda_j} |\phi_j\rangle \otimes |\eta_j\rangle$$

be a *Schmidt decomposition* of $|\psi(0)\rangle$ over $\mathcal{H}_A \otimes \mathcal{H}_B$, where $\lambda_j > 0$ with

$$\sum_{j=1}^{\mathrm{Sch}(|\psi(0)\rangle)} \lambda_j = 1$$

and

$$\{ |\phi_1\rangle, \dots, |\phi_{\mathrm{Sch}(|\psi(0)\rangle)}\rangle \}, \qquad \{ |\eta_1\rangle, \dots, |\eta_{\mathrm{Sch}(|\psi(0)\rangle)}\rangle \}$$

are orthonormal sets of states. $\mathrm{Sch}(|\psi(0)\rangle)$ denotes the Schmidt rank of $|\psi(0)\rangle$ over $\mathcal{H}_A \otimes \mathcal{H}_B$. Thus

$$
\begin{aligned}
\mathrm{tr}_B \left([\hat{H}, \rho_{AB}(0)] \right) &= \sum_{j=1}^{\mathrm{Sch}(|\psi(0)\rangle)} I \otimes \langle \eta_j | \left[\hat{H}, \rho_{AB}(0) \right] I \otimes |\eta_j\rangle \\
&= \sum_{j=1}^{\mathrm{Sch}(|\psi(0)\rangle)} I \otimes \langle \eta_j | [X_A \otimes X_B, \rho_{AB}(0)] I \otimes |\eta_j\rangle \\
&= \sum_{m=1}^{\mathrm{Sch}(|\psi(0)\rangle)} \sum_{n=1}^{\mathrm{Sch}(|\psi(0)\rangle)} \sqrt{\lambda_m \lambda_n} \langle \eta_n | X_B | \eta_m \rangle X_A |\phi_m\rangle\langle\phi_n| \\
&\quad - \sum_{m=1}^{\mathrm{Sch}(|\psi(0)\rangle)} \sum_{n=1}^{\mathrm{Sch}(|\psi(0)\rangle)} \sqrt{\lambda_m \lambda_n} |\phi_m\rangle\langle\phi_n| X_A \langle \eta_n | X_B | \eta_m \rangle \\
&= \sum_{m=1}^{\mathrm{Sch}(|\psi(0)\rangle)} \sum_{n=1}^{\mathrm{Sch}(|\psi(0)\rangle)} \sqrt{\lambda_m \lambda_n} \langle \eta_n | X_B | \eta_m \rangle [X_A, |\phi_m\rangle\langle\phi_n|]
\end{aligned}
$$

where we used the result

$$\rho_{AB}(0) = \sum_{m=1}^{\mathrm{Sch}(|\psi(0)\rangle)} \sum_{n=1}^{\mathrm{Sch}(|\psi(0)\rangle)} \sqrt{\lambda_m \lambda_n} (|\phi_m\rangle \otimes |\eta_m\rangle)(\langle\phi_n| \otimes \langle\eta_n|).$$

Since

$$\rho_A(0) = \sum_{j=1}^{\mathrm{Sch}(|\psi(0)\rangle)} \lambda_j |\phi_j\rangle\langle\phi_j|, \qquad \log_2 \rho_A(0) = \sum_{j=1}^{\mathrm{Sch}(|\psi(0)\rangle)} \log_2 \lambda_j |\phi_j\rangle\langle\phi_j|$$

we find

$$\Gamma(t)|_{t\to 0} = i \mathrm{tr}_A (\mathrm{tr}_B [\hat{H}, \rho_{AB}(0)] \log_2 \rho_A(0))$$

$$\text{Sch}(|\psi(0)\rangle)$$
$$= i \sum_{j=1} \langle \phi_j | \text{tr}_B [\hat{H}, \rho_{AB}(0)] \log_2 \rho_A(0) | \phi_j \rangle$$

$$\text{Sch}(|\psi(0)\rangle) \quad \text{Sch}(|\psi(0)\rangle)$$
$$= i \sum_{m=1} \sum_{n=1} \sqrt{\lambda_m \lambda_n} \log_2 \lambda_n \langle \eta_n | X_B | \eta_m \rangle \langle \phi_n | X_A | \phi_m \rangle$$

$$\text{Sch}(|\psi(0)\rangle) \quad \text{Sch}(|\psi(0)\rangle)$$
$$- i \sum_{m=1} \sum_{n=1} \sqrt{\lambda_m \lambda_n} \log_2 \lambda_m \langle \eta_n | X_B | \eta_m \rangle \langle \phi_n | X_A | \phi_m \rangle$$

$$\text{Sch}(|\psi(0)\rangle) \quad \text{Sch}(|\psi(0)\rangle)$$
$$= i \sum_{m=1} \sum_{n=1} \sqrt{\lambda_m \lambda_n} \log_2 \frac{\lambda_n}{\lambda_m} \langle \eta_n | X_B | \eta_m \rangle \langle \phi_n | X_A | \phi_m \rangle$$

$$\text{Sch}(|\psi(0)\rangle) \quad \text{Sch}(|\psi(0)\rangle)$$
$$\leq \sum_{m=1} \sum_{n=1} \sqrt{\lambda_m \lambda_n} \left| \log_2 \frac{\lambda_n}{\lambda_m} \right| |\langle \eta_n | X_B | \eta_m \rangle| |\langle \phi_n | X_A | \phi_m \rangle|$$

$$\text{Sch}(|\psi(0)\rangle) \quad \text{Sch}(|\psi(0)\rangle)$$
$$= \sum_{m=1} \sum_{n=1} (\lambda_m + \lambda_n) |\langle \eta_n | X_B | \eta_m \rangle| |\langle \phi_n | X_A | \phi_m \rangle|$$
$$\times \sqrt{\frac{\lambda_n}{\lambda_m + \lambda_n} \frac{\lambda_m}{\lambda_m + \lambda_n}} \left| \log_2 \frac{\lambda_n}{\lambda_m} \right|$$

$$\text{Sch}(|\psi(0)\rangle) \quad \text{Sch}(|\psi(0)\rangle)$$
$$\leq \sum_{m=1} \sum_{n=1} (\lambda_m + \lambda_n) |\langle \eta_n | X_B | \eta_m \rangle| |\langle \phi_n | X_A | \phi_m \rangle|$$
$$\times \max_{x \in (0,1)} \sqrt{x(1-x)} \log_2 \left(\frac{x}{1-x} \right)$$

$$\leq 2 \max_{x \in (0,1)} \sqrt{x(1-x)} \log_2 \left(\frac{x}{1-x} \right)$$
$$\approx 1.9123$$

where we used

$$\text{Sch}(|\psi(0)\rangle) \quad \text{Sch}(|\psi(0)\rangle)$$
$$\sum_{m=1} \sum_{n=1} \lambda_m |\langle \eta_n | X_B | \eta_m \rangle| |\langle \phi_n | X_A | \phi_m \rangle|$$

$$\text{Sch}(|\psi(0)\rangle) \quad \text{Sch}(|\psi(0)\rangle)$$
$$\leq \sum_{m=1} \sum_{n=1} |\langle \eta_n | X_B | \eta_m \rangle| |\langle \phi_n | X_A | \phi_m \rangle| \leq 1$$

since $X_A^2 = I$ and $X_B^2 = I$.

Problem 14. Consider the Hamilton operator

$$\hat{H} = \mu_x \sigma_x \otimes \sigma_x + \mu_y \sigma_y \otimes \sigma_y, \qquad \mu_x, \mu_y \in \mathbb{R}$$

where σ_x and σ_y are Pauli spin matrices.

(i) Calculate the eigenvalues and eigenvectors of $\hat{H}$.
(ii) Are the eigenvectors entangled?
(iii) Let $|\psi\rangle \in \mathbb{C}^4$. The *von Neumann entropy* is given by

$$E(|\psi\rangle) := -\text{tr}(\rho_A \log_2 \rho_A)$$

where $\rho_A := \text{tr}_{\mathbb{C}^2}(|\psi\rangle\langle\psi|)$. The *entanglement capability* of $\hat{H}$ is defined as

$$E(\hat{H}) := \max_{|\psi\rangle \in \mathbb{C}^4} \Gamma(t)|_{t \to 0}$$

where

$$\Gamma(t) = \frac{dE(\exp(-i\hat{H}t)|\psi(0)\rangle)}{dt}$$

is the *state entanglement rate*. Show that

$$E(\hat{H}) = \alpha(\mu_x + \mu_y)$$

where

$$\alpha = 2 \max_{x \in (0,1)} \sqrt{x(1-x)} \log_2 \left(\frac{x}{1-x}\right).$$

Solution 14. (i) The matrix representation of the Hamilton operator $\hat{H}$ is given by

$$\hat{H} = \begin{pmatrix} 0 & 0 & 0 & \mu_x - \mu_y \\ 0 & 0 & \mu_x + \mu_y & 0 \\ 0 & \mu_x + \mu_y & 0 & 0 \\ \mu_x - \mu_y & 0 & 0 & 0 \end{pmatrix}.$$

The eigenvalues are $\mu_x - \mu_y$ with corresponding eigenvector

$$|\phi^+\rangle = \frac{1}{\sqrt{2}}(1, 0, 0, 1)^T$$

$\mu_y - \mu_x$ with corresponding eigenvector

$$|\phi^-\rangle = \frac{1}{\sqrt{2}}(1, 0, 0, -1)^T$$

$\mu_x + \mu_y$ with corresponding eigenvector

$$|\psi^+\rangle = \frac{1}{\sqrt{2}}(0, 1, 1, 0)^T$$

and $-\mu_x - \mu_y$ with corresponding eigenvector

$$|\psi^-\rangle = \frac{1}{\sqrt{2}}(0, 1, -1, 0)^T.$$

(ii) Clearly all four eigenvectors are entangled (Bell basis).

(iii) Consider

$$|\psi_{max}\rangle := \begin{pmatrix} 0 \\ \sqrt{x_0} \\ -i\sqrt{1-x_0} \\ 0 \end{pmatrix}$$

$$= (\sqrt{x_0} - i\sqrt{1-x_0})\frac{1}{2}\begin{pmatrix} 0 \\ 1 \\ 1 \\ 0 \end{pmatrix} + (\sqrt{x_0} + i\sqrt{1-x_0})\frac{1}{2}\begin{pmatrix} 0 \\ 1 \\ -1 \\ 0 \end{pmatrix}$$

where $x_0 \in (0,1)$ satisfies

$$\alpha = 2\sqrt{x_0(1-x_0)}\log_2\left(\frac{x_0}{1-x_0}\right).$$

Now we have

$$\exp(-i\hat{H}t)|\psi_{max}\rangle = \begin{pmatrix} 0 \\ \cos(t(\mu_x + \mu_y)) - \sqrt{1-x_0}\sin(t(\mu_x + \mu_y)) \\ -i\left(\sqrt{x_0}\sin(t(\mu_x + \mu_y)) + \sqrt{1-x_0}\cos(t(\mu_x + \mu_y))\right) \\ 0 \end{pmatrix}.$$

Defining

$$a_1 := \sqrt{x_0}\cos(t(\mu_x + \mu_y)) - \sqrt{1-x_0}\sin(t(\mu_x + \mu_y))$$
$$a_2 := \sqrt{x_0}\sin(t(\mu_x + \mu_y)) + \sqrt{1-x_0}\cos(t(\mu_x + \mu_y))$$

and

$$\rho_{max}(t) := \exp(-i\hat{H}t)|\psi_{max}\rangle\langle\psi_{max}|\exp(i\hat{H}t)$$

we find

$$\rho_{max}(t) = \begin{pmatrix} 0 & 0 & 0 & 0 \\ 0 & a_1^2 & ia_1\bar{a}_2 & 0 \\ 0 & -ia_2\bar{a}_1 & a_2^2 & 0 \\ 0 & 0 & 0 & 0 \end{pmatrix}, \qquad \text{tr}_{\mathbb{C}^2}(\rho_{max}(t)) = \begin{pmatrix} a_1^2 & 0 \\ 0 & a_2^2 \end{pmatrix}.$$

Thus

$$\Gamma(t) = -\frac{d}{dt}\left(a_1^2\log_2 a_1^2 + a_2^2\log_2 a_2^2\right)$$

$$= -\left(4a_1\frac{da_1}{dt}\log_2 a_1 + \frac{2}{\ln 2}a_1\frac{da_1}{dt} + 4a_2\frac{da_2}{dt}\log_2 a_2 + \frac{2}{\ln 2}a_2\frac{da_2}{dt}\right)$$

$$= 4\mu a_1 a_2\left(\log_2\frac{a_1}{a_2}\right)$$

$$= 4\mu\left(\sqrt{x_0(1-x_0)}\cos(2t\mu) + \frac{1}{2}(2x_0 - 1)\sin(2t\mu)\right)$$

$$\times \log_2\left(\frac{\sqrt{x_0}\cos(2t\mu) - \sqrt{1-x_0}\sin(2t\mu)}{\sqrt{x_0}\sin(2t\mu) + \sqrt{1-x_0}\cos(2t\mu)}\right)$$

where $\mu \equiv \mu_x + \mu_y$ and we used that a_1 and a_2 satisfy the system of linear differential equations with constant coefficients

$$\frac{da_1}{dt} = -a_2, \qquad \frac{da_2}{dt} = a_1.$$

Since $\hat{H}$ is asymptotically equivalent to $(\mu_x + \mu_y)\sigma_x \otimes \sigma_x$ and

$$E((\mu_x + \mu_y)\sigma_x \otimes \sigma_x) \le (\mu_x + \mu_y)\alpha$$

and using

$$\Gamma(0) = (\mu_x + \mu_y)2\sqrt{x_0(1 - x_0)} \log_2 \frac{x_0}{1 - x_0} = \alpha(\mu_x + \mu_y)$$

we find $E(\hat{H}) = \alpha(\mu_x + \mu_y)$.

Problem 15.   Consider the orthonormal basis $\{\,|0\rangle, |1\rangle, \ldots, |n-1\rangle\,\}$ in the Hilbert space $\mathbb{C}^n$. We assume in the following that this is the standard basis. Consider the states (*coherent states*)

$$|\beta\rangle = \left(1 - \sum_{k=1}^{n-1} x_k\right)^{1/2} |0\rangle + \sum_{k=1}^{n-1} \sqrt{x_k} e^{i\phi_k} |k\rangle$$

where $\phi_k \in [0, 2\pi)$, $0 \le x_k \le 1$ and with the constraints

$$0 \le x_j \le 1 - \sum_{k=j+1}^{n-1} x_k, \qquad j = 1, 2, \ldots, n-2.$$

The Lebesgue measure is given by

$$d\mu(\beta) = \frac{n!}{(2\pi)^{n-1}} \prod_{j=1}^{n-1} dx_j d\phi_j.$$

(i) Let $n = 4$. Then the state $|\beta\rangle$ is given by

$$|\beta\rangle = \begin{pmatrix} (1 - x_1 - x_2 - x_3)^{1/2} \\ \sqrt{x_1} e^{i\phi_1} \\ \sqrt{x_2} e^{i\phi_2} \\ \sqrt{x_3} e^{i\phi_3} \end{pmatrix}.$$

Show that this state is normalized.
(ii) Calculate the density matrix $\rho = |\beta\rangle\langle\beta|$.
(iii) Show that the coherent states $|\beta\rangle$ satisfy

$$\int_\Omega d\mu(\beta)|\beta\rangle\langle\beta| = I_4$$

where $d\mu(\beta)$ is the uniform measure given above and Ω the domain for ϕ_j $(j = 1, 2, 3)$ and x_k $(k = 1, 2, 3)$ described above. I_4 is the 4×4 unit matrix. This equation is called the *resolution of identity* and a coherent state must satisfy this condition.

(iv) Find the reduced density matrix from $|\beta\rangle$ and a condition for entanglement.

Solution 15.   (i) Taking the scalar product we have

$$((1-x_1-x_2-x_3)^{1/2} \sqrt{x_1}e^{-i\phi_1} \sqrt{x_2}e^{-i\phi_2} \sqrt{x_3}e^{-i\phi_3}) \begin{pmatrix} (1 - x_1 - x_2 - x_3)^{1/2} \\ \sqrt{x_1}e^{i\phi_1} \\ \sqrt{x_2}e^{i\phi_2} \\ \sqrt{x_3}e^{i\phi_3} \end{pmatrix}$$

$$= (1 - x_1 - x_2 - x_3) + x_1 + x_2 + x_3 = 1.$$

Thus the state is normalized.

(ii) We find the 4×4 matrix

$$\rho = |\beta\rangle\langle\beta| =$$

$$\begin{pmatrix} d^2 & d\sqrt{x_1}e^{-i\phi_1} & d\sqrt{x_2}e^{-i\phi_2} & d\sqrt{x_3}e^{-i\phi_3} \\ d\sqrt{x_1}e^{i\phi_1} & x_1 & \sqrt{x_1}\sqrt{x_2}e^{i(\phi_1-\phi_2)} & \sqrt{x_1}\sqrt{x_3}e^{i(\phi_1-\phi_3)} \\ d\sqrt{x_2}e^{i\phi_2} & \sqrt{x_1}\sqrt{x_2}e^{i(\phi_2-\phi_1)} & x_2 & \sqrt{x_2}\sqrt{x_3}e^{i(\phi_2-\phi_3)} \\ d\sqrt{x_3}e^{i\phi_3} & \sqrt{x_3}\sqrt{x_1}e^{i(\phi_3-\phi_1)} & \sqrt{x_3}\sqrt{x_2}e^{i(\phi_3-\phi_2)} & x_3 \end{pmatrix}$$

where $d := (1 - x_1 - x_2 - x_3)^{1/2}$.

(iii) Since

$$\int_{\phi=0}^{2\pi} e^{i\phi} = 0$$

and

$$\int_{\phi_3=0}^{2\pi} \int_{\phi_2=0}^{2\pi} \int_{\phi_1=0}^{2\pi} d\phi_3 d\phi_2 d\phi_1 = (2\pi)^3$$

$$\int_{x_3=0}^{1} \int_{x_2=0}^{1-x_3} \int_{x_1=0}^{1-x_2-x_3} dx_3 dx_2 dx_1 = \frac{1}{6}$$

$$\int_{x_3=0}^{1} \int_{x_2=0}^{1-x_3} \int_{x_1=0}^{1-x_2-x_3} dx_3 dx_2 dx_1 x_1 = \frac{1}{24}$$

$$\int_{x_3=0}^{1} \int_{x_2=0}^{1-x_3} \int_{x_1=0}^{1-x_2-x_3} dx_3 dx_2 dx_1 x_2 = \frac{1}{24}$$

$$\int_{x_3=0}^{1} \int_{x_2=0}^{1-x_3} \int_{x_1=0}^{1-x_2-x_3} dx_3 dx_2 dx_1 x_3 = \frac{1}{24}$$

we find (1).

(iv) Let $|0\rangle_4$, $|1\rangle_4$, $|2\rangle_4$, $|3\rangle_4$ be the standard basis in $\mathbb{C}^4$ and $|0\rangle_2$, $|1\rangle_2$ be the standard basis in $\mathbb{C}^2$. Then we can write $|0\rangle_4 = |0\rangle_2 \otimes |0\rangle_2$, $|1\rangle_4 = |0\rangle_2 \otimes |1\rangle_2$, $|2\rangle_4 = |1\rangle_2 \otimes |0\rangle_2$ and $|3\rangle_4 = |1\rangle_2 \otimes |1\rangle_2$ with the coefficients

$$c_{00} = (1 - x_1 - x_2 - x_3)^{1/2}, \qquad c_{01} = \sqrt{x_1}e^{i\phi_1},$$

$$c_{10} = \sqrt{x_2}e^{i\phi_2}, \qquad c_{11} = \sqrt{x_3}e^{i\phi_3}$$

which leads to the 2×2 matrix

$$C = \begin{pmatrix} (1 - x_1 - x_2 - x_3)^{1/2} & \sqrt{x_1}e^{i\phi_1} \\ \sqrt{x_2}e^{i\phi_2} & \sqrt{x_3}e^{i\phi_3} \end{pmatrix}.$$

The reduced density matrix is

$$CC^\dagger = \begin{pmatrix} 1 - x_2 - x_3 & d\sqrt{x_2}e^{-i\phi_2} + \sqrt{x_1}\sqrt{x_3}e^{i\phi_1}e^{-i\phi_3} \\ d\sqrt{x_2}e^{i\phi_2} + \sqrt{x_1}\sqrt{x_3}e^{-i\phi_1}e^{i\phi_3} & x_2 + x_3 \end{pmatrix}$$

where $d := (1 - x_1 - x_2 - x_3)^{1/2}$. We obtain

$$\det(CC^\dagger) = x_3 d^2 + x_1 x_2 - 2\sqrt{x_1 x_2 x_3}d\cos(\phi_1 + \phi_2 - \phi_3).$$

The state $|\beta\rangle$ is not entangled if $\det(C^\dagger C) = 0$.

Problem 16. Consider the pure state

$$|\psi\rangle := \alpha|00\rangle + \beta|11\rangle$$

in the Hilbert space $\mathbb{C}^2 \otimes \mathbb{C}^2$, where $\alpha, \beta \in \mathbb{C}$ and $|\alpha|^2 + |\beta^2| = 1$. Let $\rho := |\psi\rangle\langle\psi|$ be the corresponding density matrix.
(i) Find $-\mathrm{tr}\rho_1 \log_2 \rho_1$ where $\rho_1 := \mathrm{tr}_{\mathbb{C}^2}\rho$.
(ii) Let $\widetilde{\rho}$ be a density matrix for a disentangled state on $\mathbb{C}^2 \otimes \mathbb{C}^2$. Find the fidelity (also called *Uhlmann's transition probability*)

$$\mathcal{F}(\rho, \widetilde{\rho}) := \left[\mathrm{tr}\sqrt{\sqrt{\rho}\widetilde{\rho}\sqrt{\rho}} \right]^2.$$

(iii) Show that the minimum over $\widetilde{\rho}$ of the *modified Bures metric*

$$D_B(\rho, \widetilde{\rho}) := 2 - 2\mathcal{F}(\rho, \widetilde{\rho})$$

is given by $4|\alpha|^2(1 - |\alpha|^2)$ at

$$\sigma := |\alpha|^2|00\rangle\langle00| + |\beta|^2|11\rangle\langle11|.$$

The *Bures metric* is defined as

$$D_{Bures}(\rho, \widetilde{\rho}) := 2 - 2\sqrt{\mathcal{F}(\rho, \widetilde{\rho})}.$$

(iv) Compare the result in (iii) with the result from (i).

Solution 16. (i) We find that

$$\rho = |\alpha|^2|00\rangle\langle00| + |\beta|^2|11\rangle\langle11| + \alpha\overline{\beta}|00\rangle\langle11| + \beta\overline{\alpha}|11\rangle\langle00|.$$

Taking the partial trace over the first qubit in $\mathbb{C}^2$ yields

$$\rho_1 = \mathrm{tr}_{\mathbb{C}^2}\rho = |\alpha|^2|0\rangle\langle0| + |\beta|^2|1\rangle\langle1|.$$

Thus

$$-\mathrm{tr}\rho_1 \log_2 \rho_1 = -|\alpha|^2 \log_2 |\alpha|^2 - (1 - |\alpha|^2) \log_2(1 - |\alpha|^2).$$

(ii) Since ρ is a pure state we have $\sqrt{\rho} = \rho$ and

$$
\begin{aligned}
\mathcal{F}(\rho, \widetilde{\rho}) &= \left[\mathrm{tr}\sqrt{\sqrt{\rho}\widetilde{\rho}\sqrt{\rho}}\right]^2 = \left[\mathrm{tr}\sqrt{\rho\widetilde{\rho}\rho}\right]^2 \\
&= \left[\mathrm{tr}\sqrt{|\psi\rangle\langle\psi|\widetilde{\rho}|\psi\rangle\langle\psi|}\right]^2 = \left[\mathrm{tr}\sqrt{\langle\psi|\widetilde{\rho}|\psi\rangle|\psi\rangle\langle\psi|}\right]^2 \\
&= |\langle\psi|\widetilde{\rho}|\psi\rangle| (\mathrm{tr}\rho)^2 = |\langle\psi|\widetilde{\rho}|\psi\rangle|.
\end{aligned}
$$

(iii) From (ii) we have

$$D_B(\rho, \sigma) = 2 - 2\mathcal{F}(\rho, \sigma) = 2 - 2|\langle\psi|\sigma|\psi\rangle|.$$

For

$$\sigma = |\alpha|^2|00\rangle\langle00| + |\beta|^2|11\rangle\langle11|$$

we find

$$D_B(\rho, \sigma) = 2 - 2(|\alpha^4| + |\beta|^4) = 2 - 2(|\alpha^4| + (1 - |\alpha|^2)^2) = 4|\alpha|^2(1 - |\alpha|^2).$$

Obviously σ is not entangled. For $|\alpha|^2 = 0$ or $|\alpha|^2 = 1$ it is immediately clear that we have a minimum. Thus consider $0 < |\alpha|^2 < 1$. Now let ν be any fixed density matrix in $\mathbb{C}^4$ and $\lambda \in [0, 1]$. Thus the convex function

$$\sigma(\lambda) := \lambda\sigma + (1 - \lambda)\nu$$

is also a density matrix. It follows that

$$
\begin{aligned}
\frac{d}{d\lambda}D_B(\rho, \sigma(\lambda))\Big|_{\lambda=1} &= -2\frac{d}{d\lambda}\left|\lambda\langle\psi|\sigma|\psi\rangle + (1 - \lambda)\langle\psi|\nu|\psi\rangle\right|\Big|_{\lambda=1} \\
&= -2\frac{d}{d\lambda}\left|\lambda(|\alpha|^4 + |\beta|^4) + (1 - \lambda)\langle\psi|\nu|\psi\rangle\right|\Big|_{\lambda=1} \\
&= \begin{cases} -2(|\alpha|^4 + |\beta|^4 - \langle\psi|\nu|\psi\rangle) & |\alpha|^4 + |\beta|^4 \geq 0 \\ +2(|\alpha|^4 + |\beta|^4 - \langle\psi|\nu|\psi\rangle) & |\alpha|^4 + |\beta|^4 < 0 \end{cases} \\
&= -2(|\alpha|^4 + |\beta|^4 - \langle\psi|\nu|\psi\rangle) \\
&= -2((|\alpha|^2 + |\beta|^2)^2 - 2|\alpha|^2|\beta|^2 - \langle\psi|\nu|\psi\rangle) \\
&= -2(-2|\alpha|^2|\beta|^2 + 1 - \langle\psi|\nu|\psi\rangle)
\end{aligned}
$$

where we used that $\langle\psi|\nu|\psi\rangle$ is real. If ν is sufficiently close to $\rho = |\psi\rangle\langle\psi|$ then

$$1 - \langle\psi|\nu|\psi\rangle < 2|\alpha|^2|\beta|^2$$

and $D_B(\rho, \sigma(\lambda))$ is increasing around σ. Thus we have found the minimum $4|\alpha|^2(1 - |\alpha|^2)$.

(iv) For $|\alpha| \in [0, 1]$ we find

$$4|\alpha|^2(1 - |\alpha|^2) \leq -|\alpha|^2 \log_2|\alpha|^2 - (1 - |\alpha|^2)\log_2(1 - |\alpha|^2).$$

Problem 17.   The two-point *Hubbard model* with cyclic boundary conditions is given by

$$\hat{H} = t(c_{1\uparrow}^\dagger c_{2\uparrow} + c_{1\downarrow}^\dagger c_{2\downarrow} + c_{2\uparrow}^\dagger c_{1\uparrow} + c_{2\downarrow}^\dagger c_{1\downarrow}) + U(n_{1\uparrow}n_{1\downarrow} + n_{2\uparrow}n_{2\downarrow})$$

where

$$n_{j\uparrow} := c_{j\uparrow}^\dagger c_{j\uparrow}, \quad n_{j\downarrow} := c_{j\downarrow}^\dagger c_{j\downarrow}, \qquad j = 1, 2.$$

The *Fermi operators* $c_{j\uparrow}^\dagger, c_{j\downarrow}^\dagger, c_{j\uparrow}, c_{j\downarrow}$ obey the *anti-commutation relations*

$$[c_{j,\sigma}^\dagger, c_{k,\sigma'}]_+ = \delta_{\sigma\sigma'}\delta_{jk}I, \quad [c_{j,\sigma}^\dagger, c_{k,\sigma'}^\dagger]_+ = [c_{j,\sigma}, c_{k,\sigma'}]_+ = 0.$$

$\hat{H}$ commutes with the total number operator $\hat{N}$, and the total spin operator $\hat{S}_z$ in the z direction

$$\hat{N} := \sum_{j=1}^{2}(c_{j\uparrow}^\dagger c_{j\uparrow} + c_{j\downarrow}^\dagger c_{j\downarrow}), \qquad \hat{S}_z := \frac{1}{2}\sum_{j=1}^{2}(c_{j\uparrow}^\dagger c_{j\uparrow} - c_{j\downarrow}^\dagger c_{j\downarrow}).$$

We consider the subspace with two electrons, $N = 2$ and $S_z = 0$. A basis for 2 particles with total spin 0 is

$$|s_1\rangle := c_{1\uparrow}^\dagger c_{1\downarrow}^\dagger|0\rangle, \quad |s_2\rangle := c_{1\uparrow}^\dagger c_{2\downarrow}^\dagger|0\rangle, \quad |s_3\rangle := c_{2\uparrow}^\dagger c_{1\downarrow}^\dagger|0\rangle, \quad |s_4\rangle := c_{2\uparrow}^\dagger c_{2\downarrow}^\dagger|0\rangle$$

where $\langle 0|0\rangle = 1$.
(i) Find the matrix representation of $\hat{H}$ in this basis.
(ii) Can the matrix representation of $\hat{H}$ be written in the form $\hat{H} = A_1 \otimes I_2 + I_2 \otimes A_2$ where A_1 and A_2 are 2×2 matrices and I_2 is the 2×2 identity matrix?

Solution 17.   (i) Applying $\hat{H}$ to the basis gives

$$\hat{H}|s_1\rangle = t|s_2\rangle + t|s_3\rangle + U|s_1\rangle$$
$$\hat{H}|s_2\rangle = t|s_1\rangle + t|s_4\rangle$$
$$\hat{H}|s_3\rangle = t|s_1\rangle + t|s_4\rangle$$
$$\hat{H}|s_4\rangle = t|s_2\rangle + t|s_3\rangle + U|s_4\rangle.$$

Identifying $|s_i\rangle$ with elements $\mathbf{e}_i$ of the standard basis in $\mathbb{C}^4$ yields the matrix representation of $\hat{H}$

$$\hat{H} = \begin{pmatrix} U & t & t & 0 \\ t & 0 & 0 & t \\ t & 0 & 0 & t \\ 0 & t & t & U \end{pmatrix}.$$

(ii) Suppose a Hamilton operator $\hat{K}$ can be written as $\hat{K} = A_1 \otimes I_2 + I_2 \otimes A_2$ where $A_1, A_2 \in M^2$ and I_2 is the 2×2 identity matrix. Then we have

$$\exp(-i\hat{K}\tau/\hbar) = \exp(-i\tau A_1/\hbar \otimes I_2 - i\tau I_2/\hbar \otimes A_2)$$
$$= \exp(-i\tau A_1/\hbar) \otimes \exp(-i\tau A_2/\hbar).$$

In this case separable states remain separable under time evolution in the model, and entangled states remain entangled under time evolution in the model. For the matrix representation of $\hat{H}$, however we have

$$\hat{H} = tV_{NOT} \otimes I_2 + tI_2 \otimes V_{NOT} + \text{diag}(U,0,0,U), \qquad V_{NOT} := \begin{pmatrix} 0 & 1 \\ 1 & 0 \end{pmatrix}.$$

The diagonal matrix $\text{diag}(U,0,0,U)$ cannot be written in the form $A_1 \otimes I_2 + I_2 \otimes A_2$. Thus we conclude that almost all initial separable states evolve into entangled states under the time evolution of the model.

Problem 18. Find the matrix representation of the two-point Hubbard model in the basis

$$\left\{ \frac{1}{\sqrt{2}}(c_{1\downarrow}^\dagger c_{1\uparrow}^\dagger|0\rangle + c_{2\downarrow}^\dagger c_{2\uparrow}^\dagger|0\rangle), \frac{1}{\sqrt{2}}(c_{1\downarrow}^\dagger c_{2\uparrow}^\dagger|0\rangle + c_{2\downarrow}^\dagger c_{1\uparrow}^\dagger|0\rangle), \right.$$

$$\left. \frac{1}{\sqrt{2}}(c_{1\downarrow}^\dagger c_{1\uparrow}^\dagger|0\rangle - c_{2\downarrow}^\dagger c_{2\uparrow}^\dagger|0\rangle), \frac{1}{\sqrt{2}}(c_{1\downarrow}^\dagger c_{2\uparrow}^\dagger|0\rangle - c_{2\downarrow}^\dagger c_{1\uparrow}^\dagger|0\rangle) \right\}.$$

Solution 18. The two-point Hubbard model admits a discrete symmetry under the change $1 \to 2$, $2 \to 1$. Thus we have a finite group with two elements. We obtain two irreducible representations. The group-theoretical reduction leads to the two invariant subspaces

$$\left\{ \frac{1}{\sqrt{2}}(c_{1\downarrow}^\dagger c_{1\uparrow}^\dagger|0\rangle + c_{2\downarrow}^\dagger c_{2\uparrow}^\dagger|0\rangle), \frac{1}{\sqrt{2}}(c_{1\downarrow}^\dagger c_{2\uparrow}^\dagger|0\rangle + c_{2\downarrow}^\dagger c_{1\uparrow}^\dagger|0\rangle) \right\},$$

$$\left\{ \frac{1}{\sqrt{2}}(c_{1\downarrow}^\dagger c_{1\uparrow}^\dagger|0\rangle - c_{2\downarrow}^\dagger c_{2\uparrow}^\dagger|0\rangle), \frac{1}{\sqrt{2}}(c_{1\downarrow}^\dagger c_{2\uparrow}^\dagger|0\rangle - c_{2\downarrow}^\dagger c_{1\uparrow}^\dagger|0\rangle) \right\}.$$

These four states can be considered as the Bell states. In the Bell basis the matrix representation of the Hubbard model is given by

$$
\begin{pmatrix} U & 2t & 0 & 0 \\ 2t & 0 & 0 & 0 \\ 0 & 0 & 0 & 0 \\ 0 & 0 & 0 & U \end{pmatrix} = \begin{pmatrix} U & 2t \\ 2t & 0 \end{pmatrix} \oplus \begin{pmatrix} 0 & 0 \\ 0 & U \end{pmatrix}
$$

where $\oplus$ denotes the direct sum.

Problem 19. The two-point Hubbard model with cyclic boundary conditions is given by

$$
\hat{H} = t(c_{1\uparrow}^\dagger c_{2\uparrow} + c_{1\downarrow}^\dagger c_{2\downarrow} + c_{2\uparrow}^\dagger c_{1\uparrow} + c_{2\downarrow}^\dagger c_{1\downarrow}) + U(n_{1\uparrow}n_{1\downarrow} + n_{2\uparrow}n_{2\downarrow}).
$$

Find the time evolution of the initial state

$$
|\psi(0)\rangle = \frac{1}{\sqrt{2}}(c_{1\uparrow}^\dagger c_{1\downarrow}^\dagger - c_{2\uparrow}^\dagger c_{2\downarrow}^\dagger)|0\rangle
$$

under the two-point Hubbard model. When is the state $|\psi(\tau)\rangle$ entangled?

Solution 19. Solving the Schrödinger equation

$$
i\hbar \frac{d}{d\tau}|\psi(\tau)\rangle = \hat{H}|\psi(\tau)\rangle
$$

we find

$$
|\psi(\tau)\rangle = e^{-iU\tau/\hbar}|\psi(0)\rangle.
$$

Consequently, the condition for separability is given by

$$
\exp\left(-2i\frac{\tau}{\hbar}U\right) = 0.
$$

Of course, this equation cannot be satisfied. Thus $|\psi(\tau)\rangle$ is entangled for all τ.

Problem 20. An arbitrary pure state in the Hilbert space $\mathbb{C}^4$ can be written as

$$
|\psi\rangle = \begin{pmatrix} \cos\theta_3 \\ \sin\theta_3 \cos\theta_2 e^{i\phi_3} \\ \sin\theta_3 \sin\theta_2 \cos\theta_1 e^{i\phi_2} \\ \sin\theta_3 \sin\theta_2 \sin\theta_1 e^{i\phi_1} \end{pmatrix}
$$

where $\theta_k \in [0, \pi/2]$, and $\phi_k \in [0, 2\pi)$ for $k = 1, 2, 3$.
(i) Find values for θ_k and ϕ_k $(k = 1, 2, 3)$ so that we obtain the unentangled state (product state)

$$
|\alpha\rangle = \begin{pmatrix} 1 \\ 0 \\ 0 \\ 0 \end{pmatrix} \equiv \begin{pmatrix} 1 \\ 0 \end{pmatrix} \otimes \begin{pmatrix} 1 \\ 0 \end{pmatrix}.
$$

(ii) Find values for θ_k and ϕ_k ($k = 1, 2, 3$) so that we obtain the entangled (Bell) state

$$|\beta\rangle = \frac{1}{\sqrt{2}} \begin{pmatrix} 1 \\ 0 \\ 0 \\ 1 \end{pmatrix} .$$

Solution 20. (i) We obtain the unentangled state by setting $\theta_3 = 0$, $\theta_2 = 0$ and $\theta_1 = 0$.
(ii) We obtain the Bell state by setting $\theta_3 = \pi/4$, $\theta_2 = \pi/2$, $\theta_1 = \pi/2$ and $\phi_3 = \phi_2 = \phi_1 = 0$.

Problem 21. A completely entangled state $|\Psi^{AB}\rangle$ of an $(N \times N)$-dimensional Hilbert space $\mathcal{H} \otimes \mathcal{H}$ can be written as

$$|\Psi^{AB}\rangle = \frac{1}{\sqrt{N}} \sum_{k=0}^{N-1} |\psi_k^A\rangle \otimes |\psi_k^B\rangle$$

where $\{\, |\psi_k\rangle \, : \, k = 0, 1, \ldots, N-1 \,\}$ is a orthonormal basis of the Hilbert space $\mathcal{H}$. We define the linear operators (*unitary depolarizers*)

$$U_{jk} := \sum_{\ell=0}^{N-1} e^{(2\pi i/N)j\ell} |\psi_{\ell \bmod N}\rangle \langle \psi_{\ell+k \bmod N}|$$

where $j, k = 0, 1, \ldots, N-1$.
(i) Calculate

$$\frac{1}{N} \sum_{j=0}^{N-1} \sum_{k=0}^{N-1} U_{jk} X U_{jk}^{\dagger}, \qquad \frac{1}{N} \mathrm{tr}(U_{jk} U_{\ell m}^{\dagger})$$

where X is an arbitrary linear operator defined on the Hilbert space $\mathcal{H}$.
(ii) Discuss the set

$$\{\, |\Psi_{jk}^{AB}\rangle = (U_{jk}^A \otimes I_B)|\Psi^{AB}\rangle \, : \, j, k = 0, 1, \ldots, N-1 \,\}.$$

Solution 21. (i) We obtain

$$\frac{1}{N} \sum_{j=0}^{N-1} \sum_{k=0}^{N-1} U_{jk} X U_{jk}^{\dagger} = (\mathrm{tr} X) I$$

where I is the identity operator in $\mathcal{H}$. We obtain

$$\frac{1}{N} \mathrm{tr}(U_{jk} U_{\ell m}^{\dagger}) = \delta_{j\ell} \delta_{km} .$$

(ii) The set is an orthonormal basis of the Hilbert space $\mathcal{H} \otimes \mathcal{H}$.

Problem 22.   Consider the *Greenberger-Horne-Zeilinger state*

$$|GHZ\rangle = \frac{1}{\sqrt{2}}(|0\rangle \otimes |0\rangle \otimes |0\rangle + |1\rangle \otimes |1\rangle \otimes |1\rangle)$$

and the *W state*

$$|W\rangle = \frac{1}{\sqrt{3}}(|0\rangle \otimes |0\rangle \otimes |1\rangle + |0\rangle \otimes |1\rangle \otimes |0\rangle + |1\rangle \otimes |0\rangle \otimes |0\rangle).$$

(i) Calculate the states

$$(\sqrt{2}\langle 0| \otimes I_2 \otimes I_2)|GHZ\rangle, \quad (\sqrt{2}I_2 \otimes \langle 0| \otimes I_2)|GHZ\rangle$$

$$(\sqrt{2}I_2 \otimes I_2 \otimes \langle 0|)|GHZ\rangle, \quad (\sqrt{2}\langle 1| \otimes I_2 \otimes I_2)|GHZ\rangle$$

$$(\sqrt{2}I_2 \otimes \langle 1| \otimes I_2)|GHZ\rangle, \quad (\sqrt{2}I_2 \otimes I_2 \otimes \langle 1|)|GHZ\rangle$$

and discuss.
(iii) Calculate

$$\left(\frac{\sqrt{3}}{\sqrt{2}}\langle 0| \otimes I_2 \otimes I_2\right)|W\rangle, \quad \left(\frac{\sqrt{3}}{\sqrt{2}}I_2 \otimes \langle 0| \otimes I_2\right)|W\rangle,$$

$$\left(\frac{\sqrt{3}}{\sqrt{2}}I_2 \otimes I_2 \otimes \langle 0|\right)|W\rangle, \quad (\sqrt{3}\langle 1| \otimes I_2 \otimes I_2)|W\rangle,$$

$$(\sqrt{3}I_2 \otimes \langle 1| \otimes I_2)|W\rangle, \quad (\sqrt{3}I_2 \otimes I_2 \otimes \langle 1|)|W\rangle$$

and discuss.

Solution 22.   (i) We find

$$(\sqrt{2}\langle 0| \otimes I_2 \otimes I_2)|GHZ\rangle = |0\rangle \otimes |0\rangle, \quad (\sqrt{2}I_2 \otimes \langle 0| \otimes I_2)|GHZ\rangle = |0\rangle \otimes |0\rangle$$

$$(\sqrt{2}I_2 \otimes I_2 \otimes \langle 0||GHZ\rangle = |0\rangle \otimes |0\rangle, \quad (\sqrt{2}\langle 1| \otimes I_2 \otimes I_2)|GHZ\rangle = |1\rangle \otimes |1\rangle$$

$$(\sqrt{2}I_2 \otimes \langle 1| \otimes I_2)|GHZ\rangle = |1\rangle \otimes |1\rangle, \quad (\sqrt{2}I_2 \otimes I_2 \otimes \langle 1|)|GHZ\rangle = |1\rangle \otimes |1\rangle.$$

Thus all the two-particle states are not entangled after measurement of the third state, although the $|GHZ\rangle$ state is entangled. The $|GHZ\rangle$ state is usually referred to as maximally entangled in several senses, e.g., it violates Bell inequalities maximally. However, from the result above we find that the state is maximally fragile, i.e., if one particle is lost or projected onto the computational basis $\{ |0\rangle, |1\rangle \}$, then all entanglement is destroyed.

(ii) We find

$$\left(\frac{\sqrt{3}}{\sqrt{2}}\langle 0| \otimes I_2 \otimes I_2\right)|W\rangle = \frac{1}{\sqrt{2}}(|0\rangle \otimes |1\rangle + |1\rangle \otimes |0\rangle)$$

$$\left(\frac{\sqrt{3}}{\sqrt{2}}I_2 \otimes \langle 0| \otimes I_2\right)|W\rangle = \frac{1}{\sqrt{2}}(|0\rangle \otimes |1\rangle + |1\rangle \otimes |0\rangle)$$

$$\left(\frac{\sqrt{3}}{\sqrt{2}}I_2 \otimes I_2 \otimes \langle 0|\right)|W\rangle = \frac{1}{\sqrt{2}}(|0\rangle \otimes |1\rangle + |1\rangle \otimes |0\rangle).$$

Thus in this case the projected states are (maximally) entangled (Bell states). However, for the other cases we find

$$(\sqrt{3}\langle 1| \otimes I_2 \otimes I_2)|W\rangle = |0\rangle \otimes |0\rangle$$
$$(\sqrt{3}I_2 \otimes \langle 1| \otimes I_2)|W\rangle = |0\rangle \otimes |0\rangle$$
$$(\sqrt{3}I_2 \otimes I_2 \otimes \langle 1|)|W\rangle = |0\rangle \otimes |0\rangle.$$

Thus these states are not entangled.

Problem 23. Let σ_x, σ_y and σ_z be the Pauli spin matrices. Consider the Hamilton operator

$$\hat{H} = \frac{1}{2}\epsilon(\sigma_z \otimes I_2 + I_2 \otimes \sigma_z) - \Delta(\sigma_x \otimes \sigma_x)$$

where $\epsilon > 0$ and $\Delta > 0$. Find the eigenvalues and normalized eigenvectors of $\hat{H}$. Are the eigenvectors entangled?

Solution 23. Let

$$|\uparrow\rangle = \begin{pmatrix} 1 \\ 0 \end{pmatrix}, \qquad |\downarrow\rangle = \begin{pmatrix} 0 \\ 1 \end{pmatrix}$$

and $|\uparrow\uparrow\rangle = |\uparrow\rangle \otimes |\uparrow\rangle$ etc.. Then the eigenvalues and normalized eigenvectors are given by

$$E_0 = -\sqrt{\epsilon^2 + \Delta^2}, \qquad |0\rangle = \frac{1}{\sqrt{1+a^2}}(|\downarrow\downarrow\rangle + a|\uparrow\uparrow\rangle)$$

$$E_1 = -\Delta, \qquad |1\rangle = \frac{1}{\sqrt{2}}(|\uparrow\downarrow\rangle + |\downarrow\uparrow\rangle)$$

$$E_2 = +\Delta, \qquad |2\rangle = \frac{1}{\sqrt{2}}(-|\uparrow\downarrow\rangle + |\downarrow\uparrow\rangle)$$

$$E_3 = +\sqrt{\epsilon^2 + \Delta^2}, \qquad |3\rangle = \frac{1}{\sqrt{1+a^2}}(-a|\downarrow\downarrow\rangle + |\uparrow\uparrow\rangle)$$

where
$$a := \frac{\sqrt{\epsilon^2 + \Delta^2} - \epsilon}{\Delta}.$$

The eigenstates are entangled.

Problem 24. Consider the Hilbert space $\mathcal{H} = \mathbb{C}^n$ and the product space $\mathcal{H} \otimes \mathcal{H}$. Let A be an arbitrary $n \times n$ matrix over $\mathbb{C}$ and I_n the $n \times n$ identity matrix. Consider the following definition. A normalized vector $\psi \in \mathcal{H} \otimes \mathcal{H}$ is called *maximally entangled*, if its reduced density matrix is maximally mixed, i.e., a multiple of I_n

$$\langle \psi | (A \otimes I_n) | \psi \rangle = \dim(\mathcal{H})^{-1} \mathrm{tr}(A). \tag{1}$$

(i) Show that the *Bell states* in $\mathbb{C}^4$

$$\Phi^+ = \frac{1}{\sqrt{2}} \begin{pmatrix} 1 \\ 0 \\ 0 \\ 1 \end{pmatrix}, \quad \Phi^- = \frac{1}{\sqrt{2}} \begin{pmatrix} 1 \\ 0 \\ 0 \\ -1 \end{pmatrix}, \quad \Psi^+ = \frac{1}{\sqrt{2}} \begin{pmatrix} 0 \\ 1 \\ 1 \\ 0 \end{pmatrix}, \quad \Psi^- = \frac{1}{\sqrt{2}} \begin{pmatrix} 0 \\ 1 \\ -1 \\ 0 \end{pmatrix}$$

satisfy equation (1), where $\mathcal{H} = \mathbb{C}^2$.
(ii) Calculate the left and right-hand side of equation (1) for the vector $(1\ 0\ 0\ 0)$. Discuss.

Solution 24. (i) We have $\dim(\mathcal{H}) = n = 2$ and $\mathrm{tr}A = a_{11} + a_{22}$. For the left-hand side we have

$$(A \otimes I_n) = \begin{pmatrix} a_{11} & 0 & a_{12} & 0 \\ 0 & a_{11} & 0 & a_{12} \\ a_{21} & 0 & a_{22} & 0 \\ 0 & a_{21} & 0 & a_{22} \end{pmatrix}.$$

Thus for the Bell state Ψ^+ we have

$$\frac{1}{2}(1\ 0\ 0\ 1) \begin{pmatrix} a_{11} & 0 & a_{12} & 0 \\ 0 & a_{11} & 0 & a_{12} \\ a_{21} & 0 & a_{22} & 0 \\ 0 & a_{21} & 0 & a_{22} \end{pmatrix} \begin{pmatrix} 1 \\ 0 \\ 0 \\ 1 \end{pmatrix} = \frac{1}{2}(a_{11} + a_{22}).$$

Analogously we prove for the other Bell states that equation (1) is satisfied.
(ii) For the vector $(1\ 0\ 0\ 0)$ we obtain for the left hand side a_{11}. Thus equation (1) is not satisfied. This state is not entangled.

Problem 25. Let $|H\rangle$ $(|V\rangle)$ indicate the state of a horizontal (vertical) polarized photon. Suppose we have the product state

$$|\psi\rangle_{1234} = \frac{1}{\sqrt{2}}(|H\rangle_1 \otimes |V\rangle_2 - |V\rangle_1 \otimes |H\rangle_2) \otimes \frac{1}{\sqrt{2}}(|H\rangle_3 \otimes |V\rangle_4 - |V\rangle_3 \otimes |H\rangle_4)$$

Thus we have a product state of two polarization entangled pairs. One photon out of each pair (2 and 3) is directed to the two inputs of a *polarizing beam splitter*. What is the output?

Solution 25. Since the polarizing beam splitter transmits horizontally polarized photons and reflects vertically polarized photons, coincidence detection between the two polarizing beam splitter outputs implies that either both photons 2 and 3 are both horizontally polarized or both vertically polarized. Thus the state (1) is projected onto a two-dimensional subspace spanned by

$$|V\rangle_1 \otimes |H\rangle_2 \otimes |H\rangle_3 \otimes |V\rangle_4, \qquad |H\rangle_1 \otimes |V\rangle_2 \otimes |V\rangle_3 \otimes |H\rangle_4.$$

After the polarizing beam splitter, the renormalized state corresponding to a fourfold coincidence is

$$|\phi\rangle_{12'3'4} = \frac{1}{\sqrt{2}}(|H\rangle_1 \otimes |V\rangle_{2'} \otimes |V\rangle_{3'} \otimes |H\rangle_4 + |V\rangle_1 \otimes |H\rangle_{2'} \otimes |H\rangle_{3'} \otimes |V\rangle_4).$$

This is a *GHZ state* of four particles, which can exhibit nonlocal behaviour according to the GHZ theorem.

Problem 26. An *entanglement witness* W on the product Hilbert space $\mathcal{H}_A \otimes \mathcal{H}_B$ is a linear operator on $\mathcal{H}_A \otimes \mathcal{H}_B$ such that W is not positive semi-definite and

$$\langle \psi | W | \psi \rangle \geq 0$$

for all separable $|\psi\rangle = |\psi_A\rangle \otimes |\psi_B\rangle$, where $|\psi_A\rangle \in \mathcal{H}_A$ and $|\psi_B\rangle \in \mathcal{H}_B$. Show that

$$W = I_4 - 2|\Phi^+\rangle\langle\Phi^+| = \begin{pmatrix} 0 & 0 & 0 & -1 \\ 0 & 1 & 0 & 0 \\ 0 & 0 & 1 & 0 \\ -1 & 0 & 0 & 0 \end{pmatrix}$$

is an entanglement witness.

Solution 26. The eigenvalues of W are -1 and 1 with multiplicity 3, with corresponding eigenvectors given by the Bell basis. Clearly W is not positive semi-definite. Using $|\psi\rangle = |\psi_A\rangle \otimes |\psi_B\rangle$, with $|\psi_A\rangle, |\psi_B\rangle \in \mathbb{C}^2$, we find

$$\begin{aligned}
\langle\psi|W|\psi\rangle &= \langle\psi_A|\psi_A\rangle\langle\psi_B|\psi_B\rangle - |\langle\psi_A|0\rangle\langle\psi_B|0\rangle + \langle\psi_A|1\rangle\langle\psi_B|1\rangle|^2 \\
&= \left(|\langle\psi_A|0\rangle|^2 + |\langle\psi_A|1\rangle|^2\right)\left(|\langle\psi_B|0\rangle|^2 + |\langle\psi_B|1\rangle|^2\right) \\
&\quad - |\langle\psi_A|0\rangle\langle\psi_B|0\rangle + \langle\psi_A|1\rangle\langle\psi_B|1\rangle|^2 \\
&\geq \left(|\langle\psi_A|0\rangle|^2 + |\langle\psi_A|1\rangle|^2\right)\left(|\langle\psi_B|0\rangle|^2 + |\langle\psi_B|1\rangle|^2\right)
\end{aligned}$$

$$- \left(|\langle \psi_A | 0 \rangle \langle \psi_B | 0 \rangle| + |\langle \psi_A | 1 \rangle \langle \psi_B | 1 \rangle| \right)^2$$
$$= \left(|\langle \psi_A | 0 \rangle \langle \psi_B | 1 \rangle| - |\langle \psi_A | 1 \rangle \langle \psi_B | 0 \rangle| \right)^2 \geq 0.$$

Thus the operator W is an entanglement witness. Furthermore,

$$\operatorname{tr}(W | \Phi^+ \rangle \langle \Phi^+ |) = \langle \Phi^+ | W | \Phi^+ \rangle = -1.$$

Consequently W is a witness for the entangled Bell state $|\Phi^+\rangle$.

Problem 27. Let L denote the space of linear operators on a Hilbert space $\mathcal{H}$. A linear map $\epsilon : L \to L$ is called a *positive map* if for all positive semidefinite $\rho \in L$ the operator $\epsilon(\rho)$ is also positive semidefinite. As an example we consider the transpose operation. Let $B = \{|1\rangle, |2\rangle, \ldots, |\dim(\mathcal{H})\rangle\}$ denote an orthonormal basis for the Hilbert space $\mathcal{H}$. The *transpose* of a linear operator

$$A := \sum_{j=1}^{\dim(\mathcal{H})} \sum_{k=1}^{\dim(\mathcal{H})} a_{jk} |j\rangle \langle k|$$

is given by

$$\epsilon_{T(B)}(A) := \sum_{j=1}^{\dim(\mathcal{H})} \sum_{k=1}^{\dim(\mathcal{H})} a_{jk} |k\rangle \langle j|.$$

The transpose operation $\epsilon_{T(B)}$ is a positive map.

Let L_A denote the space of linear operators on a Hilbert space $\mathcal{H}_A$, L_B the space of linear operators on the Hilbert space $\mathcal{H}_B$ and L_{AB} the space of linear operators on the Hilbert space $\mathcal{H}_A \otimes \mathcal{H}_B$. The extension $\epsilon \otimes I$ of the linear map $\epsilon : L_A \to L_A$ is defined by

$$(\epsilon \otimes I) \left(\sum_{k=1}^{\dim(L_A)} A_k \otimes B_k \right) = \sum_{k=1}^{\dim(L_A)} \epsilon(A_k) \otimes B_k$$

where $A_k \in L_A$ and $B_k \in L_B$ for $k = 1, 2, \ldots, \dim(L_A)$. Similarly the extension $I \otimes \mu$ of $\mu : L_B \to L_B$ can be defined. A positive map $\epsilon : L \to L$ is a *completely positive map* if all possible extensions ($\epsilon \otimes I$ or $I \otimes \epsilon$) of the map to arbitrary Hilbert spaces are positive. Every completely positive map can be written in the form

$$\epsilon(\rho) = \sum_k A_k \rho A_k^*$$

where A_k is a linear operator on the Hilbert space. Furthermore, if

$$\sum_k A_k A_k^* = I$$

then ϵ is trace preserving, i.e. $\mathrm{tr}(\epsilon(\rho)) = \mathrm{tr}(\rho)$. Is the transpose operation completely positive?

Solution 27. *Partial transposition* $\epsilon_{T(B)} \otimes I$ in the basis B of the Hilbert space $\mathcal{H}_A$ of a state ρ in the Hilbert space $\mathcal{H}_A \otimes \mathcal{H}_B$ is not completely positive. Consider the Bell state $|\Phi^+\rangle$ in the Hilbert state $\mathbb{C}^2 \otimes \mathbb{C}^2$. Then

$$|\Phi^+\rangle\langle\Phi^+| = \frac{1}{2}\begin{pmatrix} 1 & 0 & 0 & 1 \\ 0 & 0 & 0 & 0 \\ 0 & 0 & 0 & 0 \\ 1 & 0 & 0 & 1 \end{pmatrix}.$$

Thus we have

$$(\epsilon_{T(\{|0\rangle,|1\rangle\})} \otimes I)(|\Phi^+\rangle\langle\Phi^+|) = \frac{1}{2}\begin{pmatrix} 1 & 0 & 0 & 0 \\ 0 & 0 & 1 & 0 \\ 0 & 1 & 0 & 0 \\ 0 & 0 & 0 & 1 \end{pmatrix}.$$

This last matrix has as eigenvalues 1 and -1. Consequently it is not positive semidefinite.

Problem 28. Consider the distillation under 1-local operations of a pure state in the Hilbert space $\mathbb{C}^2$

$$|\psi\rangle := \alpha|00\rangle + \beta|11\rangle$$

with $|\alpha|^2 + |\beta|^2 = 1$. If this state can be distilled, it will be transformed into the Bell state

$$|\Phi^+\rangle = \frac{1}{\sqrt{2}}|00\rangle + \frac{1}{\sqrt{2}}|11\rangle.$$

Find a generalized measurement on the first qubit which can be used to distill $|\Phi^+\rangle$ from $|\psi\rangle$.

Solution 28. We could attempt to construct a separable unitary transformation to perform this transform, however one component of a generalized measurement on the first qubit will achieve this for $k \in \mathbb{C}$, $k \neq 0$

$$U_0 := \frac{k}{\alpha}|0\rangle\langle 0| + \frac{k}{\beta}|1\rangle\langle 1|$$

Problems and Solutions

which transforms the density operator $\rho := |\psi\rangle\langle\psi|$ to

$$\frac{(U_0 \otimes I)\rho(U_0 \otimes I)^*}{\text{tr}(U_0 \otimes I)\rho(U_0 \otimes I)^*} = |\Phi^+\rangle\langle\Phi^+|$$

with probability

$$p_0 := \text{tr}(U_0 \otimes I)\rho(U_0 \otimes I)^* = 2|k|^2.$$

Of course this is only possible for $|\alpha^2|, |\beta|^2 \neq 0$, i.e. for $\alpha = 0$ or $\beta = 0$ there is no distillable entanglement. To complete the generalized measurement we must find U_1 such that

$$U_0^* U_0 + U_1^* U_1 = I.$$

Since $U_0^* U_0$ is clearly positive semidefinite we can use the polar decomposition $U_1 = U_1 H_1$, where U_1 is unitary and H_1 is positive semidefinite. We obtain

$$U_1 = U_1 \sqrt{I - U_0^* U_0} = U_1 \left(\sqrt{1 - \left|\frac{k}{\alpha}\right|^2} |0\rangle\langle 0| + \sqrt{1 - \left|\frac{k}{\beta}\right|^2} |1\rangle\langle 1| \right).$$

Consequently it is necessary that $k \leq \min\{|\alpha|^2, |\beta|^2\}$.

Problem 29. Consider the GHZ state (Greenberger-Horne-Zeilinger)

$$|\psi\rangle := \frac{1}{\sqrt{2}}(|001\rangle + |110\rangle).$$

We consider the basis $\alpha = \{|L\rangle, |R\rangle\}$ and the basis $\beta = \{|H\rangle, |V\rangle\}$ described by

$$\alpha: \qquad |0\rangle = \frac{1}{\sqrt{2}}(|L\rangle + |R\rangle), \quad |1\rangle = \frac{1}{\sqrt{2}}(|L\rangle - |R\rangle),$$

$$\beta: \qquad |0\rangle = \frac{1}{\sqrt{2}}(|H\rangle + |V\rangle), \quad |1\rangle = \frac{i}{\sqrt{2}}(|H\rangle - |V\rangle).$$

(i) Express $|\psi\rangle$ in terms of the bases α and β, where one qubit is in the basis α and the remaining two in the basis β.
(ii) Use (i) to predict the measurement outcomes where all qubits are measured in the basis α. Compare with the actual outcomes.

Solution 29. (i) Expressing only one qubit of $|\psi\rangle$ in the basis α and the rest in the basis β basis yields

$$\begin{aligned}
|\psi\rangle &= \tfrac{i}{2}(|LHH\rangle + |RVH\rangle - |LVV\rangle - |RHV\rangle) & \alpha\beta\beta \\
&= \tfrac{i}{2}(|HLH\rangle + |VRH\rangle - |VLV\rangle - |HRV\rangle) & \beta\alpha\beta \\
&= \tfrac{1}{2}(|HVL\rangle + |VHL\rangle - |HHR\rangle - |VVR\rangle) & \beta\beta\alpha
\end{aligned}$$

Measuring two qubits in the β basis as given in the previous equations allows us to deduce the result of measuring the other qubit in the α basis. Let β_j denote the result after measuring qubit j in the β basis. As an example, from the first equality, if $\beta_2 = \beta_3$ then the first qubit is $|L\rangle$, and $|R\rangle$ when $\beta_2 \neq \beta_3$. Thus we construct the following table.

Outcomes in β basis	Outcomes in α basis
$\beta_1 = \beta_2 = \beta_3$	LLR
$\beta_1 = \beta_2 \neq \beta_3$	RRR
$\beta_1 = \beta_3 \neq \beta_2$	RLL
$\beta_2 = \beta_3 \neq \beta_1$	LRL

(ii) We find in the α basis

$$|\psi\rangle = \tfrac{1}{2}(|LLL\rangle + |RRL\rangle - |LRR\rangle - |RLR\rangle).$$

None of the results obtained are consistent with the outcomes in the table deduced in (i).

Problem 30. Consider the state

$$|\psi\rangle = \frac{1}{2}(|0\rangle \otimes |0\rangle + e^{i\phi_1}|0\rangle \otimes |1\rangle + e^{i\phi_2}|1\rangle \otimes |0\rangle + e^{i\phi_3}|1\rangle \otimes |1\rangle).$$

(i) Let $\phi_3 = \phi_1 + \phi_2$. Is the state $|\psi\rangle$ a product state?
(ii) Let $\phi_3 = \phi_1 + \phi_2 + \pi$. Is the state $|\psi\rangle$ a product state?

Solution 30. (i) We have a product state, i.e.

$$\frac{1}{\sqrt{2}}(|0\rangle + e^{i\phi_1}|1\rangle) \otimes \frac{1}{\sqrt{2}}(|0\rangle + e^{i\phi_2}|1\rangle).$$

(ii) We do not have a product state we have a maximally entangled state.

Problem 31. Consider the state in $\mathbb{C}^8$

$$|\psi\rangle = \sum_{j_1,j_2,j_3=0}^{1} c_{j_1 j_2 j_3} |j_1\rangle \otimes |j_2\rangle \otimes |j_3\rangle$$

where

$$|0\rangle = \begin{pmatrix} 1 \\ 0 \end{pmatrix}, \qquad |1\rangle = \begin{pmatrix} 0 \\ 1 \end{pmatrix}.$$

Let R, S, T be 2×2 matrices over $\mathbb{C}$

$$R = \begin{pmatrix} r_{11} & r_{12} \\ r_{21} & r_{22} \end{pmatrix}, \qquad S = \begin{pmatrix} s_{11} & s_{12} \\ s_{21} & s_{22} \end{pmatrix}, \qquad T = \begin{pmatrix} t_{11} & t_{12} \\ t_{21} & t_{22} \end{pmatrix}$$

where $\det R = 1$, $\det S = 1$, $\det T = 1$. This means that R, S and T are elements of the Lie group $SL(2, \mathbb{C})$. Let

$$\sum_{j_1,j_2,j_3=0}^{1} d_{j_1 j_2 j_3} |j_1\rangle \otimes |j_2\rangle \otimes |j_3\rangle = (R \otimes S \otimes T) \sum_{j_1,j_2,j_3=0}^{1} c_{j_1 j_2 j_3} |j_1\rangle \otimes |j_2\rangle \otimes |j_3\rangle.$$
(1)

We define the *hyperdeterminant* of $C = (c_{j_1 j_2 j_3})$ with $j_1, j_2, j_3 \in \{0, 1\}$ as

$$\begin{aligned}
\mathrm{Det}(C) := &\, (c_{000}^2 c_{111}^2 + c_{001}^2 c_{110}^2 + c_{010}^2 c_{101}^2 + c_{011}^2 c_{100}^2) \\
&- 2(c_{000}c_{001}c_{110}c_{111} + c_{000}c_{010}c_{101}c_{111} + c_{000}c_{011}c_{100}c_{111} \\
&+ c_{001}c_{010}c_{101}c_{110} + c_{001}c_{011}c_{110}c_{100} + c_{010}c_{011}c_{101}c_{100}) \\
&+ 4(c_{000}c_{011}c_{101}c_{110} + c_{001}c_{010}c_{100}c_{111}).
\end{aligned}$$

Show that

$$\mathrm{Det}(C) = \mathrm{Det}(D).$$
(2)

Owing to (2) the quantity $\mathrm{Det}(C)$ is called an *invariant*.

Solution 31.   Note that $\det(R \otimes S \otimes T) = 1$. We find eight equations for d_{000}, d_{001}, ..., d_{111}. From $\det R = \det S = \det T = 1$ it follows that $r_{11}r_{22} - r_{12}r_{21} = 1$, $s_{11}s_{22} - s_{12}s_{21} = 1$, $t_{11}t_{22} - t_{12}t_{21} = 1$. Inserting $d_{j_1 j_2 j_3}$ into the right-hand side of (2) and inserting these conditions it follows that $\mathrm{Det}(D) = \mathrm{Det}(C)$.

Problem 32.   The *hyperdeterminant* of a $2 \times 2 \times 2$ hypermatrix $C = (c_{ijk})$ $(i, j, k \in \{0, 1\})$ is defined by

$$\mathrm{Det}C := -\frac{1}{2} \sum_{i,j,k,m,n,p=0}^{1} \sum_{i',j',k',m',n',p'=0}^{1} \epsilon_{ii'}\epsilon_{jj'}\epsilon_{kk'}\epsilon_{mm'}\epsilon_{nn'}\epsilon_{pp'} c_{ijk} a_{i'j'm} c_{npk} c_{n'p'm'}$$

where $\epsilon_{00} = \epsilon_{11} = 0$, $\epsilon_{01} = 1$, $\epsilon_{10} = -1$.
(i) Calculate $\mathrm{Det}C$.
(ii) Consider the three qubit state

$$|\psi\rangle = \sum_{i,j,k=0}^{1} c_{ijk} |i\rangle \otimes |j\rangle \otimes |k\rangle.$$

The *three tangle* τ_3 is a measure of entanglement and is defined for the three qubit state $|\psi\rangle$ as

$$\tau_{123} := 4|\mathrm{Det}C|$$

where $C = (c_{ijk})$. Find the three tangle for the GHZ state

$$|GHZ\rangle = \frac{1}{\sqrt{2}}(|0\rangle \otimes |0\rangle \otimes |0\rangle + |1\rangle \otimes |1\rangle \otimes |1\rangle)$$

and the W state

$$|W\rangle = \frac{1}{\sqrt{3}}(|0\rangle \otimes |0\rangle \otimes |1\rangle + |0\rangle \otimes |1\rangle \otimes |0\rangle + |1\rangle \otimes |0\rangle \otimes |0\rangle).$$

Solution 32. (i) We obtain

$$\mathrm{Det}(C) := (c_{000}^2 c_{111}^2 + c_{001}^2 c_{110}^2 + c_{010}^2 c_{101}^2 + c_{011}^2 c_{100}^2)$$
$$-2(c_{000}c_{001}c_{110}c_{111} + c_{000}c_{010}c_{101}c_{111} + c_{000}c_{011}c_{100}c_{111}$$
$$+c_{001}c_{010}c_{101}c_{110} + c_{001}c_{011}c_{110}c_{100} + c_{010}c_{011}c_{101}c_{100})$$
$$+4(c_{000}c_{011}c_{101}c_{110} + c_{001}c_{010}c_{100}c_{111}).$$

(ii) For the GHZ state we have

$$c_{000} = \frac{1}{\sqrt{2}}, \qquad c_{111} = \frac{1}{\sqrt{2}}$$

all other coefficients are zero. Thus we find for the three tangle

$$\tau_3 = 4|\mathrm{Det}| = 1.$$

For the W state we have $c_{001} = c_{010} = c_{100} = \frac{1}{\sqrt{3}}$. Thus $\tau_3 = 0$. Using this measure of entanglement the W state is not entangled.

Problem 33. Let $|0\rangle$, $|1\rangle$ be an orthonormal basis in $\mathbb{C}^2$. Consider the normalized state

$$|\psi\rangle = \sum_{j,k=0}^{1} c_{jk}|j\rangle \otimes |k\rangle$$

in the Hilbert space $\mathbb{C}^4$ and the 2×2 matrix $C = (c_{jk})$. Using the 4 coefficients c_{jk} $(j,k \in \{0,1\})$ we form a multilinear polynomial p in two variables x_1, x_2

$$p(x_1, x_2) = c_{00} + c_{01}x_1 + c_{10}x_2 + c_{11}x_1x_2. \tag{1}$$

Show that determinant $\det C = c_{00}c_{11} - c_{01}c_{10}$ is the unique irreducible polynomial (up to sign) of content one in the 4 unknowns c_{jk} that vanishes whenever the system of equations

$$p = \frac{\partial p}{\partial x_1} = \frac{\partial p}{\partial x_2} = 0 \tag{2}$$

has a solution (x_1^*, x_2^*) in $\mathbb{C}^2$.

Solution 33. Inserting (2) into (1) provides the three equations

$$c_{00} + c_{01}x_1 + c_{10}x_2 + c_{11}x_1x_2 = 0$$
$$c_{01} + c_{11}x_2 = 0$$
$$c_{10} + c_{11}x_1 = 0 \,.$$

Multiplication of the first equation with c_{11} and inserting the second and third equation yields

$$c_{00}c_{11} + c_{01}c_{11}x_1 + c_{10}c_{11}x_2 + c_{11}x_1c_{11}x_2 = c_{00}c_{11} - c_{01}c_{10} - c_{10}c_{01} + c_{10}c_{10}$$
$$= c_{00}c_{11} - c_{01}c_{10}$$
$$= 0 \,.$$

Problem 34. Can we find 2×2 matrices S_1 and S_2 such that

$$(S_1 \otimes S_2)\left(\begin{pmatrix} 1 \\ 0 \end{pmatrix} \otimes \begin{pmatrix} 1 \\ 0 \end{pmatrix} \right) = \frac{1}{\sqrt{2}} \begin{pmatrix} 1 \\ 0 \\ 0 \\ 1 \end{pmatrix} \,. \tag{1}$$

Solution 34. From (1) we find

$$\left(S_1 \begin{pmatrix} 1 \\ 0 \end{pmatrix} \right) \otimes \left(S_2 \begin{pmatrix} 1 \\ 0 \end{pmatrix} \right) = \frac{1}{\sqrt{2}} \begin{pmatrix} 1 \\ 0 \\ 0 \\ 1 \end{pmatrix} \,.$$

Thus

$$\begin{pmatrix} s_{11}^{(1)} \\ s_{21}^{(1)} \end{pmatrix} \otimes \begin{pmatrix} s_{11}^{(2)} \\ s_{21}^{(2)} \end{pmatrix} = \frac{1}{\sqrt{2}} \begin{pmatrix} 1 \\ 0 \\ 0 \\ 1 \end{pmatrix}$$

or

$$\begin{pmatrix} s_{11}^{(1)} s_{11}^{(2)} \\ s_{11}^{(1)} s_{21}^{(2)} \\ s_{21}^{(1)} s_{11}^{(2)} \\ s_{21}^{(1)} s_{21}^{(2)} \end{pmatrix} = \frac{1}{\sqrt{2}} \begin{pmatrix} 1 \\ 0 \\ 0 \\ 1 \end{pmatrix} \,.$$

Thus we have the four conditions

$$s_{11}^{(1)} s_{11}^{(2)} = \frac{1}{\sqrt{2}}, \quad s_{11}^{(1)} s_{21}^{(2)} = 0, \quad s_{21}^{(1)} s_{11}^{(2)} = 0, \quad s_{21}^{(1)} s_{21}^{(2)} = \frac{1}{\sqrt{2}}$$

which are not compatible. Thus no S_1 and S_2 exist such that (1) is satisfied.

Problem 35. Consider the state

$$|\psi\rangle = \cos(\alpha)|00\rangle + \sin(\alpha)|11\rangle, \qquad 0 < \alpha < \pi/4$$

where α is called the Schmidt angle.
(i) Find the eigenvalues of the density matrix $|\psi\rangle\langle\psi|$.
(ii) Find the partially traced density matrix (we find when we trace over one of the subsystems).
(iii) Show that the partially traced density matrix has two unequal and non-zero eigenvalues $\lambda_1 = \cos^2(\alpha)$ and $\lambda_2 = \sin^2(\alpha)$.
(iv) Calculate the von Neumann entropy for the corresponding density matrix. Show that the entropy grows monotonically with the Schmidt angle.

Solution 35. (i) We obtain for the density matrix

$$\rho(\alpha) = \begin{pmatrix} \cos^2\alpha & 0 & 0 & \sin(2\alpha)/2 \\ 0 & 0 & 0 & 0 \\ 0 & 0 & 0 & 0 \\ \sin(2\alpha)/2 & 0 & 0 & \sin^2\alpha \end{pmatrix}$$

with the eigenvalues 0 (triple) and 1. Thus the eigenvalues are independent of α.
(ii) We obtain the diagonal matrix

$$\rho_2 = \begin{pmatrix} \cos^2\alpha & 0 \\ 0 & \sin^2\alpha \end{pmatrix}.$$

(iii) From (ii) we see that the eigenvalues are $\cos^2(\alpha)$ and $\sin^2(\alpha)$.
(iv) Thus for the entropy we find

$$S(\rho_2) = -\cos^2(\alpha)\log_2\cos^2\alpha - \sin^2\alpha\log_2\sin^2\alpha$$

which is monotonically increasing for $0 < \cos^2\alpha < 1/2$, i.e. $0 < \alpha < \pi/4$.

Problem 36. Let $\sigma_1 = \sigma_x$, $\sigma_2 = \sigma_y$, $\sigma_3 = \sigma_z$ be the Pauli spin matrices. We form the nine 4×4 matrices

$$\Sigma_{jk} := \sigma_j \otimes \sigma_k, \qquad j, k = 1, 2, 3.$$

Note that $[\Sigma_{jk}, \Sigma_{mn}] = 0$. The *variance* of an hermitian operator $\hat{O}$ and a wave vector $|\phi\rangle$ is defined by

$$V_{\hat{O}}(|\phi\rangle) := \langle\phi|(\hat{O})^2|\phi\rangle - (\langle\phi|\hat{O}|\phi\rangle)^2.$$

The *remoteness* for a given normalized state $|\psi\rangle$ in $\mathbb{C}^4$ is defined by

$$R(|\psi\rangle) = \sum_{j=1}^{3}\sum_{k=1}^{3}\left(\langle\psi|(\Sigma_{jk})^2|\psi\rangle - (\langle\psi|\Sigma_{jk}|\psi\rangle)^2\right).$$

Find the remoteness for the Bell states

$$|\phi^+\rangle = \frac{1}{\sqrt{2}}(|0\rangle \otimes |0\rangle + |1\rangle \otimes |1\rangle), \qquad |\phi^-\rangle = \frac{1}{\sqrt{2}}(|0\rangle \otimes |0\rangle - |1\rangle \otimes |1\rangle)$$

$$|\psi^+\rangle = \frac{1}{\sqrt{2}}(|0\rangle \otimes |1\rangle + |1\rangle \otimes |0\rangle), \qquad |\psi^-\rangle = \frac{1}{\sqrt{2}}(|0\rangle \otimes |1\rangle - |1\rangle \otimes |0\rangle).$$

Solution 36.   We have for all $j, k = 1, 2, 3$

$$\Sigma_{jk}^2 = I_2 \otimes I_2 = I_4.$$

Thus

$$\langle \psi | (\Sigma_{jk})^2 | \psi \rangle = 1$$

where $|\psi\rangle$ is one of the Bell states. Now

$$\langle \psi | (\sigma_x \otimes \sigma_x) | \psi \rangle = \langle \psi | (\sigma_y \otimes \sigma_y) | \psi \rangle = \langle \psi | (\sigma_z \otimes \sigma_z) | \psi \rangle = 1.$$

All the other matrices of Σ_{jk} yield 0 for all Bell states. Thus we find for the remoteness of all Bell states $R = 6$.

Problem 37.   We consider the finite-dimensional Hilbert space $\mathcal{H} = \mathbb{C}^{2^n}$ and the normalized state

$$|\psi\rangle = \sum_{j_1, j_2, \dots, j_n = 0}^{1} c_{j_1, j_2, \dots, j_n} |j_1\rangle \otimes |j_2\rangle \otimes \cdots \otimes |j_n\rangle$$

in this Hilbert space. Here $|0\rangle$, $|1\rangle$ denotes the standard basis. Let ϵ_{jk} $(j, k = 0, 1)$ be defined by $\epsilon_{00} = \epsilon_{11} = 0$, $\epsilon_{01} = 1$, $\epsilon_{10} = -1$. Let n be even or $n = 3$. Then an n-tangle can be introduced by

$$\tau_{1\dots n} = 2 \left| \sum_{\substack{\alpha_1, \dots, \alpha_n = 0 \\ \delta_1, \dots, \delta_n = 0}}^{1} c_{\alpha_1 \dots \alpha_n} c_{\beta_1 \dots \beta_n} c_{\gamma_1 \dots \gamma_n} c_{\delta_1 \dots \delta_n} \right.$$

$$\left. \times \epsilon_{\alpha_1 \beta_1} \epsilon_{\alpha_2 \beta_2} \cdots \epsilon_{\alpha_{n-1} \beta_{n-1}} \epsilon_{\gamma_1 \delta_1} \epsilon_{\gamma_2 \delta_2} \cdots \epsilon_{\gamma_{n-1} \delta_{n-1}} \epsilon_{\alpha_n \gamma_n} \epsilon_{\beta_n \delta_n} \right|.$$

Consider the case $n = 4$ and a state $|\psi\rangle$ with $c_{0000} = 1/\sqrt{2}$, $c_{1111} = 1/\sqrt{2}$ and all other coefficients are 0. Find τ_{1234}.

Solution 37.   We find $\tau_{1234} = 1$.

Problem 38.   A general pure state $|\Psi\rangle$ of two qubits can be written as

$$|\Psi\rangle = e^{i\phi_0} \cos\theta_0 |00\rangle + e^{i\phi_1} \sin\theta_0 \cos\theta_1 |01\rangle$$

$$\tag{1}$$

$$+e^{i\phi_2}\sin\theta_0\sin\theta_1\cos\theta_2|10\rangle + e^{i\phi_3}\sin\theta_0\sin\theta_1\sin\theta_2|11\rangle$$

where ϕ_j and θ_k are chosen uniformly according to the *Haar measure*

$$d\mu = \frac{1}{(2\pi)^4}d(\sin\theta_0)^6 d(\sin\theta_1)^4 d(\sin\theta_2)^2 d\phi_0 d\phi_1 d\phi_2 d\phi_3 \qquad (2)$$

with

$$0 \le \phi_j < 2\pi, \qquad 0 \le \theta_k < \frac{\pi}{2} \qquad (3)$$

where $j = 0, 1, 2, 3$ and $k = 0, 1, 2$. An extra overall random phase $e^{i\phi_0}$ is included to maintain consistency with $SU(n)$, where $n = 4$. For a pure state of two qubits the *tangle* τ, is defined as

$$\tau := 4\det\rho_A \qquad (4)$$

where ρ_A is the *reduced density matrix* obtained when qubit B has been traced over (or vice versa, permuting A and B). The tangle τ is an entanglement measure.
(i) Find τ for $|\psi\rangle$. Then find τ for the four Bell states and the unentangled state $|00\rangle$.
(ii) Using the Haar measure find τ for a randomly selected pure state.

Solution 38. (i) From the state $|\Psi\rangle$ we obtain the 4×4 density matrix

$$\rho = |\Psi\rangle\langle\Psi| = \begin{pmatrix} \psi_0\psi_0^* & \psi_0\psi_1^* & \psi_0\psi_2^* & \psi_0\psi_3^* \\ \psi_1\psi_0^* & \psi_1\psi_1^* & \psi_1\psi_2^* & \psi_1\psi_3^* \\ \psi_2\psi_0^* & \psi_2\psi_1^* & \psi_2\psi_2^* & \psi_2\psi_3^* \\ \psi_3\psi_0^* & \psi_3\psi_1^* & \psi_3\psi_2^* & \psi_3\psi_3^* \end{pmatrix} \qquad (5)$$

where

$$\psi_0 = e^{i\phi_0}\cos\theta_0, \qquad \psi_1 = e^{i\phi_1}\sin\theta_0\cos\theta_1$$

$$\psi_2 = e^{i\phi_2}\sin\theta_0\sin\theta_1\cos\theta_2, \qquad \psi_3 = e^{i\phi_3}\sin\theta_0\sin\theta_1\sin\theta_2 . \qquad (6)$$

Using the basis

$$\begin{pmatrix} 1 \\ 0 \end{pmatrix} \otimes I_2, \qquad \begin{pmatrix} 0 \\ 1 \end{pmatrix} \otimes I_2 \qquad (7)$$

where I_2 is the 2×2 unit matrix we find the 2×2 matrix

$$\rho_A = \begin{pmatrix} \psi_0\psi_0^* + \psi_2\psi_2^* & \psi_0\psi_1^* + \psi_2\psi_3^* \\ \psi_1\psi_0^* + \psi_3\psi_2^* & \psi_1\psi_1^* + \psi_3\psi_3^* \end{pmatrix} . \qquad (8)$$

Thus the determinant of (8) is given by

$$\det\rho_A = (\psi_0\psi_0^* + \psi_2\psi_2^*)(\psi_1\psi_1^* + \psi_3\psi_3^*) - (\psi_1\psi_0^* + \psi_3\psi_2^*)(\psi_0\psi_1^* + \psi_2\psi_3^*) . \qquad (9)$$

Therefore

$$\det \rho_A = \psi_0 \psi_0^* \psi_3 \psi_3^* + \psi_1 \psi_1^* \psi_2 \psi_2^* - \psi_0 \psi_1^* \psi_2^* \psi_3 - \psi_0^* \psi_1 \psi_2 \psi_3^* . \qquad (10)$$

Inserting (6) into (10) we get

$$\det \rho_A = \cos^2 \theta_0 \sin^2 \theta_0 \sin^2 \theta_1 \sin^2 \theta_2 + \sin^4 \theta_0 \cos^2 \theta_1 \sin^2 \theta_1 \cos^2 \theta_2$$
$$- (e^{i(\phi_0 - \phi_1 - \phi_2 - \phi_3)} + e^{i(-\phi_0 + \phi_1 + \phi_2 - \phi_3)})$$
$$\times \sin^3 \theta_0 \cos \theta_0 \sin^2 \theta_1 \cos \theta_1 \sin \theta_2 \cos \theta_2 .$$

It follows that the four Bell states have the maximum possible entanglement, i.e., $\tau = 1$. The product state $|00\rangle$ has $\tau = 0$.
(ii) From (2) we find

$$\frac{48}{(2\pi)^4} \cos \theta_0 (\sin \theta_0)^5 \cos \theta_1 (\sin \theta_1)^3 \cos \theta_2 \sin \theta_2 d\theta_0 d\theta_1 d\theta_2 d\phi_0 d\phi_1 d\phi_2 d\phi_3$$

and

$$\int_{SU(4)} d\mu = 1$$

i.e., the Haar measure is normalized. Here we made use of

$$\int_0^{2\pi} d\phi = 2\pi, \quad \text{and} \quad \int_0^{\pi/2} \sin^k(x) \cos(x) dx = \frac{1}{k+1}$$

where $k = 1, 2, \ldots$. Integrating $\det \rho_A$ (or $\det \rho_B$) over the Haar measure gives $\langle \tau \rangle = \frac{2}{5}$, where we used

$$\int_0^{\pi/2} \sin^m(x) \cos^n(x) dx = \frac{m-1}{m+n} \int_0^{\pi/2} \sin^{m-2}(x) \cos^n(x) dx$$

$$\int_0^{\pi/2} \sin^m(x) \cos^n(x) dx = \frac{n-1}{m+n} \int_0^{\pi/2} \sin^m(x) \cos^{n-2}(x) dx$$

and

$$\int_0^{\pi/2} \sin(x) \cos(x) dx = \frac{1}{2} .$$

A randomly selected pure state of two qubits might thus be expected to have 0.4 tangle units of entanglement.

Chapter 11

Bell Inequality

Bell's theorem states that, according to quantum mechanics, the value of a certain combination of correlations for experiments of two distant systems can be higher than the highest value allowed by any local-realistic theory of the type proposed by Einstein, Podolsky and Rosen, in which local properties of a system determine the result of any experiment on that system. The most discussed Bell inequality, the Clauser-Horne-Shimony-Holt (CHSH) inequality states that in any local-realistic theory the absolute value of a combination of four correlations is bounded by 2. Consider a system with two distant particles i and j. Let A and a (B and b) be physical observables taking values -1 and 1 referring to local experiments on particle i (j). The correlation $C(A, B)$ of A and B is defined as

$$C(A, B) := P_{AB}(1, 1) - P_{AB}(1, -1) - P_{AB}(-1, 1) + P_{AB}(-1, -1)$$

where $P_{AB}(1, -1)$ denotes the joint probability of obtaining $A = 1$ and $B = -1$ when A and B are measured. In any theory in which local variables of particle i (j) determine the result of local experiments on particle i (j), the absolute value of a particular combination of correlations is bounded by 2

$$|C(A, B) - mC(A, b) - nC(a, B) - mnC(a, b)| \leq 2$$

where m and n can either be -1 or 1. The CHSH inequality holds for any local-realistic theory, where $m, n \in \{-1, 1\}$. For a two particle system in a quantum pure state $|\psi\rangle$, the quantum correlation of A and B is defined $C_Q(A, B) := \langle\psi|\hat{A}\hat{B}|\psi\rangle$, where $\hat{A}$ and $\hat{B}$ are the self-adjoint operators which represent the observables A and B. For certain choices of $\hat{A}$, $\hat{a}$, $\hat{B}$, $\hat{b}$, and $|\psi\rangle$, the quantum correlation violates the CHSH inequality.

Problem 1.   Let $\mathcal{H}_A$ and $\mathcal{H}_B$ be finite-dimensional Hilbert spaces. Let $\mathcal{H}$ be the Hilbert space $\mathcal{H} = \mathcal{H}_A \otimes \mathcal{H}_B$, i.e., $\mathcal{H}$ is the tensor product of the two Hilbert spaces $\mathcal{H}_A$ and $\mathcal{H}_B$. Let $|\psi\rangle$ be a normalized vector (pure state) in $\mathcal{H}$. Let X be an observable (described as a hermitian matrix $\hat{X}$) in $\mathcal{H}$. Then $\langle\psi|\hat{X}|\psi\rangle$ defines the expectation values. The following three conditions are equivalent when applied to pure states.

1. *Factorisability:* $|\psi\rangle = |\alpha\rangle \otimes |\beta\rangle$, where $|\alpha\rangle \in \mathcal{H}_A$ and $|\beta\rangle \in \mathcal{H}_B$ with $|\alpha\rangle$ and $|\beta\rangle$ normalized.

2. The generalized *Bell inequality:* Let $\hat{A}_1$, $\hat{A}_2$ be hermitian operators (matrices) in $\mathcal{H}_A$ with

$$\hat{A}_1^2 = I_A, \qquad \hat{A}_2^2 = I_A$$

where I_A is the identity operator in $\mathcal{H}_A$. Let $\hat{B}_1$, $\hat{B}_2$ be hermitian operators (matrices) in $\mathcal{H}_B$ with

$$\hat{B}_1^2 = I_B, \qquad \hat{B}_2^2 = I_B$$

where I_B is the identity operator in $\mathcal{H}_B$. Thus the eigenvalues of $\hat{A}_1$, $\hat{A}_2$, $\hat{B}_1$ and $\hat{B}_2$ can only be ± 1. The generalized Bell inequality is

$$|\langle\psi|\hat{A}_1 \otimes \hat{B}_1|\psi\rangle + \langle\psi|\hat{A}_1 \otimes \hat{B}_2|\psi\rangle + \langle\psi|\hat{A}_2 \otimes \hat{B}_1|\psi\rangle - \langle\psi|\hat{A}_2 \otimes \hat{B}_2|\psi\rangle| \leq 2.$$

3. *Statistical independence:* For all hermitian operators $\hat{A}$ on $\mathcal{H}_A$ and $\hat{B}$ on $\mathcal{H}_B$ with the conditions given above

$$\langle\psi|\hat{A} \otimes \hat{B}|\psi\rangle = \langle\psi|\hat{A} \otimes I_B|\psi\rangle\langle\psi|I_A \otimes \hat{B}|\psi\rangle.$$

(i) Show that condition 3 follows from condition 1.
(ii) Show that condition 2 follows from condition 3.

Solution 1.   (i) Consider the product state

$$|\psi\rangle = |\alpha\rangle \otimes |\beta\rangle.$$

Then

$$\begin{aligned}\langle\psi|(\hat{A} \otimes \hat{B})|\psi\rangle &= ((\langle\beta| \otimes \langle\alpha|)(\hat{A} \otimes \hat{B})(|\alpha\rangle \otimes |\beta\rangle)) \\ &= \langle\alpha|\hat{A}|\alpha\rangle\langle\beta|\hat{B}|\beta\rangle \\ &= \langle\psi|\hat{A} \otimes I_B|\psi\rangle\langle\psi|I_A \otimes \hat{B}|\psi\rangle.\end{aligned}$$

(ii) We use the shortcut notation

$$\langle\hat{A}_1 \otimes \hat{B}_1\rangle \equiv \langle\psi|\hat{A}_1 \otimes \hat{B}_1|\psi\rangle$$

etc. Using statistical independence we have

$$|\langle \hat{A}_1 \otimes \hat{B}_1 \rangle + \langle \hat{A}_1 \otimes \hat{B}_2 \rangle + \langle \hat{A}_2 \otimes \hat{B}_1 \rangle - \langle \hat{A}_2 \otimes \hat{B}_2 \rangle| =$$

$$|\langle \hat{A}_1 \otimes I_B \rangle (\langle \hat{I}_A \otimes \hat{B}_1 \rangle + \langle \hat{I}_A \otimes \hat{B}_2 \rangle) + \langle \hat{A}_2 \otimes I_B \rangle (\langle I_A \otimes \hat{B}_1 \rangle - \langle I_A \otimes \hat{B}_2 \rangle)| \,.$$

Using the fact that $|\langle \hat{A}_1 \otimes I_B \rangle| \leq 1$ and $|\langle \hat{A}_2 \otimes I_B \rangle| \leq 1$ we have

$$|\langle \hat{A}_1 \otimes \hat{B}_1 \rangle + \langle \hat{A}_1 \otimes \hat{B}_2 \rangle + \langle \hat{A}_2 \otimes \hat{B}_1 \rangle - \langle \hat{A}_2 \otimes \hat{B}_2 \rangle|$$
$$\leq |\langle I_A \otimes \hat{B}_1 \rangle + \langle I_A \otimes \hat{B}_2 \rangle| + |\langle I_A \otimes \hat{B}_1 \rangle - \langle I_A \otimes \hat{B}_2 \rangle|$$
$$\leq \max(((\langle I_A \otimes \hat{B}_1 \rangle + \langle I_A \otimes \hat{B}_2 \rangle) + (\langle I_A \otimes \hat{B}_1 \rangle - (\langle I_A \otimes \hat{B}_2 \rangle)),$$
$$((\langle I_A \otimes \hat{B}_1 \rangle + \langle I_A \otimes \hat{B}_2 \rangle) - (\langle I_A \otimes \hat{B}_1 \rangle - \langle I_A \otimes \hat{B}_2 \rangle)),$$
$$-(\langle \hat{I}_A \otimes \hat{B}_1 \rangle + \langle I_A \otimes \hat{B}_1 \rangle) + (\langle I_A \otimes \hat{B}_1 \rangle - \langle I_A \otimes \hat{B}_2 \rangle)),$$
$$-(\langle I_A \otimes \hat{B}_1 \rangle + \langle I_A \otimes \hat{B}_2 \rangle) - (\langle I_A \otimes \hat{B}_1 \rangle - \langle I_A \otimes \hat{B}_1 \rangle)))$$
$$= \max(2\langle I_A \otimes \hat{B}_1 \rangle, 2\langle I_A \otimes \hat{B}_2 \rangle, -2\langle I_A \otimes \hat{B}_2 \rangle, -2\langle I_A \otimes \hat{B}_1 \rangle) \leq 2$$

where we also used

$$|\langle I_A \otimes \hat{B}_1 \rangle| \leq 1, \qquad |\langle I_A \otimes \hat{B}_2 \rangle| \leq 1 \,.$$

Problem 2. Let $\mathcal{H}_A$ and $\mathcal{H}_B$ be finite-dimensional Hilbert spaces. Let $\mathcal{H}$ be the Hilbert space $\mathcal{H} = \mathcal{H}_A \otimes \mathcal{H}_B$, i.e., $\mathcal{H}$ is the tensor product of the two Hilbert spaces $\mathcal{H}_A$ and $\mathcal{H}_B$.

Let $\hat{A}_1$, $\hat{A}_2$ be hermitian operators (matrices) in $\mathcal{H}_A$ with

$$\hat{A}_1^2 = I_A, \qquad \hat{A}_2^2 = I_A \,.$$

Let $\hat{B}_1$, $\hat{B}_2$ be hermitian operators (matrices) in $\mathcal{H}_B$ with

$$\hat{B}_1^2 = I_B, \qquad \hat{B}_2^2 = I_B \,.$$

The generalized *Bell inequality* is given by

$$|\langle \psi | \hat{A}_1 \otimes \hat{B}_1 | \psi \rangle + \langle \psi | \hat{A}_1 \otimes \hat{B}_2 | \psi \rangle + \langle \psi | \hat{A}_2 \otimes \hat{B}_1 | \psi \rangle - \langle \psi | \hat{A}_2 \otimes \hat{B}_2 | \psi \rangle| \leq 2 \,.$$

Let $\mathcal{H}_A = \mathcal{H}_B = \mathbb{C}^2$. Let $\{\, |0\rangle, |1\rangle \,\}$ be the standard basis in $\mathbb{C}^2$. Consider the entangled state in $\mathcal{H} = \mathbb{C}^4$ (*EPR state*)

$$|\psi\rangle = \frac{1}{\sqrt{2}} (|0\rangle \otimes |1\rangle - |1\rangle \otimes |0\rangle)$$

which is one of the Bell states. Show that this state and the operators

$$\hat{A}_1 := \sigma_x, \qquad \hat{A}_2 := \sigma_y$$

$$\hat{B}_1 := \frac{1}{\sqrt{2}}(\sigma_x + \sigma_y), \qquad \hat{B}_2 := \frac{1}{\sqrt{2}}(\sigma_x - \sigma_y)$$

violate the Bell inequality.

Solution 2.   We have

$$\hat{A}_1|0\rangle = |1\rangle, \qquad \hat{A}_1|1\rangle = |0\rangle$$

$$\hat{A}_2|0\rangle = i|1\rangle, \qquad \hat{A}_2|1\rangle = -i|0\rangle$$

$$\hat{B}_1|0\rangle = \frac{1}{\sqrt{2}}(|1\rangle + i|1\rangle), \qquad \hat{B}_1|1\rangle = \frac{1}{\sqrt{2}}(|0\rangle - i|0\rangle)$$

$$\hat{B}_2|0\rangle = \frac{1}{\sqrt{2}}(|1\rangle - i|1\rangle), \qquad \hat{B}_2|1\rangle = \frac{1}{\sqrt{2}}(|0\rangle - i|0\rangle).$$

Using $\langle 0|0\rangle = \langle 1|1\rangle = 1$ and $\langle 0|1\rangle = \langle 1|0\rangle = 0$, we find

$$|\langle\psi|\hat{A}_1 \otimes \hat{B}_1|\psi\rangle + \langle\psi|\hat{A}_1 \otimes \hat{B}_2|\psi\rangle + \langle\psi|\hat{A}_2 \otimes \hat{B}_1|\psi\rangle - \langle\psi|\hat{A}_2 \otimes \hat{B}_2|\psi\rangle| = 2\sqrt{2}.$$

Thus the Bell inequality is violated since $2\sqrt{2} > 2$.

Problem 3.   Let σ_1, σ_2 and σ_3 be the Pauli spin matrices. Consider the *Bell operator*

$$B := \frac{1}{2}\sum_{j,k=1}^{3}(a_j(c_k + d_k) + b_j(c_k - d_k))\sigma_j \otimes \sigma_k$$

where $\mathbf{a}, \mathbf{b}, \mathbf{c}, \mathbf{d}$ are real unit vectors in $\mathbb{R}^3$.
(i) Calculate the matrix B^2.
(ii) Let

$$|\Phi^+\rangle = \frac{1}{\sqrt{2}}\begin{pmatrix} 1 \\ 0 \\ 0 \\ 1 \end{pmatrix}.$$

Calculate the density operator $\rho = |\Phi^+\rangle\langle\Phi^+|$ and then $\mathrm{tr}(\rho B)$. Discuss.
(iii) Let

$$|\xi\rangle = \begin{pmatrix} 1 \\ 0 \\ 0 \\ 0 \end{pmatrix}.$$

Calculate the density operator $\rho = |\xi\rangle\langle\xi|$ and then $\mathrm{tr}(\rho B)$. Compare to (ii).

Solution 3.   (i) Since

$$(\sigma_j \otimes \sigma_k)(\sigma_m \otimes \sigma_n) \equiv (\sigma_j\sigma_m) \otimes (\sigma_k\sigma_n)$$

we have

$$B^2 = \frac{1}{4} \sum_{j,k=1}^{3} \sum_{m,n=1}^{3} (a_j(c_k + d_k) + b_j(c_k - d_k))(a_m(c_n + d_n)$$
$$+ b_m(c_n - d_n))(\sigma_j \sigma_m) \otimes (\sigma_k \sigma_n).$$

Since $\sigma_j^2 = I_2$,

$$\sigma_1 \sigma_2 = -\sigma_2 \sigma_1 = i\sigma_3, \quad \sigma_2 \sigma_3 = -\sigma_3 \sigma_2 = i\sigma_1, \quad \sigma_3 \sigma_1 = -\sigma_1 \sigma_3 = i\sigma_2$$

and B^2 is hermitian (since B is hermitian) we obtain

$$\begin{aligned}
B^2 = {}& I_2 \otimes I_2 + (a_2 b_3 - a_3 b_2)(c_2 d_3 - c_3 d_2)\sigma_1 \otimes \sigma_1 \\
& + (a_1 b_3 - b_1 a_3)(c_1 d_3 - d_1 c_3)\sigma_2 \otimes \sigma_2 \\
& + (a_1 b_2 - a_2 b_1)(c_1 d_2 - c_2 d_1)\sigma_3 \otimes \sigma_3 \\
& + (a_2 b_3 - b_2 a_3)(c_3 d_1 - c_1 d_3)\sigma_1 \otimes \sigma_2 \\
& + (a_1 b_3 - b_1 a_3)(c_3 d_2 - c_2 d_3)\sigma_2 \otimes \sigma_1 \\
& + (a_2 b_3 - a_3 b_2)(c_1 d_2 - c_2 d_1)\sigma_1 \otimes \sigma_3 \\
& + (a_1 b_2 - a_2 b_1)(c_2 d_3 - c_3 d_2)\sigma_3 \otimes \sigma_1 \\
& + (a_1 b_3 - a_3 b_1)(c_2 d_1 - c_1 d_2)\sigma_2 \otimes \sigma_3 \\
& + (a_1 b_2 - a_2 b_1)(c_3 d_1 - c_1 d_3)\sigma_3 \otimes \sigma_2.
\end{aligned}$$

(ii) We obtain

$$\rho = |\Phi^+\rangle\langle\Phi^+| = \frac{1}{2} \begin{pmatrix} 1 & 0 & 0 & 1 \\ 0 & 0 & 0 & 0 \\ 0 & 0 & 0 & 0 \\ 1 & 0 & 0 & 1 \end{pmatrix}.$$

Thus ρ can be written as a sum of Kronecker products

$$\rho = \frac{1}{2} \left(\begin{pmatrix} 1 & 0 \\ 0 & 0 \end{pmatrix} \otimes \begin{pmatrix} 1 & 0 \\ 0 & 0 \end{pmatrix} + \begin{pmatrix} 0 & 0 \\ 0 & 1 \end{pmatrix} \otimes \begin{pmatrix} 0 & 0 \\ 0 & 1 \end{pmatrix} \right.$$
$$\left. + \begin{pmatrix} 0 & 1 \\ 0 & 0 \end{pmatrix} \otimes \begin{pmatrix} 0 & 1 \\ 0 & 0 \end{pmatrix} + \begin{pmatrix} 0 & 0 \\ 1 & 0 \end{pmatrix} \otimes \begin{pmatrix} 0 & 0 \\ 1 & 0 \end{pmatrix} \right).$$

Since

$$\text{tr}\left(\begin{pmatrix} 1 & 0 \\ 0 & 0 \end{pmatrix} \sigma_j \right) = 0, \qquad \text{tr}\left(\begin{pmatrix} 0 & 0 \\ 0 & 1 \end{pmatrix} \sigma_j \right) = 0$$

for $j = 1, 2$ and

$$\text{tr}\left(\begin{pmatrix} 1 & 0 \\ 0 & 0 \end{pmatrix} \sigma_3 \right) = 1, \qquad \text{tr}\left(\begin{pmatrix} 0 & 0 \\ 0 & 1 \end{pmatrix} \sigma_3 \right) = -1,$$

$$\text{tr}\left(\begin{pmatrix} 0 & 1 \\ 0 & 0 \end{pmatrix} \sigma_1 \right) = 1, \qquad \text{tr}\left(\begin{pmatrix} 0 & 0 \\ 1 & 0 \end{pmatrix} \sigma_1 \right) = 1,$$

$$\text{tr}\left(\begin{pmatrix} 0 & 1 \\ 0 & 0 \end{pmatrix}\sigma_2\right) = i, \qquad \text{tr}\left(\begin{pmatrix} 0 & 0 \\ 1 & 0 \end{pmatrix}\sigma_2\right) = -i$$

$$\text{tr}\left(\begin{pmatrix} 0 & 1 \\ 0 & 0 \end{pmatrix}\sigma_3\right) = 0, \qquad \text{tr}\left(\begin{pmatrix} 0 & 0 \\ 1 & 0 \end{pmatrix}\sigma_3\right) = 0$$

we find

$$\text{tr}(\rho B) = \frac{1}{2}\left(a_1(c_1 + d_1) + b_1(c_1 - d_1) - a_2(c_2 + d_2) - b_2(c_2 - d_2)\right)$$
$$+ a_3(c_3 + d_3) + b_3(c_3 - d_3)) .$$

(iii) We have

$$\rho = |\xi\rangle\langle\xi| = \begin{pmatrix} 1 & 0 & 0 & 0 \\ 0 & 0 & 0 & 0 \\ 0 & 0 & 0 & 0 \\ 0 & 0 & 0 & 0 \end{pmatrix} = \begin{pmatrix} 1 & 0 \\ 0 & 0 \end{pmatrix} \otimes \begin{pmatrix} 1 & 0 \\ 0 & 0 \end{pmatrix} .$$

Thus

$$\text{tr}(\rho B) = \frac{1}{2}(a_3(c_3 + d_3) + b_3(c_3 - d_3)) .$$

Problem 4. Consider the Pauli spin matrices $\boldsymbol{\sigma} = (\sigma_1, \sigma_2, \sigma_3)$. Let $\mathbf{q}$, $\mathbf{r}$, $\mathbf{s}$, $\mathbf{t}$ be unit vectors in $\mathbb{R}^3$. We define

$$Q := \mathbf{q} \cdot \boldsymbol{\sigma}, \quad R := \mathbf{r} \cdot \boldsymbol{\sigma}, \quad S := \mathbf{s} \cdot \boldsymbol{\sigma}, \quad T := \mathbf{t} \cdot \boldsymbol{\sigma} .$$

Calculate the matrix

$$(Q \otimes S + R \otimes S + R \otimes T - Q \otimes T)^2$$

and express the result using commutators.

Solution 4. Using that for $j, k = 1, 2, 3$ we have

$$\sigma_j \sigma_k = \delta_{jk} I_2 + i \sum_{\ell=1}^{3} \epsilon_{jk\ell} \sigma_\ell$$

with $\epsilon_{123} = \epsilon_{231} = \epsilon_{312} = 1$, $\epsilon_{321} = \epsilon_{213} = \epsilon_{132} = -1$ and 0 otherwise we obtain

$$(Q \otimes S + R \otimes S + R \otimes T - Q \otimes T)^2 \equiv 4I_2 \otimes I_2 + [Q, R] \otimes [S, T] .$$

Problem 5. Let X_1 and X_2 be $m \times m$ hermitian matrices with

$$X_1^2 = X_2^2 = I_m .$$

Let Y_1 and Y_2 be $n \times n$ hermitian matrices with

$$Y_1^2 = Y_2^2 = I_n \, .$$

(i) What can be said about the eigenvalues of X_1, X_2, Y_1 and Y_2?
(ii) Consider the so-called *Bell operator*

$$B := X_1 \otimes (Y_1 + Y_2) + X_2 \otimes (Y_1 - Y_2) \, .$$

Calculate B^2. Express the result using commutators.
(iii) What can be said about the eigenvalues of B^2 and B?

Solution 5.   (i) From the eigenvalue equation $X_1 \mathbf{x}_1 = \lambda \mathbf{x}_1$ we obtain

$$X_1(X_1 \mathbf{x}_1) = X_1(\lambda \mathbf{x}_1) = \lambda X_1 \mathbf{x}_1 = \lambda^2 \mathbf{x}_1 \, .$$

Consequently

$$I_m \mathbf{x}_1 = \lambda^2 \mathbf{x}_1$$

and therefore $1 = \lambda^2$. Thus the eigenvalues can only be ± 1. Analogously, we find that the eigenvalues for the matrices X_2, Y_1 and Y_2 can only be ± 1.
(ii) Since $X_1^2 = X_2^2 = I_m$ and $Y_1^2 = Y_2^2 = I_n$ we have

$$
\begin{aligned}
B^2 &= I_m \otimes (Y_1 + Y_2)(Y_1 + Y_2) + (X_1 X_2) \otimes (Y_1 + Y_2)(Y_1 - Y_2) \\
&\quad + (X_2 X_1) \otimes (Y_1 - Y_2)(Y_1 + Y_2) + I_m \otimes (Y_1 - Y_2)(Y_1 - Y_2) \\
&= I_m \otimes (2I_n + Y_1 Y_2 + Y_2 Y_1) + X_1 X_2 \otimes (Y_2 Y_1 - Y_1 Y_2) \\
&\quad + X_2 X_1 \otimes (Y_1 Y_2 - Y_2 Y_1) + I_m \otimes (2I_n - Y_1 Y_2 - Y_2 Y_1) \\
&= 4I_m \otimes I_n - (X_1 X_2) \otimes [Y_1, Y_2] + (X_2 X_1) \otimes [Y_1, Y_2] \\
&= 4I_m \otimes I_n - [X_1, X_2] \otimes [Y_1, Y_2] \\
&= 4I_m \otimes I_n + (i[X_1, X_2]) \otimes (i[Y_1, Y_2]) \, .
\end{aligned}
$$

(iii) Note that the commutator $[A, B]$ of two hermitian $d \times d$ matrices A, B is in general not hermitian. However $i[A, B]$ is hermitian. Thus we can find an $m \times m$ unitary matrix U and an $n \times n$ unitary matrix V such that

$$U(i[X_1, X_2])U^* = \mathrm{diag}(\alpha_1, \alpha_2, \ldots, \alpha_m)$$

$$V(i[Y_1, Y_2])V^* = \mathrm{diag}(\beta_1, \beta_2, \ldots, \beta_n)$$

with $\alpha_j, \beta_j \in \mathbb{R}$. It follows that

$$
\begin{aligned}
(U \otimes V)B^2(U^* \otimes V^*) &= 4I_m \otimes I_n + (U(i[X_1, X_2])U^*) \otimes (V(i[Y_1, Y_2])V^*) \\
&= 4I_m \otimes I_n + \mathrm{diag}(\alpha_1, \ldots, \alpha_m) \otimes \mathrm{diag}(\beta_1, \ldots, \beta_n) \, .
\end{aligned}
$$

The real eigenvalues α_j of the hermitian matrix $i[X_1, X_2]$ are restricted by $-2 \le \alpha_j \le 2$ for $j = 1, 2, \ldots, m$. Analogously, the real eigenvalues β_k of the hermitian matrix $i[Y_1, Y_2]$ are restricted by $-2 \leq \beta_k \leq 2$ for $k = 1, 2, \ldots, n$. Thus we have $-4 \leq \alpha_j \beta_k \leq 4$. The eigenvalues of B^2 are therefore given by

$$4 + \alpha_j \beta_k, \qquad j = 1, 2, \ldots, m, \quad k = 1, 2, \ldots, n.$$

It follows that the eigenvalues of B are

$$|\lambda_{jk}| = \sqrt{4 + \alpha_j \beta_k}$$

with $j = 1, 2, \ldots, m$ and $k = 1, 2, \ldots, n$.

Problem 6.   Let $\mathbf{n}$, $\mathbf{m}$ be unit vectors in $\mathbb{R}^3$. Consider the spin singlet state (entangled state)

$$|\psi\rangle = \frac{1}{\sqrt{2}} \left(\begin{pmatrix} 1 \\ 0 \end{pmatrix} \otimes \begin{pmatrix} 0 \\ 1 \end{pmatrix} - \begin{pmatrix} 0 \\ 1 \end{pmatrix} \otimes \begin{pmatrix} 1 \\ 0 \end{pmatrix} \right).$$

(i) Show that the quantum mechanical expectation values $E(\mathbf{n}, \mathbf{m})$

$$E(\mathbf{n}, \mathbf{m}) = \langle \psi | (\boldsymbol{\sigma} \cdot \mathbf{n}) \otimes (\boldsymbol{\sigma} \cdot \mathbf{m}) | \psi \rangle$$

is given by

$$E(\mathbf{n}, \mathbf{m}) = -\mathbf{m} \cdot \mathbf{n} = -\|\mathbf{n}\| \cdot \|\mathbf{m}\| \cos\phi = -\cos\phi_{\mathbf{n},\mathbf{m}} \tag{1}$$

where $\phi_{\mathbf{n},\mathbf{m}}$ are the angles between the two quantization directions $\mathbf{n}$ and $\mathbf{m}$. We write $\cos\phi_{\mathbf{n},\mathbf{m}}$ to indicate that $\phi_{\mathbf{m},\mathbf{n}}$ is the angle between $\mathbf{m}$ and $\mathbf{n}$.
(ii) The *CHSH inequality* is given by

$$|E(\mathbf{n}, \mathbf{m}) - E(\mathbf{n}, \mathbf{m}')| + |E(\mathbf{n}', \mathbf{m}') + E(\mathbf{n}', \mathbf{m})| \leq 2. \tag{2}$$

Insert (1) into (2) and then find the angles, where the inequality is maximally violated.

Solution 6.   (i) Note that

$$\boldsymbol{\sigma} \cdot \mathbf{n} = n_1 \sigma_1 + n_2 \sigma_2 + n_3 \sigma_3.$$

Straightforward calculation yields the result.
(ii) We obtain

$$|\cos\phi_{\mathbf{n},\mathbf{m}} - \cos\phi_{\mathbf{n},\mathbf{m}'}| + |\cos\phi_{\mathbf{n}',\mathbf{m}'} + \cos\phi_{\mathbf{n}',\mathbf{m}}| \leq 2.$$

The maximal violation is $2\sqrt{2}$, achieved by the angles

$$\phi_{\mathbf{n},\mathbf{m}'} = 3\pi/4, \qquad \phi_{\mathbf{n},\mathbf{m}} = \phi_{\mathbf{n}',\mathbf{m}'} = \phi_{\mathbf{n}',\mathbf{m}} = \pi/4$$

where $\cos(3\pi/4) = -1/\sqrt{2}$ and $\cos(\pi/4) = 1/\sqrt{2}$. Angles which violate the inequality (2) are called *Bell angles*.

Problem 7. Consider four observers: Alice (A), Bob (B), Charlie (C) and Dora (D) each having one qubit. Every observer is allowed to choose between two dichotomic observables. Denote the outcome of observer X's measurement by X_i ($X = A, B, C, D$) with $i = 1, 2$. Under the assumption of local realism, each outcome can either take the value $+1$ or -1. The correlations between the measurement outcomes of all four observers can be represented by the product $A_i B_j C_k D_l$, where $i, j, k, l = 1, 2$. In a local realistic theory, the correlation function of the measurement performed by all four observers is the average of $A_i B_j C_k D_l$ over many runs of the experiment

$$Q(A_i B_j C_k D_l) := \langle \psi | A_i B_j C_k D_l | \psi \rangle$$

The Mermin-Ardehali-Belinskii-Klyshko inequality is given by

$$
\begin{aligned}
& Q(A_1 B_1 C_1 D_1) - Q(A_1 B_1 C_1 D_2) - Q(A_1 B_1 C_2 D_1) - Q(A_1 B_2 C_1 D_1) \\
& - Q(A_2 B_1 C_1 D_1) - Q(A_1 B_1 C_2 D_2) - Q(A_1 B_2 C_1 D_2) - Q(A_2 B_1 C_1 D_2) \\
& - Q(A_1 B_2 C_2 D_1) - Q(A_2 B_1 C_2 D_1) - Q(A_2 B_2 C_1 D_1) + Q(A_2 B_2 C_2 D_2) \\
& + Q(A_2 B_2 C_2 D_1) + Q(A_2 B_2 C_1 D_2) + Q(A_2 B_1 C_2 D_2) + Q(A_1 B_2 C_2 D_2) \le 4 \, .
\end{aligned}
$$

Each observer X measures the spin of each qubit by projecting it either along $\mathbf{n}_1^X$ or $\mathbf{n}_2^X$. Every observer can independently choose between two arbitrary directions. For a four qubit state $|\psi\rangle$, the correlation functions are thus given by

$$Q(A_i B_j C_k D_l) = \langle \psi | (\mathbf{n}_i^A \cdot \boldsymbol{\sigma}) \otimes (\mathbf{n}_j^B \cdot \boldsymbol{\sigma}) \otimes (\mathbf{n}_k^C \cdot \boldsymbol{\sigma}) \otimes (\mathbf{n}_l^D \cdot \boldsymbol{\sigma}) | \psi \rangle \, .$$

where $\cdot$ denotes the scalar product, i.e.

$$\mathbf{n}_j^X \cdot \boldsymbol{\sigma} := n_{j1}^X \sigma_1 + n_{j2}^X \sigma_2 + n_{j3}^X \sigma_3 \, .$$

Let

$$
\mathbf{n}_1^A = \begin{pmatrix} 1 \\ 0 \\ 0 \end{pmatrix}, \quad
\mathbf{n}_2^A = \begin{pmatrix} 0 \\ 0 \\ 1 \end{pmatrix}, \quad
\mathbf{n}_1^B = \begin{pmatrix} 0 \\ 1 \\ 0 \end{pmatrix}, \quad
\mathbf{n}_2^B = \begin{pmatrix} 0 \\ 0 \\ 1 \end{pmatrix}
$$

$$
\mathbf{n}_1^C = \begin{pmatrix} 0 \\ 1 \\ 0 \end{pmatrix}, \quad
\mathbf{n}_2^C = \begin{pmatrix} 0 \\ 0 \\ 1 \end{pmatrix}, \quad
\mathbf{n}_1^D = \frac{1}{\sqrt{2}}\begin{pmatrix} -1 \\ 0 \\ 1 \end{pmatrix}, \quad
\mathbf{n}_2^D = \frac{1}{\sqrt{2}}\begin{pmatrix} 1 \\ 0 \\ 1 \end{pmatrix} \, .
$$

Show that the Mermin-Ardehali-Belinskii-Klyshko inequality is violated for the state

$$|\psi\rangle = \frac{1}{2\sqrt{2}}(|0000\rangle - |0011\rangle - |0101\rangle + |0110\rangle + |1001\rangle + |1010\rangle + |1100\rangle + |1111\rangle)$$

where

$$|0\rangle = \begin{pmatrix} 1 \\ 0 \end{pmatrix}, \qquad |1\rangle = \begin{pmatrix} 0 \\ 1 \end{pmatrix}.$$

and $|0000\rangle \equiv |0\rangle \otimes |0\rangle \otimes |0\rangle \otimes |0\rangle$ etc..

Solution 7.   For the first term we have

$$
\begin{aligned}
Q(A_1 B_1 C_1 D_1) &= \frac{1}{\sqrt{2}} \langle \psi | \sigma_x \otimes \sigma_y \otimes \sigma_y \otimes (-\sigma_x + \sigma_z) | \psi \rangle \\
&= -\frac{1}{\sqrt{2}} \langle \psi | (\sigma_x \otimes \sigma_y \otimes \sigma_y \otimes \sigma_x) | \psi \rangle \\
&\quad + \frac{1}{\sqrt{2}} \langle \psi | (\sigma_x \otimes \sigma_y \otimes \sigma_y \otimes \sigma_z) | \psi \rangle .
\end{aligned}
$$

Using

$$\langle \psi | (\sigma_x \otimes \sigma_y \otimes \sigma_y \otimes \sigma_x) | \psi \rangle = -1, \qquad \langle \psi | (\sigma_x \otimes \sigma_y \otimes \sigma_y \otimes \sigma_z) | \psi \rangle = 0$$

we obtain for the first term

$$Q(A_1 B_1 C_1 D_1) = \frac{1}{\sqrt{2}}.$$

Analogously we calculate the other terms. Summing up the terms we find the value $4\sqrt{2}$. Since $4\sqrt{2} > 4$ the Mermin-Ardehali-Belinskii-Klyshko inequality is violated by the state $|\psi\rangle$.

Chapter 12

Teleportation

Teleportation is the transmission of quantum information using a classical channel and entanglement. It demonstrates the use of entanglement as a communication resource. The simplest case is to consider the teleportation of a single qubit using two bits of classical communication and one entangled pair (EPR-pair). Quantum teleportation is the disembodied transport of an unknown quantum state from one place to another. The key idea is that two distant operators, Alice at a sending station and Bob at a receiving terminal, share an entangled quantum bipartite state and exploit its nonlocal character as a quantum resource. The resource state can be the *singlet state* of a pair of spin-$\frac{1}{2}$ particles (Bell state)

$$|\psi\rangle = \frac{1}{\sqrt{2}}(|0\rangle \otimes |1\rangle - |1\rangle \otimes |0\rangle).$$

Particle 1 is given to Alice and particle 2 is given to Bob. Alice intends to transport an unknown state of a third spin-$\frac{1}{2}$ particle to Bob. She performs a complete projective measurement on the joint system consisting of particle 1 and 3 and then conveys its outcome to Bob via a classical communication channel. As a consequence of Alice's measurement, the total-spin state of the three-particle system collapses. Owing to the entanglement, this involves a breakdown of the spin-$\frac{1}{2}$ state of Bob's particle 2. Nevertheless, Bob makes use of the information transmitted classically by Alice to transform his reduced state into an output that is an accurate replica of the original unknown input.

Problem 1.   Consider the following states $(a, b \in \mathbb{C})$

$$|\psi\rangle := a|0\rangle + b|1\rangle, \quad |a|^2 + |b|^2 = 1 \tag{1}$$

$$|\phi\rangle := |\psi\rangle \otimes \frac{1}{\sqrt{2}}(|00\rangle + |11\rangle). \tag{2}$$

(i) Show that the state $|\phi\rangle$ can be written as

$$|\phi\rangle = \frac{1}{2\sqrt{2}}(|00\rangle + |11\rangle) \otimes (a|0\rangle + b|1\rangle) + \frac{1}{2\sqrt{2}}(|00\rangle - |11\rangle) \otimes (a|0\rangle - b|1\rangle)$$
$$+ \frac{1}{2\sqrt{2}}(|01\rangle + |10\rangle) \otimes (a|1\rangle + b|0\rangle) + \frac{1}{2\sqrt{2}}(|01\rangle - |10\rangle) \otimes (a|1\rangle - b|0\rangle).$$

(ii) Describe how measurement of the first two qubits of $|\phi\rangle$ can be used to obtain $|\psi\rangle$ as the last qubit. Alice has the first qubit of $|\phi\rangle$ and Alice and Bob share the second and third qubits of $|\phi\rangle$ (an EPR-pair).

Solution 1.   (i) Inserting (1) into (2) we obtain

$$|\phi\rangle = \frac{1}{\sqrt{2}}(a|000\rangle + a|011\rangle + b|100\rangle + b|111\rangle).$$

On the other hand we have

$$\frac{1}{2\sqrt{2}}(|00\rangle + |11\rangle) \otimes (a|0\rangle + b|1\rangle) + \frac{1}{2\sqrt{2}}(|00\rangle - |11\rangle) \otimes (a|0\rangle - b|1\rangle)$$
$$+ \frac{1}{2\sqrt{2}}(|01\rangle + |10\rangle) \otimes (a|1\rangle + b|0\rangle) + \frac{1}{2\sqrt{2}}(|01\rangle - |10\rangle) \otimes (a|1\rangle - b|0\rangle)$$
$$= \frac{1}{2\sqrt{2}}(a|000\rangle + a|110\rangle + b|001\rangle + b|111\rangle)$$
$$+ \frac{1}{2\sqrt{2}}(a|000\rangle - a|110\rangle - b|001\rangle + b|111\rangle)$$
$$+ \frac{1}{2\sqrt{2}}(a|011\rangle + a|101\rangle + b|010\rangle + b|100\rangle)$$
$$+ \frac{1}{2\sqrt{2}}(a|011\rangle - a|101\rangle - b|010\rangle + b|100\rangle)$$
$$= \frac{1}{\sqrt{2}}(a|000\rangle + a|011\rangle + b|100\rangle + b|111\rangle)$$
$$= |\phi\rangle.$$

(ii) We measure in the *Bell basis*

$$\left\{ \frac{1}{\sqrt{2}}(|00\rangle + |11\rangle), \ \frac{1}{\sqrt{2}}(|00\rangle - |11\rangle), \ \frac{1}{\sqrt{2}}(|01\rangle + |10\rangle), \ \frac{1}{\sqrt{2}}(|01\rangle - |10\rangle) \right\}.$$

From the state $|\phi\rangle$ we can see that the first two qubits are in each of the Bell states with equal probability. Thus if we measure the first two qubits in the Bell basis we obtain a result corresponding to each of the Bell states and can perform a transform to obtain $|\psi\rangle$ in the last qubit as follows

Bell State	Transform
$\frac{1}{\sqrt{2}}(|00\rangle + |11\rangle)$	I_2
$\frac{1}{\sqrt{2}}(|00\rangle - |11\rangle)$	$|0\rangle\langle 0| - |1\rangle\langle 1|$
$\frac{1}{\sqrt{2}}(|01\rangle + |10\rangle)$	U_{NOT}
$\frac{1}{\sqrt{2}}(|01\rangle - |10\rangle)$	$|0\rangle\langle 1| - |1\rangle\langle 0|$

After measurement and applying the corresponding transform we obtain $|\psi\rangle$ as the last qubit. So if Alice and Bob initially share the entangled pair

$$\frac{1}{\sqrt{2}}(|00\rangle + |11\rangle)$$

Alice can perform a measurement in the Bell basis on her qubit and her part of the entangled pair and sends the result (two bits) to Bob who applies the corresponding transform to his part of the entangled pair. The state $|\psi\rangle$ is thus *teleported* from Alice's qubit to Bob's qubit. Note that the Bell basis is obtained by applying $U_{CNOT}(U_H \otimes I_2)$ to the computational basis $\{|00\rangle, |01\rangle, |10\rangle, |11\rangle\}$. The transforms are unitary and therefore invertible. Thus we can also measure the first two qubits in the computational basis after applying

$$(U_H \otimes I)U_{CNOT}.$$

Problem 2. Let $|\psi\rangle := a|0\rangle + b|1\rangle$ be an arbitrary qubit state. Let $|\phi\rangle$ be another arbitrary qubit state. Let U be a unitary operator which acts on two qubits.
(i) Determine the implications of measuring the first two qubits of

$$|\theta\rangle := |\psi\rangle \otimes \frac{1}{\sqrt{2}}(I_2 \otimes U)\left((|00\rangle + |11\rangle) \otimes |\phi\rangle\right)$$

with respect to the Bell basis. How can we obtain $U(|\psi\rangle \otimes |\phi\rangle)$ as the last two qubits?
(ii) Alice has $|\psi\rangle$ and Bob has $|\phi\rangle$. Describe how U can be applied to $|\psi\rangle \otimes |\phi\rangle$ using only classical communication and prior shared entanglement. After the computation, Alice must still have the first qubit of $U(|\psi\rangle \otimes |\phi\rangle)$ and Bob must still have the second qubit of $U(|\psi\rangle \otimes |\phi\rangle)$.

Solution 2. (i) We have

$$|\theta\rangle = a|0\rangle \otimes \frac{1}{\sqrt{2}}(I_2 \otimes U)\left((|00\rangle + |11\rangle) \otimes |\phi\rangle\right)$$

$$+ b|1\rangle \otimes \frac{1}{\sqrt{2}} (I_2 \otimes U) \left((|00\rangle + |11\rangle) \otimes |\phi\rangle \right)$$

$$= a|0\rangle \otimes \frac{1}{\sqrt{2}} \left(|0\rangle \otimes U(|0\rangle \otimes |\phi\rangle) + |1\rangle \otimes U(|1\rangle \otimes |\phi\rangle) \right)$$

$$+ b|1\rangle \otimes \frac{1}{\sqrt{2}} \left(|0\rangle \otimes U(|0\rangle \otimes |\phi\rangle) + |1\rangle \otimes U(|1\rangle \otimes |\phi\rangle) \right)$$

$$= \frac{1}{\sqrt{2}} \left(|00\rangle \otimes U(a|0\rangle \otimes |\phi\rangle) + |01\rangle \otimes U(a|1\rangle \otimes |\phi\rangle) \right)$$

$$+ \frac{1}{\sqrt{2}} \left(|10\rangle \otimes U(b|0\rangle \otimes |\phi\rangle) + |11\rangle \otimes U(b|1\rangle \otimes |\phi\rangle) \right).$$

Expanding $|00\rangle$, $|01\rangle$, $|10\rangle$ and $|11\rangle$ in the Bell basis for the first two qubits yields the state

$$|\theta\rangle = \frac{1}{2\sqrt{2}} (|00\rangle + |11\rangle) \otimes U\left((a|0\rangle + b|1\rangle) \otimes |\phi\rangle \right)$$

$$+ \frac{1}{2\sqrt{2}} (|00\rangle - |11\rangle) \otimes U\left((a|0\rangle - b|1\rangle) \otimes |\phi\rangle \right)$$

$$+ \frac{1}{2\sqrt{2}} (|01\rangle + |10\rangle) \otimes U\left((a|1\rangle + b|0\rangle) \otimes |\phi\rangle \right)$$

$$+ \frac{1}{2\sqrt{2}} (|01\rangle - |10\rangle) \otimes U\left((a|1\rangle - b|0\rangle) \otimes |\phi\rangle \right).$$

We measure in the *Bell basis*

$$\left\{ \frac{1}{\sqrt{2}}(|00\rangle + |11\rangle), \; \frac{1}{\sqrt{2}}(|00\rangle - |11\rangle), \; \frac{1}{\sqrt{2}}(|01\rangle + |10\rangle), \; \frac{1}{\sqrt{2}}(|01\rangle - |10\rangle) \right\}.$$

From $|\theta\rangle$ we can see that the first two qubits are in each of the Bell states with equal probability. If we make a measurement we obtain a result corresponding to each of the Bell states and can perform a transform to obtain $U(|\psi\rangle \otimes |\phi\rangle)$ in the last two qubits as follows

Bell State	Transform
$\frac{1}{\sqrt{2}}(|00\rangle + |11\rangle)$	I_2
$\frac{1}{\sqrt{2}}(|00\rangle - |11\rangle)$	$U\left((|0\rangle\langle 0| - |1\rangle\langle 1|) \otimes I_2 \right) U^*$
$\frac{1}{\sqrt{2}}(|01\rangle + |10\rangle)$	$U(U_{NOT} \otimes I_2)U^*$
$\frac{1}{\sqrt{2}}(|01\rangle - |10\rangle)$	$U\left((|0\rangle\langle 1| - |1\rangle\langle 0|) \otimes I_2 \right) U^*$

Thus after measurement and applying the corresponding transform we obtain $U(|\psi\rangle \otimes |\phi\rangle)$ as the last two qubits. Thus if Alice and Bob initially share the entangled state

$$\frac{1}{\sqrt{2}}(|00\rangle + |11\rangle)$$

Bob applies U to his two qubits. Then Alice can perform a measurement in the Bell basis on her qubit and her part of the entangled pair and sends the result (two bits) to Bob who applies the corresponding transform from the table to his part of the entangled pair. Thus with probability $\frac{1}{4}$ Bob can begin the computation $U(|\psi\rangle \otimes |\phi\rangle)$ without knowing the state $|\psi\rangle$ and still obtain the correct result after Alice measures her two qubits. With probability $\frac{3}{4}$ he still has to apply a transform which is independent of $|\psi\rangle$. (ii) Alice teleports $|\psi\rangle$ to Bob with one entangled pair, Bob performs the computation $U(|\psi\rangle \otimes |\phi\rangle)$ on his two qubits and then teleports the first qubit back to Alice with a second entangled pair. Thus 4 bits of communication are used in this scheme (Alice sends two to Bob, and then Bob sends two to Alice). Alice and Bob can perform U_{CNOT} even though their qubits are spatially separated if they have prior entanglement.

Problem 3. In quantum teleportation we start with the following state in the Hilbert space $\mathbb{C}^8$

$$|\psi\rangle \otimes |0\rangle \otimes |0\rangle \equiv (a|0\rangle + b|1\rangle) \otimes |0\rangle \otimes |0\rangle \equiv |\psi 00\rangle$$

where $|a|^2 + |b|^2 = 1$. The quantum circuit for teleportation is given by

where A is the input $|\psi\rangle$, B the input $|0\rangle$ and C the input $|0\rangle$. Study what happens when we feed the product state $|\psi 00\rangle$ into the quantum circuit. From the circuit we have the following eight 8×8 unitary matrices (left to right)

$$U_1 = I_2 \otimes U_H \otimes I_2, \quad U_2 = I_2 \otimes U_{XOR},$$
$$U_3 = U_{XOR} \otimes I_2, \quad U_4 = U_H \otimes I_2 \otimes I_2,$$
$$U_5 = I_2 \otimes U_{XOR}, \quad U_6 = I_2 \otimes I_2 \otimes U_H,$$
$$U_7 = I_4 \oplus U_{NOT} \oplus U_{NOT}, \quad U_8 = I_2 \otimes I_2 \otimes U_H$$

where $\oplus$ denotes the direct sum of matrices, U_H denotes the Hadamard gate, U_{XOR} denotes the XOR-gate and

$$U_{NOT} := \begin{pmatrix} 0 & 1 \\ 1 & 0 \end{pmatrix}.$$

(i) Find $U_8U_7U_6U_5U_4U_3U_2U_1|\psi00\rangle$.

(ii) Write a program which implements and verifies the teleportation algorithm.

Solution 3.   (i) Applying the first four unitary matrices to the input state we obtain

$$U_4U_3U_2U_1|\psi00\rangle$$

$$= \frac{a}{2}(|000\rangle + |100\rangle + |011\rangle + |111\rangle) + \frac{b}{2}(|010\rangle - |110\rangle + |001\rangle - |101\rangle).$$

This state can be rewritten as

$$U_4U_3U_2U_1|\psi00\rangle = \frac{1}{\sqrt{2}}(|0\rangle + |1\rangle) \otimes (\frac{a}{\sqrt{2}}(|0\rangle \otimes |0\rangle + |1\rangle \otimes |1\rangle))$$

$$+ \frac{1}{\sqrt{2}}(|0\rangle - |1\rangle) \otimes (\frac{b}{\sqrt{2}}(|0\rangle \otimes |1\rangle + |1\rangle \otimes |0\rangle)).$$

Applying all eight unitary matrices to the input state we obtain

$$U_8U_7U_6U_5U_4U_3U_2U_1|\psi00\rangle$$

$$= \frac{a}{2}(|000\rangle + |100\rangle + |010\rangle + |110\rangle) + \frac{b}{2}(|011\rangle + |111\rangle + |001\rangle + |101\rangle).$$

This state can be rewritten as

$$\left(\frac{1}{\sqrt{2}}(|0\rangle + |1\rangle)\right) \otimes \left(\frac{1}{\sqrt{2}}(|0\rangle + |1\rangle)\right) \otimes |\psi\rangle.$$

The state $|\psi\rangle$ will be transferred to the lower output, where both other outputs will come out in the state $(|0\rangle + |1\rangle)/\sqrt{2}$. If the two upper outputs are measured in the standard basis ($|0\rangle$ versus $|1\rangle$), two random classical bits will be obtained in addition to the quantum state $|\psi\rangle$ on the lower output.

(ii) The implementation in SymbolicC++ is as follows. The **Matrix** class of SymbolicC++ includes the method **kron** for the Kronecker product of two matrices and the method **dsum** for the direct sum of two matrices. The overloaded operators * and + are used for matrix multiplication and addition. The identity matrix is also implemented. Thus the code for the quantum circuit is as follows.

```
// teleport.cpp

#include <iostream>
#include "symbolicc++.h"
using namespace std;
```

```
Symbolic Hadamard(const Symbolic &v)
{
 Symbolic H("",2,2);
 Symbolic sqrt12 = sqrt(1/Symbolic(2));
 H(0,0) = sqrt12; H(0,1) =  sqrt12;
 H(1,0) = sqrt12; H(1,1) = -sqrt12;
 return (H*v);
}


Symbolic XOR(const Symbolic &v)
{
 Symbolic X("",4,4);
 X(0,0) = 1; X(0,1) = 0; X(0,2) = 0; X(0,3) = 0;
 X(1,0) = 0; X(1,1) = 1; X(1,2) = 0; X(1,3) = 0;
 X(2,0) = 0; X(2,1) = 0; X(2,2) = 0; X(2,3) = 1;
 X(3,0) = 0; X(3,1) = 0; X(3,2) = 1; X(3,3) = 0;
 return (X*v);
}


Symbolic Bell(const Symbolic &v)
{
 Symbolic I("",2,2), H("",2,2), X("",4,4);
 Symbolic sqrt12 = sqrt(1/Symbolic(2));
 I = I.identity();
 H(0,0) = sqrt12; H(0,1) =  sqrt12;
 H(1,0) = sqrt12; H(1,1) = -sqrt12;
 Symbolic UH = kron(H,I);
 X(0,0) = 1; X(0,1) = 0; X(0,2) = 0; X(0,3) = 0;
 X(1,0) = 0; X(1,1) = 1; X(1,2) = 0; X(1,3) = 0;
 X(2,0) = 0; X(2,1) = 0; X(2,2) = 0; X(2,3) = 1;
 X(3,0) = 0; X(3,1) = 0; X(3,2) = 1; X(3,3) = 0;
 return (X*(UH*v));
}


Symbolic Swap(const Symbolic &v)
{
 Symbolic S("",4,4);
 S(0,0) = 1; S(0,1) = 0; S(0,2) = 0; S(0,3) = 0;
 S(1,0) = 0; S(1,1) = 0; S(1,2) = 0; S(1,3) = 1;
 S(2,0) = 0; S(2,1) = 0; S(2,2) = 1; S(2,3) = 0;
 S(3,0) = 0; S(3,1) = 1; S(3,2) = 0; S(3,3) = 0;
 return XOR(S*XOR(v));
}
```

```
Symbolic Teleport(const Symbolic &v)
{
 Symbolic result;
 Symbolic NOT("",2,2),H("",2,2),I("",2,2),X("",4,4);
 Symbolic sqrt12 = sqrt(1/Symbolic(2));
 NOT(0,0) = 0; NOT(0,1) = 1; NOT(1,0) = 1; NOT(1,1) = 0;
 H(0,0) = sqrt12; H(0,1) =  sqrt12;
 H(1,0) = sqrt12; H(1,1) = -sqrt12;
 I = I.identity();
 X(0,0) = 1; X(0,1) = 0; X(0,2) = 0; X(0,3) = 0;
 X(1,0) = 0; X(1,1) = 1; X(1,2) = 0; X(1,3) = 0;
 X(2,0) = 0; X(2,1) = 0; X(2,2) = 0; X(2,3) = 1;
 X(3,0) = 0; X(3,1) = 0; X(3,2) = 1; X(3,3) = 0;
 Symbolic U1 = kron(I,kron(H,I)); Symbolic U2 = kron(I,X);
 Symbolic U3 = kron(X,I); Symbolic U4 = kron(H,kron(I,I));
 Symbolic U5 = kron(I,X); Symbolic U6 = kron(I,kron(I,H));
 Symbolic U7 = dsum(I,dsum(I,dsum(NOT,NOT)));
 Symbolic U8 = kron(I,kron(I,H));
 result = U8*(U7*(U6*(U5*(U4*(U3*(U2*(U1*v)))))));
 return result;
}

// The outcome after measuring value for qubit.
// Since the probabilities may be symbolic this function
// cannot simulate a measurement where random outcomes
// have the correct distribution
Symbolic Measure(const Symbolic &v,unsigned int qubit,
                 unsigned int value)
{
 int i,len,skip = 1-value;
 Symbolic result(v);
 Symbolic D;
 len = v.rows()/int(pow(2.0,qubit+1.0));
 for(i=0;i<v.rows();i++)
 {
  if(!(i%len)) skip = 1-skip;
  if(skip) result(i) = 0;
  else D += result(i)*result(i);
 }
 return result/sqrt(D);
}

// for output clarity
```

```
ostream &print(ostream &o, const Symbolic &v)
{
 char *b2[2]={"|0>","|1>"};
 char *b4[4]={"|00>","|01>","|10>","|11>"};
 char *b8[8]={"|000>","|001>","|010>","|011>",
              "|100>","|101>","|110>","|111>"};
 char **b;
 if(v.rows()==2) b=b2;
 if(v.rows()==4) b=b4;
 if(v.rows()==8) b=b8;
 for(int i=0;i<v.rows();i++)
  if(v(i)!=0) o << "+(" << v(i) << ")" << b[i];
 return o;
}

int main(void)
{
 Symbolic zero("", 2),one("", 2);
 Symbolic zz("", 4),zo("", 4),oz("", 4),oo("", 4),qreg;
 Symbolic tp00,tp01,tp10,tp11,psiGHZ;
 Symbolic a("a"), b("b");
 Symbolic sqrt12 = sqrt(1/Symbolic(2));
 zero(0) = 1; zero(1) = 0; one(0)  = 0; one(1)  = 1;
 zz = kron(zero,zero); zo = kron(zero,one);
 oz = kron(one,zero);  oo = kron(one,one);
 cout << "UH|0> = "; print(cout,Hadamard(zero))<< endl;
 cout << "UH|1> = "; print(cout,Hadamard(one)) << endl;
 cout << endl;
 cout << "UXOR|00> = "; print(cout,XOR(zz)) << endl;
 cout << "UXOR|01> = "; print(cout,XOR(zo)) << endl;
 cout << "UXOR|10> = "; print(cout,XOR(oz)) << endl;
 cout << "UXOR|11> = "; print(cout,XOR(oo)) << endl;
 cout << endl;
 cout << "UBELL|00> = "; print(cout,Bell(zz)) << endl;
 cout << "UBELL|01> = "; print(cout,Bell(zo)) << endl;
 cout << "UBELL|10> = "; print(cout,Bell(oz)) << endl;
 cout << "UBELL|11> = "; print(cout,Bell(oo)) << endl;
 cout << endl;
 cout << "USWAP|00> = "; print(cout,Swap(zz)) << endl;
 cout << "USWAP|01> = "; print(cout,Swap(zo)) << endl;
 cout << "USWAP|10> = "; print(cout,Swap(oz)) << endl;
 cout << "USWAP|11> = "; print(cout,Swap(oo)) << endl;
 cout << endl;
```

```
qreg=kron(a*zero+b*one,kron(zero,zero));
cout << "UTELEPORT("; print(cout,qreg) << ") = ";
 print(cout,qreg=Teleport(qreg)) << endl;
cout << "Results after measurement of first 2 qubits:" << endl;
tp00 = Measure(Measure(qreg,0,0),1,0);
tp01 = Measure(Measure(qreg,0,0),1,1);
tp10 = Measure(Measure(qreg,0,1),1,0);
tp11 = Measure(Measure(qreg,0,1),1,1);

Equations simplify = (a*a==1-b*b,1/sqrt(1/Symbolic(4))==2);
tp00 = tp00.subst_all(simplify);
tp01 = tp01.subst_all(simplify);
tp10 = tp10.subst_all(simplify);
tp11 = tp11.subst_all(simplify);

cout << "  |00> : " ; print(cout,tp00) << endl;
cout << "  |01> : " ; print(cout,tp01) << endl;
cout << "  |10> : " ; print(cout,tp10) << endl;
cout << "  |11> : " ; print(cout,tp11) << endl;
cout << endl;

psiGHZ=kron(zz,zero)*sqrt12+kron(oo,one)*sqrt12;
cout << "Greenberger-Horne-Zeilinger state : ";
print(cout,psiGHZ) << endl;
cout << "Measuring qubit 0 as 1 yields : ";
print(cout,Measure(psiGHZ,0,1)) << endl;
cout << "Measuring qubit 1 as 1 yields : ";
print(cout,Measure(psiGHZ,1,1)) << endl;
cout << "Measuring qubit 2 as 0 yields : ";
print(cout,Measure(psiGHZ,2,0)) << endl;
 return 0;
}
```

The program generates the following output

```
UTELEPORT(+(a)|000>+(b)|100>) =
    +(1/2*a)|000>+(1/2*b)|001>+(1/2*a)|010>
    +(1/2*b)|011>+(1/2*a)|100>+(1/2*b)|101>
    +(1/2*a)|110>+(1/2*b)|111>
Results after measurement of first 2 qubits:
 |00> : +(a)|000>+(b)|001>
 |01> : +(a)|010>+(b)|011>
 |10> : +(a)|100>+(b)|101>
 |11> : +(a)|110>+(b)|111>
```

Chapter 13

Cloning

Cloning is the duplication of information. Cloning is necessarily a physical process. Obviously we have to find unitary transformations for the cloning process. The *no cloning theorem* is a result of quantum mechanics which forbids the creation of identical copies of an arbitrary unknown quantum state. However, approximate copies still have many uses in quantum computing. Owing to the superposition principle of quantum mechanics it it not possible in general to clone an arbitrary quantum state. As a consequence any successful attempt to clone a state will destroy the original state in the process. Quantum cloning machines are devices for approximately cloning arbitrary quantum states. Suppose we want to clone the state $|\psi\rangle \in \mathcal{H}$ to obtain

$$|\psi\rangle \otimes |\psi\rangle .$$

The approximate cloning operation yields some mixed state ρ in $\mathcal{H} \otimes \mathcal{H}$. The fidelity of the cloning process for $|\psi\rangle$ is given by

$$((\langle\psi| \otimes \langle\psi|)\rho(|\psi\rangle \otimes |\psi\rangle)).$$

Optimal cloning attempts to maximize the average fidelity (closeness to $|\psi\rangle \otimes |\psi\rangle$). In the case of symmetric cloning we assume

$$\langle\psi|\mathrm{tr}_{\mathcal{H}}\rho|\psi\rangle$$

is independent of the system on which the partial trace is performed. Thus for symmetric cloning we need only use $\langle\psi|\mathrm{tr}_{\mathcal{H}}\rho|\psi\rangle$ as the measure of fidelity. We provide exercises describing what types of information can be cloned accurately and techniques for cloning certain types of information.

Problem 1.   The *CNOT gate* maps $(a, b \in \{0, 1\})$

$$|a\rangle \otimes |b\rangle \rightarrow |a\rangle \otimes |a \oplus b\rangle$$

where $\oplus$ is the XOR operation. Show that the CNOT gate can be used to clone a bit.

Solution 1.   Setting $b = 0$ we obtain from the CNOT gate

$$|a\rangle \otimes |0\rangle \rightarrow |a\rangle \otimes |a\rangle$$

since $a \oplus 0 = a$ for all a. Thus we have cloned a bit.

Problem 2.   Let

$$\mathbf{x} = \begin{pmatrix} x_1 \\ x_2 \end{pmatrix}, \qquad x_1 x_1^* + x_2 x_2^* = 1$$

be an arbitrary normalized vector in $\mathbb{C}^2$. Can we construct a 4×4 unitary matrix U such that

$$U\left(\begin{pmatrix} x_1 \\ x_2 \end{pmatrix} \otimes \begin{pmatrix} 1 \\ 0 \end{pmatrix} \right) = \begin{pmatrix} x_1 \\ x_2 \end{pmatrix} \otimes \begin{pmatrix} x_1 \\ x_2 \end{pmatrix}? \tag{1}$$

Prove or disprove this equation.

Solution 2.   Such a matrix does not exist. This can be seen as follows. From the right-hand side of (1) we have

$$\begin{pmatrix} x_1 \\ x_2 \end{pmatrix} \otimes \begin{pmatrix} x_1 \\ x_2 \end{pmatrix} = \left(\begin{pmatrix} x_1 \\ 0 \end{pmatrix} + \begin{pmatrix} 0 \\ x_2 \end{pmatrix} \right) \otimes \left(\begin{pmatrix} x_1 \\ 0 \end{pmatrix} + \begin{pmatrix} 0 \\ x_2 \end{pmatrix} \right) =$$

$$\begin{pmatrix} x_1 \\ 0 \end{pmatrix} \otimes \begin{pmatrix} x_1 \\ 0 \end{pmatrix} + \begin{pmatrix} x_1 \\ 0 \end{pmatrix} \otimes \begin{pmatrix} 0 \\ x_2 \end{pmatrix} + \begin{pmatrix} 0 \\ x_2 \end{pmatrix} \otimes \begin{pmatrix} x_1 \\ 0 \end{pmatrix} + \begin{pmatrix} 0 \\ x_2 \end{pmatrix} \otimes \begin{pmatrix} 0 \\ x_2 \end{pmatrix}.$$

On the other hand, from the left-hand side of (1) we find

$$U\left(\begin{pmatrix} x_1 \\ x_2 \end{pmatrix} \otimes \begin{pmatrix} 1 \\ 0 \end{pmatrix} \right) = U\left(\begin{pmatrix} x_1 \\ 0 \end{pmatrix} \otimes \begin{pmatrix} 1 \\ 0 \end{pmatrix} + \begin{pmatrix} 0 \\ x_2 \end{pmatrix} \otimes \begin{pmatrix} 1 \\ 0 \end{pmatrix} \right)$$

$$= \begin{pmatrix} x_1 \\ 0 \end{pmatrix} \otimes \begin{pmatrix} x_1 \\ 0 \end{pmatrix} + \begin{pmatrix} 0 \\ x_2 \end{pmatrix} \otimes \begin{pmatrix} 0 \\ x_2 \end{pmatrix}$$

where we used the linearity of the unitary matrix U. Comparing these two equations we find a contradiction. This is the *no cloning theorem*.

However equation (1) does hold when

$$\begin{pmatrix} x_1 \\ 0 \end{pmatrix} \otimes \begin{pmatrix} 0 \\ x_2 \end{pmatrix} + \begin{pmatrix} 0 \\ x_2 \end{pmatrix} \otimes \begin{pmatrix} x_1 \\ 0 \end{pmatrix} = \begin{pmatrix} 0 \\ 0 \end{pmatrix} \otimes \begin{pmatrix} 0 \\ 0 \end{pmatrix} .$$

Therefore $x_1 x_2 = 0$. Thus at least one of x_1 and x_2 must be zero. It is still possible to clone elements of a known orthonormal basis.

Problem 3. Let $\{ |0\rangle, |1\rangle \}$ be a basis in $\mathbb{C}^2$. Let $|\psi\rangle$ be an arbitrary qubit. Is there a unitary transformation such that

$$|\psi\rangle \otimes |\psi\rangle \to |\psi\rangle \otimes |0\rangle \ ?$$

Solution 3. Such a unitary transformation does not exist. For an arbitrary qubit $|\psi\rangle$ the product states $|\psi\rangle \otimes |\psi\rangle$ span a three-dimensional subspace of the four-dimensional Hilbert space $\mathbb{C}^4$ of two qubits. However, the product states $|\psi\rangle \otimes |0\rangle$ span only a two-dimensional subspace, as $|0\rangle$ is a fixed state. Thus the unitary transform would take a system with *von Neumann entropy* $\log_2 3$ to one with von Neumann entropy $\log_2 2$. Since the system is closed (we have a unitary transformation), this decrease of entropy is therefore a violation of the *second law of thermodynamics*. Thus the second law of thermodynamics implies that such a unitary transformation does not exist.

Problem 4. Consider the approximate cloning of a qubit $|\psi\rangle \in \mathbb{C}^2$ by simply measuring the qubit with respect to a randomly chosen orthonormal basis $\{ |0\rangle, |1\rangle \}$, and then using the state $|0\rangle \otimes |0\rangle$ or $|1\rangle \otimes |1\rangle$ corresponding to the measurement outcome. Determine the average fidelity of this cloning process.

Solution 4. The state $|\psi\rangle$ is represented by a density operator

$$|\psi\rangle\langle\psi| = \frac{1}{2}(I_2 + \mathbf{n} \cdot \boldsymbol{\sigma})$$

where $\mathbf{n}$ is a unit vector in $\mathbb{R}^3$ and $\mathbf{n} \cdot \boldsymbol{\sigma} := n_1\sigma_1 + n_2\sigma_2 + n_3\sigma_3$. The randomly chosen orthonormal basis can be represented by the density operators

$$|0\rangle\langle 0| = \frac{1}{2}(I_2 + \mathbf{m} \cdot \boldsymbol{\sigma})$$

$$|1\rangle\langle 1| = I_2 - |0\rangle\langle 0| = \frac{1}{2}(I_2 - \mathbf{m} \cdot \boldsymbol{\sigma})$$

where $\mathbf{m}$ is a unit vector in $\mathbb{R}^3$. Measuring $|\psi\rangle$ in the basis $\{ |0\rangle, |1\rangle \}$ yields $|0\rangle$ with probability

$$|\langle 0|\psi\rangle|^2 = \frac{1}{2}(1 + \mathbf{n} \cdot \mathbf{m})$$

and $|1\rangle$ with probability

$$|\langle 1|\psi\rangle|^2 = \frac{1}{2}(1 - \mathbf{n} \cdot \mathbf{m})$$

where $|\langle 0|\psi\rangle|^2 = \langle 0|\psi\rangle\langle\psi|0\rangle$. Thus we construct a two qubit system described by the density operator

$$\frac{1}{2}(1 + \mathbf{n} \cdot \mathbf{m})|0\rangle\langle 0| \otimes |0\rangle\langle 0| + \frac{1}{2}(1 - \mathbf{n} \cdot \mathbf{m})|1\rangle\langle 1| \otimes |1\rangle\langle 1|.$$

If we trace out either qubit we are left with the same density operator, i.e. this is symmetric cloning. The fidelity for $|\psi\rangle$ (symmetric cloning) is given by

$$\left[\frac{1}{2}(1 + \mathbf{n} \cdot \mathbf{m})\right]^2 + \left[\frac{1}{2}(1 - \mathbf{n} \cdot \mathbf{m})\right]^2 = \frac{1}{2}(1 + (\mathbf{n} \cdot \mathbf{m})^2).$$

To determine the average fidelity we integrate over all $\mathbf{m} \in \mathbb{R}^3$. Thus we use spherical coordinates

$$\mathbf{m} = (\sin\theta\cos\phi, \sin\theta\sin\phi, \cos\theta), \qquad \phi \in [0, 2\pi], \quad \theta \in [0, \pi].$$

Consequently the average fidelity is given by

$$\frac{1}{4\pi}\int_{\|\mathbf{m}\|=1} \frac{1}{2}(1 + (\mathbf{n} \cdot \mathbf{m})^2)\, d\mathbf{m}$$

$$= \frac{1}{4\pi}\int_0^\pi \int_0^{2\pi} \frac{1}{2}(1 + (n_1\sin\theta\cos\phi + n_2\sin\theta\sin\phi + n_3\cos\theta)^2)\sin\theta\, d\phi\, d\theta$$

$$= \frac{1}{2} + \frac{1}{8}\int_0^\pi (n_1^2\sin^3\theta + n_2^2\sin^3\theta + 2n_3^2\cos^2\theta\sin\theta)\, d\theta$$

$$= \frac{1}{2} + \frac{1}{8}\int_0^\pi (2\cos^2\theta\sin\theta + (n_1^2 + n_2^2)(\sin\theta - 3\cos^2\theta\sin\theta))\, d\theta$$

$$= \frac{2}{3}$$

where we used $n_3^2 = 1 - n_1^2 - n_2^2$. The average fidelity is independent of $|\psi\rangle$, i.e. the cloning process in universal. However the cloning process is not optimal.

Chapter 14

Quantum Algorithms

An *algorithm* is a precise description of how to realize a given objective, for example solving a computational problem. We distinguish between *classical* and *quantum* algorithms where quantum physical resources are used. Deutsch's problem leads to the simplest quantum algorithm. Consider the Boolean functions f that map $\{\, 0,\ 1\,\}$ to $\{\, 0,\ 1\,\}$. There are exactly four such functions: two constant functions $(f(0) = f(1) = 0$ and $f(0) = f(1) = 1)$ and two balanced functions $(f(0) = 0, f(1) = 1$ and $f(0) = 1, f(1) = 0)$. In Deutsch's problem one is allowed to evaluate the function only once and we are required to deduce from the result whether f is constant or balanced. Thus we are asked for the global property of f. A generalization is the Deutsch-Josza problem. One considers the Boolean functions $f : \{\, 0,1\,\}^n \to \{\, 0,1\,\}$ in the following way. Assume that, for one of these functions, it is promised that it either constant or balanced (i.e. has an equal number of 0's outputs as 1's) and consider the goal of determining which of the two properties the function actually has. The *quantum Fourier transform* plays an important role in a number of quantum algorithms. The quantum Fourier transform on the additive group of integers modulo 2^m is the mapping

$$|a\rangle \overset{F_{2^m}}{\rightarrow} \sum_{y=0}^{2^m-1} \exp(2\pi i a y/2^m)|y\rangle$$

where $a \in \{\, 0,1,\ldots,2^m - 1\,\}$. The state on the right-hand side is unentangled. The quantum Fourier transform is used in Shor's algorithm. Given positive integers a and N which are relatively prime and such that $a < N$. The goal is to find the minimum positive integer r such that $a^r \bmod N = 1$.

Problem 1. In classical communication complexity Alice is provided with a binary string

$$\mathbf{x} = x_0 x_1 \cdots x_{n-1}$$

of length n and Bob is provided with a binary string

$$\mathbf{y} = y_0 y_1 \cdots y_{n-1}$$

of length n. Alice has to determine a boolean function

$$f : \{0, 1\}^n \times \{0, 1\}^n \to \{0, 1\}$$

with the least communication between herself and Bob.
(i) Consider the *parity function*

$$f(\mathbf{x}, \mathbf{y}) = x_0 \oplus x_1 \oplus \cdots \oplus x_{n-1} \oplus y_0 \oplus y_1 \oplus \cdots \oplus y_{n-1}$$

where $\oplus$ is the *XOR-operation*, i.e.,

$$0 \oplus 0 = 0, \quad 0 \oplus 1 = 1, \quad 1 \oplus 0 = 1, \quad 1 \oplus 1 = 0.$$

How many bits has Bob to send to Alice so that she can determine f?
(ii) Consider the inner product modulo-2 function

$$f(\mathbf{x}, \mathbf{y}) = (x_0 \cdot y_0) \oplus (x_1 \cdot y_1) \oplus \cdots \oplus (x_{n-1} \cdot y_{n-1})$$

where $\cdot$ denotes the *AND operation*, i.e.,

$$0 \cdot 0 = 0, \quad 0 \cdot 1 = 0, \quad 1 \cdot 0 = 0, \quad 1 \cdot 1 = 1.$$

What is the minimum number of bits Bob has to send to Alice so that she can compute this function?

Solution 1. (i) Obviously Bob has to send only one bit, the one he finds by computing $y_0 \oplus y_1 \oplus \cdots \oplus y_{n-1}$.

(ii) Bob must send all n bits in order for Alice to compute f.

Problem 2. Find all $x_A, x_B, x_C \in \{0, 1\}$ such that $x_A + x_B + x_C = 1 \bmod 2$. We use the mapping $f_1 : \{0, 1\} \to U(2)$

$$f_1(0) := U_H, \qquad f_1(1) := I_2$$

where U_H is the Walsh-Hadamard transform and $U(2)$ denotes the unitary group over $\mathbb{C}^2$. Thus we can map from the triple (x_A, x_B, x_C) to linear operators acting on three qubits

$$f_3(x_A, x_B, x_C) := f_1(x_A) \otimes f_1(x_B) \otimes f_1(x_C).$$

Let

$$|\psi\rangle := \frac{1}{2}(|001\rangle + |010\rangle + |100\rangle - |111\rangle).$$

For each triple (x_A, x_B, x_C) found in the first part of the problem, calculate

$$|\phi\rangle := f_3(x_A, x_B, x_C)|\psi\rangle.$$

Let s_A, s_B, s_C denote the result (0 or 1) of measuring the first, second and third qubit, respectively of $|\phi\rangle$ in the computational basis. In each case determine

$$s_A + s_B + s_C \bmod 2, \qquad x_A \cdot x_B \cdot x_C.$$

Solution 2. We have

$$(x_A, x_B, x_C) \in \{ (0,0,1), (0,1,0), (1,0,0), (1,1,1) \}.$$

We note the symmetry of the state $|\psi\rangle$ with respect to qubit ordering. Thus we need only to calculate the transform for $(0,0,1)$ and $(1,1,1)$. For $(1,1,1)$ we have $f_3(1,1,1)|\psi\rangle = (I_2 \otimes I_2 \otimes I_2)|\psi\rangle = |\psi\rangle$. Measuring the qubits yields

$$(s_A, s_B, s_C) \in \{ (0,0,1), (0,1,0), (1,0,0), (1,1,1) \}$$

with equal probability. In each case we find $s_A + s_B + s_C = 1 \bmod 2$. For $(0,0,1)$ we have $f_3(0,0,1) = U_H \otimes U_H \otimes I_2$. Since

$$|\psi\rangle = \frac{1}{2}(|01\rangle + |10\rangle) \otimes |0\rangle + \frac{1}{2}(|00\rangle - |11\rangle) \otimes |1\rangle$$

we obtain

$$f_3(0,0,1)|\psi\rangle = \frac{1}{2}(|00\rangle - |11\rangle) \otimes |0\rangle + \frac{1}{2}(|01\rangle + |10\rangle) \otimes |1\rangle.$$

We find that measuring the qubits yields

$$(s_A, s_B, s_C) \in \{ (0,0,0), (0,1,1), (1,0,1), (1,1,0) \}$$

with equal probability. In each case we find $s_A + s_B + s_C = 0 \bmod 2$.

(x_A, x_B, x_C)	$x_A \cdot x_B \cdot x_C$	$s_A + s_B + s_C \bmod 2$
(0,0,1)	0	0
(0,1,0)	0	0
(1,0,0)	0	0
(1,1,1)	1	1

We find that $s_A + s_B + s_C = x_A \cdot x_B \cdot x_C$ mod 2. Suppose Alice, Bob and Carol each have a bit string $(x_{A,1}, \ldots, x_{A,n})$, $(x_{B,1}, \ldots, x_{B,n})$ and $(x_{C,1}, \ldots, x_{C,n})$, respectively. They want to calculate

$$f(\mathbf{x}_A, \mathbf{x}_B, \mathbf{x}_C) = \sum_{j=1}^{n} (x_{A,j} \cdot x_{B,j} \cdot x_{C,j}) \text{ mod } 2$$

sharing (communicating) as little information as possible. If Alice, Bob and Carol share n triplets of qubits in the state $|\psi\rangle$ they can calculate $s_{A,1}, \ldots, s_{A,n}, s_{B,1}, \ldots, s_{B,n}$ and $s_{C,1}, \ldots, s_{C,n}$ respectively as above. Thus

$$f(\mathbf{x}_A, \mathbf{x}_B, \mathbf{x}_C) = \sum_{j=1}^{n} (s_{A,j} + s_{B,j} + s_{C,j}) \text{ mod } 2.$$

If Alice, Bob and Carol calculate

$$S_{A|B|C} = \sum_{j=1}^{n} S_{A|B|C,j} \text{ mod } 2$$

Bob and Carol need only to send one bit each (S_B and S_C) to Alice for Alice to compute $f(\mathbf{x}_A, \mathbf{x}_B, \mathbf{x}_C) = S_A + S_B + S_C$, for any n. In other words the *communication complexity* is 2. Classically, for $n \geq 3$, three bits of communication are required.

Problem 3.   (i) Find all $x, y, z \in \{0, 1, 2, 3\}$ such that

$$x + y + z = 0 \text{ mod } 2. \tag{1}$$

What are the possible values of the function

$$f(x, y, z) := \frac{(x + y + z) \text{ mod } 4}{2}$$

when the condition (1) holds?
(ii) Now use the binary representation for $x = x_1 x_0$, $y = y_1 y_0$ and $z = z_1 z_0$ where $x_0, x_1, y_0, y_1, z_0, z_1 \in \{0, 1\}$. Describe the condition $x + y + z = 0$ mod 2 in terms of x_0, x_1, y_0, y_1, z_0 and z_1.
(iii) We use the map

$$f_1(0) = I_2, \qquad f_1(1) = U_H.$$

Thus we can map from the triple (x_0, y_0, z_0) to linear operators acting on three qubits

$$f_3(x_0, y_0, z_0) = f_1(x_0) \otimes f_1(y_0) \otimes f_1(z_0).$$

Let

$$|\psi\rangle := \frac{1}{2}(|000\rangle - |011\rangle - |101\rangle - |110\rangle).$$

For each triple (x_0, y_0, z_0) found in part (i) calculate

$$|\phi\rangle := f_3(x_0, y_0, z_0)|\psi\rangle.$$

Let s_x, s_y, s_z denote the result (0 or 1) of measuring the first, second and third qubit, respectively of $|\phi\rangle$ in the computational basis. In each case determine

$$s_x + s_y + s_z \bmod 2, \qquad x_0 + y_0 + z_0.$$

Solution 3.   (i) Obviously $x + y + z$ must be even. Thus the sum includes only an even number (0 or 2) of odd numbers. Thus we have the nine combinations

$$(0,0,0), (0,1,1), (0,0,2), (1,1,2), (0,2,2), (0,1,3), (2,2,2), (1,2,3), (0,3,3).$$

(ii) Let (x, y, z) be an element of the set of all permutations of elements of the above set. When $x + y + z$ is even, $(x + y + z) \bmod 4 \in \{0, 2\}$. Now when $x+y+z = 0 \bmod 2$ then $f(x, y, z) \in \{0, 1\}$. Since $x+y+z = 0 \bmod 2$ the least significant bit of the sum must be zero. The least significant bit is given by $x_0 \oplus y_0 \oplus z_0 = 0$. We find that

$$f(x, y, z) = x_1 \oplus y_1 \oplus z_1 \oplus (x_0 + y_0 + z_0).$$

XOR is denoted by "$\oplus$" and OR is denoted by "$+$". Thus we have

$$(x_0, y_0, z_0) \in \{ (0,0,0), (0,1,1), (1,0,1), (1,1,0) \}.$$

(iii) We note the symmetry of the state $|\psi\rangle$ with respect to the qubit ordering. Thus we need only calculate the transform for $(0,0,0)$ and $(0,1,1)$. For $(0,0,0)$ we have

$$f_3(0,0,0)|\psi\rangle = I_2 \otimes I_2 \otimes I_2 |\psi\rangle = |\psi\rangle.$$

Measuring the qubits yields

$$(s_x, s_y, s_z) \in \{ (0,0,0), (0,1,1), (1,0,1), (1,1,0) \}$$

with equal probability. In each case we find $s_x + s_y + s_z = 0 \bmod 2$. For $(0,1,1)$ we have $f_3(0,1,1) = I_2 \otimes U_H \otimes U_H$. Note that $|\psi\rangle$ can be written as

$$|\psi\rangle = \frac{1}{2}|0\rangle \otimes (|00\rangle - |11\rangle) - \frac{1}{2}|1\rangle \otimes (|01\rangle + |10\rangle).$$

Therefore

$$f_3(0,1,1)|\psi\rangle = \frac{1}{2}|0\rangle \otimes (|01\rangle + |10\rangle) - \frac{1}{2}|1\rangle \otimes (|00\rangle - |11\rangle).$$

We find that measuring the qubits yields

$$(s_x, s_y, s_z) \in \{\, (0,1,0), (1,0,0), (0,0,1), (1,1,1) \,\}$$

with equal probability. In each case $s_x + s_y + s_z = 1 \bmod 2$.

(x_0, y_0, z_0)	$x_0 + y_0 + z_0$	$s_x + s_y + s_z \bmod 2$
(0,0,0)	0	0
(0,1,1)	1	1
(1,0,1)	1	1
(1,1,0)	1	1

We find that $(s_x + s_y + s_z \bmod 2) = x_0 + y_0 + z_0$. Thus for three parties to calculate $f(x, y, z)$, where each party has one of the x, y and z, it is sufficient for each party to send one bit ($x_1 \oplus s_x$ or $y_1 \oplus s_y$ or $z_1 \oplus s_z$) to the other parties to calculate $f(x, y, z)$. In other words each party can calculate

$$x_1 \oplus s_x \oplus y_1 \oplus s_y \oplus z_1 \oplus s_z = x_1 \oplus y_1 \oplus z_1 \oplus (x_0 + y_0 + z_0) = f(x, y, z)$$

after communication. In other words three bits broadcast to all parties are sufficient to calculate $f(x, y, z)$, the *communication complexity* is 3 bits. Classically it is necessary that 4 bits be broadcast.

Problem 4.   (i) Determine the eigenvalues and eigenvectors of

$$A(x) := (1 - x)I_2 + xU_{NOT}, \qquad x \in \{0, 1\}.$$

(ii) Show that the unitary transform

$$U_f = |0f(0)\rangle\langle 00| + |0\overline{f(0)}\rangle\langle 01| + |1f(1)\rangle\langle 10| + |1\overline{f(1)}\rangle\langle 11|$$

where $f : \{0, 1\} \to \{0, 1\}$ is a boolean function and $\overline{x}$ denotes the boolean negation of x, can be written as

$$U_f = |0\rangle\langle 0| \otimes A(f(0)) + |1\rangle\langle 1| \otimes A(f(1)).$$

(iii) Calculate

$$U_f \left(I_2 \otimes \frac{1}{\sqrt{2}} (|0\rangle - |1\rangle) \right).$$

Consider the cases $f(0) = f(1)$ and $f(0) \neq f(1)$.

Solution 4.   (i) We have $A(0) = I_2$ and $A(1) = U_{NOT}$. Thus A_0 has eigenvalues 1 (twice), and $A(1)$ has eigenvalues 1 and -1. We tabulate the eigenvalues and corresponding eigenvectors of $A(x)$

eigenvalue	eigenvector
1	$\frac{1}{\sqrt{2}}(|0\rangle + |1\rangle)$
$(-1)^x$	$\frac{1}{\sqrt{2}}(|0\rangle - |1\rangle)$

(ii) We have

$$
\begin{aligned}
U_f &= (1 - f(0))|00\rangle\langle00| + f(0)|01\rangle\langle00| \\
&\quad + f(0)|00\rangle\langle01| + (1 - f(0))|01\rangle\langle01| \\
&\quad + (1 - f(1))|10\rangle\langle10| + f(1)|11\rangle\langle10| \\
&\quad + f(1)|10\rangle\langle11| + (1 - f(1))|11\rangle\langle11| \\
&= |0\rangle\langle0| \otimes ((1 - f(0))|0\rangle\langle0| + f(0)|1\rangle\langle0|) \\
&\quad + |0\rangle\langle0| \otimes (f(0)|0\rangle\langle1| + (1 - f(0))|1\rangle\langle1|) \\
&\quad + |1\rangle\langle1| \otimes ((1 - f(1))|0\rangle\langle0| + f(1)|1\rangle\langle0|) \\
&\quad + |1\rangle\langle1| \otimes (f(1)|0\rangle\langle1| + (1 - f(1))|1\rangle\langle1|) \\
&= |0\rangle\langle0| \otimes ((1 - f(0))(|0\rangle\langle0| + |1\rangle\langle1|) + f(0)(|0\rangle\langle1| + |1\rangle\langle0|)) \\
&\quad + |1\rangle\langle1| \otimes ((1 - f(1))(|0\rangle\langle0| + |1\rangle\langle1|) + f(1)(|0\rangle\langle1| + |1\rangle\langle0|)) \\
&= |0\rangle\langle0| \otimes A(f(0)) + |1\rangle\langle1| \otimes A(f(1)) \,.
\end{aligned}
$$

(iii) We find

$$
\begin{aligned}
U_f \left(I_2 \otimes \frac{1}{\sqrt{2}}(|0\rangle - |1\rangle) \right) &= |0\rangle\langle0| \otimes A(f(0))\frac{1}{\sqrt{2}}(|0\rangle - |1\rangle) \\
&\quad + |1\rangle\langle1| \otimes A(f(1))\frac{1}{\sqrt{2}}(|0\rangle - |1\rangle) \\
&= |0\rangle\langle0| \otimes (-1)^{f(0)}\frac{1}{\sqrt{2}}(|0\rangle - |1\rangle) \\
&\quad + |1\rangle\langle1| \otimes (-1)^{f(1)}\frac{1}{\sqrt{2}}(|0\rangle - |1\rangle) \\
&= (-1)^{f(0)}|0\rangle\langle0| \otimes \frac{1}{\sqrt{2}}(|0\rangle - |1\rangle) \\
&\quad + (-1)^{f(1)}|1\rangle\langle1| \otimes \frac{1}{\sqrt{2}}(|0\rangle - |1\rangle) \\
&= \left((-1)^{f(0)} \left(|0\rangle\langle0| + (-1)^{f(0)+f(1)}|1\rangle\langle1| \right) \otimes I_2 \right) \\
&\quad \times \left(I_2 \otimes \frac{1}{\sqrt{2}}(|0\rangle - |1\rangle) \right) \,.
\end{aligned}
$$

Thus when $f(0) = f(1)$ we apply the identity operator to the first qubit and when $f(0) \neq f(1)$ we apply a phase change to the first qubit. The eigenvalues $(-1)^{f(0)}$ and $(-1)^{f(1)}$ are said to *kick back* to the first qubit. A phase change combined with two Walsh-Hadamard transforms in the appropriate order implements a NOT gate.

Problem 5.   (i) Alice and Bob share n entangled pairs of the form $\frac{1}{\sqrt{2}}(|00\rangle + |11\rangle)$. We can write their shared state of $2n$ qubits in the form of the generalized Bell state

$$|\psi\rangle = \frac{1}{\sqrt{2^n}} \sum_{j=0}^{2^n-1} |j\rangle \otimes |j\rangle \tag{1}$$

where the first n qubits belong to Alice and the second n qubits belong to Bob. Furthermore Alice has 2^n bits $a_0, \ldots, a_{2^n-1}$ and Bob has 2^n bits $b_0, \ldots, b_{2^n-1}$. Let the unitary operators U_{PA} and U_{PB} act on the computational basis as follows

$$U_{PA}|j\rangle = (-1)^{a_j}|j\rangle, \qquad j = 0, 1, \ldots, 2^n - 1$$

$$U_{PB}|j\rangle = (-1)^{b_j}|j\rangle, \qquad j = 0, 1, \ldots, 2^n - 1.$$

Let

$$|\phi\rangle := (U_{PA} \otimes U_{PB})|\psi\rangle. \tag{2}$$

Calculate the state

$$\left(\bigotimes_n U_H\right) \otimes \left(\bigotimes_n U_H\right)|\phi\rangle. \tag{3}$$

(ii) For each of the cases

$$(a) \quad a_0 = b_0, \, a_1 = b_1, \ldots, \, a_{2^n-1} = b_{2^n-1}$$

$$(b) \quad \sum_{k=0}^{2^n-1} |a_k - b_k| = 2^{n-1}$$

determine when measurement of the first n qubits in the computational basis yields the same result as measurement of the second n qubits in the computational basis.

Solution 5.   (i) From (1) and (2) we obtain

$$|\phi\rangle = \frac{1}{\sqrt{2^n}} \sum_{j=0}^{2^n-1} (-1)^{a_j+b_j}|j\rangle \otimes |j\rangle.$$

Thus we find for (3)

$$\left(\bigotimes_n U_H\right) \otimes \left(\bigotimes_n U_H\right)|\phi\rangle = \frac{1}{(2\sqrt{2})^n} \sum_{j=0}^{2^n-1} \sum_{k=0}^{2^n-1} \sum_{l=0}^{2^n-1} (-1)^{a_j+b_j+j*k+j*l}|k\rangle \otimes |l\rangle$$

since

$$\left(\bigotimes_n U_H\right)|j\rangle = \bigotimes_{s=0}^{n-1} U_H |j_s\rangle$$

$$= \bigotimes_{s=0}^{n-1} U_H \frac{1}{\sqrt{2}}(|0\rangle + (-1)^{j_s}|1\rangle)$$

$$= \sum_{k=0}^{2^n-1} (-1)^{j_0 k_0 + j_1 k_1 + \cdots + j_{n-1} k_{n-1}}|k\rangle$$

where we decompose j and k as follows

$$j = j_0 + j_1 2 + j_2 4 + \cdots + j_{n-1} 2^{n-1}, \qquad k = k_0 + k_1 2 + k_2 4 + \cdots + k_{n-1} 2^{n-1}$$

and

$$j * k := (j_0 \cdot k_0) \oplus (j_1 \cdot k_1) \oplus \cdots \oplus (j_{n-1} \cdot k_{n-1})$$
$$= j_0 k_0 + j_1 k_1 + \cdots + j_{n-1} k_{n-1} \mod 2.$$

(ii) For the case (a) we have for $k = l$

$$\frac{1}{(2\sqrt{2})^n} \sum_{j=0}^{2^n-1} (-1)^{a_j + b_j + j*k + j*l} = \frac{1}{(2\sqrt{2})^n} \sum_{j=0}^{2^n-1} (-1)^{j*k+j*l}$$

$$= \frac{1}{(2\sqrt{2})^n} \sum_{j=0}^{2^n-1} (-1)^{j*(k+l)}$$

$$= \frac{1}{(2\sqrt{2})^n} 2^n$$

$$= 2^{-n/2}.$$

In other words the probability of measuring $|k\rangle \otimes |k\rangle$ for a given k is 2^{-n}. Furthermore

$$\sum_{k=0}^{2^n-1} 2^{-n} = 2^{-n} \sum_{k=0}^{2^n-1} 1 = 1.$$

For the case (b) we find when $k = l$

$$\sum_{j=0}^{2^n-1} (-1)^{a_j + b_j + j*k + j*l} = \sum_{j=0}^{2^n-1} (-1)^{a_j + b_j} = 0.$$

Thus if condition (a) holds measuring the $2n$ qubits in the computational basis always yields $|j\rangle$ and $|j\rangle$, i.e., the first n qubits always yield exactly the same result as the second n qubits.

If condition (b) holds then measuring the $2n$ qubits in the computational basis yields $|j\rangle$ and $|k\rangle$ where $j \neq k$, i.e., the first n qubits never yield the same result as the second n qubits.

Problem 6.   (i) Show that the vectors

$$|0_H\rangle := \frac{1}{\sqrt{2}}(|0\rangle + |1\rangle), \qquad |1_H\rangle := \frac{1}{\sqrt{2}}(|0\rangle - |1\rangle)$$

form an orthonormal basis for $\mathbb{C}^2$.
(ii) Determine the probabilities associated with finding $|0\rangle$ in the states $|0_H\rangle$ and $|1_H\rangle$.
(iii) Determine how to obtain $|0_H\rangle$ and $|1_H\rangle$ using only measurement and the phase change operation

$$U_{PS} := |0\rangle\langle 0| - |1\rangle\langle 1|.$$

(iv) Let $f : \{0,1\} \to \{0,1\}$ be a boolean function and

$$U_f := |0f(0)\rangle\langle 00| + |0\overline{f(0)}\rangle\langle 01| + |1f(1)\rangle\langle 10| + |1\overline{f(1)}\rangle\langle 11|.$$

Determine in terms of $|0_H\rangle$ and $|1_H\rangle$

$$(a)\ U_f|0_H\rangle \otimes |0_H\rangle, \qquad (b)\ U_f|0_H\rangle \otimes |1_H\rangle.$$

These techniques are used to solve *Deutsch's problem*.

Solution 6.   (i) First we demonstrate the linear independence of the vectors

$$a|0\rangle + b|1\rangle = \frac{1}{\sqrt{2}}(a + b)|0_H\rangle + \frac{1}{\sqrt{2}}(a - b)|1_H\rangle$$

$$a|0_H\rangle + b|1_H\rangle = \frac{1}{\sqrt{2}}(a + b)|0\rangle + \frac{1}{\sqrt{2}}(a - b)|1\rangle.$$

Thus for $a|0_H\rangle + b|1_H\rangle = 0$ it follows that $a = b = 0$.
(ii) We find

$$\langle 0_H|0_H\rangle = \frac{1}{2}(\langle 0|0\rangle + \langle 0|1\rangle + \langle 1|0\rangle + \langle 1|1\rangle) = 1$$

$$\langle 1_H|1_H\rangle = \frac{1}{2}(\langle 0|0\rangle - \langle 0|1\rangle - \langle 1|0\rangle + \langle 1|1\rangle) = 1$$

$$\langle 0_H|1_H\rangle = \frac{1}{2}(\langle 0|0\rangle - \langle 0|1\rangle + \langle 1|0\rangle - \langle 1|1\rangle) = 0$$

$$|\langle 0|0_H\rangle|^2 = \frac{1}{2}|(\langle 0|0\rangle + \langle 0|1\rangle|^2 = \frac{1}{2}$$

$$|\langle 0|1_H\rangle|^2 = \frac{1}{2}|(\langle 0|0\rangle - \langle 0|1\rangle|^2 = \frac{1}{2}.$$

Thus measurement projects the state $|0\rangle$ onto $|0_H\rangle$ and $|1_H\rangle$ with equal probability.

(iii) Starting with $|0\rangle$, we can obtain $|0_H\rangle$ and $|1_H\rangle$ by measurement in the $|0_H\rangle$ and $|1_H\rangle$ basis and applying U_{PS} as follows

Desired state	Measure	Transform
$|0_H\rangle$	$|0_H\rangle$	I_2
$|0_H\rangle$	$|1_H\rangle$	U_{PS}
$|1_H\rangle$	$|0_H\rangle$	U_{PS}
$|1_H\rangle$	$|1_H\rangle$	I_2

(iv) For (a) we have

$$|0_H\rangle \otimes |0_H\rangle = \frac{1}{2}(|00\rangle + |01\rangle + |10\rangle + |11\rangle).$$

Thus

$$U_f|0_H\rangle \otimes |0_H\rangle = \frac{1}{2}(|0f(0)\rangle + |0\overline{f(0)}\rangle + |1f(1)\rangle + |1\overline{f(1)}\rangle)$$

$$= \frac{1}{2}\Big((1-f(0))|00\rangle + f(0)|01\rangle + f(0)|00\rangle + (1-f(0))|01\rangle$$

$$+(1-f(1))|10\rangle + f(1)|11\rangle + f(1)|10\rangle + (1-f(1))|11\rangle\Big)$$

$$= |0_H\rangle \otimes |0_H\rangle.$$

For (b) we have

$$|0_H\rangle \otimes |1_H\rangle = \frac{1}{2}(|00\rangle - |01\rangle + |10\rangle - |11\rangle).$$

Thus

$$U_f|0_H\rangle \otimes |1_H\rangle = \frac{1}{2}(|0f(0)\rangle - |0\overline{f(0)}\rangle + |1f(1)\rangle - |1\overline{f(1)}\rangle)$$

$$= \frac{1}{2}((1-f(0))|00\rangle + f(0)|01\rangle - f(0)|00\rangle - (1-f(0))|01\rangle$$

$$+(1-f(1))|10\rangle + f(1)|11\rangle - f(1)|10\rangle - (1-f(1))|11\rangle)$$

$$= \frac{1}{2}((1-2f(0))|00\rangle - (1-2f(0))|01\rangle$$

$$+(1-2f(1))|10\rangle - (1-2f(1))|11\rangle)$$

$$= \frac{1}{2}((-1)^{f(0)}|00\rangle - (-1)^{f(0)}|01\rangle$$

$$-(-1)^{f(1)}|10\rangle - (-1)^{f(1)}|11\rangle)$$
$$= \frac{1}{2}\left((-1)^{f(0)}|0\rangle \otimes (|0\rangle - |1\rangle) + (-1)^{f(1)}|1\rangle(|0\rangle - |1\rangle)\right)$$
$$= \frac{1}{\sqrt{2}}(-1)^{f(0)}((|0\rangle - (-1)^{f(0)+f(1)}|1\rangle) \otimes |1_H\rangle)$$
$$= \frac{1}{\sqrt{2}}(-1)^{f(0)}((|0\rangle - (-1)^{f(0)\oplus f(1)}|1\rangle) \otimes |1_H\rangle)$$
$$= (-1)^{f(0)}|f(0) \oplus f(1)_H\rangle \otimes |1_H\rangle.$$

Note that $f(0) \oplus f(1)$ is 0 when f is constant, and 1 when f is balanced. Thus by determining $f(0) \oplus f(1)$ we have solved *Deutsch's problem* .

Problem 7.   Consider the following quantum game G_n with $n \geq 3$ players. Each player P_j ($j = 0, 1, \ldots, n-1$) receives a single input bit x_j and has to produce a single output bit y_j. It is known that there is an even number of 1s among the inputs. The players are not allowed to communicate after receiving their inputs. Then they are challenged to produce a collective output that contains an even number of 1s if and only if the number of 1s in the input is divisible by 4. Therefore, we require that

$$\sum_{j=0}^{n-1} y_j \equiv \frac{1}{2}\sum_{j=0}^{n-1} x_j \quad (\text{mod})\, 2$$

provided that

$$\sum_{j=0}^{n-1} x_j \equiv 0 \,(\text{mod})\, 2\,.$$

We call $\mathbf{x} = x_0 x_1 \cdots x_{n-1}$ the question and $\mathbf{y} = y_0 y_1 \cdots y_{n-1}$ the answer. Show that if the n-players are allowed to share prior entanglement, then they can always win the game G_n.

Solution 7.   We define the following n-qubit entangled state in the Hilbert space $\mathbb{C}^{2^n}$

$$|\psi_+\rangle := \frac{1}{\sqrt{2}}(|00\cdots0\rangle + |11\cdots1\rangle)$$

$$|\psi_-\rangle := \frac{1}{\sqrt{2}}(|00\cdots0\rangle - |11\cdots1\rangle)\,.$$

The *Walsh-Hadamard transform* is given by

$$U_H|0\rangle \to \frac{1}{\sqrt{2}}|0\rangle + \frac{1}{\sqrt{2}}|1\rangle, \qquad U_H|1\rangle \to \frac{1}{\sqrt{2}}|0\rangle - \frac{1}{\sqrt{2}}|1\rangle\,.$$

Furthermore consider the unitary transformation

$$U_S|0\rangle \to |0\rangle, \qquad U_S|1\rangle \to e^{i\pi/2}|1\rangle$$

where $e^{i\pi/2} = i$. If the unitary transformation U_S is applied to any two qubits of $|\psi_+\rangle$, while other qubits are left undisturbed, then

$$U_S|\psi_+\rangle = |\psi_-\rangle$$

and if U_S is applied to any two qubits of $|\psi_-\rangle$, then

$$U_S|\psi_-\rangle = |\psi_+\rangle.$$

Therefore, if the qubits of $|\psi_+\rangle$ are distributed among n players, and if exactly m of them apply S to their qubit, the resulting state will be $|\psi_+\rangle$ if $m \equiv 0 \pmod 4$ and $|\psi_-\rangle$ if $m \equiv 2 \pmod 4$. The effect of applying the Walsh-Hadamard transform to each qubit in $|\psi_+\rangle$ is to produce an equal superposition of all classical n-bit strings that contain an even number of 1s, whereas the effect of applying the Walsh-Hadamard transform to each qubit in $|\psi_-\rangle$ is to produce an equal superposition of all classical n-bits that contain an odd number of 1s. Thus

$$(U_H \otimes U_H \otimes \cdots \otimes U_H)|\psi_+\rangle = \frac{1}{\sqrt{2^{n-1}}} \sum_{\Delta(\mathbf{y})=0 \,(mod\, 2)} |y_0 y_1 \cdots y_{n-1}\rangle$$

$$(U_H \otimes U_H \otimes \cdots \otimes U_H)|\psi_-\rangle = \frac{1}{\sqrt{2^{n-1}}} \sum_{\Delta(\mathbf{y})=1 \,(mod\, 2)} |y_0 y_1 \cdots y_{n-1}\rangle$$

where

$$\Delta(\mathbf{y}) := \sum_{j=0}^{n-1} y_j$$

denotes the *Hamming weight* of $\mathbf{y}$. Consequently the strategy is as follows: At the beginning the state $|\psi_+\rangle$ is produced and its n-qubits are distributed among the n players. After the separation each player A_j receives input bit x_j and does the following

1. If $x_j = 1$, A_j applies the unitary transformation U_S to his qubit; otherwise he/she does nothing.

2. He/she applies U_H to this qubit.

3. He/she measures his/her qubit in order to obtain y_j.

4. He/she produces y_j as his/her output.

An even number of players will apply U_S to their qubit. If that number is divisible by 4, which means that $\frac{1}{2}\sum_{j=0}^{n-1} x_j$ is even, then the states reverts to $|\psi_+\rangle$ after step 1 and therefore to a superposition of all $|y_0 y_1 \cdots y_{n-1}\rangle$

such that $\Delta(\mathbf{y}) \equiv 0 \pmod 2$ after step 2. It follows that $\sum_{j=0}^{n-1} y_i$, the number of players who measure and output 1, is even. If the number of players who apply S to their qubit is congruent to 2 modulo 4, which means that $\frac{1}{2} \sum_{j=0}^{n-1} x_j$ is odd, then the state evolves to $|\psi_-\rangle$ after step 1 and therefore to a superposition of all $|\mathbf{y}\rangle \equiv |y_0 y_1 \cdots y_{n-1}\rangle$ such that $\Delta(\mathbf{y}) \equiv 1 \pmod 2$ after step 2. In this case $\sum_{j=0}^{n-1} y_j$ is odd. In either case, (1) is satisfied at the end of the protocol.

Problem 8. Let $x_0, x_1, y_0, y_1 \in \{0, 1\}$ where Alice has x_0 and x_1 and Bob has y_0 and y_1. Alice and Bob want to calculate the boolean function

$$g(x_0, x_1, y_0, y_1) := x_1 \oplus y_1 \oplus (x_0 \cdot y_0)$$

where $\oplus$ denotes the XOR operation and $\cdot$ denotes the AND operation. Furthermore Alice and Bob share an EPR pair (Bell state)

$$\frac{1}{\sqrt{2}}(|00\rangle - |11\rangle).$$

Alice applies the unitary matrix

$$U_R\left(-\frac{\pi}{16} + x_0 \frac{\pi}{4}\right) \otimes I_2$$

to her qubit of the EPR pair and Bob applies the unitary matrix

$$I_2 \otimes U_R\left(-\frac{\pi}{16} + y_0 \frac{\pi}{4}\right)$$

to his qubit of the EPR pair, where

$$U_R(\theta) := \begin{pmatrix} \cos\theta & -\sin\theta \\ \sin\theta & \cos\theta \end{pmatrix}.$$

Let a denote the result of Alice measuring her qubit of the EPR pair and let b denote the result of Bob measuring his qubit of the EPR pair. Find the probability that $a \oplus b = x_0 \cdot y_0$, where $\oplus$ denotes the boolean XOR operation and $\cdot$ denotes the boolean AND operation.

Solution 8. We define $|\psi\rangle$ to be the state of the EPR-pair after Alice and Bob apply their transforms. Consequently

$$|\psi\rangle := U_R\left(-\frac{\pi}{16} + x_0 \frac{\pi}{4}\right) \otimes U_R\left(-\frac{\pi}{16} + y_0 \frac{\pi}{4}\right) \frac{1}{\sqrt{2}}(|00\rangle - |11\rangle).$$

Thus

$$|\psi\rangle = \frac{1}{\sqrt{2}}\left(\left(\cos\left(-\frac{\pi}{16} + x_0 \frac{\pi}{4}\right)|0\rangle + \sin\left(-\frac{\pi}{16} + x_0 \frac{\pi}{4}\right)|1\rangle\right)\right.$$

$$\otimes \left(\cos\left(-\frac{\pi}{16} + y_0\frac{\pi}{4} \right) |0\rangle + \sin\left(-\frac{\pi}{16} + y_0\frac{\pi}{4} \right) |1\rangle \right)$$

$$- \left(-\sin\left(-\frac{\pi}{16} + x_0\frac{\pi}{4} \right) |0\rangle + \cos\left(-\frac{\pi}{16} + x_0\frac{\pi}{4} \right) |1\rangle \right)$$

$$\otimes \left(-\sin\left(-\frac{\pi}{16} + y_0\frac{\pi}{4} \right) |0\rangle + \cos\left(-\frac{\pi}{16} + y_0\frac{\pi}{4} \right) |1\rangle \right) \Big)$$

$$= \frac{1}{\sqrt{2}} \left(\cos\left(-\frac{\pi}{8} + (x_0 + y_0)\frac{\pi}{4} \right) |00\rangle + \sin\left(-\frac{\pi}{8} + (x_0 + y_0)\frac{\pi}{4} \right) |01\rangle \right.$$

$$\left. + \sin\left(-\frac{\pi}{8} + (x_0 + y_0)\frac{\pi}{4} \right) |10\rangle - \cos\left(-\frac{\pi}{8} + (x_0 + y_0)\frac{\pi}{4} \right) |11\rangle \right).$$

Thus we find for the probabilities of obtaining a and b

a	b	$a \oplus b$	$P(a,b)$
0	0	0	$\frac{1}{2}\cos^2\left(-\frac{\pi}{8} + (x_0 + y_0)\frac{\pi}{4} \right)$
0	1	1	$\frac{1}{2}\sin^2\left(-\frac{\pi}{8} + (x_0 + y_0)\frac{\pi}{4} \right)$
1	0	1	$\frac{1}{2}\sin^2\left(-\frac{\pi}{8} + (x_0 + y_0)\frac{\pi}{4} \right)$
1	1	0	$\frac{1}{2}\cos^2\left(-\frac{\pi}{8} + (x_0 + y_0)\frac{\pi}{4} \right)$

Next we find the probability that

$$a \oplus b = x_0 \cdot y_0$$

for given x_0 and y_0

x_0	y_0	$x_0 \cdot y_0$	$P(a \oplus b = x_0 \cdot y_0)$
0	0	0	$P(a=0, b=0) + P(a=1, b=1) = \cos^2\frac{\pi}{8}$
0	1	0	$P(a=0, b=0) + P(a=1, b=1) = \cos^2\frac{\pi}{8}$
1	0	0	$P(a=0, b=0) + P(a=1, b=1) = \cos^2\frac{\pi}{8}$
1	1	1	$P(a=1, b=0) + P(a=0, b=1) = \cos^2\frac{\pi}{8}$

We find the probability

$$P(a \oplus b = x_0 \cdot y_0) = \cos^2\frac{\pi}{8}.$$

Problem 9. Let G be a finite group and let $\rho : G \to GL(n, \mathbb{C})$ be a representation of G where $n \in \mathbb{N}$. Show that

$$C := \sum_{g \in G} \rho(g)^* \rho(g)$$

is positive definite. Thus the positive definite matrices C^{-1}, $\sqrt{C}$ and $\sqrt{C}^{-1}$ exist. Is $\rho_C : G \to GL(n, \mathbb{C})$ defined by

$$\rho_C(g) := \sqrt{C}\rho(g)\sqrt{C}^{-1}$$

a unitary representation of G?

Solution 9.   Clearly C is hermitian. Since $\rho(g)$ is invertible, we have $\rho(g)\mathbf{x} \neq 0$ for all $\mathbf{x} \neq \mathbf{0}$, $\mathbf{x} \in \mathbb{C}^n$. Thus

$$\mathbf{x}^* C \mathbf{x} = \sum_{g \in G} \mathbf{x}^* \rho(g)^* \rho(g) \mathbf{x} = \sum_{g \in G} \|\rho(g)\mathbf{x}\|^2 > 0.$$

It follows that C is positive definite and the positive definite matrices C^{-1}, $\sqrt{C}$ and $\sqrt{C}^{-1}$ exist. Obviously $\rho_C(g)$ is invertible, then the inverse is given by $\rho_C(g)^{-1} = \sqrt{C}\rho(g)^{-1}\sqrt{C}^{-1}$. Let $g_1, g_2 \in \mathbb{G}$ then

$$\rho_C(g_1)\rho_C(g_2) = \sqrt{C}\rho(g_1)\sqrt{C}^{-1}\sqrt{C}\rho(g_2)\sqrt{C}^{-1} = \sqrt{C}\rho(g_1)\rho(g_2)\sqrt{C}^{-1}$$
$$= \sqrt{C}\rho(g_1 \cdot g_2)\sqrt{C}^{-1}$$

where $\cdot$ is the group operation. Thus ρ_C provides a representation for G. We also have for all $g' \in G$

$$\rho(g')^* C \rho(g') = \sum_{g \in G} \rho(g')^* \rho(g)^* \rho(g) \rho(g') = \sum_{g \in G} (\rho(g)\rho(g'))^* \rho(g)\rho(g')$$
$$= \sum_{g \in G} (\rho(g \cdot g'))^* \rho(g \cdot g') = \sum_{g \in G} \rho(g)^* \rho(g) = C$$

where we used that $\{g' \cdot g : g \in G\} = G$. Now we find

$$\rho_C(g)^* \rho_C(g) = \sqrt{C}^{-1*}\rho(g)^*\sqrt{C}^*\sqrt{C}\rho(g)\sqrt{C}^{-1} = \sqrt{C}^{-1}\rho(g)^* C \rho(g)\sqrt{C}^{-1}$$
$$= \sqrt{C}^{-1} C \sqrt{C}^{-1} = I_n$$

where I_n is the identity operator on $\mathbb{C}^n$. Thus ρ_C is a unitary representation for G.

Problem 10.   Let G be a finite abelian group with identity 0 and let $f : G \to S$ for some finite set S. We use the orthonormal bases

$$\{\,|g\rangle_G : g \in G\,\} \subset \mathbb{C}^{|G|} \quad \text{and} \quad \{\,|s\rangle_S : s \in S\,\} \subset \mathbb{C}^{|S|}.$$

Find a unitary operator U_f such that

$$U_f(|g\rangle_G \otimes |s_0\rangle_S) = |g\rangle_G \otimes |f(g)\rangle_S$$

where $s_0 \in S$.

Solution 10.   Let $h_g : S \setminus \{s_0\} \to S \setminus \{f(g)\}$ be arbitrary one to one functions for each $g \in G$. One solution is

$$U_f = \sum_{g \in G} |g\rangle_G\langle g|_G \otimes \left(|f(g)\rangle_S\langle s_0|_S + \sum_{s \in S \setminus \{s_0\}} |h_g(s)\rangle_S\langle s|_S \right).$$

Problem 11. Let G be a finite abelian group with identity 0 and let $f : G \to S$ for some finite set S. We use the orthonormal bases

$$\{ |g\rangle \ : \ g \in G \} \subset \mathbb{C}^{|G|}.$$

Discuss the quantum Fourier transform over the group G.

Solution 11. Since G is a finite abelian group the irreducible representations of G are of the form $\chi : G \to \mathbb{C}$. There are $|G|$ such representations labeled χ_g for $g \in G$. Since χ is a group homomorphism we have $\chi(g_1 \cdot g_2) = \chi(g_1)\chi(g_2)$ for all $g_1, g_2 \in G$. Irreducible representations $\chi_1 \neq \chi_2$ have the property that

$$\sum_{g \in G} \chi_1(g)\overline{\chi_2(g)} = 0.$$

The trivial irreducible representation $\chi_1(g) = 1$ for all $g \in G$ provides

$$\sum_{g \in G} \chi(g)\overline{\chi_1(g)} = \sum_{g \in G} \chi(g) = 0$$

for nontrivial χ. The transform over the group structure is

$$U_F(G) := \frac{1}{\sqrt{|G|}} \sum_{u \in G} \sum_{v \in G} \overline{\chi_u(v)} |u\rangle\langle v|$$

where the $\chi_l : G \to \mathbb{C}$ are the $|G|$ irreducible representations of G, $\chi_l(g)$ is a $|G|$-th root of unity. For $G = \mathbb{Z}_N$, the group of integers $\{0, 1, \dots, N-1\}$ with addition modulo N, we obtain the quantum Fourier transform.

Problem 12. Let G be a finite abelian group with identity 0, and K be a subgroup of G. The cosets of K, i.e. $g \cdot K := \{ g \cdot k \ : \ k \in K \}$ for $g \in G$, partition G. Now suppose $f : G \to X$, for some finite set X, with the property $f(g_1) = f(g_2)$ if and only if $g_1 \cdot K = g_2 \cdot K$. From f and G we wish to find the hidden subgroup K. As the computational basis we use the orthonormal basis $B_G = \{ |g\rangle_G \ : \ g \in G \} \subset \mathbb{C}^{|G|}$ and the orthonormal basis $B_X = \{ |x\rangle_X \ : \ x \in X \} \subset \mathbb{C}^{|X|}$. We begin with the state

$$\frac{1}{\sqrt{|G|}} \sum_{g \in G} |g\rangle_G \otimes |f(g)\rangle_X.$$

Show how measurement of the second system in the basis B_X and then performing the quantum Fourier transform over the group structure for the first system can be used to find K.

Solution 12.   We can factor out the second state according to the cosets of K

$$\frac{1}{\sqrt{|G|}} \sum_{g \cdot K} \left(\sum_{k \in K} |g \cdot k\rangle_G \right) \otimes |f(g)\rangle_X.$$

Measuring the second state in the basis B_X projects the first state, for some g, onto

$$\frac{1}{\sqrt{|K|}} \sum_{k \in K} |g \cdot k\rangle_G.$$

Applying $U_F(G)$ yields

$$U_F(G) \frac{1}{\sqrt{|K|}} \sum_{k \in K} |g \cdot k\rangle_G = \frac{1}{\sqrt{|G| \cdot |K|}} \sum_{u \in G} \left(\sum_{k \in K} \overline{\chi_u(g \cdot k)} \right) |u\rangle_G$$

$$= \frac{1}{\sqrt{|G| \cdot |K|}} \sum_{u \in G} \overline{\chi_u(g)} \left(\sum_{k \in K} \overline{\chi_u(k)} \right) |u\rangle_G$$

since the irreducible representations are group homomorphisms. Due to the fact that the irreducible representation χ_u of G can also be considered as an irreducible representation of the subgroup K we have

$$\sum_{k \in K} \chi(k) = 0$$

for nontrivial χ so that

$$U_F(G) \frac{1}{\sqrt{|K|}} \sum_{k \in K} |g \cdot k\rangle_G = \frac{1}{\sqrt{|G| \cdot |K|}} \sum_{u \in G} \overline{\chi_u(g)} \left(\sum_{k \in K} \overline{\chi_u(k)} \right) |u\rangle_G$$

$$= \sqrt{\frac{|K|}{|G|}} \sum_{\substack{u \in G \\ \chi_u(k)=1 \ \forall k \in K}} \overline{\chi_u(g)} |u\rangle_G.$$

In the last result we find non-zero probability amplitudes for $|u\rangle_G$ when $\chi_u(k) = 1$ for all $k \in K$. Measurement of the state yields $u \in G$ with $\chi_u(k) = 1 \ \forall k \in K$. We can consequently test the different K against χ_u. If K has a generator k_0, then $\chi_u(k_0) = 1$ so that measurement provides the possibility of recovering the generator of K. In general, the process must be repeated since we may have found a generator for a subgroup.

Chapter 15

Quantum Error Correction

In classical communication theory where bits are communicated, the only possible type of error that can occur is a bit flip. In the quantum case any rotation or phase change in the Hilbert space of the quantum state is an error. Thus there are an infinite number of different errors that could occur just for a single qubit. Fortunately the measurement process involves the projection of the quantum state into a compatible subspace. Thus measurement to determine the occurrence of an error reduces the error to one compatible with the measurement. Suppose the data is contained in the state $|\psi\rangle$, and the environment is described by the state $|E\rangle$. The initial state of the entire system is described by the tensor product of the states $|\psi\rangle \otimes |E\rangle$, which evolves according to some unitary operation U. The state $|\psi\rangle$ evolves according to the unitary operation U_ψ which describes a quantum algorithm. In classical error correction codes, all that needs to be corrected are bit flips. In the quantum case errors such as bit flips, phase changes and rotations complicate the error correction techniques. Since arbitrary errors in an encoding of information cannot be corrected, only certain types of errors are assumed to occur. The types of errors depend on the implementation. For example, suppose the types of errors (which we assume are distinguishable due to an encoding) are described by the unitary basis $E_1, \ldots, E_n$ so that all errors are described by a linear combination

$$E = c_1 E_1 + \cdots + c_n E_n, \qquad E^\dagger E = I$$

where I is the identity operator. We use the product state $|\psi\rangle \otimes |0\rangle$, where

265

$|\psi\rangle$ is an encoded quantum state with the necessary property that it can be used to determine if any error of $E_1, \ldots, E_n$ has occurred, and the second quantum register will hold the number of the type of error which occurred. Let S denote the operator for the *error syndrome*

$$S(E_j \otimes I)|\psi\rangle \otimes |0\rangle := |\psi\rangle \otimes |j\rangle.$$

Now the encoded state with errors is given by

$$(E \otimes I)|\psi\rangle \otimes |0\rangle = \sum_{j=1}^{n} E_j|\psi\rangle \otimes |0\rangle.$$

Applying the operator for the error syndrome gives

$$S(E \otimes I)|\psi\rangle \otimes |0\rangle = \sum_{j=1}^{n} E_j|\psi\rangle \otimes |j\rangle.$$

Measuring the second register identifies the error. Suppose the measurement corresponds to $|k\rangle$, then the error can be repaired since

$$(E_k^{-1} \otimes I)((E_k \otimes I)|\psi\rangle \otimes |k\rangle) = |\psi\rangle \otimes |k\rangle.$$

Problem 1. Calculate the following in terms of I_2, X, Y, Z

(i) XZ

(ii) ZX

(iii) $U_{CNOT}(X \otimes I_2)U_{CNOT}$

(iv) $U_{CNOT}(I_2 \otimes X)U_{CNOT}$

(v) $U_{CNOT}(Z \otimes I_2)U_{CNOT}$

(vi) $U_{CNOT}(I_2 \otimes Z)U_{CNOT}$

(vii) $U_{CNOT}(X \otimes X)U_{CNOT}$

(viii) $U_{CNOT}(Z \otimes Z)U_{CNOT}$

(ix) $U_{CNOT}U_{CNOT}$

where

$$I_2 := |0\rangle\langle 0| + |1\rangle\langle 1|$$
$$X := |0\rangle\langle 1| + |1\rangle\langle 0|$$
$$Y := |0\rangle\langle 1| - |1\rangle\langle 0|$$
$$Z := |0\rangle\langle 0| - |1\rangle\langle 1|$$
$$U_{CNOT} := |0\rangle\langle 0| \otimes I_2 + |1\rangle\langle 1| \otimes X.$$

Solution 1.   Straightforward calculation yields

$$(i)\ XZ = -Y$$
$$(ii)\ ZX = Y$$
$$(iii)\ U_{CNOT}(X \otimes I_2)U_{CNOT} = X \otimes X$$
$$(iv)\ U_{CNOT}(I_2 \otimes X)U_{CNOT} = I_2 \otimes X$$
$$(v)\ U_{CNOT}(Z \otimes I_2)U_{CNOT} = Z \otimes I_2$$
$$(vi)\ U_{CNOT}(I_2 \otimes Z)U_{CNOT} = Z \otimes Z$$
$$(vii)\ U_{CNOT}(X \otimes X)U_{CNOT} = X \otimes I_2$$
$$(viii)\ U_{CNOT}(Z \otimes Z)U_{CNOT} = I_2 \otimes Z$$
$$(ix)\ U_{CNOT}U_{CNOT} = I_2 \otimes I_2 .$$

Problem 2.   Suppose that the only errors which can occur to three qubits are described by the set of unitary matrices

$$\{ I_2 \otimes I_2 \otimes I_2, I_2 \otimes U_{NOT} \otimes U_{NOT}, I_2 \otimes U_P \otimes U_P, I_2 \otimes (U_P U_{NOT}) \otimes (U_P U_{NOT}) \}$$

where

$$U_P := |0\rangle\langle 0| - |1\rangle\langle 1|, \quad U_{NOT} := |0\rangle\langle 1| + |1\rangle\langle 0| .$$

A linear combination of these unitary matrices is given by

$$E := \alpha I_2 \otimes I_2 \otimes I_2 + \beta I_2 \otimes U_{NOT} \otimes U_{NOT} + \delta I_2 \otimes U_P \otimes U_P$$
$$+ \gamma I_2 \otimes (U_P U_{NOT}) \otimes (U_P U_{NOT})$$

where $\alpha, \beta, \delta, \gamma \in \mathbb{C}$. Describe how an arbitrary error E on the three-qubit state

$$\frac{1}{\sqrt{2}}(|00\rangle + |11\rangle) \otimes |\psi\rangle$$

can be corrected to obtain the correct $|\psi\rangle$ as the last qubit, where

$$|\psi\rangle := a|0\rangle + b|1\rangle, \quad |a|^2 + |b|^2 = 1, \quad a, b \in \mathbb{C} .$$

Solution 2.   Applying the matrix $(I_8 \equiv I_2 \otimes I_2 \otimes I_2)$

$$\alpha I_8 + \beta I_2 \otimes U_{NOT} \otimes U_{NOT} + \delta I_2 \otimes U_P \otimes U_P + \gamma I_2 \otimes (U_P U_{NOT}) \otimes (U_P U_{NOT})$$

to the state

$$\frac{1}{\sqrt{2}}(|00\rangle + |11\rangle) \otimes |\psi\rangle$$

yields the state

$$\alpha \frac{1}{\sqrt{2}}(|00\rangle + |11\rangle) \otimes |\psi\rangle$$

$$+ \beta \frac{1}{\sqrt{2}}(|01\rangle + |10\rangle) \otimes (a|1\rangle + b|0\rangle)$$

$$+ \delta \frac{1}{\sqrt{2}}(|00\rangle - |11\rangle) \otimes (a|0\rangle - b|1\rangle)$$

$$+ \gamma \frac{1}{\sqrt{2}}(|01\rangle - |10\rangle) \otimes (a|1\rangle - b|0\rangle).$$

Thus we measure the first two qubits in the Bell basis and apply the corresponding transform to the last qubit to obtain $|\psi\rangle$.

measure	transform
$\frac{1}{\sqrt{2}}(|00\rangle + |11\rangle)$	I_2
$\frac{1}{\sqrt{2}}(|01\rangle + |10\rangle)$	U_{NOT}
$\frac{1}{\sqrt{2}}(|00\rangle - |11\rangle)$	U_P
$\frac{1}{\sqrt{2}}(|01\rangle - |10\rangle)$	$U_{NOT}U_P$

Problem 3.   Assume that the only errors that occur in a system of qubits are isolated to individual qubits, i.e., the error in one qubit state is independent of the error in another qubit state. Hence the error for each qubit can be expressed as a linear operator E on the Hilbert space $\mathbb{C}^2$. Furthermore E can be expressed as a linear combination of the 2×2 identity matrix and the Pauli spin matrices σ_x, σ_y and σ_z. Now consider a non-degenerate n-qubit code representing a single qubit state which can correct errors in up to k qubits.
(i) Find a lower bound describing n.
(ii) Find the lower bound for $k = 1$.

Hint. The n-qubit states representing qubits with errors should be distinct (orthogonal) for distinct errors and distinct from the case where there are no errors.

Solution 3.   (i) We have 3 distinct errors on a single qubit described by the Pauli matrices. Thus there are

$$3^l \binom{n}{l} \equiv 3^l \frac{n!}{l!(n-l)!}$$

distinct errors in l qubits of n qubits. The total number of ways to have at most k errors in n qubits is then given by

$$\sum_{l=0}^{k} 3^l \binom{n}{l}.$$

There are 2^n orthogonal states in a Hilbert space describing n qubits. Since the states representing qubits ($|0\rangle$ or $|1\rangle$) with distinct errors should be orthogonal, we find

$$2 \sum_{l=0}^{k} 3^l \binom{n}{l} \leq 2^n.$$

(ii) For $k = 1$ we have the bound

$$2(1 + 3n) \leq 2^n.$$

In other words, for $k = 1$ we find $n \geq 5$.

Problem 4.   (i) Do the Pauli matrices (including the 2×2 unit matrix) form a group under matrix multiplication? If so, provide a proof. If not, what set with minimal cardinality includes the Pauli matrices and forms a group under matrix multiplication? The Pauli matrices are

$$I_2 = \begin{pmatrix} 1 & 0 \\ 0 & 1 \end{pmatrix}, \quad \sigma_x = \begin{pmatrix} 0 & 1 \\ 1 & 0 \end{pmatrix}, \quad \sigma_y = \begin{pmatrix} 0 & -i \\ i & 0 \end{pmatrix}, \quad \sigma_z = \begin{pmatrix} 1 & 0 \\ 0 & -1 \end{pmatrix}.$$

(ii) Show that the Pauli matrices form a basis for the 2×2 matrices over $\mathbb{C}$.

Solution 4.   (i) Obviously the Pauli matrices do not form a group since $\sigma_x \sigma_y = i\sigma_z$ which is not a Pauli matrix. Since the factor i is introduced, all factors which are powers of i must be included i.e. ± 1 and $\pm i$:

$$\{ \pm\sigma, \, \pm i\sigma \, | \, \sigma \in \{ I, \sigma_x, \sigma_y, \sigma_z \} \}.$$

Thus we have a group of cardinality 16.
(ii) We have

$$\begin{pmatrix} a & b \\ c & d \end{pmatrix} = \frac{a+d}{2} I + \frac{a-d}{2} \sigma_z + \frac{b+c}{2} \sigma_x + \frac{b-c}{2} i\sigma_y.$$

Alternatively

$$c_1 I + c_2 \sigma_x + c_3 \sigma_y + c_4 \sigma_z = 0$$

leads to the four linear equations

$$c_1 + c_4 = 0, \qquad c_1 - c_4 = 0, \qquad c_2 - ic_3 = 0, \qquad c_2 + ic_3 = 0.$$

These equations obviously only have the solution $c_1 = c_2 = c_3 = c_4 = 0$. These properties can be used in quantum error corrections.

Problem 5.   Let S be a set of operators closed under the hermitian conjugate (adjoint) such that the quantum code C_S of the set of states

$$C_S := \{ \, |\psi\rangle \, : \, U|\psi\rangle = |\psi\rangle, \quad \forall U \in S \, \}$$

is non trivial (does not consist only of the zero state). The set S is called the *stabilizer* of the code C_S.
(i) Show that

$$[M, N]|\psi\rangle = 0, \qquad \forall M, N \in S, \quad |\psi\rangle \in C_S.$$

(ii) Let $[E, M] = 0$ for some $M \in S$. What can be said about $E|\psi\rangle$ when $|\psi\rangle \in C_S$?
(iii) Let $[E, M]_+ = 0$ for some $M \in S$. What can be said about $E|\psi\rangle$ when $|\psi\rangle \in C_S$?

Solution 5.   (i) Since

$$MN|\psi\rangle = M|\psi\rangle = |\psi\rangle$$

$$NM|\psi\rangle = N|\psi\rangle = |\psi\rangle$$

it follows that

$$[M, N]|\psi\rangle = 0.$$

(ii) For $|\psi\rangle \in C_S$ we have

$$ME|\psi\rangle = EM|\psi\rangle = E|\psi\rangle \,.$$

Thus $E|\psi\rangle$ is an eigenstate of M corresponding to the eigenvalue 1.
(iii) For $|\phi\rangle, |\psi\rangle \in C_S$ we have

$$\langle\phi|E|\psi\rangle = \langle\phi|EM|\psi\rangle = -\langle\phi|ME|\psi\rangle = -\langle\phi|M^*E|\psi\rangle = -\langle\phi|E|\psi\rangle.$$

Thus $\langle\phi|E|\psi\rangle = 0$. In other words, $E|\psi\rangle \notin C_S$. Furthermore

$$ME|\psi\rangle = -EM|\psi\rangle = -E|\psi\rangle$$

thus $E|\psi\rangle$ is an eigenstate of M corresponding to the eigenvalue -1.

Problem 6.   Consider the application of the controlled NOT gate to the state $|10\rangle \equiv |1\rangle \otimes |0\rangle$. Suppose that a single qubit error $\sigma_x \otimes I_2$ occurs before the controlled NOT operation is performed. How many errors does the resulting state have after performing the control NOT operation? Also discuss the case when the error is given by

$$\frac{1}{\sqrt{2}}(I_2 + \sigma_x) \otimes I_2 .$$

Solution 6. The intended operation is

$$U_{CNOT}|10\rangle = |11\rangle .$$

The operation with error is

$$U_{CNOT}(\sigma_x \otimes I_2)|10\rangle = U_{CNOT}|00\rangle = |00\rangle.$$

Thus we have two single qubit errors $\sigma_x \otimes I_2$ and $I_2 \otimes \sigma_x$. For the second type of single qubit error we have

$$U_{CNOT}\frac{1}{\sqrt{2}}(I_2 + \sigma_x) \otimes I_2|10\rangle = U_{CNOT}\frac{1}{\sqrt{2}}(|10\rangle + |00\rangle) = \frac{1}{\sqrt{2}}(|00\rangle + |11\rangle).$$

Here we find that the error cannot be expressed in terms of single qubit errors.

Problem 7. Consider the application of the controlled NOT gate to corresponding pairs of qubits of the state $|\tilde{1}\tilde{0}\rangle$ where $|\tilde{0}\rangle$ is a code word for $|0\rangle$ and $|\tilde{1}\rangle$ is a code word for $|1\rangle$ in a quantum error correction code that can correct single qubit errors. In other words the controlled NOT gate is applied to the i-th qubit of $|\tilde{1}\rangle$ as control and the i-th qubit of $|\tilde{0}\rangle$ as target. Suppose that a single qubit error

$$\frac{1}{\sqrt{2}}(I_2 + \sigma_x)$$

on one of the qubits of $|\tilde{1}\rangle$ occurs before the controlled NOT operation is performed. Discuss the error correction of this computation.

Solution 7. Suppose the single qubit error occurs on the j-th qubit, and that the encoding $|\tilde{1}\rangle$ and $|\tilde{0}\rangle$ is n qubits long. Thus the error is

$$E := \frac{1}{\sqrt{2}}\left(\bigotimes_{k=1}^{j-1} I_2\right) \otimes (I_2 + \sigma_x) \otimes \left(\bigotimes_{k=j+1}^{2n} I_2\right) .$$

Let $\tilde{U}_{CNOT}$ denote the pairwise application of the controlled NOT operation. Then

$$\tilde{U}_{CNOT}E|\tilde{1}\tilde{0}\rangle = \tilde{U}_{CNOT}E\tilde{U}_{CNOT}^*\tilde{U}_{CNOT}|\tilde{1}\tilde{0}\rangle.$$

The controlled NOT operation is given by

$$U_{CNOT} = \frac{1}{2}(I_2 + \sigma_z) \otimes I_2 + \frac{1}{2}(I_2 - \sigma_z) \otimes \sigma_x.$$

We find the matrix

$$\frac{1}{\sqrt{2}}U_{CNOT}(I_2 + \sigma_x)U_{CNOT}^* = \frac{1}{\sqrt{2}}(I_2 \otimes I_2 + \sigma_x \otimes \sigma_x).$$

Consequently we obtain the normalized state

$$\tilde{U}_{CNOT}E|\tilde{1}\tilde{0}\rangle = \left(\frac{1}{\sqrt{2}} \left(\bigotimes_{k=1}^{2n} I_2 \right) \right.$$

$$\left. + \frac{1}{\sqrt{2}} \left(\bigotimes_{k=1}^{j-1} I_2 \right) \otimes \sigma_x \otimes \left(\bigotimes_{k=j+1}^{n+j-1} I_2 \right) \otimes \sigma_x \otimes \left(\bigotimes_{k=n+j+1}^{2n} I_2 \right) \right)$$

$$\times \tilde{U}_{CNOT}|\tilde{1}\tilde{0}\rangle.$$

Thus there is a linear combination of no error, or a single qubit error in the first block of n qubits and a single qubit error in the second block of n qubits. Both errors can be independently corrected.

Chapter 16

Quantum Cryptography

Cryptography usually involves a key or keys to be used in encryption and decryption algorithms. Classical cryptography generally relies on maps that are perceived to be very difficult to invert with incomplete information. One popular algorithm due to Rivest, Shamir and Adelman is the RSA algorithm

$$n = pq, \qquad M \in \{0, 1, \ldots, n-1\},$$

$$ed = 1 \bmod (p-1)(q-1)$$

$$C := (M^e \bmod n), \qquad M \equiv (C^d \bmod n).$$

where p and q are large prime numbers, M is the message and C is the encrypted message. If p, q and d are unknown then, in general, C cannot easily be obtained from M. However, Shor found that p and q can be obtained with relative ease (and consequently also d) using the quantum Fourier transform.

Quantum cryptography is concerned with the secure distribution of keys using quantum communication channels. Another application is hiding classical data in quantum states. Quantum cryptographic techniques rely on physics to supply secure communication in the sense that it is possible to determine if someone has intercepted the message. This is due to the fact that measurement in quantum mechanics is associated with a disturbance (projection) of the quantum state. Entanglement can also be used to detect whether a message has been intercepted. For example, when one qubit of an EPR pair is measured the correlation is destroyed which can be tested with Bell's inequality.

Problem 1.   Let p, q be large prime numbers, (e, n) a public key, and (d, n) a private key for the RSA cryptosystem where $n = pq$. Let $M, C \in \{0, 1, \ldots, n - 1\}$ with

$$C \equiv M^e \bmod n, \qquad M \equiv C^d \bmod n.$$

(i) Let $gcd(a, n) = 1$. Show that p and q can be determined from even $r \in \mathbb{N}$ if

$$a^r \equiv 1 \bmod n, \quad a^{r/2} \not\equiv 1 \bmod n, \quad a^{r/2} \not\equiv 1 \bmod n.$$

(ii) Illustrate (i) with $p = 5$, $q = 11$ and $a = 6$.

(iii) Show that M can be determined from C and $t \in \mathbb{N}$, where (t minimal)

$$C^{et} \equiv C \bmod n.$$

(iv) Illustrate (iii) with $p = 5$, $q = 11$, $e = 9$ and $C = 48$.

Solution 1.   (i) We have

$$a^r - 1 \equiv 0 \bmod n$$

$$(a^{\frac{r}{2}} - 1)(a^{\frac{r}{2}} + 1) \equiv 0 \bmod n$$

Consequently

$$[(a^{\frac{r}{2}} - 1) \bmod n][(a^{\frac{r}{2}} + 1) \bmod n] = kn = kpq$$

for some $k \in \mathbb{Z}$. Consequently one of p and q is given by

$$gcd(a^{\frac{r}{2}} - 1 \bmod n, n).$$

The second is found by division of n. The gcd can be determined efficiently using the Eulidean algorithm.

(ii) The powers (from 0) of a in modulo $5 \cdot 11 = 55$ arithmetic are

$$1, 6, 36, 216 \equiv 51, \equiv 306 \equiv 31, \equiv 186 \equiv 21, \equiv 126 \equiv 16, \equiv 96 \equiv 41,$$

$$\equiv 246 \equiv 26, \equiv 156 \equiv 46, \equiv 276 \equiv 1$$

where the last power is 10. Thus we use $r = 10$. Thus the periodicity of this sequence is also 10. Consequently

$$6^5 - 1 = 7775 \equiv 20 \bmod n.$$

Using the Euclidean algorithm

$$55 = 2 \cdot 20 + 15, \quad 20 = 1 \cdot 15 + 5, \quad 15 = 3 \cdot 5 + 0$$

Thus we find the $gcd(20, 55) = 5$. Consequently $p = 5$ and $q = n/5 = 11$.
(iii) We find

$$C^t \equiv M^{et} \equiv C^{det} \equiv (C^{et})^d \equiv C^d \equiv M \bmod n.$$

(iv) We find

$$C^e = 48^9 = 4^9 4^9 3^9 \equiv 14 \cdot 14 \cdot 48 \equiv 3.$$

The powers (from 0) of C^e in modulo $5 \cdot 11 = 55$ arithmetic are

$$1, 3, 9, 27, 26, 23, 14, 42, 16, 48$$

i.e. we use $t = 9$. Now

$$C^t \equiv 3 \equiv M \bmod n.$$

Problem 2. Apply the quantum Fourier transform to the first register of

$$|\psi\rangle := \frac{1}{\sqrt{6}}(|0\rangle \otimes |0\rangle + |1\rangle \otimes |1\rangle + |2\rangle \otimes |2\rangle + |3\rangle \otimes |0\rangle + |4\rangle \otimes |1\rangle + |5\rangle \otimes |2\rangle)$$

to find the underlying periodicity in the second register. In this case the quantum Fourier transform is given by

$$U_{QFT} = \frac{1}{\sqrt{6}} \sum_{j,k=0}^{5} e^{-i2\pi jk/6} |j\rangle\langle k|.$$

Solution 2. Applying the quantum Fourier transform yields

$$(U_{QFT} \otimes I)|\psi\rangle = \frac{1}{6} \sum_{j=0}^{5} |j\rangle \otimes \left(|0\rangle + e^{-i2\pi j/6}|1\rangle + e^{-i2\pi j2/6}|2\rangle \right.$$

$$\left. + e^{-i2\pi j3/6}|0\rangle + e^{-i2\pi j4/6}|1\rangle + e^{-i2\pi j5/6}|2\rangle \right)$$

$$= \frac{1}{6} \sum_{j=0}^{5} \left[1 + e^{-i\pi j}\right] |j\rangle \otimes \left(|0\rangle + e^{-i\pi j/3}|1\rangle + e^{-i2\pi j/3}|2\rangle \right)$$

$$= \frac{1}{3} \sum_{j=0}^{2} |2j\rangle \otimes \left(|0\rangle + e^{-i\pi 2j/3}|1\rangle + e^{-i4\pi j/3}|2\rangle \right).$$

Thus measuring the first register yields 0, 2 or 4. The minimum positive value is 2. The period is $6/2 = 3$.

Problem 3. Let $B_1 = \{|0\rangle, |1\rangle\}$ and $B_2 = \{|0_H\rangle, |1_H\rangle\}$ denote two orthonormal bases in $\mathbb{C}^2$ where

$$|0_H\rangle := \frac{1}{\sqrt{2}}(|0\rangle + |1\rangle), \qquad |1_H\rangle := \frac{1}{\sqrt{2}}(|0\rangle - |1\rangle).$$

Show that

$$|\psi\rangle := \frac{1}{\sqrt{2}}(|0\rangle \otimes |1\rangle - |1\rangle \otimes |0\rangle) = -\frac{1}{\sqrt{2}}(|0_H\rangle \otimes |1_H\rangle - |1_H\rangle \otimes |0_H\rangle).$$

Solution 3. A simple calculation yields

$$|\psi\rangle := \frac{1}{\sqrt{2}}(|0\rangle \otimes |1\rangle - |1\rangle \otimes |0\rangle)$$

$$= \frac{1}{2\sqrt{2}}([|0_H\rangle + |1_H\rangle] \otimes [|0_H\rangle - |1_H\rangle] - [|0_H\rangle - |1_H\rangle] \otimes [|0_H\rangle + |1_H\rangle])$$

$$= -\frac{1}{\sqrt{2}}(|0_H\rangle \otimes |1_H\rangle - |1_H\rangle \otimes |0_H\rangle).$$

Problem 4. Let

$$B_1 := \{ \ |\psi_0\rangle := |H\rangle, \ |\psi_1\rangle := |V\rangle \ \}$$

denote an orthonormal basis in the Hilbert space $\mathbb{C}^2$. The states $|H\rangle$ and $|V\rangle$ can be identified with the horizontal and vertical polarization of a photon. Let

$$B_2 := \left\{ \ |\phi_0\rangle := \frac{1}{\sqrt{2}}(|H\rangle + |V\rangle), \ |\phi_1\rangle := \frac{1}{\sqrt{2}}(|H\rangle - |V\rangle) \ \right\}$$

denote a second orthonormal basis in $\mathbb{C}^2$. These states are identified with the 45° and -45° polarization of a photon. Alice sends photons randomly prepared in one of the four states $|H\rangle$, $|V\rangle$, $|\phi_0\rangle$ and $|\phi_1\rangle$ to Bob. Bob then randomly chooses a basis B_1 or B_2 to measure the polarization of the photon. All random decisions follow the uniform distribution. Alice and Bob interpret $|\psi_0\rangle$ as binary 0 and $|\psi_1\rangle$ as binary 1 in the basis B_1. They interpret $|\phi_0\rangle$ as binary 0 and $|\phi_1\rangle$ as binary 1 in the basis B_2.
(i) What is the probability that Bob measures the photon in the state prepared by Alice, i.e., what is the probability that the binary interpretation is identical for Alice and Bob?
(ii) An eavesdropper (named Eve) intercepts the photons sent to Bob and then resends a photon to Bob. Eve also detects the photon polarization in one of the bases B_1 or B_2 before resending. What is the probability that the binary interpretation is identical for Alice and Bob?

Solution 4. (i) The probability that Alice chooses to prepare a state from the basis B_1 is $\frac{1}{2}$ and from B_2 is $\frac{1}{2}$. Similarly the probabilities that Bob chooses to measure in the basis B_1 and B_2 are also $\frac{1}{2}$. Thus the probability that Alice and Bob measure in the same basis is $\frac{1}{4} + \frac{1}{4} = \frac{1}{2}$. To determine the correlations in the binary interpretation we consider the two cases (a) Alice and Bob use the same basis and (b) Alice and Bob use a different basis. The cases (a) and (b) have equal probability of $\frac{1}{2}$. For the case (a) Alice and Bob have the same binary interpretation. For the case (b) we note that

$$|\langle \psi_0 | \phi_0 \rangle|^2 = |\langle \psi_0 | \phi_1 \rangle|^2 = |\langle \psi_1 | \phi_0 \rangle|^2 = |\langle \psi_1 | \phi_1 \rangle|^2 = \frac{1}{2}.$$

In other words, if Bob uses the wrong basis he obtains the correct binary interpretation with probability $\frac{1}{2}$. Therefore the total probability that Alice and Bob have the same binary interpretation is

$$\frac{1}{2} \cdot 1 + \frac{1}{2} \cdot \frac{1}{2} = \frac{3}{4}.$$

Thus 75% of the photons sent by Alice have an identical binary interpretation shared by Alice and Bob.

(ii) From (i) the probability that Alice and Eve, Eve and Bob, as well as Alice and Bob measure in the same basis are all $\frac{1}{4} + \frac{1}{4} = \frac{1}{2}$. Also from (i) we find that if Alice and Eve work in the same basis Bob has a 75% chance of obtaining the correct result since Eve does not perturb the state of the photon. Similarly if Bob and Eve work in the same basis Bob has a 75% chance of obtaining the correct result since Bob does not perturb the state of the photon after Eve resends it. Now we consider the case when Eve uses a different basis from that of Alice and Bob. Suppose Alice sends $|\psi_0\rangle$ from B_1, and Eve measures in B_2. Thus Eve will obtain $|\phi_0\rangle$ or $|\phi_1\rangle$ with equal probability $\frac{1}{2}$. Now Bob measures in the basis B_1 and obtains $|\psi_0\rangle$ with probability $\frac{1}{2}$ or $|\psi_1\rangle$ with probability $\frac{1}{2}$. Thus we can construct the following table where P_1 is the probability that Eve obtains Alice's binary interpretation of the state correctly and P_2 is the probability that Bob obtains Alice's binary interpretation of the state correctly.

Alice's basis	Eve's basis	Bob's basis	P_1	P_2
B_1	B_1	B_1	1	1
B_1	B_1	B_2	1	1/2
B_1	B_2	B_1	1/2	1/2
B_1	B_2	B_2	1/2	1/2
B_2	B_1	B_1	1/2	1/2
B_2	B_1	B_2	1/2	1/2
B_2	B_2	B_1	1	1/2
B_2	B_2	B_2	1	1

The total probability that Bob's binary interpretation corresponds to Alice's binary interpretation is

$$\frac{1}{8}\left(1 + \frac{1}{2} + \frac{1}{2} + \frac{1}{2} + \frac{1}{2} + \frac{1}{2} + \frac{1}{2} + 1\right) = \frac{5}{8}.$$

i.e., 62.5%.

Problem 5.   (i) Consider the two-qubit singlet state in the Hilbert space $\mathbb{C}^4$

$$|\psi\rangle = \frac{1}{\sqrt{2}}(|01\rangle - |10\rangle) \equiv \frac{1}{\sqrt{2}}(|0\rangle \otimes |1\rangle - |1\rangle \otimes |0\rangle).$$

Let U be a 2×2 unitary matrix with $\det(U) = 1$. Find the state $(U \otimes U)|\psi\rangle$.
(ii) Consider the state

$$|\psi\rangle = \frac{1}{2\sqrt{3}}(2|0011\rangle - |0101\rangle - |0110\rangle - |1001\rangle - |1010\rangle + 2|1100\rangle)$$

in the Hilbert space $\mathbb{C}^{16}$. This state is an extension of the two-qubit singlet state given in (i). Calculate the state $(U \otimes U \otimes U \otimes U)|\psi\rangle$.
(iii) The state given in (i) and (ii) can be extended to arbitrary N ($N =$ even) as follows

$$|\psi\rangle = \frac{1}{(N/2)!\sqrt{N/2+1}} \sum_{\substack{permutations \\ 0...01...1}} p!\left(\frac{N}{2} - p\right)!(-1)^{N/2-p}|j_1\, j_2 \cdots j_N\rangle$$

where the sum is extended over all the states obtained by permuting the state

$$|0 \ldots 0 1 \ldots 1\rangle \equiv |0\rangle \otimes \cdots \otimes |0\rangle \otimes |1\rangle \otimes \cdots \otimes |1\rangle$$

which contains the same number of 0s and 1s and p is the number of 0s in the first $N/2$ positions. Thus the state is a singlet state. Let

$$U^{\otimes N} \equiv U \otimes \cdots \otimes U \qquad N - \text{times}$$

Find the state $U^{\otimes N}|\psi\rangle$.

Solution 5.   (i) A unitary transformation for 2×2 matrices is given by

$$|0\rangle \to a|0\rangle + b|1\rangle, \qquad |1\rangle \to c|0\rangle + d|1\rangle$$

where $ad - bc = e^{i\phi}$ ($\phi \in \mathbb{R}$). We obtain

$$(U \otimes U)|\psi\rangle = \frac{e^{i\phi}}{\sqrt{2}}(|0\rangle \otimes |1\rangle - |1\rangle \otimes |0\rangle).$$

For $\phi = 0$ ($\det(U) = 1$), we obtain the eigenvalue equation $(U \otimes U)|\psi\rangle = |\psi\rangle$.
(ii) Using the results from (i) and $\det(U) = 1$, we find

$$(U \otimes U \otimes U \otimes U)|\psi\rangle = |\psi\rangle.$$

(iii) Using the result from (i), we also find $U^{\otimes N}|\psi\rangle = |\psi\rangle$. The state $|\psi\rangle$ given in (iii) can be used to distribute cryptographic keys, encode quantum information in decoherence-free subspaces, perform secret sharing, teleclone quantum states, and also for solving the liar detection and Byzantine generals problems.

Problem 6. Let ρ denote an arbitrary 4×4 density matrix. Consider the unitary operators (bilateral rotations)

$$B_x := \frac{1}{2}\begin{pmatrix} 1 & i \\ i & 1 \end{pmatrix} \otimes \begin{pmatrix} 1 & i \\ i & 1 \end{pmatrix}, \quad B_y := \frac{1}{2}\begin{pmatrix} 1 & 1 \\ -1 & 1 \end{pmatrix} \otimes \begin{pmatrix} 1 & 1 \\ -1 & 1 \end{pmatrix},$$

$$B_z := i\begin{pmatrix} 1 & 0 \\ 0 & -i \end{pmatrix} \otimes \begin{pmatrix} 1 & 0 \\ 0 & -i \end{pmatrix}$$

and the unitary operators

$$U_1 := I_4, \ U_2 := B_x^2, \ U_3 := B_y^2, \ U_4 := B_z^2, \ U_5 := B_x B_y, \ U_6 := B_y B_z,$$

$$U_7 := B_z B_x, \ U_8 := B_y B_x, \ U_9 := U_5^2, \ U_{10} := U_6^2, \ U_{11} := U_7^2, \ U_{12} := U_8^2.$$

The mixed state ρ_W is prepared by transforming 12 systems each described by the mixed state ρ according to each of the operators $U_1, \ldots, U_{12}$. Calculate the density matrix

$$\rho_W := \frac{1}{12}\sum_{j=1}^{12} U_j \rho U_j^*.$$

Express ρ_W in terms of the Bell basis.

Solution 6. An arbitrary 4×4 density matrix ρ can be written in the form

$$\begin{pmatrix} a_{11} & a_{12} & a_{13} & a_{14} \\ \overline{a_{12}} & a_{22} & a_{23} & a_{24} \\ \overline{a_{13}} & \overline{a_{23}} & a_{33} & a_{34} \\ \overline{a_{14}} & \overline{a_{24}} & \overline{a_{34}} & 1 - a_{11} - a_{22} - a_{33} \end{pmatrix}$$

where $a_{11}, a_{12}, a_{13} \in \mathbb{R}$. We find

$$\rho_W = \frac{1}{6}\begin{pmatrix} 2 - 2F & 0 & 0 & 0 \\ 0 & 1 + 2F & 1 - 4F & 0 \\ 0 & 1 - 4F & 1 + 2F & 0 \\ 0 & 0 & 0 & 2 - 2F \end{pmatrix}$$

where $F := (a_{22} + a_{33} - 2\Re a_{23})/2$. In terms of the Bell basis we have

$$\rho_W = F|\psi^-\rangle\langle\psi^-| + \frac{1 - F}{3}(|\psi^+\rangle\langle\psi^+| + |\phi^+\rangle\langle\phi^+| + |\phi^-\rangle\langle\phi^-|).$$

Problem 7.   Let

$$\rho_W := F|\phi^+\rangle\langle\phi^+| + \frac{1-F}{3}(|\psi^+\rangle\langle\psi^+| + |\psi^-\rangle\langle\psi^-| + |\phi^-\rangle\langle\phi^-|)$$

where $F \in [0,1]$ is the fidelity. Consider the unitary matrices

$$U_{BXOR1} := \begin{pmatrix} 1 & 0 \\ 0 & 0 \end{pmatrix} \otimes I_2 \otimes I_2 \otimes I_2 + \begin{pmatrix} 0 & 0 \\ 0 & 1 \end{pmatrix} \otimes I_2 \otimes \begin{pmatrix} 0 & 1 \\ 1 & 0 \end{pmatrix} \otimes I_2,$$

$$U_{BXOR2} := I_2 \otimes \begin{pmatrix} 1 & 0 \\ 0 & 0 \end{pmatrix} \otimes I_2 \otimes I_2. + I_2 \otimes \begin{pmatrix} 0 & 0 \\ 0 & 1 \end{pmatrix} \otimes I_2 \otimes \begin{pmatrix} 0 & 1 \\ 1 & 0 \end{pmatrix}$$

and $U_{BXOR} = U_{BXOR1}U_{BXOR2}$ (*bilateral exclusive or*). Let

$$\rho := U_{BXOR}(\rho_W \otimes \rho_W)U_{BXOR}^*.$$

Calculate the probability p_c that the last two qubits of a system described by ρ are found in the same state when measured with respect to the standard basis, i.e. the last two qubits are in one of the states

$$\begin{pmatrix} 1 \\ 0 \end{pmatrix} \otimes \begin{pmatrix} 1 \\ 0 \end{pmatrix}, \qquad \begin{pmatrix} 0 \\ 1 \end{pmatrix} \otimes \begin{pmatrix} 0 \\ 1 \end{pmatrix}.$$

Determine the fidelity

$$F' := \frac{\operatorname{tr}\left(\rho\left(|\phi^+\rangle\langle\phi^+| \otimes \Pi\right)\right)}{p_c}, \qquad \Pi = \begin{pmatrix} 1 & 0 & 0 & 0 \\ 0 & 0 & 0 & 0 \\ 0 & 0 & 0 & 0 \\ 0 & 0 & 0 & 1 \end{pmatrix}.$$

It is the projection onto the space of states compatible with the measurement outcomes above. Discuss the case $1/2 < F < 1$.

Solution 7.   The probability p_c is given by

$$p_c = \operatorname{tr}\left(\rho\left(I_4 \otimes \Pi\right)\right) = \frac{8F^2 - 4F + 5}{9}.$$

Thus the new fidelity F' is given by

$$F' = \frac{10F^2 - 2F + 1}{8F^2 - 4F + 5}.$$

For $F = 1/2$ we find $F' = 1/2$ and for $F = 1$ we find $F' = 1$. We have

$$F' - F = \frac{(2F - 1)(4F - 1)(1 - F)}{4F(2F - 1) + 5}.$$

Obviously for $1/2 < F < 1$ we have $F > 0$, $2F - 1 > 0$, $4F - 1 > 0$ and $1 - F > 0$. Thus $F' > F$ when $1/2 < F < 1$.

Part II

Infinite-Dimensional
Hilbert Spaces

Chapter 17

Harmonic Oscillator and Bose Operators

Besides qubit-based quantum computing and quantum algorithms, quantum information over continuous variables is also applied and used in fields such as quantum teleportation and quantum cryptography. For continuous systems Bose operators play the central role. Consider a family of linear operators b_j, $b_j^\dagger$, $j = 1, 2, \ldots, m$ on an inner product space V, satisfying the commutation relations (*Heisenberg algebra*)

$$[b_j, b_k] = [b_j^\dagger, b_k^\dagger] = 0, \qquad [b_j, b_k^\dagger] = \delta_{jk} I \qquad (1)$$

where I is the identity operator. The operator $b_j^\dagger$ is called a *Bose creation operator* and the operator b_j is called an *Bose annihilation operator*. The inner product space must be infinite dimensional for (1) to hold. For, if A and B are $n \times n$ matrices such that $[A, B] = \lambda I$, then $\text{tr}([A, B]) = 0$ implies $\lambda = 0$. Let $|\mathbf{0}\rangle \equiv |00...0\rangle$ be the *vacuum state*, i.e., $b_j|\mathbf{0}\rangle = 0$, $\langle \mathbf{0}|\mathbf{0}\rangle = 1$, $j = 1, 2, \ldots, m$. A normalized state is given by

$$\frac{1}{\sqrt{n_1! n_2! \cdots n_k!}} (b_{j_1}^\dagger)^{n_1} (b_{j_2}^\dagger)^{n_2} \cdots (b_{j_k}^\dagger)^{n_k} |\mathbf{0}\rangle \equiv |n_1, n_2, \ldots, n_k\rangle$$

where $j_1, j_2, \ldots, j_k \in \{1, 2, \ldots, m\}$ and $n_1, n_2, \ldots, n_k \in \{0, 1, 2, \ldots\}$. For b_j we also use the notation $I \otimes I \otimes \cdots \otimes b \otimes I \otimes I \otimes \cdots \otimes I$, where b is in the jth position.

Problem 1.   Consider the Hamilton operator for the one dimensional *harmonic oscillator*

$$\hat{H} = \frac{1}{2m}\hat{p}^2 + \frac{1}{2}m\omega^2\hat{q}^2$$

where ω is the frequency and m the mass. We introduce the *characteristic length*

$$\ell_0 := \sqrt{\frac{\hbar}{m\omega}}$$

and define the dimensionless linear operators (Bose operators)

$$b := \frac{1}{\sqrt{2}}\left(\frac{\hat{q}}{\ell_0} + i\frac{\hat{p}}{\hbar/\ell_0}\right), \qquad b^\dagger := \frac{1}{\sqrt{2}}\left(\frac{\hat{q}}{\ell_0} - i\frac{\hat{p}}{\hbar/\ell_0}\right).$$

(i) Find the commutator $[b, b^\dagger]$.
(ii) Express $\hat{q}$ and $\hat{p}$ in terms of b and $b^\dagger$, i.e., find the inverse transform.
(iii) Express $\hat{H}$ in terms of b and $b^\dagger$. Find $\hat{H}|n\rangle$. Discuss

Solution 1.   (i) Since

$$\hat{p} := -i\hbar\frac{\partial}{\partial q}$$

we obtain $[b, b^\dagger] = I$, where I is the identity operator.
(ii) We find

$$\hat{q} = \frac{1}{\sqrt{2}}\ell_0(b + b^\dagger), \qquad \hat{p} = \frac{\hbar/\ell_0}{\sqrt{2}i}(b - b^\dagger).$$

(iii) We find

$$\hat{H} = \hbar\omega(b^\dagger b + \frac{1}{2}I)$$

and $\hat{H}|n\rangle = \hbar\omega(n + 1/2)|n\rangle$ with $n = 0, 1, \ldots$.

Problem 2.   Consider the Hamilton operator for the one dimensional *harmonic oscillator* in the form ($\hbar = 1, m = 1, \omega = 1$)

$$\hat{H} = \frac{1}{2}\hat{p}^2 + \frac{1}{2}\hat{q}^2$$

and the unitary operator

$$U := e^{i\hat{H}\pi/4}.$$

Find

$$U\begin{pmatrix}\hat{p}\\\hat{q}\end{pmatrix}U^\dagger \equiv \begin{pmatrix}U\hat{p}U^\dagger\\U\hat{q}U^\dagger\end{pmatrix}.$$

Solution 2. From $\hat{p} = -i\frac{\partial}{\partial q}$ and $[\hat{p}, \hat{q}] = -iI$ it follows that

$$[\hat{H}, \hat{q}] = -i\hat{p}, \quad [\hat{H}, \hat{p}] = i\hat{q}, \quad [\hat{H}, [\hat{H}, \hat{q}]] = \hat{q}, \quad [\hat{H}, [\hat{H}, \hat{p}]] = \hat{p}.$$

Using this result and the expansions

$$U\hat{p}U^{\dagger} = \sum_{j=0}^{\infty} \frac{\left(i\frac{\pi}{4}\right)^j}{j!} \overbrace{[\hat{H}, [\hat{H}, \ldots, [\hat{H}, \hat{p}]\ldots]]}^{j\,\text{times}}$$

$$U\hat{q}U^{\dagger} = \sum_{j=0}^{\infty} \frac{\left(i\frac{\pi}{4}\right)^j}{j!} \overbrace{[\hat{H}, [\hat{H}, \ldots, [\hat{H}, \hat{q}]\ldots]]}^{j\,\text{times}}$$

we obtain

$$U\begin{pmatrix}\hat{p}\\\hat{q}\end{pmatrix}U^{\dagger} = \frac{1}{\sqrt{2}}\begin{pmatrix}\hat{p}-\hat{q}\\\hat{p}+\hat{q}\end{pmatrix} = \frac{1}{\sqrt{2}}\begin{pmatrix}1&-1\\1&1\end{pmatrix}\begin{pmatrix}\hat{p}\\\hat{q}\end{pmatrix}$$

i.e., the unitary operator U is a *Walsh-Hadamard transform*.

Problem 3. Let b, $b^{\dagger}$ be Bose annihilation and creation operators. Find the commutator of the operators

$$\widetilde{b} = \epsilon_1 I + b, \qquad \widetilde{b}^{\dagger} = \epsilon_2 I + b^{\dagger}$$

where I is the identity operator and ϵ_1, ϵ_2 are constants.

Solution 3. We have

$$\begin{aligned}[\widetilde{b}, \widetilde{b}^{\dagger}] &= [\epsilon_1 I + b, \epsilon_2 I + b^{\dagger}]\\ &= [\epsilon_1 I, \epsilon_2 I] + [\epsilon_1 I, b^{\dagger}] + [b, \epsilon_2 I] + [b, b^{\dagger}]\\ &= [b, b^{\dagger}]\\ &= I.\end{aligned}$$

Problem 4. Let $\beta \in \mathbb{C}$. Calculate the commutators

$$[b, \beta b^{\dagger} - \beta^* b], \qquad [b^{\dagger}, \beta b^{\dagger} - \beta^* b].$$

Solution 4. We find

$$[b, \beta b^{\dagger} - \beta^* b] = \beta I, \qquad [b^{\dagger}, \beta b^{\dagger} - \beta^* b] = \beta^* I.$$

Problem 5.   (i) Calculate the commutators

$$[b^2, b^\dagger b], \qquad [b^2, b^{\dagger 2}].$$

(ii) Using the commutation relations and $b|0\rangle = 0$ calculate the state

$$bbb^\dagger b^\dagger |0\rangle.$$

Is the state normalized?
(iii) Let

$$|n\rangle := \frac{1}{\sqrt{n!}}(b^\dagger)^n|0\rangle, \quad n = 0, 1, 2, \dots$$

be the *number states* (*Fock state*). Find the operator

$$\sum_{n=0}^{\infty} |n\rangle\langle n|.$$

Solution 5.   (i) Using the commutation relations we obtain

$$[b^2, b^\dagger b] = 2b^2, \qquad [b^2, b^{\dagger 2}] = 2I + 4b^\dagger b.$$

(ii) Using the commutation relations and $b|0\rangle = 0$ we find

$$bbb^\dagger b^\dagger |0\rangle = 2|0\rangle.$$

(iii) We find

$$\sum_{n=0}^{\infty} |n\rangle\langle n| = I$$

where I is the identity operator, i.e. we have the infinite-dimensional unit matrix. This is the *completeness relation*.

Problem 6.   Let $n \in \mathbb{N}$ and $\beta \in \mathbb{C}$. Calculate the commutator

$$[b^\dagger, (\beta b^\dagger - \beta^* b)^n].$$

Solution 6.   If f is an analytic function we have

$$[b^\dagger, f(b)] = -\frac{df(b)}{db}.$$

Thus

$$[b^\dagger, (\beta b^\dagger - \beta^* b)^n] = n\beta^*(\beta b^\dagger - \beta^* b)^{n-1}.$$

For $n = 1$ we have $[b^\dagger, (\beta b^\dagger - \beta^* n)] = \beta^* I$.

Problem 7. Using the number states $|n\rangle$ find the matrix representation of the unbounded operators $b^\dagger b$.

Solution 7. Since $b^\dagger b|n\rangle = n|n\rangle$ we obtain the infinite dimensional unbounded matrix

$$\text{diag}(0, 1, 2, \ldots).$$

Problem 8. Using the number states $|n\rangle$ find the matrix representation of the unbounded operators $b^\dagger + b$.

Solution 8. Since $b|n\rangle = \sqrt{n}|n-1\rangle$, $b^\dagger|n\rangle = \sqrt{n+1}|n+1\rangle$ we obtain the infinite dimensional diagonal matrix

$$\begin{pmatrix} 0 & 1 & 0 & 0 & \cdots \\ 1 & 0 & \sqrt{2} & 0 & \cdots \\ 0 & \sqrt{2} & 0 & \sqrt{3} & \cdots \\ 0 & 0 & \sqrt{3} & 0 & \cdots \\ \vdots & \vdots & \vdots & & \ddots \end{pmatrix}.$$

Problem 9. Let $b^\dagger$, b be Bose creation and annihilation operators, respectively.
(i) Calculate the commutator $[b^\dagger + b, b^\dagger b]$.
(ii) Consider the matrices

$$A = \begin{pmatrix} 0 & 0 & 0 & 0 \\ 0 & 1 & 0 & 0 \\ 0 & 0 & 2 & 0 \\ 0 & 0 & 0 & 3 \end{pmatrix}, \quad B = \begin{pmatrix} 0 & 1 & 0 & 0 \\ 1 & 0 & \sqrt{2} & 0 \\ 0 & \sqrt{2} & 0 & \sqrt{3} \\ 0 & 0 & \sqrt{3} & 0 \end{pmatrix}.$$

Find the commutator $[A, B]$. These matrices appear when we truncate the infinite dimensional unbounded matrices $b^\dagger b$ and $b^\dagger + b$.

Solution 9. (i) Using $bb^\dagger = I + b^\dagger b$ we obtain

$$[b^\dagger + b, b^\dagger b] = -b^\dagger + b.$$

(ii) We obtain the skew-symmetric matrix

$$[A, B] = \begin{pmatrix} 0 & -1 & 0 & 0 \\ 1 & 0 & -\sqrt{2} & 0 \\ 0 & \sqrt{2} & 0 & -\sqrt{3} \\ 0 & 0 & \sqrt{3} & 0 \end{pmatrix}$$

which is a truncation of the infinite dimensional unbounded matrix $-b^\dagger + b$.

Problem 10.   Consider the operators

$$\hat{A} = b_k e^{-i\phi_k} + b_k^\dagger e^{i\phi_k}, \qquad \hat{B} = b_\ell e^{-i\phi_\ell} + b_\ell^\dagger e^{i\phi_\ell}.$$

Find the commutator $[\hat{A}, \hat{B}]$. What is the condition on the phases ϕ_k and ϕ_ℓ such that $[\hat{A}, \hat{B}] = 0$ for $k = \ell$?

Solution 10.   We have

$$\begin{aligned}
[\hat{A}, \hat{B}] &= [b_k e^{-i\phi_k} + b_k^\dagger e^{i\phi_k}, b_\ell e^{-i\phi_\ell} + b_\ell^\dagger e^{i\phi_\ell}] \\
&= [b_k e^{-i\phi_k}, b_\ell^\dagger e^{i\phi_\ell}] + [b_k^\dagger e^{i\phi_k}, b_\ell e^{-i\phi_\ell}] \\
&= (e^{i(\phi_\ell - \phi_k)} - e^{i(\phi_k - \phi_\ell)}) \delta_{\ell k} \\
&= 2i \sin(\phi_\ell - \phi_k) \delta_{\ell k}.
\end{aligned}$$

Thus the commutator is zero for $k = \ell$ if $\sin(\phi_\ell - \phi_k) = 0$ i.e. $\phi_\ell - \phi_k = n\pi$ and $n \in \mathbb{Z}$.

Problem 11.   Let f be an analytic function in z and $z \in \mathbb{C}$. Let $\hat{n} := b^\dagger b$ be the number operator. Calculate

$$e^{z\hat{n}} f(b), \quad e^{z\hat{n}} f(b^\dagger), \quad e^{zb^\dagger} f(b), \quad e^{zb} f(b^\dagger).$$

Solution 11.   Since $\hat{n}b = b(\hat{n} - I)$ and $\hat{n}b^\dagger = b^\dagger(I + \hat{n})$ we find

$$\begin{aligned}
e^{z\hat{n}} f(b) &= f(e^{-z}b) e^{z\hat{n}} \\
e^{z\hat{n}} f(b^\dagger) &= f(e^z b^\dagger) e^{z\hat{n}} \\
e^{zb^\dagger} f(b) &= f(b - zI) e^{zb^\dagger} \\
e^{zb} f(b^\dagger) &= f(b^\dagger + zI) e^{zb}.
\end{aligned}$$

Problem 12.   Let $b^\dagger$, b be Bose creation and annihilation operator with the commutator relation $[b, b^\dagger] = I$, where I is the identity operator. Show that one has a representation

$$b^\dagger \mapsto z, \qquad b \mapsto \frac{d}{dz}.$$

Solution 12.   We calculate the commutator

$$[d/dz, z] f(z)$$

where f is an analytic functions. Using the product rule we obtain

$$[d/dz, z]f(z) = \frac{d}{dz}(zf(z)) - z\frac{df(z)}{dz} = f(z) + z\frac{df(z)}{dz} - z\frac{df(z)}{dz} = f(z).$$

Problem 13. Consider the differential operators

$$b = \frac{1}{\sqrt{2}}\left(x + \frac{d}{dx}\right), \qquad b^\dagger = \frac{1}{\sqrt{2}}\left(x - \frac{d}{dx}\right)$$

acting the the vector space $S(\mathbb{R})$. Find the commutator $[b, b^\dagger]$. Find the operator $\hat{N} = b^\dagger b$.

Solution 13. Let $f \in S(\mathbb{R})$. We have

$$[b, b^\dagger]f = bb^\dagger f - b^\dagger b f$$
$$= \frac{1}{2}\left(x^2 f - \frac{d}{dx}(xf) - x\frac{df}{dx} - \frac{d^2 f}{dx^2}\right)$$
$$- \frac{1}{2}\left(x^2 f - \frac{d}{dx}(xf) + x\frac{df}{dx} - \frac{d^2 f}{dx^2}\right)$$
$$= \frac{d}{dx}(xf) - x\frac{df}{dx}$$
$$= f.$$

We obtain the differential operator

$$\hat{N} = \frac{1}{2}\left(x^2 - 1 - \frac{d^2}{dx^2}\right).$$

Problem 14. Let $\hat{n} := b^\dagger b$ be the *number operator*. Calculate the commutators

$$[\hat{n}, b], \quad [\hat{n}, [\hat{n}, b]], \quad [\hat{n}, [\hat{n}, [\hat{n}, b]]].$$

Discuss the general case for m commutators.

Solution 14. We have

$$[\hat{n}, b] = -b, \quad [\hat{n}, [\hat{n}, b]] = (-1)^2 b, \quad [\hat{n}, [\hat{n}, [\hat{n}, b]]] = (-1)^3 b.$$

Obviously, for the general case with m commutators we find $(-1)^m b$.

Problem 15. Let $\epsilon \in \mathbb{R}$ and $\epsilon > 0$, $\hat{n} = b^\dagger b$ and $\{\,|n\rangle : n = 0, 1, 2, \dots\}$ denote the number states (Fock states).

(i) Using the number states calculate the *trace*

$$\text{tr}(e^{-\epsilon\hat{n}}) \equiv \sum_{n=0}^{\infty} \langle n|e^{-\epsilon\hat{n}}|n\rangle \,.$$

(ii) Using the number states calculate the trace

$$\text{tr}(b^{\dagger}be^{-\epsilon b^{\dagger}b}) \equiv \text{tr}(b^{\dagger}be^{-\epsilon b^{\dagger}b}) = \sum_{n=0}^{\infty} \langle n|b^{\dagger}be^{-\epsilon b^{\dagger}b}|n\rangle \,.$$

Solution 15. (i) Since $\hat{n}|n\rangle = n|n\rangle$ we have $e^{-\epsilon b^{\dagger}b}|n\rangle = e^{-\epsilon n}|n\rangle$. Thus

$$\text{tr}(e^{-\epsilon\hat{n}}) = \sum_{n=0}^{\infty} e^{-\epsilon n} = \frac{e^{\epsilon}}{e^{\epsilon} - 1} \,.$$

(ii) Using $\langle n|b^{\dagger}b = \langle n|n$ we obtain

$$\text{tr}(b^{\dagger}be^{-\epsilon b^{\dagger}b}) = \sum_{n=0}^{\infty} ne^{-\epsilon n} = \frac{e^{\epsilon}}{(e^{\epsilon} - 1)^2} \,.$$

Problem 16. Let b and $b^{\dagger}$ be Bose annihilation and creation operators, respectively. Consider the general one-mode canonical *Bogolubov transform*

$$\widetilde{b} := e^{i\phi} \cosh(r)b + e^{i\psi} \sinh(r)b^{\dagger}$$
$$\widetilde{b}^{\dagger} := e^{-i\phi} \cosh(r)b^{\dagger} + e^{-i\psi} \sinh(r)b$$

where r is a real parameter (*squeezing parameter*).
(i) Show that the operators $\widetilde{b}$ and $\widetilde{b}^{\dagger}$ satisfy the Bose commutation relations.
(ii) Find the inverse Bogolubov transform.

Solution 16. (i) Since $\cosh^2(r) - \sinh^2(r) = 1$ we find

$$[\widetilde{b}, \widetilde{b}^{\dagger}] = bb^{\dagger} - b^{\dagger}b = I \,.$$

(ii) The transformation can be written in the matrix form

$$\begin{pmatrix} \widetilde{b} \\ \widetilde{b}^{\dagger} \end{pmatrix} = \begin{pmatrix} e^{i\phi} \cosh(r) & e^{i\psi} \sinh(r) \\ e^{-i\psi} \sinh(r) & e^{-i\phi} \cosh(r) \end{pmatrix} \begin{pmatrix} b \\ b^{\dagger} \end{pmatrix} \,.$$

The determinant of this matrix is $+1$. Thus the inverse transformation is given by

$$\begin{pmatrix} b \\ b^{\dagger} \end{pmatrix} = \begin{pmatrix} e^{-i\phi} \cosh(r) & -e^{i\psi} \sinh(r) \\ -e^{-i\psi} \sinh(r) & e^{i\phi} \cosh(r) \end{pmatrix} \begin{pmatrix} \widetilde{b} \\ \widetilde{b}^{\dagger} \end{pmatrix} \,.$$

Problem 17. Let $\epsilon \in \mathbb{R}$ and $f : \mathbb{C}^2 \to \mathbb{C}$ be an *entire analytic function*. If a function f is analytic on the whole complex plane then f is said to be entire. Show that

$$e^{\epsilon b} f(b, b^\dagger) e^{-\epsilon b} = f(b, b^\dagger + \epsilon I) \tag{1a}$$

$$e^{-\epsilon b^\dagger} f(b, b^\dagger) e^{\epsilon b^\dagger} = f(b + \epsilon I, b^\dagger). \tag{1b}$$

Solution 17. We have

$$e^{\epsilon b} f(b, b^\dagger) e^{-\epsilon b} = f(e^{\epsilon b} b e^{-\epsilon b}, e^{\epsilon b} b^\dagger e^{-\epsilon b}) = f(b, e^{\epsilon b} b^\dagger e^{-\epsilon b}).$$

Since

$$e^{\epsilon b} b^\dagger e^{-\epsilon b} = b^\dagger + \epsilon I$$

we find (1a), where we used $[b, b^\dagger] = I$. A similar proof holds for (1b).

Problem 18. Let $f : \mathbb{C} \to \mathbb{C}$ be an entire analytic function. Show that

$$f(b^\dagger b) = \sum_{n=0}^{\infty} f(n) |n\rangle \langle n|$$

where $|n\rangle$ is the number state (Fock state).

Solution 18. The completeness relation is given by

$$\sum_{n=0}^{\infty} |n\rangle \langle n| = I.$$

Since $b^\dagger b |n\rangle = n|n\rangle$, we have

$$f(b^\dagger b)|n\rangle = f(n)|n\rangle.$$

It follows that

$$f(b^\dagger b) = f(b^\dagger b)I = f(b^\dagger b) \sum_{n=0}^{\infty} |n\rangle \langle n| = \sum_{n=0}^{\infty} f(b^\dagger b)|n\rangle \langle n| = \sum_{n=0}^{\infty} f(n)|n\rangle \langle n|.$$

Problem 19. Let f be an analytic function in x and y. Let $b^\dagger$ and b be Bose creation and annihilation operators, respectively. We can define $f(b, b^\dagger)$ by its power series expansion

$$f(b, b^\dagger) := \sum_{j_1=0}^{\infty} \sum_{j_2=0}^{\infty} \cdots \sum_{j_n=0}^{\infty} f(j_1, j_2, \ldots, j_n)(b^\dagger)^{j_1} b^{j_2} \ldots b^{j_n}.$$

We can use the commutation relation for Bose operators repeatedly to rearrange the operators b, $b^\dagger$ so that

$$f(b, b^\dagger) = \sum_{m=0}^{\infty} \sum_{n=0}^{\infty} f_{mn}^{(n)} (b^\dagger)^m b^n .$$

We say that the function $f(b, b^\dagger)$ is in *normal order form*.
(i) Consider the functions

$$f(b, b^\dagger) = b^\dagger b b^\dagger b, \qquad g(b, b^\dagger) = b^\dagger b b^\dagger b b^\dagger b .$$

Find the normal order form for these functions.
(ii) Consider the operator

$$e^{-\epsilon b^\dagger b}$$

where ϵ is a real positive parameter. Find the *normal order form*.

Solution 19. (i) From the commutation relations for Bose operators we find $bb^\dagger = I + b^\dagger b$. Thus

$$f(b, b^\dagger) = b^\dagger b + b^\dagger b^\dagger bb, \quad g(b, b^\dagger) = b^\dagger b^\dagger b^\dagger bbb + 3b^\dagger b^\dagger bb + b^\dagger b .$$

(ii) Using the results from (i) and $b^\dagger b (b^\dagger b)^j \equiv b^\dagger (I + b^\dagger b)^j b$ we find

$$e^{-\epsilon b^\dagger b} = \sum_{j=0}^{\infty} \frac{1}{j!} (e^{-\epsilon} - 1)^j (b^\dagger)^j b^j .$$

Problem 20. Let $\{\, |n\rangle \,:\, n = 0, 1, 2, \dots \,\}$ be the number states (Fock states). We define the linear operators

$$\hat{E} := \sum_{n=0}^{\infty} |n\rangle\langle n+1|, \qquad \hat{E}^\dagger := \sum_{n=0}^{\infty} |n+1\rangle\langle n| .$$

Obviously, $\hat{E}^\dagger$ follows from $\hat{E}$.
(i) Find $\hat{E}\hat{E}^\dagger$ and $\hat{E}^\dagger \hat{E}$.
(ii) Let f be an analytic function. Calculate $\hat{E} f(\hat{n}) \hat{E}^\dagger$ and $\hat{E}^\dagger f(\hat{n} + I)\hat{E}$, where $\hat{n}$ is the photon number operator and I is the identity operator.

Solution 20. (i) Using $\langle m|n\rangle = \delta_{mn}$ and the completeness relation we find

$$\hat{E}\hat{E}^\dagger = \sum_{m=0}^{\infty} |m\rangle\langle m+1| \sum_{n=0}^{\infty} |n+1\rangle\langle n| = \sum_{m=0}^{\infty} \sum_{n=0}^{\infty} |m\rangle \delta_{m+1,n+1} \langle n|$$

$$= \sum_{m=0}^{\infty} |m\rangle\langle m| = I .$$

Analogously, we find

$$\hat{E}^\dagger \hat{E} = I - |0\rangle\langle 0|\,.$$

(ii) Using the Taylor expansion around 0 of an analytic function we have

$$\hat{E}f(\hat{n}) = \sum_{m=0}^{\infty} |m\rangle\langle m+1| \sum_{j=0}^{\infty} \frac{f^{(j)}(0)}{j!}\hat{n}^j\,.$$

Applying $\hat{n}|m\rangle = m|m\rangle$, $\hat{n}^j|m\rangle = m^j|m\rangle$ and

$$(\hat{n}+I)|m\rangle = (m+1)|m\rangle, \quad (\hat{n}+I)^j|m\rangle = (m+1)^j|m\rangle$$

we obtain

$$\hat{E}f(\hat{n}) = f(\hat{n}+I)\hat{E}\,.$$

Thus

$$\hat{E}f(\hat{n})\hat{E}^\dagger = f(\hat{n}+I)\,.$$

Analogously

$$\hat{E}^\dagger f(\hat{n}+I)\hat{E} = f(\hat{n})\,.$$

Problem 21. Consider the *Susskind-Glogower canonical phase states*

$$|\phi\rangle := \sum_{n=0}^{\infty} e^{i\phi n}|n\rangle$$

where $|n\rangle$ are the number states. Let

$$\hat{L} := \sum_{n=1}^{\infty} |n-1\rangle\langle n|$$

be the non unitary number-lowering operator. Find $\hat{L}|\phi\rangle$.

Solution 21. Since $\langle m|n\rangle = \delta_{mn}$ we have

$$\hat{L}|\phi\rangle = e^{i\phi}|\phi\rangle\,.$$

This means that $|\phi\rangle$ is an eigenstate of the operator $\hat{L}$.

Problem 22. Let $b^\dagger$ and b be Bose creation and annihilation operators, respectively. Consider the operator

$$e^{\alpha_1 b^2 + \alpha_2 (b^\dagger)^2 + \alpha_3 (bb^\dagger + b^\dagger b)} \tag{1}$$

where $\alpha_1, \alpha_2, \alpha_3 \in \mathbb{R}$. Let $\epsilon \in \mathbb{R}$ be an arbitrary real parameter. Find the smooth functions f_0, f_1, f_2 and f_3, depending on ϵ, such that

$$e^{\epsilon(\alpha_1 b^2 + \alpha_2 (b^\dagger)^2 + \alpha_3 (bb^\dagger + b^\dagger b))} = e^{f_0(\epsilon)I + f_1(\epsilon)(b^\dagger)^2} e^{f_2(\epsilon)b^\dagger b} e^{f_3(\epsilon)b^2} \qquad (2)$$

where I denotes the identity operator. Then set $\epsilon = 1$. Solve the problem using *parameter differentiation* with respect to ϵ. We find a system of ordinary differential equations for the functions f_0, f_1, f_2 and f_3.

Solution 22.   Differentiating the left-hand side of (2) with respect to ϵ yields

$$(\alpha_1 b^2 + \alpha_2 (b^\dagger)^2 + \alpha_3 (bb^\dagger + b^\dagger b)) e^{\epsilon(\alpha_1 b^2 + \alpha_2 (b^\dagger)^2 + \alpha_3 (bb^\dagger + b^\dagger b))} =$$

$$e^{f_0(\epsilon)I + f_1(\epsilon)(b^\dagger)^2} \left(\frac{df_0}{d\epsilon} I + \frac{df_1}{d\epsilon} (b^\dagger)^2 \right) e^{f_2(\epsilon)b^\dagger b} e^{f_3(\epsilon)b^2}$$

$$+ e^{f_0(\epsilon)I + f_1(\epsilon)(b^\dagger)^2} e^{f_2(\epsilon)b^\dagger b} \left(\frac{df_2}{d\epsilon} b^\dagger b \right) e^{f_3(\epsilon)b^2}$$

$$+ e^{f_0(\epsilon)I + f_1(\epsilon)(b^\dagger)^2} e^{f_2(\epsilon)b^\dagger b} e^{f_3(\epsilon)b^2} \left(\frac{df_3}{d\epsilon} b^2 \right).$$

Owing to the identity $e^{\epsilon \hat{X}} e^{-\epsilon \hat{X}} = I$, we have

$$e^{-\epsilon(\alpha_1 b^2 + \alpha_2 (b^\dagger)^2 + \alpha_3 (bb^\dagger + b^\dagger b))} = e^{-f_3(\epsilon)b^2} e^{-f_2(\epsilon)b^\dagger b} e^{-f_0(\epsilon)I - f_1(\epsilon)(b^\dagger)^2}.$$

From the last two equation we obtain

$$\alpha_1 b^2 + \alpha_2 (b^\dagger)^2 + \alpha_3 (bb^\dagger + b^\dagger b) = \frac{df_0}{d\epsilon} I + \frac{df_1}{d\epsilon} e^{-f_3 b^2} e^{-f_2 b^\dagger b} (b^\dagger)^2 e^{f_2 b^\dagger b} e^{f_3 b^2}$$

$$+ \frac{df_2}{d\epsilon} e^{-f_3 b^2} b^\dagger b e^{f_3 b^2} + \frac{df_3}{d\epsilon} b^2.$$

Since

$$e^{-f_2(\epsilon)b^\dagger b} (b^\dagger)^2 e^{f_2(\epsilon)b^\dagger b} = (b^\dagger)^2 e^{-2f_2(\epsilon)}$$

$$e^{-f_3(\epsilon)b^2} (b^\dagger)^2 e^{f_3(\epsilon)b^2} = (b^\dagger)^2 + 4f_3^2(\epsilon)b^2 - 2f_3(\epsilon)(I + 2b^\dagger b)$$

$$e^{-f_3(\epsilon)b^2} b^\dagger b e^{f_3(\epsilon)b^2} = b^\dagger b - 2f_3(\epsilon)b^2$$

we find

$$\alpha_1 b^2 + \alpha_2 (b^\dagger)^2 + 2\alpha_3 b^\dagger b + \alpha_3 I$$

$$= \frac{df_0}{d\epsilon} I + \frac{df_1}{d\epsilon} e^{-2f_2} ((b^\dagger)^2 + 4f_3^2 b^2 - f_3(2I + 4b^\dagger b)) + \frac{df_2}{d\epsilon} (b^\dagger b - 2f_3 b^2) + \frac{df_3}{d\epsilon} b^2$$

where we used

$$e^{-\mu \hat{A}} \hat{B} e^{\mu \hat{A}} = \hat{B} - \mu[\hat{A}, \hat{B}] + \frac{1}{2!} \mu^2 [\hat{A}, [\hat{A}, \hat{B}]] + \cdots$$

and $bb^\dagger = I + b^\dagger b$. Separating out terms with I, b^2, $(b^\dagger)^2$, and $b^\dagger b$ we find

$$\frac{df_0}{d\epsilon} - \alpha_3 - 2f_3 \frac{df_1}{d\epsilon} e^{-2f_2} = 0$$

$$4f_3^2 \frac{df_1}{d\epsilon} e^{-2f_2} + \frac{df_3}{d\epsilon} - 2\frac{df_2}{d\epsilon} f_3 - \alpha_1 = 0$$

$$\frac{df_1}{d\epsilon} - \alpha_2 e^{2f_2} = 0$$

$$4\frac{df_1}{d\epsilon} f_3 e^{-2f_2} - \frac{df_2}{d\epsilon} + 2\alpha_3 = 0 .$$

Using these system of equations we can cast the system of nonlinear differential equations in the form

$$\frac{df_0}{d\epsilon} = \alpha_3 + 2\alpha_2 f_3$$

$$\frac{df_1}{d\epsilon} = \alpha_2 e^{2f_2}$$

$$\frac{df_2}{d\epsilon} = 2\alpha_3 + 4\alpha_2 f_3$$

$$\frac{df_3}{d\epsilon} = \alpha_1 + 4\alpha_3 f_3 + 4\alpha_2 f_3^2$$

with the initial conditions $f_j(0) = 0$ for $j = 0, 1, 2, 3$. We first solve the fourth equation which is a *Riccati equation* and then insert it into the third and second equation to find f_2 and f_0. Finally we solve for f_1. The integration yields

$$f_0(\epsilon) = \frac{1}{2} \ln(\cosh(2\lambda\epsilon) - (\alpha_3/\lambda) \sinh(2\lambda\epsilon))$$

$$f_1(\epsilon) = \frac{(\alpha_2/2\lambda) \sinh(2\lambda\epsilon)}{\cosh(2\lambda\epsilon) - (\alpha_3/\lambda) \sinh(2\lambda\epsilon)}$$

$$f_2(\epsilon) = -\ln(\cosh(2\lambda\epsilon) - (\alpha_3/\lambda) \sinh(2\lambda\epsilon))$$

$$f_3(\epsilon) = \frac{(\alpha_1/2\lambda) \sinh(2\lambda\epsilon)}{\cosh(2\lambda\epsilon) - (\alpha_3/\lambda) \sinh(2\lambda\epsilon)}$$

where $\lambda := \sqrt{\alpha_3 - \alpha_1 \alpha_2}$. Setting $\epsilon = 1$ we have

$$e^{\alpha_1 b^2 + \alpha_2 (b^\dagger)^2 + \alpha_3 (bb^\dagger + b^\dagger b)} = \frac{1}{\sqrt{\cosh(2\lambda) - (\alpha_3/\lambda) \sinh(2\lambda)}}$$

$$\times \exp\left(\frac{(\alpha_2/\lambda) \sinh(2\lambda)}{\cosh(2\lambda) - (\alpha_3/\lambda) \sinh(2\lambda)} (b^\dagger)^2\right)$$

$$\times \exp(\ln(\cosh(2\lambda) - (\alpha_3/\lambda) \sinh(2\lambda))^{-1} b^\dagger b)$$

$$\times \exp\left(\frac{(\alpha_1/2\lambda) \sinh(2\lambda)}{\cosh(2\lambda) - (\alpha_3/\lambda) \sinh(2\lambda)} b^2\right) .$$

Problem 23. The homogeneous *Bogolubov transform* of the Bose creation operator $b^\dagger$ and Bose annihilation operator b

$$\widetilde{b} = \mu b + \nu b^\dagger, \qquad \mu, \nu \in \mathbb{C}$$

for a pair of complex parameters

$$\mu = |\mu| \exp(i\phi), \qquad \nu = |\nu| \exp(i\theta)$$

obeying additionally $|\mu|^2 - |\nu|^2 = 1$ is *canonical* since it leaves the commutator invariant

$$[b, b^\dagger] = [\widetilde{b}, \widetilde{b}^\dagger] = I.$$

Every canonical transform can be represented as a unitary transformation

$$\widetilde{b} = B(\mu, \nu) b B^\dagger(\mu, \nu).$$

The *Bogolubov unitary operator* $B(\mu, \nu)$ is defined by this relation up to an arbitrary phase factor. One choice is the normal form

$$B(\mu, \nu) = \mu^{-1/2} \exp\left(-\frac{\nu}{2\mu} b^{\dagger 2}\right) \exp\left(-\ln(\mu) b^\dagger b\right) \exp\left(\frac{\nu^*}{2\mu} b^2\right).$$

Show that the Bogolubov transform forms a continuous non-commutative group.

Solution 23. Let

$$|\mu'|^2 - |\nu'|^2 = 1, \qquad |\mu''|^2 - |\nu''|^2 = 1.$$

Then we have

$$B(\mu', \nu') B(\mu'', \nu'') = B(\mu, \nu)$$

and

$$\mu = \mu' \mu'' + \nu'^* \nu'', \qquad \nu = \mu'^* \nu'' + \nu' \mu''$$

with $|\mu|^2 - |\nu|^2 = 1$. The identity element of the group is given by $B(1,0)$, where we used that $\ln(1) = 0$. The inverse element of $B(\mu, \nu)$ is given by

$$B^{-1}(\mu, \nu) = B^\dagger(\mu, \nu) = B(\mu^*, -\nu).$$

Obviously, the associative law also holds.

Problem 24. The semi-simple Lie algebra $su(1,1)$ is given by the commutation relations

$$[k_1, k_2] = -ik_3, \qquad [k_3, k_1] = ik_2, \qquad [k_2, k_3] = -ik_1$$

where k_1, k_2 and k_3 are the basis elements of the Lie algebra. Show that an infinite-dimensional matrix representation is given by

$$k_1 = \frac{1}{2}\begin{pmatrix} 0 & 1 & 0 & 0 & 0 & \cdots \\ 1 & 0 & 2 & 0 & 0 & \cdots \\ 0 & 2 & 0 & 3 & 0 & \cdots \\ 0 & 0 & 3 & 0 & 4 & \cdots \\ \vdots & \vdots & \vdots & \vdots & \vdots & \ddots \end{pmatrix}$$

$$k_2 = \frac{1}{2}\begin{pmatrix} 0 & i & 0 & 0 & \cdots \\ -i & 0 & 2i & 0 & \cdots \\ 0 & -2i & 0 & 3i & \cdots \\ 0 & 0 & -3i & 0 & \cdots \\ \vdots & \vdots & \vdots & \vdots & \ddots \end{pmatrix}$$

$$k_3 = \frac{1}{2}\mathrm{diag}(1,3,5,7,\ldots).$$

Solution 24. Straightforward calculation yields the commutation relations given above.

Problem 25. Consider the operators

$$K_+ := b_1^\dagger b_2^\dagger, \qquad K_- := b_2 b_1, \qquad K_3 := \frac{1}{2}(b_1^\dagger b_1 + b_2^\dagger b_2 + I)$$

where I is the identity operator. Find the commutators

$$[K_+, K_-], \quad [K_3, K_+], \quad [K_3, K_-].$$

The operators K_+, K_-, K_3 form a representation of the Lie algebra $su(1,1)$.

Solution 25. Using the commutation relations given above we find

$$[K_+, K_-] = b_1^\dagger b_2^\dagger b_2 b_1 - b_2 b_1 b_1^\dagger b_2^\dagger = b_1^\dagger b_2^\dagger b_2 b_1 - b_2 b_2^\dagger - b_1^\dagger b_1 b_2 b_2^\dagger$$
$$= -b_2 b_2^\dagger - b_1^\dagger b_1 = -I - b_2^\dagger b_2 - b_1^\dagger b_1$$
$$= -2K_3.$$

Analogously $[K_3, K_-] = -K_-$, $[K_3, K_+] = K_+$.

Problem 26. Consider the linear operators

$$J_+ := b_1^\dagger b_2, \quad J_- := b_2^\dagger b_1, \quad J_3 := \frac{1}{2}(b_1^\dagger b_1 - b_2^\dagger b_2)$$

where $b_1^\dagger, b_2^\dagger$ are Bose creation operators and b_1, b_2 are Bose annihilation operators and I is the identity operator. Find the commutators

$$[J_+, J_-], \quad [J_3, J_+], \quad [J_3, J_-].$$

The operators J_+, J_-, J_3 form a representation of the semisimple Lie algebra $su(2)$.

Solution 26.   Using the commutation relation given above we find

$$[J_+, J_-] = b_1^\dagger b_2 b_2^\dagger b_1 - b_2^\dagger b_1 b_1^\dagger b_2 = b_1^\dagger b_2 b_2^\dagger b_1 - b_2^\dagger b_2 - b_2^\dagger b_1^\dagger b_1 b_2$$
$$= -b_2^\dagger b_2 + b_1^\dagger b_1 = 2J_3.$$

Analogously $[J_3, J_-] = -J_-$, $[J_3, J_+] = J_+$.

Problem 27.   Consider the Bose creation operators $b_1^\dagger, b_2^\dagger$ and Bose annihilation operators b_1, b_2 with $b_1 = b \otimes I$, $b_2 = I \otimes b$ and $b_1|00\rangle = 0$, $b_2|00\rangle = 0$, where $|00\rangle \equiv |0\rangle \otimes |0\rangle$ is the vacuum state. Consider the linear transformation

$$\tilde{b}_1 = u_{11}b_1 + u_{12}b_2 + v_{11}b_1^\dagger + v_{12}b_2^\dagger$$
$$\tilde{b}_2 = u_{21}b_1 + u_{22}b_2 + v_{21}b_1^\dagger + v_{22}b_2^\dagger$$
$$\tilde{b}_1^\dagger = v_{11}^*b_1 + v_{12}^*b_2 + u_{11}^*b_1^\dagger + u_{12}^*b_2^\dagger$$
$$\tilde{b}_2^\dagger = v_{21}^*b_1 + v_{22}^*b_2 + u_{21}^*b_1^\dagger + u_{22}^*b_2^\dagger$$

where $u_{jk}, v_{jk} \in \mathbb{C}$.
(i) Find the condition that the operators $\tilde{b}_1$, $\tilde{b}_2$, $\tilde{b}_1^\dagger$, $\tilde{b}_2^\dagger$ also satisfy the commutation relation for Bose operators.
(ii) For the vacuum state of the Bose fields $\tilde{b}_1$, $\tilde{b}_2$ we can write

$$|\tilde{0}\rangle \equiv |\widetilde{00}\rangle \equiv |\tilde{0}\rangle \otimes |\tilde{0}\rangle = \sum_{m=0}^{\infty}\sum_{n=0}^{\infty} c_{mn}|m\rangle \otimes |n\rangle.$$

Find the recurrence relation for c_{mn} from the condition

$$\tilde{b}_1|\tilde{0}\rangle \otimes |\tilde{0}\rangle = 0, \qquad \tilde{b}_2|\tilde{0}\rangle \otimes |\tilde{0}\rangle = 0. \tag{1}$$

Solution 27.   (i) From the conditions

$$[\tilde{b}_1, \tilde{b}_2] = 0, \qquad [\tilde{b}_1, \tilde{b}_2^\dagger] = 0, \qquad [\tilde{b}_1, \tilde{b}_1^\dagger] = I, \qquad [\tilde{b}_2, \tilde{b}_2^\dagger] = I$$

we find

$$u_{11}v_{21} + u_{12}v_{22} - u_{21}v_{11} - u_{22}v_{12} = 0 \tag{2a}$$
$$u_{11}u_{21}^* + u_{12}u_{22}^* - v_{11}v_{21}^* - v_{12}v_{22}^* = 0 \tag{2b}$$
$$u_{11}u_{11}^* + u_{12}u_{12}^* - v_{11}v_{11}^* - v_{12}v_{12}^* = 1 \tag{2c}$$
$$u_{21}u_{21}^* + u_{22}u_{22}^* - v_{21}v_{21}^* - v_{22}v_{22}^* = 1. \tag{2d}$$

(ii) From the conditions (1) we find

$$c_{(m+1)n}u_{11}\sqrt{m+1} + c_{m(n+1)}u_{12}\sqrt{n+1}$$

$$+c_{(m-1)n}v_{11}\sqrt{m} + c_{m(n-1)}v_{12}\sqrt{n} = 0 \tag{3a}$$

and

$$c_{(m+1)n}u_{21}\sqrt{m+1} + c_{m(n+1)}u_{22}\sqrt{n+1}$$

$$+c_{(m-1)n}v_{21}\sqrt{m} + c_{m(n-1)}v_{22}\sqrt{n} = 0. \tag{3b}$$

Let

$$\Delta_1 := u_{11}u_{22} - u_{12}u_{21}, \qquad \Delta_2 := u_{11}v_{21} - u_{21}v_{11}, \qquad \Delta_3 := u_{11}v_{22} - u_{21}v_{12}$$

$$\Delta_4 := u_{22}v_{11} - u_{12}v_{21}, \qquad \Delta_5 := u_{22}v_{12} - u_{12}v_{22}.$$

Multiplication of (3a) with u_{21} and (3b) with u_{11} and subtracting yields

$$c_{m(n+1)}\Delta_1\sqrt{n+1} = -c_{(m-1)n}\Delta_2\sqrt{m} - c_{m(n-1)}\Delta_3\sqrt{n}.$$

Multiplication of (3a) with u_{22} and (3b) with u_{12} and subtracting yields

$$c_{(m+1)n}\Delta_1\sqrt{m+1} = -c_{(m-1)n}\Delta_4\sqrt{m} - c_{m(n-1)}\Delta_5\sqrt{n}.$$

We assumed that $\Delta_1 \neq 0$. From (2a) we see that $\Delta_2 = \Delta_5$. Thus we have

$$c_{(2k)(2n+1)} = c_{(2k+1)(2n)} = 0$$

$$c_{(2k)(2n)} = (-1)^{n+k}\sqrt{(2n)!}\sqrt{(2k)!}$$

$$\times \sum_{\substack{0 \le s \le n \\ s \le k}} \left(\frac{\Delta_2}{\Delta_1}\right)^{2s} \left(\frac{\Delta_3}{2\Delta_1}\right)^{n-s} \left(\frac{\Delta_4}{2\Delta_1}\right)^{k-s} \frac{1}{(n-s)!(k-s)!(2s)!}c_0$$

and

$$c_{(2k+1)(2n+1)} = (-1)^{n+k+1}\sqrt{(2n+1)!}\sqrt{(2k+1)!}$$

$$\times \sum_{\substack{0 \le s \le n \\ s \le k}} \left(\frac{\Delta_2}{\Delta_1}\right)^{2s+1} \left(\frac{\Delta_3}{2\Delta_1}\right)^{n-s} \left(\frac{\Delta_4}{2\Delta_1}\right)^{k-s} \frac{1}{(n-s)!(k-s)!(2s+1)!}c_0.$$

Consequently, for the vacuum state of Bose operators $\widetilde{b}_1$ and $\widetilde{b}_2$ we find

$$|\widetilde{00}\rangle = \sum_{k=0,n=0}^{\infty} \left(c_{(2k)(2n)}|2k\rangle \otimes |2n\rangle + c_{(2k+1)(2n+1)}|2k+1\rangle \otimes |2n+1\rangle\right).$$

In operator form this can be written as

$$|\widetilde{00}\rangle = c_0 \exp\left(-\frac{\Delta_4}{2\Delta_1}(b_1^\dagger)^2 - \frac{\Delta_2}{\Delta_1}b_1^\dagger b_2^\dagger - \frac{\Delta_3}{2\Delta_1}(b_2^\dagger)^2\right)|0\rangle \otimes |0\rangle \,.$$

Thus the unitary operator

$$U = \exp\left(-\frac{\Delta_4}{2\Delta_1}(b_1^\dagger)^2 - \frac{\Delta_2}{\Delta_1}b_1^\dagger b_2^\dagger - \frac{\Delta_3}{2\Delta_1}(b_2^\dagger)^2\right)$$

is the operator of transformation of the vacuum states for the most general two-dimensional Bogolubov transformation. Thus we also have

$$|\widetilde{m}\rangle \otimes |\widetilde{n}\rangle = U|m\rangle \otimes |n\rangle \,.$$

Problem 28.   Quantum mechanically, a *phase shift* δ induced by a linear optical element on a single-mode optical field is described by the unitary operator

$$U := \exp(i\delta\hat{n})$$

where $\hat{n} := b^\dagger b$ is the *number operator* and b the annihilation operator for the optical mode. Assume the optical field is in the state $|\psi\rangle$.
(i) Express $|\psi\rangle$ in the basis of photon number Fock state representation.
(ii) Find the state

$$|\psi'\rangle := U|\psi\rangle \,.$$

(iii) Find $|\Delta\psi\rangle := |\psi'\rangle - |\psi\rangle$ and the norm $\|\,|\Delta\psi\rangle\|$.

Solution 28.   (i) We can write

$$|\psi\rangle = \sum_{m=0}^{\infty} c_m|m\rangle$$

in the basis of photon number Fock state representation, where c_m are the expansion coefficients.
(ii) The phase-shifted state $|\psi'\rangle$ can be written as

$$|\psi'\rangle = \exp(i\delta\hat{n})\sum_{m=0}^{\infty} c_m|m\rangle = \sum_{m=0}^{\infty} c_m e^{i\delta m}|m\rangle \,.$$

(iii) Thus for the difference we find

$$|\Delta\psi\rangle = |\psi'\rangle - |\psi\rangle = \sum_{m=0}^{\infty} c_m(e^{i\delta m} - 1)|m\rangle$$

and therefore

$$\| \, |\Delta\psi\rangle \, \|^2 = \langle\Delta\psi|\Delta\psi\rangle = 4\sum_{m=0}^{\infty} |c_m|^2 \sin^2(\delta m/2) = 4\sum_{m=0}^{\infty} P_m \sin^2(\delta m/2)$$

where $P_m = |c_m|^2$ is the photon number distribution for the input field.

Problem 29. The generator of displacements for numbers is formally defined by

$$D(k) := \int_{-\pi}^{\pi} d\phi \, e^{ik\phi} |\phi\rangle\langle\phi|$$

where

$$|\phi\rangle = \sum_{n=0}^{\infty} e^{in\phi}|n\rangle, \qquad \phi \in \mathbb{R}.$$

Show that these basis states are not normalized.

Solution 29. Since

$$\langle\phi| = \sum_{m=0}^{\infty} \langle m|e^{-im\phi}$$

and $\langle m|n\rangle = \delta_{mn}$ we find

$$\langle\phi|\phi\rangle = \sum_{m=0}^{\infty} 1 \, .$$

Problem 30. Let b_1, b_2 be Bose annihilation operators. Show that

$$e^{\mu b_1 b_2} e^{\nu b_1^\dagger b_2^\dagger}|00\rangle = \frac{1}{1-\mu\nu} e^{\nu b_1^\dagger b_2^\dagger/(1-\mu\nu)}|00\rangle, \qquad \mu, \nu \in \mathbb{R} \qquad (1)$$

where $|00\rangle \equiv |0\rangle \otimes |0\rangle$.

Solution 30. We solve the problem by considering the expression

$$e^{\mu b_1 b_2} e^{\nu b_1^\dagger b_2^\dagger}|00\rangle = e^{f(\mu, b_1^\dagger, b_2^\dagger)}|00\rangle$$

where f is an analytic function. Differentiating both sides with respect to μ yields

$$b_1 b_2 e^{\mu b_1 b_2} e^{\nu b_1^\dagger b_2^\dagger}|00\rangle = e^f \frac{\partial f}{\partial \mu}|00\rangle \, .$$

Thus

$$b_1 b_2 e^f|00\rangle = e^f \frac{\partial f}{\partial \mu}|00\rangle \, .$$

Note that $\partial f/\partial \mu$ commutes with $\exp(f)$ since f is a function of $b_1^\dagger$ and $b_2^\dagger$ only. If we multiply from the left by $\exp(-f)$ we find

$$e^{-f} b_1 b_2 e^f |00\rangle = \frac{\partial f}{\partial \mu} |00\rangle \,.$$

If follows that

$$e^{-f} b_1 e^f e^{-f} b_2 e^f |00\rangle = \frac{\partial f}{\partial \mu} |00\rangle \,.$$

Using

$$[b, g(b, b^\dagger)] = \frac{\partial g}{\partial b^\dagger} \tag{2}$$

with $g = e^f$, we obtain

$$e^{-f} b_1 e^f = e^{-f} \left(e^f b_1 + \frac{\partial e^f}{\partial b_1^\dagger} \right) = b_1 + \frac{\partial f}{\partial b_1^\dagger}$$

since e^f commutes with $\partial f/\partial b_1^\dagger$. Similarly

$$e^{-f} b_2 e^f = b_2 + \frac{\partial f}{\partial b_2^\dagger} \,.$$

Thus we have

$$\left(b_1 + \frac{\partial f}{\partial b_1^\dagger} \right) \left(b_2 + \frac{\partial f}{\partial b_2^\dagger} \right) |00\rangle = \frac{\partial f}{\partial \mu} |00\rangle \,.$$

Since $b_2 |00\rangle = 0$ we arrive at

$$\left(b_1 \frac{\partial f}{\partial b_2^\dagger} + \frac{\partial f}{\partial b_1^\dagger} \frac{\partial f}{\partial b_2^\dagger} \right) |00\rangle = \frac{\partial f}{\partial \mu} |00\rangle \,.$$

Using (2) again with $g = \partial f/\partial b_2^\dagger$ we obtain

$$b_1 \frac{\partial f}{\partial b_2^\dagger} = \frac{\partial f}{\partial b_2^\dagger} b_1 + \frac{\partial^2 f}{\partial b_1^\dagger \partial b_2^\dagger} \,.$$

Since $b_1 |00\rangle = 0$ we obtain

$$\left(\frac{\partial^2 f}{\partial b_1^\dagger \partial b_2^\dagger} + \frac{\partial f}{\partial b_1^\dagger} \frac{\partial f}{\partial b_2^\dagger} \right) |00\rangle = \frac{\partial f}{\partial \mu} |00\rangle \,.$$

Since f contains only $b_1^\dagger$ and $b_2^\dagger$ which commute, the solution of this partial differential equation must be of the form

$$f(\mu, b_1^\dagger, b_2^\dagger) = h_1(\mu) I + h_2(\mu) b_1^\dagger b_2^\dagger \,.$$

Thus

$$f(0, b_1^\dagger, b_2^\dagger) = \nu b_1^\dagger b_2^\dagger$$

or $h_1(0) = 0$, $h_2(0) = \nu$ owing to (1). Inserting this ansatz into the partial differential equation and equating equal powers of $b_1^\dagger b_2^\dagger$, we find that h_1 and h_2 satisfy the nonlinear system of ordinary differential equations

$$\frac{dh_2}{d\mu} = h_2^2, \qquad \frac{dh_1}{d\mu} = h_2$$

with the solution of the initial value problem

$$h_2(\mu) = \frac{\nu}{1 - \mu\nu}, \qquad h_1(\mu) = -\ln(1 - \mu\nu)$$

and thus we find (1).

Problem 31. The standard *Pauli group* for continuous variable quantum computing of n coupled oscillator systems is the Heisenberg-Weyl group which consists of phase-space displacement operators for n harmonic oscillators. This group is a continuous Lie group and can therefore only be generated by a set of continuously parameterized operators. The *Lie algebra* that generates this group is spanned by the $2n$ canonical operators $\hat{p}_j$, $\hat{q}_j$, $j = 1, 2, \ldots, n$ along with the commutation relation

$$[\hat{q}_j, \hat{p}_k] = i\hbar\delta_{jk}I .$$

For a single oscillator ($n = 1$) the algebra is spanned by the canonical operators $\{\hat{q}, \hat{p}, I\}$. We define the operators

$$X(q) := e^{-(i/\hbar)q\hat{p}}, \qquad Z(p) := e^{(i/\hbar)p\hat{q}}$$

where $q, p \in \mathbb{R}$. Let $\{\,|s\rangle \,:\, s \in \mathbb{R}\}$ be position eigenstates (in the sense of generalized functions).
(i) Calculate the states

$$X(q)|s\rangle, \qquad Z(p)|s\rangle .$$

(ii) Find the commutator $[X(q), Z(p)]$.

Solution 31. (i) We find in the sense of generalized functions that

$$X(q)|s\rangle = |s + q\rangle, \qquad Z(p)|s\rangle = \exp((i/\hbar)ps)|s\rangle .$$

Thus the operator $X(q)$ is a position translation operator. The operator $Z(p)$ is a momentum boost operator.
(ii) We obtain

$$X(q)Z(p) = e^{-(i/\hbar)qp}Z(p)X(q) .$$

Thus
$$[X(q), Z(p)] = (I - e^{(i/\hbar)qp})X(q)Z(p).$$

Problem 32.   Let $r \in \mathbb{R}$. Find ϵ_1, ϵ_2 and ϵ_3 such that
$$e^{r(b_1^\dagger b_2^\dagger - b_1 b_2)} \equiv e^{\epsilon_1 b_1^\dagger b_2^\dagger} e^{\epsilon_2 (b_1^\dagger b_1 + b_2^\dagger b_2 + I)} e^{\epsilon_3 b_1 b_2}.$$

Solution 32.   Using the fact that the operators
$$K_+ := -b_1 b_2, \qquad K_- := b_1^\dagger b_2^\dagger, \qquad K_3 := -\frac{1}{2}(b_1^\dagger b_1 + b_2^\dagger b_2 + I)$$

form the semi-simple Lie algebra
$$[K_3, K_+] = K_+, \qquad [K_3, K_-] = -K_-, \qquad [K_+, K_-] = 2K_3$$

and
$$e^{r(K_+ + K_-)} \equiv e^{K_- \tanh(r)} e^{2\ln(\cosh(r)) K_3} e^{K_+ \tanh(r)}$$

we find $\epsilon_1 = \tanh(r)$, $\epsilon_2 = -\ln(\cosh(r))$, $\epsilon_3 = -\tanh(r)$.

Problem 33.   To build a simple quantum computer one could use the following optical gates

$$U_S := \exp(i\pi b^\dagger b) \qquad\qquad \textit{phase modulator}$$
$$U_B := \exp\left(\frac{\pi}{4}(b_1^\dagger b_2 - b_1 b_2^\dagger)\right) \qquad \textit{quantum beam splitter}$$
$$U_F := \exp\left(\frac{\chi}{2}b_3^\dagger b_3(b_1^\dagger b_2 - b_1 b_2^\dagger)\right) \qquad \textit{Fredkin gate}$$

(i) Calculate the state
$$U_S|n\rangle.$$

(ii) Calculate the state
$$U_B|01\rangle.$$

(iii) Calculate the states
$$U_F|011\rangle, \qquad U_F|101\rangle, \qquad U_F|xy0\rangle$$

with $\chi = \pi$ and $x, y \in \{0, 1\}$.

Solution 33.   (i) Since $b^\dagger b|n\rangle = n|n\rangle$ we obtain
$$e^{i\pi b^\dagger b}|n\rangle = e^{i\pi n}|n\rangle = (-1)^n|n\rangle.$$

(ii) Since

$$(b_1^\dagger b_2 - b_1 b_2^\dagger)|01\rangle = |10\rangle, \qquad (b_1^\dagger b_2 - b_1 b_2^\dagger)|10\rangle = -|01\rangle$$

we find

$$U_B|01\rangle = \frac{1}{\sqrt{2}}\left(|01\rangle + |10\rangle\right)$$

where we used $\sin(\pi/4) = 1/\sqrt{2}$ and $\cos(\pi/4) = 1/\sqrt{2}$.

(iii) Since

$$b_3^\dagger b_3 (b_1^\dagger b_2 - b_1 b_2^\dagger)|011\rangle = |101\rangle, \qquad b_3^\dagger b_3 (b_1^\dagger b_2 - b_1 b_2^\dagger)|101\rangle = -|011\rangle$$

and

$$b_3^\dagger b_3 (b_1^\dagger b_2 - b_1 b_2^\dagger)|xy0\rangle = 0$$

we find the states

$$U_F|101\rangle = -|011\rangle, \qquad U_F|011\rangle = |101\rangle, \qquad U_F|xy0\rangle = |xy0\rangle$$

where we used that $b|0\rangle = 0$ and $b|1\rangle = |0\rangle$. Thus $b_3^\dagger b_3$ plays the role of a control operator.

Problem 34. Consider the Hamilton operator

$$\hat{H} = \hbar\omega b^\dagger b.$$

(i) Calculate

$$\operatorname{tr} e^{-\hat{H}/k_B T}.$$

(ii) Consider the *density operator*

$$\rho = \frac{\exp(-\hat{H}/k_B T)}{\operatorname{tr}\exp(-\hat{H}/k_B T)}.$$

Let $\hat{n} := b^\dagger b$. Calculate

$$\bar{n} \equiv \langle \hat{n} \rangle \equiv \langle b^\dagger b \rangle \equiv \operatorname{tr}(b^\dagger b \rho)$$

and $\langle \hat{H} \rangle$.

(iii) Calculate $\operatorname{tr}(\rho b)$, $\operatorname{tr}(\rho b^\dagger)$.

Solution 34. (i) We set $\lambda := \hbar\omega/k_B T$. Then using the completeness relation

$$I = \sum_{n=0}^{\infty} |n\rangle\langle n|,$$

$\hat{n}|n\rangle = n|n\rangle$ and $e^{-\lambda \hat{n}}|n\rangle = e^{-\lambda n}|n\rangle$ we find

$$e^{-\hat{H}/k_B T} = e^{-\lambda b^\dagger b} = \sum_{n=0}^{\infty} e^{-\lambda b^\dagger b}|n\rangle\langle n| = \sum_{n=0}^{\infty} e^{-\lambda n}|n\rangle\langle n|.$$

Thus

$$\operatorname{tr} e^{-\hat{H}/k_B T} = \sum_{n=0}^{\infty} (e^{-\lambda})^n = \frac{1}{1 - e^{-\lambda}}.$$

(ii) Using the result from (i) we obtain

$$\bar{n} = \langle b^\dagger b \rangle = \frac{1}{e^\lambda - 1}$$

and therefore

$$\langle \hat{H} \rangle = \frac{\hbar\omega}{e^{\hbar\omega/k_B T} - 1}.$$

(iii) Since $\langle n|b|n\rangle = 0$ and $\langle n|b^\dagger|n\rangle = 0$ we obtain

$$\langle b \rangle = \operatorname{tr}(\rho b) = 0, \qquad \langle b^\dagger \rangle = \operatorname{tr}(\rho b^\dagger) = 0.$$

Problem 35.   Let ϵ be a real parameter, σ_z the Pauli matrix and I_2 the 2×2 identity matrix. Calculate

$$f(\epsilon) = e^{\epsilon \sigma_z \otimes (b - b^\dagger)}(I_2 \otimes b^\dagger b)e^{-\epsilon \sigma_z \otimes (b - b^\dagger)} \tag{1}$$

using *parameter differentiation* and then solving the differential equation with the corresponding initial values (operators), i.e. $f(0) = I_2 \otimes b^\dagger b$.

Solution 35.   From (1) we obtain $f(0) = I_2 \otimes b^\dagger b$. Now

$$\frac{df}{d\epsilon} = e^{\epsilon \sigma_z \otimes (b - b^\dagger)}(\sigma_z \otimes ((b - b^\dagger)b^\dagger b - b^\dagger b(b - b^\dagger)))e^{-\epsilon \sigma_z \otimes (b - b^\dagger)}$$

$$= e^{\epsilon \sigma_z \otimes (b - b^\dagger)}(\sigma_z \otimes (b + b^\dagger))e^{-\epsilon \sigma_z \otimes (b - b^\dagger)}.$$

Thus

$$\frac{df(0)}{d\epsilon} = \sigma_z \otimes (b + b^\dagger).$$

Since

$$(b - b^\dagger)(b + b^\dagger) - (b + b^\dagger)(b - b^\dagger) = 2I$$

and $\sigma_z^2 = I_2$ we obtain for the second order derivative

$$\frac{d^2 f}{d\epsilon^2} = 2I_2 \otimes I.$$

Thus the solution of this second order linear differential equation is

$$f(\epsilon) = \epsilon^2 I_2 \otimes I + C_1 \epsilon + C_2 \, .$$

Inserting the two initial values (operators) yields

$$C_1 = \sigma_z \otimes (b + b^\dagger), \qquad C_2 = I_2 \otimes b^\dagger b \, .$$

It follows that

$$f(\epsilon) = \epsilon^2 I_2 \otimes I + \epsilon \sigma_z \otimes (b + b^\dagger) + I_2 \otimes b^\dagger b \, .$$

Problem 36. Let $\hat{n} := b^\dagger b$ and $\alpha \in \mathbb{R}$. Calculate the state $\exp(\alpha \hat{n})|0\rangle$.

Solution 36. Since $\hat{n}|0\rangle = 0$ we obtain $\exp(\alpha \hat{n})|0\rangle = |0\rangle$.

Problem 37. Let b, $b^\dagger$ be Bose operators. Find the eigenvalues of the operator $(-1)^{b^\dagger b}$. This unitary operator is defined by $e^{i\pi b^\dagger b}$. Why is this operator called the *parity operator*?

Solution 37. Let $|n\rangle$ be a Fock state. Then $b^\dagger b|n\rangle = n|n\rangle$, $b^\dagger b b^\dagger b|n\rangle = n^2|n\rangle$ etc. Thus

$$e^{i\pi b^\dagger b}|n\rangle = e^{i\pi n}|n\rangle$$

where $n = 0, 1, 2, \ldots, \infty$. It follows that

$$\langle m|e^{i\pi n}|n\rangle = e^{i\pi n}\langle m|n\rangle = e^{i\pi n}\delta_{mn} \, .$$

Since $e^{i\pi n} = +1$ if n is even and $e^{i\pi n} = -1$ if n is odd we find that the eigenvalues of $(-1)^{b^\dagger b}$ are $+1$ and -1 both infinitely degenerate.

Problem 38. The hermitian cosine $\hat{C}_{SG}$ and sine $\hat{S}_{SG}$ operators introduced by Susskind and Glogower are given in the Fock basis by

$$\hat{C}_{SG} := \frac{1}{2} \sum_{n=0}^{\infty} (|n\rangle\langle n+1| + |n+1\rangle\langle n|)$$

$$\hat{S}_{SG} := \frac{1}{2i} \sum_{n=0}^{\infty} (|n\rangle\langle n+1| - |n+1\rangle\langle n|) \, .$$

Solve the eigenvalue equations $\hat{C}_{SG}|c\rangle_{SG} = c|c\rangle_{SG}$, $\hat{S}_{SG}|s\rangle_{SG} = s|s\rangle_{SG}$.

Solution 38. We find

$$|c\rangle_{SG} = \sqrt{\frac{2}{\pi}} \sqrt[4]{1 - c^2} \sum_{n=0}^{\infty} U_n(c)|n\rangle, \quad |s\rangle_{SG} = \sqrt{\frac{2}{\pi}} \sqrt[4]{1 - s^2} \sum_{n=0}^{\infty} i^n U_n(s)|n\rangle$$

where $U_n(x)$ for $x = c$ or $x = s$ are the *Chebyshev polynomials* of second kind and $x \in [-1, 1]$. They obey the recursion formula

$$U_{n+1}(x) - 2xU_n(x) + U_{n-1}(x) = 0$$

where $U_0(x) = 1$ and $U_1(x) = 2x$.

Problem 39.   Let $|0\rangle$ be the vacuum state, i.e. $b|0\rangle = 0$, and $\epsilon \in \mathbb{R}$. Calculate the state

$$\exp(\epsilon b^\dagger \otimes b^\dagger)(|0\rangle \otimes |0\rangle).$$

Sometimes one also writes

$$\exp(\epsilon b_1^\dagger b_2^\dagger)|0\rangle|0\rangle.$$

Solution 39.   Since $b^\dagger|n\rangle = \sqrt{n+1}|n+1\rangle$ for $n = 0, 1, 2, \dots$ we obtain

$$(b^\dagger \otimes b^\dagger)(|n\rangle \otimes |n\rangle) = (b^\dagger|n\rangle) \otimes (b^\dagger|n\rangle) = (n+1)|n+1\rangle \otimes |n+1\rangle.$$

Now

$$
\begin{aligned}
\exp(\epsilon b^\dagger \otimes b^\dagger)|0\rangle \otimes |0\rangle &= \left(I \otimes I + \frac{\epsilon}{1!} b^\dagger \otimes b^\dagger + \frac{\epsilon^2}{2!}(b^\dagger)^2 \otimes (b^\dagger)^2 + \cdots\right)|0\rangle \otimes |0\rangle \\
&= |0\rangle \otimes |0\rangle + \epsilon|1\rangle \otimes |1\rangle + \epsilon^2|2\rangle \otimes |2\rangle + \cdots \\
&= \sum_{j=0}^{\infty} \epsilon^j |j\rangle \otimes |j\rangle.
\end{aligned}
$$

What is the condition on ϵ so that the series converges? This means: what is the condition that the state can be normalized?

Problem 40.   Let $b, b^\dagger$ be Bose annihilation and creation operators, respectively. Let $\epsilon \in \mathbb{R}$ and $|0\rangle$ be the vacuum state, i.e. $b|0\rangle = 0$.
(i) Calculate the state

$$b \exp(\epsilon b^\dagger)|0\rangle.$$

(ii) Calculate the state

$$b^n \exp(\epsilon b^\dagger)|0\rangle, \qquad n = 2, 3, \dots.$$

Solution 40.   (i) Since

$$bb^\dagger|0\rangle = |0\rangle, \qquad b(b^\dagger)^2|0\rangle = 2b^\dagger|0\rangle, \qquad b(b^\dagger)^3|0\rangle = 3(b^\dagger)^2|0\rangle$$

and in general $b(b^\dagger)^n|0\rangle = n(b^\dagger)^{n-1}|0\rangle$ we obtain

$$b\exp(\epsilon b^\dagger)|0\rangle = \epsilon|0\rangle + \frac{\epsilon^2}{2!}2b^\dagger|0\rangle + \frac{\epsilon^3}{3!}3(b^\dagger)^2|0\rangle + \cdots$$

$$= \epsilon(I + \epsilon b^\dagger + \frac{\epsilon^2}{2!}(b^\dagger)^2 + \cdots)|0\rangle$$

$$= \epsilon\exp(\epsilon b^\dagger)|0\rangle .$$

(ii) Using the result from (i) we have

$$b^n\exp(\epsilon b^\dagger)|0\rangle = \epsilon^n\exp(\epsilon b^\dagger)|0\rangle .$$

Problem 41. Consider the *Bose-Einstein density operator*

$$\rho = \frac{1}{\overline{n}+1}\sum_{n=0}^{\infty}\left(\frac{\overline{n}}{\overline{n}+1}\right)^n|n\rangle\langle n|$$

where $|n\rangle$, $n = 0, 1, 2, \ldots$ are the number states and

$$\overline{n} := \left(\exp\left(\frac{\hbar\omega}{k_BT}\right) - 1\right)^{-1} .$$

Show that $\mathrm{tr}\rho = 1$.

Solution 41. We set $\alpha = \hbar\omega/k_BT$ and use the number states as an orthonormal basis. Thus

$$\mathrm{tr}\rho = \sum_{m=0}^{\infty}\langle m|\rho|m\rangle = \frac{1}{\overline{n}+1}\sum_{n=0}^{\infty}\left(\frac{\overline{n}}{\overline{n}+1}\right)^n$$

where we used $\langle m|n\rangle = \langle n|m\rangle = \delta_{mn}$. Now we have

$$\frac{1}{\overline{n}+1} = 1 - e^{-\alpha}, \qquad \frac{\overline{n}}{\overline{n}+1} = e^{-\alpha} .$$

Since the sum is a *geometric series* we have

$$\sum_{n=0}^{\infty}(e^{-\alpha})^n = \frac{1}{1-e^{-\alpha}} .$$

Thus $\mathrm{tr}\rho = 1$.

Problem 42. Let $|0\rangle$ be the vacuum state. Calculate the state

$$\frac{1}{2}(b^\dagger \otimes I - I \otimes b^\dagger)(b^\dagger \otimes I + I \otimes b^\dagger)(|0\rangle \otimes |0\rangle)$$

where I is the identity operator. Discuss.

Solution 42.   Since $(b^\dagger \otimes I)(I \otimes b^\dagger) = (I \otimes b^\dagger)(b^\dagger \otimes I)$ we obtain the state

$$\frac{1}{\sqrt{2}}\left(|2\rangle \otimes |0\rangle - |0\rangle \otimes |2\rangle\right).$$

The photons interfere constructively or destructively. Complete destructive interference implies that the affected outgoing mode is in the vacuum state.

Problem 43.   Consider the operators

$$J_x = \frac{1}{2}(b_1^\dagger b_2 + b_2^\dagger b_1), \qquad J_y = \frac{1}{2}i(b_1^\dagger b_2 - b_2^\dagger b_1), \qquad J_z = \frac{1}{2}(b_1^\dagger b_1 - b_2^\dagger b_2)$$

and $J_+ := J_x + iJ_y = b_1^\dagger b_2$, $J_- := J_x - iJ_y = b_2^\dagger b_1$. Let $\phi \in \mathbb{R}$. Find

$$e^{-iJ_z\phi}b_1^\dagger e^{iJ_z\phi}, \quad e^{-iJ_z\phi}b_2^\dagger e^{iJ_z\phi}, \quad e^{-iJ_y\phi}b_1^\dagger e^{iJ_y\phi}, \qquad e^{-iJ_y\phi}b_2^\dagger e^{iJ_y\phi}.$$

Solution 43.   We obtain

$$\exp(-iJ_z\phi)b_1^\dagger \exp(iJ_z\phi) = b_1^\dagger e^{-i\phi/2}$$
$$\exp(-iJ_z\phi)b_2^\dagger \exp(iJ_z\phi) = b_2^\dagger e^{-i\phi/2}$$
$$\exp(-iJ_y\phi)b_1^\dagger \exp(iJ_y\phi) = b_1^\dagger \cos(\phi/2) + b_2^\dagger \sin(\phi/2)$$
$$\exp(-iJ_y\phi)b_2^\dagger \exp(iJ_y\phi) = b_2^\dagger \cos(\phi/2) - b_1^\dagger \sin(\phi/2).$$

Problem 44.   Let $\{\,|n\rangle \,:\, n = 0, 1, 2, \ldots\,\}$ be Fock states (number states). Consider the linear operator

$$T_{13} := \sum_{n=0}^{\infty}(|n\rangle \otimes I \otimes I)(I \otimes I \otimes \langle n|)$$

in the product (infinite-dimensional) Hilbert space $\mathcal{H} = \mathcal{H}_1 \otimes \mathcal{H}_2 \otimes \mathcal{H}_3$ with $\mathcal{H}_1 = \mathcal{H}_2 = \mathcal{H}_3$. Here I denotes the identity operator. Apply the operator T_{13} to the state

$$I \otimes I \otimes |\psi\rangle.$$

Solution 44.   We find

$$T_{13}(I \otimes I \otimes |\psi\rangle) = \sum_{n=0}^{\infty}(|n\rangle \otimes I \otimes I)(I \otimes I \otimes \langle n|\psi\rangle I)$$

$$= \sum_{n=0}^{\infty}(\langle n|\psi\rangle|n\rangle) \otimes I \otimes I$$

$$= |\psi\rangle \otimes I \otimes I$$

where we used

$$|\psi\rangle = \sum_{n=0}^{\infty} \langle n|\psi\rangle |n\rangle .$$

The operator T_{13} can be considered as a *transfer operator*.

Problem 45. A *beam splitter* can be realized by means of a linear medium where the *polarization vector* is proportional to the incoming *electric field*

$$\hat{\mathbf{P}} = \chi \hat{\mathbf{E}}$$

with $\chi \equiv \chi^{(1)}$ denoting the first order (linear) susceptibility. We consider the incoming field excited only in the relevant spatial modes b_1 and b_2 (at the same frequency ω)

$$\hat{\mathbf{E}}(\mathbf{r}, t) = i\sqrt{\frac{\hbar\omega}{2\epsilon_0 V}} \left((b_1 + b_2)e^{i(\mathbf{k}\cdot\mathbf{r}-\omega t)} + h.c. \right)$$

where $h.c.$ denotes the hermitian conjugate. The interaction Hamilton operator contains only the resonant terms

$$\hat{H}_I = -\hat{\mathbf{P}} \cdot \hat{\mathbf{E}} = -\chi\hat{\mathbf{E}}^2 = \frac{\chi\hbar\omega}{2\epsilon_0 V}(b_1^\dagger b_2 + b_1 b_2^\dagger)$$

where $\cdot$ denotes the scalar product. The evolution operator (in the interaction picture) of the whole device is expressed as

$$U := \exp\left(i \arctan\left(\sqrt{\frac{1-\tau}{\tau}} \right) \left(b_1^\dagger b_2 + b_1 b_2^\dagger \right) \right)$$

where τ, given by

$$\tau = \left(1 + \tan^2\left(\frac{\chi\hbar\omega}{2\epsilon_0 V} \right) \right)^{-1}$$

represents the the *transmissivity* of the beam splitter.
(i) Calculate

$$\tilde{b}_1 = U^\dagger b_1 U, \qquad \tilde{b}_2 = U^\dagger b_2 U .$$

(ii) Find a rotation of the phase frame by $3\pi/2$.

Solution 45. (i) Straightforward calculation yields

$$\tilde{b}_1 = U^\dagger b_1 U = -i\tau^{1/2} b_1 + (1-\tau)^{1/2} b_2$$

$$\tilde{b}_2 = U^\dagger b_2 U = i(1-\tau)^{1/2} b_1 + \tau^{1/2} b_2 .$$

(ii) A rotation of the phase frame can be obtained by the substitution $b_1 \to -ib_1$. Then we obtain

$$\widetilde{b}_1 = \tau^{1/2}b_1 + (1-\tau)^{1/2}b_2, \qquad \widetilde{b}_2 = \tau^{1/2}b_2 - (1-\tau)^{1/2}b_1.$$

Problem 46. Owing to their helical wave fronts the electromagnetic field of photons having an orbital angular momentum has a phase singularity. There the intensity has to vanish resulting in a doughnut-like intensity distribution. These light fields can be described using Laguerre Gaussian (LG_{pl}) modes with two indices p and l. The p-index $(p = 0,1,2,\ldots)$ identifies the non-axial radial nodes observed in the transversal plane and the l-index $(l = 0,\pm1,\pm2,\ldots)$ the number of the 2π-phase shifts along a closed path around the beam center. The index l is also called the topological winding number. It describes the helical structure of the wave front around a wave front singularity or dislocation. The index l also determines the amount of orbital angular momentum in units of $\hbar$ carried by one photon. When the pump beam is a $LG_{l_0 p_0}$ mode, under conditions of collinear phase-matching, the two-photon state at the output of the nonlinear crystal can be written as a coherent superposition of eigenstates of the orbital angular-momentum operator that are correlated in orbital angular momentum, i.e., $l_1 + l_2 = l_0$, where l_1 and l_2 refer to the orbital angular momentum eigenvalues for the signal and idler photons. A photon state described by a LG mode can be written as

$$|lp\rangle := \int dq \, LG_{lp}(q) b^\dagger(q)|0\rangle$$

where the mode function in the spatial frequency domain is given by

$$LG_{lp}(\rho,\phi) := \left(\frac{\omega_0 p!}{2\pi(|l|+p)!}\right)^{1/2}\left(\frac{\omega_0 \rho_k}{\sqrt{2}}\right)^{|l|} L_p^{|l|}\left(\frac{\rho_k^2 \omega_0^2}{2}\right)\exp\left(-\frac{\rho_k^2 \omega_0^2}{4}\right)$$
$$\times \exp\left(il\phi_k + i\left(p - \frac{|l|}{2}\right)\pi\right)$$

with ρ_k and ϕ_k being the modulus and phase, respectively, of the transverse coordinate q. The functions $L_p^{|l|}$ are the *associated Laguerre polynomials* and ω_0 is the beam width. Find the state $|lp\rangle$ for $l = p = 0$.

Solution 46. Since the associated Laguerre polynomial L_0^0 is given by $L_0^0(x) = 1$ we obtain

$$LG_{00} = \left(\frac{\omega_0}{2\pi}\right)^{1/2}\exp\left(-\frac{\rho_k^2 \omega_0^2}{4}\right).$$

Thus for LG_{00} we find a Gaussian.

Chapter 18

Coherent States

Quantum coherent states are the closest quantum-mechanical analogue to a classical particle oscillating in a harmonic potential. Coherent states are minimum uncertainty states. Quantum computation circuits with coherent states as the logical qubits can be constructed using simple linear networks, conditional measurements and coherent superposition resource states. Coherent states are very sensitive to their environment. The output of a single mode stabilised laser can be described by a coherent state $|\beta\rangle$, where β is a complex number which determines the average field amplitude. Harmonic oscillator coherent states can be defined in three different equivalent ways. Firstly, the coherent states are the eigenstates of the Bose annihilation operator

$$b|\beta\rangle = \beta|\beta\rangle, \qquad \beta \in \mathbb{C}.$$

Thus the spectrum of the operator b fills the entire complex plane. Secondly, they are displaced vacuum states

$$|\beta\rangle = \exp(-|\beta|^2/2)\exp(\beta b^\dagger)\exp(-\beta^* b)|0\rangle$$

where $|0\rangle$ is the vacuum state with $\langle 0|0\rangle = 1$. Since $b|0\rangle = 0$ we have

$$|\beta\rangle = \exp(-|\beta|^2/2)\exp(\beta b^\dagger)|0\rangle.$$

Thirdly, coherent states are states of minimum uncertainty

$$\Delta p \Delta x = \frac{\hbar}{2}$$

and are thus most classical within the quantum framework.

Problem 1.   Bose creation $b^\dagger$ and annihilation b operators obey the *Heisenberg algebra* $[b, b^\dagger] = I$, $[b, b] = [b^\dagger, b^\dagger] = 0$ with $b|0\rangle = 0$, where $|0\rangle$ is the vacuum state. The *coherent states* $|\beta\rangle$ can be obtained by applying the unitary *displacement operator*

$$D(\beta) := \exp(\beta b^\dagger - \beta^* b)$$

on the vacuum state $|0\rangle$, i.e.,

$$|\beta\rangle := D(\beta)|0\rangle = \exp(\beta b^\dagger - \beta^* b)|0\rangle = \exp\left(-\frac{1}{2}|\beta|^2\right) \sum_{n=0}^{\infty} \frac{\beta^n}{\sqrt{n!}}|n\rangle, \qquad \beta \in \mathbb{C}.$$

Show that from this definition the coherent states can also be obtained as the eigenstates of the annihilation (destruction) operator b, i.e., $b|\beta\rangle = \beta|\beta\rangle$.

Solution 1.   We find for the commutator

$$[b, (\beta b^\dagger - \beta^* b)^n] = \beta n (\beta b^\dagger - \beta^* b)^{n-1}, \qquad n = 1, 2, \dots$$

and therefore we have the commutation relation $[b, D(\beta)] = \beta D(\beta)$. Since $b|0\rangle = 0$ we have

$$0 = D(\beta)b|0\rangle = (b - \beta I)D(\beta)|0\rangle = (b - \beta I)|\beta\rangle$$

where we used the above commutation relation.

Problem 2.   Let $D(\beta)$ be the displacement operator. Find the operators $D^\dagger(\beta)$ and $D^{-1}(\beta)$.

Solution 2.   We find $D^\dagger(\beta) = D^{-1}(\beta) = D(-\beta)$.

Problem 3.   Let $|\beta\rangle$ and $|\gamma\rangle$ be coherent states.
(i) Calculate $\langle\gamma|\beta\rangle$.
(ii) Calculate $\langle 0|D(\beta)|0\rangle \equiv \langle 0|\beta\rangle$.
(iii) Find the probability $|\langle\gamma|\beta\rangle|^2$.

Solution 3.   (i) Since

$$|\beta\rangle = \exp\left(-\frac{1}{2}|\beta|^2\right) \sum_{n=0}^{\infty} \frac{\beta^n}{\sqrt{n!}}|n\rangle, \qquad |\gamma\rangle = \exp\left(-\frac{1}{2}|\gamma|^2\right) \sum_{m=0}^{\infty} \frac{\gamma^m}{\sqrt{m!}}|m\rangle$$

we find

$$\langle\gamma|\beta\rangle = \exp\left(-\frac{1}{2}(|\beta|^2 + |\gamma|^2)\right) \sum_{m=0}^{\infty}\sum_{n=0}^{\infty} \frac{\gamma^{*m}}{\sqrt{m!}} \frac{\beta^n}{\sqrt{n!}} \langle m|n\rangle$$

$$= \exp\left(-\frac{1}{2}(|\beta|^2 + |\gamma|^2)\right) \sum_{n=0}^{\infty} \frac{(\beta\gamma^*)^n}{n!}$$

$$= \exp\left(-\frac{1}{2}(|\beta|^2 + |\gamma|^2) + \beta\gamma^*\right)$$

where we used $\langle m|n \rangle = \delta_{mn}$.

(ii) Using (i) we find

$$\langle 0|\beta \rangle = \exp\left(-\frac{1}{2}|\beta|^2\right).$$

(iii) From (i) we obtain

$$|\langle \gamma|\beta \rangle|^2 = \exp(-|\beta - \gamma|^2).$$

If $\gamma = \beta$ we have $|\langle \beta|\beta \rangle|^2 = 1$. If $\gamma = -\beta$ we obtain $|\langle \beta| - \beta \rangle|^2 = e^{-4|\beta|^2}$.

Problem 4. Let $|\beta\rangle$ be a coherent state. Let $\hat{n} := b^\dagger b$. Calculate

$$\langle \beta|\hat{n}|\beta \rangle.$$

Solution 4. Since $b|\beta\rangle = \beta|\beta\rangle$, $\langle\beta|b^\dagger = \langle\beta|\beta^*$ we obtain

$$\langle \beta|\hat{n}|\beta \rangle = \beta\beta^* = |\beta|^2.$$

Problem 5. Let b, $b^\dagger$ be Bose annihilation and creation operators and $\hat{x}$, $\hat{p}$ be canonical position and momentum operators which are related by the equations

$$b = \frac{1}{\sqrt{2}}(\hat{x} + i\hat{p}), \qquad b^\dagger = \frac{1}{\sqrt{2}}(\hat{x} - i\hat{p}).$$

Express the displacement operator with $\hat{x}$ and $\hat{p}$.

Solution 5. We obtain

$$D(\hat{x}, \hat{p}) = \exp(i(p\hat{x} - x\hat{p}))$$

where the real and complex parameters are also related by

$$\beta = \frac{1}{\sqrt{2}}(x + ip), \qquad \bar{\beta} = \frac{1}{\sqrt{2}}(x - ip).$$

Problem 6. Consider the Hamilton operator $\hat{H} = \hbar\omega b^\dagger b$. Let

$$U(t) := \exp(-it\hat{H}/\hbar).$$

Find the state $U(t)|\beta\rangle$, where $|\beta\rangle$ is a coherent state.

Solution 6.   Since

$$|\beta\rangle = e^{-|\beta|^2/2} \sum_{n=0}^{\infty} \frac{\beta^n}{\sqrt{n!}} |n\rangle$$

and $b^\dagger b|n\rangle = n|n\rangle$ we find

$$U(t)|\beta\rangle = |\beta e^{-i\omega t}\rangle.$$

Thus the linear evolution of $|\beta\rangle$ is a rotation in phase space. The initial state will be revived at $\omega t = 2\pi, 4\pi, \ldots$ as expected.

Problem 7.   Let $D(\beta) := \exp(\beta b^\dagger - \beta^* b)$ be the displacement operator.
(i) Find the operators

$$D(\beta)bD(-\beta), \qquad D(\beta)b^\dagger D(-\beta).$$

(ii) Find the operators

$$D^\dagger(\beta)bD(\beta), \qquad D^\dagger(\beta)b^\dagger D(\beta).$$

Solution 7.   (i) Since

$$[b^\dagger, b] = -I, \qquad [b^\dagger, [b^\dagger, b]] = 0$$

we have

$$D(\beta)bD(-\beta) = b - \beta I, \qquad D(\beta)b^\dagger D(-\beta) = b^\dagger - \beta^* I.$$

Note that $D(-\beta) = D^\dagger(\beta) = D^{-1}(\beta)$.
(ii) Using parameter differentiation yields

$$D^\dagger(\beta)bD(\beta) = b + \beta I, \qquad D^\dagger(\beta)b^\dagger D(\beta) = b^\dagger + \beta^* I.$$

Problem 8.   Coherent states are defined as

$$|\beta\rangle := e^{-\beta\beta^*/2} \sum_{m=0}^{\infty} \frac{\beta^m}{\sqrt{m!}} |m\rangle$$

where β is a complex number. We have $b|\beta\rangle = \beta|\beta\rangle$ (eigenvalue equation),

$$\sum_{m=0}^{\infty} |m\rangle\langle m| = I$$

(completeness relation) and $b^\dagger b |n\rangle = n|n\rangle$.
(i) Calculate $\langle n|\beta\rangle$ and then the probability $P_n(\beta) := |\langle n|\beta\rangle|^2$.
(ii) Let $\hat{n} := b^\dagger b$. Calculate

$$\langle \hat{n} \rangle := \langle \beta | \hat{n} | \beta \rangle, \quad \langle \hat{n}^2 \rangle := \langle \beta | \hat{n}^2 | \beta \rangle \,.$$

(iii) Calculate the *variance*

$$\langle (\Delta \hat{n})^2 \rangle := \langle (\hat{n} - \langle \hat{n} \rangle I)^2 \rangle \,.$$

Solution 8. (i) Using $\langle n|m \rangle = \delta_{nm}$ we have

$$\langle n|\beta \rangle = \exp\left(-\frac{1}{2}|\beta|^2 \right) \frac{\beta^n}{\sqrt{n!}} \,.$$

Thus we obtain

$$P_n(\beta) = \frac{(\beta\beta^*)^n \exp(-\beta\beta^*)}{n!} \,.$$

This is a *Poisson distribution*.
(ii) Using $b|\beta\rangle = \beta|\beta\rangle$ and therefore $\langle \beta|b^\dagger = \langle \beta|\beta^*$ we find

$$\langle \hat{n} \rangle = \langle \beta|b^\dagger b|\beta \rangle = \beta\beta^*, \quad \langle \hat{n}^2 \rangle = (\beta\beta^*)^2 + \beta\beta^*$$

where we used $bb^\dagger = b^\dagger b + I$.
(iii) Applying the results from (ii) we obtain

$$\langle (\Delta \hat{n})^2 \rangle := \langle (\hat{n} - \langle \hat{n} \rangle I)^2 \rangle = \beta\beta^* \,.$$

Problem 9. Let $|0\rangle$, $|1\rangle$, ... be the number states. An arbitrary normalized state $|g\rangle$ can be expanded as

$$|g\rangle = \sum_{j=0}^{\infty} c_j |j\rangle, \quad \sum_{j=0}^{\infty} c_j^* c_j = 1 \,.$$

Express the state using coherent states $|\beta\rangle$. Consider the special case that $c_j = 0$ for all j except for $c_n = 1$.

Solution 9. We find

$$|g\rangle = \frac{1}{\pi} \int_{\mathbb{C}} e^{-|\beta|^2/2} g(\beta^*) |\beta\rangle d^2\beta$$

where $d^2\beta = d(\Re\beta)d(\Im\beta)$ and

$$g(\beta^*) = \sum_{j=0}^{\infty} \frac{c_j (\beta^*)^j}{\sqrt{j!}} \,.$$

If all c_j are equal to 0 except for $c_n = 1$ we obtain

$$|g\rangle = |n\rangle = \frac{1}{\pi} \int_{\mathbb{C}} e^{-|\beta|^2/2} \frac{(\beta^*)^n}{\sqrt{n!}} |\beta\rangle d^2\beta \,.$$

Problem 10.   Coherent states are defined as

$$|\beta\rangle := e^{-\beta\beta^*/2} \sum_{m=0}^{\infty} \frac{\beta^m}{\sqrt{m!}} |m\rangle$$

where β is a complex number. Let $|\psi\rangle$ be an arbitrary state in the Hilbert space containing $|\beta\rangle$. Show that

$$|\psi(\beta)| \le \exp\left(\frac{1}{2}|\beta|^2\right) \,.$$

Solution 10.   We have the following identity (*completeness relation*)

$$\frac{1}{\pi} \int_{\mathbb{C}} d^2\beta |\beta\rangle\langle\beta| = \sum_{m=0}^{\infty} |m\rangle\langle m| = I$$

from which it follows that the system of coherent states is complete. Using this equation we can expand an arbitrary state $|\psi\rangle$ with respect to the state $|\beta\rangle$

$$|\psi\rangle = \frac{1}{\pi} \int_{\mathbb{C}} d^2\beta \langle\beta|\psi\rangle |\beta\rangle \,.$$

If the coherent state $|\beta\rangle$ is taken as $|\psi\rangle$, then this equation defines a linear dependence between the different coherent states. It follows that the system of coherent states is *supercomplete*, i.e. it contains subsystems which are complete. Using the definition for the coherent state given above we obtain

$$\langle\beta|\psi\rangle = \exp\left(-\frac{1}{2}|\beta|^2\right)\psi(\beta^*), \qquad \psi(\beta) = \sum_{n=0}^{\infty} \frac{\beta^n}{\sqrt{n!}} \langle n|\psi\rangle \,.$$

The inequality $|\langle n|\psi\rangle| \le 1$ means that the function $\psi(\beta)$ for the normalization state $|\psi\rangle$ is an *entire analytic function* of the complex variables β. We also have $|\langle\beta|\psi\rangle| \le 1$. Therefore we find a bound on the growth of $\psi(\beta)$

$$|\psi(\beta)| \le \exp\left(\frac{1}{2}|\beta|^2\right) \,.$$

The normalization condition can now be written as

$$\frac{1}{\pi} \int_{\mathbb{C}} d^2\beta \exp(-|\beta|^2)|\psi(\beta)|^2 = \langle\psi|\psi\rangle \,.$$

The expansion of an arbitrary state $|\psi\rangle$ with respect to coherent states now takes the form

$$|\psi\rangle = \frac{1}{\pi} \int_{\mathbb{C}} d^2\beta \exp\left(-\frac{1}{2}|\beta|^2\right) \psi(\beta^*)|\beta\rangle.$$

Problem 11. Coherent states $|\beta\rangle$ can be written as $|\beta\rangle = D(\beta)|0\rangle$, where $D(\beta)$ is the displacement operator and $|0\rangle$ denotes the vacuum state. Show that

$$D(\beta)D(\gamma) = \exp(i\Im(\beta\gamma^*))D(\beta + \gamma). \tag{1}$$

Solution 11. Since $[b, b^\dagger] = I$ we have

$$[\beta b^\dagger - \beta^* b, \gamma b^\dagger - \gamma^* b] = -[\beta b^\dagger, \gamma^* b] - [\beta^* b, \gamma b^\dagger] = (\beta\gamma^* - \beta^*\gamma)I = 2i\Im(\beta\gamma^*)I.$$

Using the *Baker-Campbell-Hausdorff formula*

$$e^A e^B = e^{A+B} e^{[A,B]/2}$$

for $[[A, B], A] = 0$ and $[[A, B], B] = 0$, we find (1). As a consequence we have

$$D(\beta)D(\gamma)D(-\beta) = e^{2i\Im(\beta\gamma^*)}D(\gamma).$$

It follows that the operators $\exp(2\pi it)D(\beta)$ form a group. An element g of this group is defined by the real number t and a complex number β: $g(t, \beta)$. The product of two group elements $g = g_1 g_2$ is given by $g(t, \beta)$ with

$$t = t_1 + t_2 + \frac{1}{2\pi}\Im(\beta_2\beta_1^*)$$

and $\beta = \beta_1 + \beta_2$.

Problem 12. Consider the displacement operator $D(\beta)$. Show that

$$D(\beta)D(\gamma) = e^{\beta\gamma^* - \beta^*\gamma}D(\gamma)D(\beta). \tag{1}$$

Solution 12. The Baker-Campbell-Hausdorff formula

$$e^{A+B} = e^A e^B e^{-[A,B]/2} = e^B e^A e^{[A,B]/2}$$

for $[A, [A, B]] = [B, [A, B]] = 0$ can be applied since

$$[\beta b^\dagger - \beta^* b, \gamma b^\dagger - \gamma^* b] = (\beta\gamma^* - \beta^*\gamma)I$$

where I is the identity operator. Thus (1) follows.

Problem 13.   Let $|\beta\rangle$ be a coherent state. Show that

$$\frac{1}{\pi} \int_{\mathbb{C}} |\beta\rangle\langle\beta| d^2\beta = I$$

where I is the identity operator and the integration is over the entire complex plane. Set $\beta = r\exp(i\phi)$ with $0 \leq r < \infty$ and $0 \leq \phi < 2\pi$.

Solution 13.   We have

$$\frac{1}{\pi} \int_{\mathbb{C}} |\beta\rangle\langle\beta| d^2\beta = \frac{1}{\pi} \sum_{n=0}^{\infty} \sum_{m=0}^{\infty} \frac{|n\rangle\langle m|}{\sqrt{n!\, m!}} \int_{\mathbb{C}} e^{-|\beta|^2} \beta^{*m} \beta^n d^2\beta.$$

Using $\beta = r\exp(i\phi)$ we arrive at

$$\frac{1}{\pi} \int_{\mathbb{C}} |\beta\rangle\langle\beta| d^2\beta = \frac{1}{\pi} \sum_{n=0}^{\infty} \sum_{m=0}^{\infty} \frac{|n\rangle\langle m|}{\sqrt{n!\, m!}} \int_0^{\infty} re^{-r^2} r^{n+m} dr \int_0^{2\pi} e^{i(n-m)\phi} d\phi.$$

Since

$$\int_0^{2\pi} e^{i(n-m)\phi} d\phi = 2\pi\delta_{nm}$$

we have

$$\frac{1}{\pi} \int_{\mathbb{C}} |\beta\rangle\langle\beta| d^2\beta = \sum_{n=0}^{\infty} \frac{|n\rangle\langle n|}{n!} \int_0^{\infty} e^{-s} s^n ds$$

where we set $s = r^2$ and therefore $ds = 2r dr$. Thus

$$\frac{1}{\pi} \int_{\mathbb{C}} |\beta\rangle\langle\beta| d^2\beta = \sum_{n=0}^{\infty} |n\rangle\langle n| = I$$

where we used the completeness relation for the number states.

Problem 14.   (i) The *Husimi distribution* of a coherent state γ is given by

$$\rho_\gamma^H(\beta) := |\langle\beta|\gamma\rangle|^2.$$

Calculate $\rho_\gamma^H(\beta)$.

(ii) The Husimi distribution of the number state (Fock state) $|n\rangle$ is given by

$$\rho_{|n\rangle}^H(\beta) := |\langle\beta|n\rangle|^2.$$

Calculate $\rho_{|n\rangle}^H(\beta)$.

(iii) Consider the state $|n_1\rangle \otimes |n_2\rangle$. Find

$$\rho_{|n_1\rangle \otimes |n_2\rangle}^H(\beta) = |((\langle\beta_1| \otimes \langle\beta_2|)(|n_1\rangle \otimes |n_2\rangle))|^2.$$

Solution 14. (i) Since

$$|\beta\rangle = e^{-|\beta|^2/2} \sum_{n=0}^{\infty} \frac{\beta^n}{\sqrt{n!}} |n\rangle, \qquad |\gamma\rangle = e^{-|\gamma|^2/2} \sum_{n=0}^{\infty} \frac{\gamma^n}{\sqrt{n!}} |n\rangle$$

we find

$$\langle\beta|\gamma\rangle = e^{-|\beta|^2/2} e^{-|\gamma|^2/2} e^{\beta^*\gamma}$$

where we used $\langle m|n\rangle = \delta_{mn}$. Thus

$$|\langle\beta|\gamma\rangle|^2 = e^{-|\beta-\gamma|^2}$$

and the Husimi distribution of a coherent state is *Gaussian*.
(ii) Since $\langle m|n\rangle = \delta_{mn}$ we find

$$\langle\beta|n\rangle = e^{-|\beta|^2/2} \frac{\beta^{*n}}{\sqrt{n!}}$$

and hence

$$|\langle\beta|n\rangle|^2 = \frac{e^{-|\beta|^2}(|\beta|^2)^n}{n!}.$$

The Husimi distribution represents a *Poisson distribution* over the photon number states.
(iii) Since

$$((\langle\beta_1| \otimes \langle\beta_2|)(|n_1\rangle \otimes |n_2\rangle)) = \langle\beta_1|n_1\rangle\langle\beta_2|n_2\rangle$$

we obtain

$$\rho_{|n_1\rangle\otimes|n_2\rangle}^{H}(\beta) = \frac{e^{-|\beta_1|^2}(|\beta_1|^2)^{n_1}}{n_1!} \frac{e^{-|\beta_2|^2}(|\beta_2|^2)^{n_2}}{n_2!}.$$

Problem 15. In a *Kerr medium* the state evolution is governed by the interaction Hamilton operator

$$\hat{H} = \kappa(b^\dagger b)^2$$

where κ is a coupling constant proportional to the nonlinear susceptibility of the medium. A coherent input signal state $|\beta\rangle$ evolves according to the solution of the Schrödinger equation

$$|\psi_c(t)\rangle = \exp(-i\hat{H}t)|\beta\rangle.$$

Calculate $|\psi_c(t)\rangle$ for $t = \pi/(2\kappa)$. Discuss.

Solution 15.   Straightforward calculation yields

$$|\psi_c(t = \pi/(2\kappa)))\rangle = \frac{1}{\sqrt{2}}(e^{-i\pi/4}|\beta\rangle + e^{i\pi/4}|-\beta\rangle)$$

$$= \frac{1}{\sqrt{2}}(e^{-i\pi/4}D(\beta) + e^{i\pi/4}D(-\beta))|0\rangle$$

where $D(\beta)$ is the displacement operator. The state describes a superposition of two coherent states with opposite phases. When $|\beta|$ becomes large the two components become mesoscopically distinguishable states of the radiation field. Realistic values of the Kerr nonlinear susceptibilities are quite small, thus requiring a long interaction time, or equivalently a large interaction length. Thus losses become significant and the resulting decoherence may destroy the quantum superposition.

Problem 16.   Let

$$|\beta\rangle = e^{-|\beta|^2/2}\sum_{n=0}^{\infty}\frac{\beta^n}{\sqrt{n!}}|n\rangle.$$

Find the operator

$$\rho = \frac{1}{2\pi}\int_0^{2\pi}d\phi||\beta|e^{-i\phi}\rangle\langle|\beta|e^{-i\phi}|.$$

Solution 16.   We obtain

$$\rho = e^{-|\beta|^2}\sum_{n=0}^{\infty}\frac{|\beta|^{2n}}{n!}|n\rangle\langle n|.$$

This is a collection of number states.

Problem 17.   Let $b^\dagger$, b be Bose creation and annihilation operators and $\hat{n} = b^\dagger b$. We can introduce *nonlinear coherent states* B and $B^\dagger$ with an intensity dependent function $f(\hat{n})$ (which is an operator valued analytic function)

$$B := bf(\hat{n}) = f(\hat{n} + I)b, \qquad B^\dagger := f^\dagger(\hat{n})b^\dagger = b^\dagger f^\dagger(\hat{n} + I)$$

We consider f to be real and non-negative, i.e., $f^\dagger(\hat{n}) = f(\hat{n})$.
(i) Calculate the commutator $[B, B^\dagger]$. Then consider the special case $f(\hat{n}) = I$ and $f(\hat{n}) = \hat{n}$.
(ii) Express the harmonic oscillator for $B^\dagger$ and B

$$\hat{H} = \frac{1}{2}(BB^\dagger + B^\dagger B)$$

using $\hat{n}$ and $f(\hat{n})$.

(iii) Find the eigenstate of the operator B, i.e., find the nonlinear coherent state

$$B|z\rangle_{NL} = \lambda|z\rangle_{NL} \,.$$

What is the condition that the state belongs to the Fock space?

Solution 17. (i) Since f is analytic we have a Taylor expansion. We obtain

$$[B, B^\dagger] = (\hat{n} + I)f(\hat{n} + I)f(\hat{n} + I) - \hat{n}f(\hat{n})f(\hat{n}) \,.$$

With $f(\hat{n}) = I$ we obtain the result for the Bose operators b and $b^\dagger$. For the case $f(\hat{n}) = \hat{n}$ we obtain

$$[B, B^\dagger] = 3\hat{n}^2 + 3\hat{n} + I \,.$$

(ii) We obtain

$$\hat{H} = \frac{1}{2}((\hat{n} + I)f(\hat{n} + I)f(\hat{n} + I) + \hat{n}f(\hat{n})f(\hat{n})) \,.$$

(iii) Using the expansion with respect to the Fock basis $\{\,|0\rangle, |1\rangle, \ldots\,\}$ and the fact that f is an analytic function we obtain

$$|z\rangle_{NL} = \mathcal{N}_f(|z|^2)^{-1/2} \sum_{n=0}^{\infty} C_n z^n |n\rangle$$

where the coefficients C_n are given by

$$C_n = \frac{1}{\sqrt{[nf(n)f(n)]!}}, \qquad C_0 = 1, \qquad [f(n)]! := f(n)f(n-1)\cdots f(1)$$

and the normalization constant is given by

$$\mathcal{N}_f(|z|^2) = \sum_{n=0}^{\infty} |C_n|^2 |z|^{2n} \,.$$

In order to have states belonging to the Fock space, it is required that $0 < \mathcal{N}_f(|z|^2) < \infty$. This implies that

$$|z| \leq \lim_{n\to\infty} n[f(n)]^2 \,.$$

The function $f(n)$ corresponding to any nonlinear coherent state is found to be

$$f(n) = \frac{C_{n-1}}{\sqrt{n}C_n} \,.$$

Problem 18.   Let $D(\beta)$ be the displacement operator. Is

$$D(\beta)D(\beta') = D(\beta + \beta') ?$$

Prove or disprove.

Solution 18.   The answer is no. We find

$$D(\beta)D(\beta') = \exp\left(\frac{1}{2}(\beta\beta'^* - \beta^*\beta')\right) D(\beta + \beta').$$

Note that $\beta \in \mathbb{C}$. If $\beta, \beta' \in \mathbb{R}$ then we have $D(\beta)D(\beta') = D(\beta + \beta')$.

Problem 19.   Consider the coherent state $|\beta\rangle$. We define the *Schrödinger cat states*

$$|\beta_+\rangle := N_+(|\beta\rangle + |-\beta\rangle), \qquad |\beta_-\rangle := N_-(|\beta\rangle - |-\beta\rangle).$$

(i) Normalize the two states.
(ii) Calculate the probabilities

$$|\langle\beta_+|n\rangle|^2, \qquad |\langle\beta_-|n\rangle|^2$$

and discuss.

Solution 19.   (i) From $\langle\beta_+|\beta_+\rangle = 1$, $\langle\beta_-|\beta_-\rangle = 1$ and

$$\langle\gamma|\beta\rangle = \exp(-|\beta - \gamma|^2/2)$$

we obtain

$$N_+ = \frac{\exp(|\beta|^2/2)}{2\sqrt{\cosh|\beta|^2}}, \qquad N_- = \frac{\exp(|\beta|^2/2)}{2\sqrt{\sinh|\beta|^2}}.$$

(ii) Since

$$\langle n|\beta\rangle = e^{-|\beta|^2/2}\frac{\beta^n}{\sqrt{n!}}$$

we obtain the probabilities

$$|\langle\beta_+|n\rangle|^2 = \frac{1}{\cosh|\beta|^2}\frac{|\beta|^{2n}}{n!}(1 + (-1)^n)$$

$$|\langle\beta_-|n\rangle|^2 = \frac{1}{\sinh|\beta|^2}\frac{|\beta|^{2n}}{n!}(1 - (-1)^n).$$

Problem 20.   Let $D(\beta)$ be the displacement operator.
(i) Find the normally ordered form of $D(\beta)$.

(ii) Find the antinormally ordered form of $D(\beta)$.

Solution 20. (i) We find

$$D(\beta) = \exp\left(-\frac{1}{2}|\beta|^2\right)\exp(\beta b^\dagger)\exp(-\beta^* b).$$

(ii) We find

$$D(\beta) = \exp\left(\frac{1}{2}|\beta|^2\right)\exp(-\beta^* b)\exp(\beta b^\dagger).$$

Problem 21. Consider the linear operator

$$Z := b_1 + b_2^\dagger \equiv b \otimes I + I \otimes b \tag{1}$$

where $b_1 = b \otimes I$ and $b_2^\dagger = I \otimes b^\dagger$. Let $D_b(z) := e^{zb^\dagger - z^* b}$ ($z \in \mathbb{C}$) be the displacement operator and

$$|0\rangle\rangle := \frac{1}{\sqrt{\pi}}\sum_{n=0}^{\infty}(-1)^n|n\rangle \otimes |n\rangle.$$

The states $|z\rangle\rangle$ are defined by

$$|z\rangle\rangle := D_{b_1}(z)|0\rangle\rangle = D_{b_2}(z^*)|0\rangle\rangle.$$

Find the state $Z|z\rangle\rangle$. Discuss.

Solution 21. We have $[Z, Z^\dagger] = 0$ and

$$Z|z\rangle\rangle = z|z\rangle\rangle, \qquad z \in \mathbb{C}.$$

Thus $|z\rangle\rangle$ is an eigenstate of Z. For $z = 0$ the state $|0\rangle\rangle$ can be approximated by a physical (normalizable) state called the *twin beam state* - corresponding to the output of a non-degenerate optical parametric amplifier in the limit of infinite gain.

Problem 22. Consider the *beam splitter interaction* given by the unitary transformation

$$U_{BS} = \exp(i\theta(b_1 b_2^\dagger + b_1^\dagger b_2))$$

where b_1 and b_2 are the Bose annihilation operators. Let $|\beta\rangle$, $|\gamma\rangle$ be coherent states. Calculate

$$U_{BS}|\gamma\rangle \otimes |\beta\rangle.$$

Solution 22. We obtain

$$U_{BS}|\gamma\rangle \otimes |\beta\rangle = |\cos(\theta)\gamma + i\sin(\theta)\beta\rangle \otimes |\cos(\theta)\beta + i\sin(\theta)\gamma\rangle$$

where $\cos^2(\theta)$ $(\sin^2(\theta))$ is the reflectivity (transmissivity) of the beam splitter.

Problem 23. Bose creation $(\mathbf{b}^\dagger)$ and annihilation $(\mathbf{b})$ operators, where

$$\mathbf{b}^\dagger = (b_1^\dagger, b_2^\dagger, \dots, b_N^\dagger), \qquad \mathbf{b} = (b_1, b_2, \dots, b_N)$$

obey the *Heisenberg algebra*

$$[b_j, b_k^\dagger] = \delta_{jk} I, \quad [b_j, b_k] = [b_j^\dagger, b_k^\dagger] = 0, \quad j, k = 1, 2, \dots, N.$$

Coherent states, where $\mathbf{z} \in \mathbb{C}^n$, are defined as eigenvectors of the annihilation operators, that is

$$\mathbf{b}|\mathbf{z}\rangle = \mathbf{z}|\mathbf{z}\rangle.$$

(i) Show that the normalized coherent states are given by

$$|\mathbf{z}\rangle = \exp\left(-\frac{1}{2}|\mathbf{z}|^2\right) \exp(\mathbf{z} \cdot \mathbf{b}^\dagger)|\mathbf{0}\rangle \tag{1}$$

where

$$|\mathbf{z}|^2 = \sum_{j=1}^N |z_j|^2, \qquad \mathbf{z} \cdot \mathbf{b}^\dagger = \sum_{j=1}^N z_j b_j^\dagger$$

and $|\mathbf{0}\rangle = |0\,0\dots0\rangle$ is the vacuum vector satisfying $\mathbf{b}|\mathbf{0}\rangle = \mathbf{0}$.

(ii) Let $|\mathbf{w}\rangle$ be a coherent state. Find

$$\langle \mathbf{z}|\mathbf{w}\rangle, \qquad |\langle \mathbf{z}|\mathbf{w}\rangle|^2.$$

(iii) Calculate

$$\int_{\mathbb{R}^{2N}} d\mu(\mathbf{z})|\mathbf{z}\rangle\langle \mathbf{z}|.$$

Solution 23. (i) Consider the number representation

$$|\mathbf{n}\rangle \equiv |n_1, n_2, \dots, n_N\rangle = \frac{(b_1^\dagger)^{n_1}}{\sqrt{n_1!}} \frac{(b_2^\dagger)^{n_2}}{\sqrt{n_2!}} \cdots \frac{(b_N^\dagger)^{n_N}}{\sqrt{n_N!}} |\mathbf{0}\rangle.$$

We expand $|\mathbf{z}\rangle$ with respect to $|\mathbf{n}\rangle$ and apply

$$b_j|\mathbf{n}\rangle = \sqrt{n_j}|n_1, \cdots, n_j - 1, \cdots, n_N\rangle$$

$$b_j^\dagger|\mathbf{n}\rangle = \sqrt{n_j + 1}|n_1, \cdots, n_j + 1, \cdots, n_N\rangle$$

we find, after normalization, that $|\mathbf{z}\rangle$ is given by (1).

(ii) We find

$$\langle \mathbf{z}|\mathbf{w}\rangle = \exp\left(-\frac{1}{2}(|\mathbf{z}|^2 + |\mathbf{w}|^2 - 2\mathbf{z}^* \cdot \mathbf{w})\right)$$

and

$$|\langle \mathbf{z}|\mathbf{w}\rangle|^2 = \exp(-|\mathbf{z} - \mathbf{w}|^2).$$

(iii) Since

$$d\mu(\mathbf{z}) = \frac{1}{\pi} \prod_{j=1}^{N} d(\Re z_j) d(\Im z_j)$$

we find

$$\int_{\mathbb{R}^{2N}} d\mu(\mathbf{z})|\mathbf{z}\rangle\langle \mathbf{z}| = I$$

where I is the identity operator.

Problem 24. We know that

$$
\begin{aligned}
|\beta\rangle = D(\beta)|0\rangle &= e^{\beta b^\dagger - \bar{\beta} b}|0\rangle \\
&= e^{\beta b^\dagger} e^{-\bar{\beta} b} e^{-\beta\bar{\beta} I/2}|0\rangle = e^{-\beta\bar{\beta}/2} e^{\beta b^\dagger}|0\rangle \\
&= e^{-\beta\bar{\beta}/2} \sum_{n=0}^{\infty} \frac{\beta^n}{\sqrt{n!}}|n\rangle
\end{aligned}
$$

where

$$|n\rangle = \frac{(b^\dagger)^n}{\sqrt{n!}}|0\rangle.$$

Let σ_x be the Pauli spin matrix. Extend the calculation to the state

$$\tilde{D}(\beta)\left(|0\rangle \otimes \begin{pmatrix} 1 \\ 0 \end{pmatrix}\right)$$

where

$$\tilde{D}(\beta) := \exp(\beta b^\dagger \otimes \sigma_x - \bar{\beta} b \otimes \sigma_x).$$

Solution 24. First we note that since $\sigma_x^2 = I_2$ and $[b, b^\dagger] = I$

$$-[\beta b^\dagger \otimes \sigma_x, -\bar{\beta} b \otimes \sigma_x] = -\beta\bar{\beta} I_B \otimes I_2.$$

Thus we obtain

$$
\begin{aligned}
\tilde{D}(\beta)|0\rangle \otimes \begin{pmatrix} 1 \\ 0 \end{pmatrix} &= e^{\beta b^\dagger \otimes \sigma_x} e^{-\bar{\beta} b \otimes \sigma_x} e^{-\beta\bar{\beta} I_B \otimes I_2/2}|0\rangle \otimes \begin{pmatrix} 1 \\ 0 \end{pmatrix} \\
&= e^{-\beta\bar{\beta}/2} e^{\beta b^\dagger \otimes \sigma_x} e^{-\bar{\beta} b \otimes \sigma_x}|0\rangle \otimes \begin{pmatrix} 1 \\ 0 \end{pmatrix}
\end{aligned}
$$

$$= e^{-\beta\bar{\beta}/2} e^{\beta b^\dagger \otimes \sigma_x} |0\rangle \otimes \begin{pmatrix} 1 \\ 0 \end{pmatrix}$$

$$= e^{-\beta\bar{\beta}/2} \left(\left(\sum_{n=0}^{\infty} \frac{\beta^{2n}}{\sqrt{(2n)!}} |2n\rangle \right) \otimes \begin{pmatrix} 1 \\ 0 \end{pmatrix} \right.$$

$$\left. + \left(\sum_{n=0}^{\infty} \frac{\beta^{2n+1}}{\sqrt{(2n+1)!}} |2n+1\rangle \right) \otimes \begin{pmatrix} 0 \\ 1 \end{pmatrix} \right).$$

Problem 25. The normalized *Schrödinger cat state* of a single-mode radiation field is given by

$$|\beta, \theta\rangle = \frac{1}{\sqrt{2 + 2\cos\theta e^{-2|\beta|^2}}} (|\beta\rangle + e^{i\theta} |-\beta\rangle).$$

Discuss the special cases $\theta = 0$, $\theta = \pi$, $\theta = \pi/2$.

Solution 25. For $\theta = 0$ we obtain the even coherent states

$$|\beta, 0\rangle = \frac{1}{\sqrt{2 + 2e^{-2|\beta|^2}}} (|\beta\rangle + |-\beta\rangle).$$

For $\theta = \pi$ we obtain the odd states

$$|\beta, \pi\rangle = \frac{1}{\sqrt{2 - 2e^{-2|\beta|^2}}} (|\beta\rangle - |-\beta\rangle).$$

For $\theta = \pi/2$ we obtain the *Yurke-Stoler states*

$$|\beta, \pi/2\rangle = \frac{1}{\sqrt{2}} (|\beta\rangle + i|-\beta\rangle).$$

Chapter 19

Squeezed States

19.1 Single Mode Squeezing

Coherent states are not the most general kind of Gaussian wave packet. They are also not the most general kind of minimum-uncertainty wave packets, since the minimum uncertainty wave packet satisfies $\Delta q \Delta p = \hbar/2$ which only constrains the product of the dispersions Δq and Δp, whereas for coherent states we have that $(\Delta q)^2 = \hbar/(2\omega)$ and $(\Delta p)^2 = \hbar\omega/2$. For squeezed states one does not have this restriction. Unlike a coherent state, an initial squeezed state does not remain a minimum-uncertainty state in the course of time under the harmonic oscillator evolution. The product $\Delta q \Delta p$ oscillates at twice the harmonic oscillator frequency between a maximum value and a minimum value. Squeezed states possess the property that one quadrature phase has reduced fluctuations compared to the ordinary vacuum. Squeezed states of the electromagnetic field are generated by degenerate parametric down conversion in an optical cavity. The ideal squeezed state is defined as

$$|\zeta\rangle := S(\zeta)|0\rangle$$

where

$$S(\zeta) := \exp\left(\frac{1}{2}\zeta^* b^2 - \frac{1}{2}\zeta b^{\dagger 2}\right)$$

is the one-mode *squeezing operator* with $\zeta \in \mathbb{C}$. If we set $\zeta = re^{i\phi}$ $(r \geq 0)$ then r is called the *squeezing parameter*.

Problem 1.   Let $\zeta \in \mathbb{C}$. The *squeezing operator* is defined by

$$S(\zeta) := \exp\left(-\frac{\zeta}{2}(b^\dagger)^2 + \frac{\zeta^*}{2}b^2\right)$$

where $\zeta \in \mathbb{C}$. We can set $\zeta = re^{i\phi}$, where $r \geq 0$ and $\phi \in \mathbb{R}$. The *single mode squeezed states* $|\zeta\rangle$ are defined by

$$|\zeta\rangle := S(\zeta)|0\rangle.$$

(i) Show that $S(\zeta)$ is unitary.
(ii) Find $S(\zeta)$ in the disentangled form.
(iii) Calculate the state $|\zeta\rangle = S(\zeta)|0\rangle$ using the number states.

Solution 1.   (i) We obtain

$$S^\dagger(\zeta) = \exp\left(\frac{\zeta}{2}(b^\dagger)^2 - \frac{\zeta^*}{2}b^2\right) = S(-\zeta).$$

Thus we have $S^\dagger(\zeta)S(\zeta) = I$.
(ii) We obtain

$$S(\zeta) = \exp\left(-\frac{1}{2}(b^\dagger)^2 e^{i\phi}\tanh(r)\right)\exp\left(-\frac{1}{2}(b^\dagger b + bb^\dagger)\ln(\cosh(r))\right)$$
$$\times \exp\left(\frac{1}{2}b^2 e^{-i\phi}\tanh(r)\right).$$

(iii) Using the result from (ii) we obtain

$$|\zeta\rangle = \sqrt{(\mathrm{sech}(r))}\sum_{n=0}^{\infty}\frac{\sqrt{((2n)!)}}{n!}\left(-\frac{1}{2}\exp(i\phi)\tanh(r)\right)^n|2n\rangle$$

where we used that $(b^\dagger b + bb^\dagger)|0\rangle = |0\rangle$.

Problem 2.   Consider the one-mode squeezing operator

$$S(\zeta) = e^{1/2(\zeta^* b^2 - \zeta(b^\dagger)^2)}$$

where $\zeta \in \mathbb{C}$. Calculate

$$\widetilde{b} = S^\dagger(\zeta)bS(\zeta)$$

and

$$\widetilde{b}^\dagger = S^\dagger(\zeta)b^\dagger S(\zeta).$$

Solution 2. We obtain

$$\tilde{b} = \cosh(|\zeta|)b - \frac{\zeta}{|\zeta|}\sinh(|\zeta|)b^\dagger$$

and

$$\tilde{b}^\dagger = \cosh(|\zeta|)b^\dagger - \frac{\zeta^*}{|\zeta|}\sinh(|\zeta|)b \,.$$

Problem 3. Consider the single-mode squeezing operator

$$S(\zeta) = \exp\left(\frac{1}{2}\zeta(b^\dagger)^2 - \frac{1}{2}\zeta^* b^2\right)\,.$$

Let $|0\rangle$ be the vacuum state. Find the state $S(\zeta)|0\rangle \equiv |\zeta\rangle$ using number states $|n\rangle$. Use this result to find the expectation value $\langle\zeta|b^\dagger b|\zeta\rangle$.

Solution 3. Expansion over the number state basis contains only even components, i.e.

$$S(\zeta)|0\rangle \equiv |\zeta\rangle = \frac{1}{\sqrt{\mu}}\sum_{k=0}^{\infty}\left(\frac{\nu}{2\mu}\right)^k \frac{\sqrt{(2k)!}}{k!}|2k\rangle$$

where $\zeta = re^{i\phi}$, $\mu = \cosh(r)$, $\nu = e^{i\phi}\sinh(r)$. Using this result and

$$b^\dagger b|2k\rangle = 2k|2k\rangle$$

we obtain

$$\langle\zeta|b^\dagger b|\zeta\rangle = |\nu|^2\,.$$

Problem 4. Let $S(\zeta)$ be the one-mode *squeezing operator* $(\zeta \in \mathbb{C})$

$$S(\zeta) := \exp\left(\frac{1}{2}\zeta^* b^2 - \frac{1}{2}\zeta(b^\dagger)^2\right)\,.$$

Calculate

$$S(\chi)\begin{pmatrix} b \\ b^\dagger \end{pmatrix} S^\dagger(\zeta) \equiv \begin{pmatrix} S(\zeta)bS^\dagger(\zeta) \\ S(\zeta)b^\dagger S^\dagger(\zeta) \end{pmatrix}$$

and show that we can write

$$\begin{pmatrix} S(\zeta)bS^\dagger(\zeta) \\ S(\zeta)b^\dagger S^\dagger(\zeta) \end{pmatrix} = T(\zeta,\zeta^*)\begin{pmatrix} b \\ b^\dagger \end{pmatrix}$$

where the 2×2 matrix $T(\zeta,\zeta^*)$ depends only on ζ and ζ^*.

Solution 4.   We have $S^\dagger(\zeta) = S(-\zeta)$. Let $\epsilon \in \mathbb{R}$. We set

$$f_1(\epsilon) = e^{\frac{1}{2}\epsilon(\zeta^* b^2 - \zeta(b^\dagger)^2)} b \, e^{-\frac{1}{2}\epsilon(\zeta^* b^2 - \zeta(b^\dagger)^2)}$$

and

$$f_2(\epsilon) = e^{\frac{1}{2}\epsilon(\zeta^* b^2 - \zeta(b^\dagger)^2)} b^\dagger \, e^{-\frac{1}{2}\epsilon(\zeta^* b^2 - \zeta(b^\dagger)^2)}$$

with the initial conditions $f_1(\epsilon = 0) = b$ and $f_2(\epsilon = 0) = b^\dagger$. We find the system of differential equations for f_1 and f_2. Using $bb^\dagger = I + b^\dagger b$ we have

$$\frac{df_1}{d\epsilon} = -\frac{1}{2}\zeta e^{\frac{1}{2}\epsilon(\zeta^* b^2 - \zeta(b^\dagger)^2)}((b^\dagger)^2 b - b(b^\dagger)^2) e^{-\frac{1}{2}\epsilon(\zeta^* b^2 - \zeta(b^\dagger)^2)}$$
$$= \zeta f_2(\epsilon).$$

Analogously, we obtain

$$\frac{df_2}{d\epsilon} = \zeta^* f_1(\epsilon).$$

In matrix notation we have the system

$$\begin{pmatrix} df_1/d\epsilon \\ df_2/d\epsilon \end{pmatrix} = \begin{pmatrix} 0 & \zeta \\ \zeta^* & 0 \end{pmatrix} \begin{pmatrix} f_1 \\ f_2 \end{pmatrix}.$$

Let

$$A = \begin{pmatrix} 0 & \zeta \\ \zeta^* & 0 \end{pmatrix}.$$

We find

$$\exp(\epsilon A) = \begin{pmatrix} \cosh(\epsilon r) & \zeta \sinh(\epsilon r)/r \\ \zeta^* \sinh(\epsilon r)/r & \cosh(\epsilon r) \end{pmatrix} = \begin{pmatrix} \cosh(\epsilon r) & e^{i\phi}\sinh(\epsilon r) \\ e^{-i\phi}\sinh(\epsilon r) & \cosh(\epsilon r) \end{pmatrix}$$

where we used $\chi = r\exp(i\phi)$. Taking into account the initial conditions we obtain

$$\begin{pmatrix} S(\zeta) b S^\dagger(\zeta) \\ S(\zeta) b^\dagger S^\dagger(\zeta) \end{pmatrix} = \begin{pmatrix} \cosh(r) & e^{i\phi}\sinh(r) \\ e^{-i\phi}\sinh(r) & \cosh(r) \end{pmatrix} \begin{pmatrix} b \\ b^\dagger \end{pmatrix}.$$

The matrix on the right-hand side can be decomposed as

$$\begin{pmatrix} \cosh(r) & e^{i\phi}\sinh(r) \\ e^{-i\phi}\sinh(r) & \cosh(r) \end{pmatrix} \equiv$$

$$\begin{pmatrix} 0 & e^{i\phi/2} \\ e^{-i\phi/2} & 0 \end{pmatrix} \begin{pmatrix} \cosh(r) & \sinh(r) \\ \sinh(r) & \cosh(r) \end{pmatrix} \begin{pmatrix} 0 & e^{i\phi/2} \\ e^{-i\phi/2} & 0 \end{pmatrix}.$$

Problem 5.   For generating a squeezed state of one mode we start from the Hamilton operator

$$\hat{H} = \hbar\omega b^\dagger b + i\hbar\Lambda(b^2 e^{2i\omega t} - (b^\dagger)^2 e^{-2i\omega t}).$$

A photon of the driven mode, with frequency 2ω, splits into two photons of the mode of interest, each with frequency ω. Solve the Heisenberg equation of motion for b and $b^\dagger$.

Solution 5. The Heisenberg equation of motion yields

$$\frac{db}{dt} = \frac{1}{i\hbar}[b, \hat{H}](t) = -i\omega b(t) - 2\Lambda b^\dagger(t)e^{-2i\omega t}$$

$$\frac{db^\dagger}{dt} = \frac{1}{i\hbar}[b^\dagger, \hat{H}](t) = i\omega b^\dagger - 2\Lambda be^{2i\omega t}.$$

Using $b = \tilde{b}\exp(-i\omega t)$ we obtain

$$\frac{d\tilde{b}}{dt} = -2\Lambda\tilde{b}^\dagger(t), \qquad \frac{d\tilde{b}^\dagger}{dt} = -2\Lambda\tilde{b}(t).$$

Introducing the operators

$$\tilde{b}_P := \frac{1}{2}(\tilde{b} + \tilde{b}^\dagger), \qquad \tilde{b}_Q := \frac{1}{2i}(\tilde{b} - \tilde{b}^\dagger)$$

i.e.

$$\tilde{b} = \tilde{b}_P + i\tilde{b}_Q, \qquad \tilde{b}^\dagger = \tilde{b}_P - i\tilde{b}_Q$$

we finally arrive at the system of differential equations

$$\frac{db_P}{dt} = -2\Lambda b_P(t), \qquad \frac{db_Q}{dt} = 2\Lambda b_Q(t)$$

with the solution of the initial value problem

$$b_P(t) = b_P(0)e^{-2\Lambda t}, \qquad b_Q(t) = b_Q(0)e^{2\Lambda t}.$$

If the state is the vacuum state $|0\rangle$ we obtain a squeezed state. For the *electric field* $E(\mathbf{r}, t)$ we have

$$
\begin{aligned}
E(\mathbf{r}, t) &= i\mathcal{E}(b(t)e^{i\mathbf{k}\cdot\mathbf{r}} - b^\dagger(t)e^{-i\mathbf{k}\cdot\mathbf{r}}) \\
&= i\mathcal{E}(\tilde{b}(t)e^{i\mathbf{k}\cdot\mathbf{r}-i\omega t} - \tilde{b}^\dagger(t)e^{-i\mathbf{k}\cdot\mathbf{r}-i\omega t}) \\
&= -2\mathcal{E}(b_P(t)\sin(\mathbf{k}\cdot\mathbf{r} - \omega t) + b_Q(t)\cos(\mathbf{k}\cdot\mathbf{r} - \omega t)).
\end{aligned}
$$

Thus b_P and b_Q are the amplitudes of two quadrature components of the electric field. They are measurable by phase sensitive detection.

Problem 6. Consider the *squeezing operator*

$$S(r) := \exp\left(\frac{1}{2}r(b^2 - b^{\dagger 2})\right)$$

where $r \in \mathbb{R}$. Find the operators

$$S(r)\hat{q}S^\dagger(r), \qquad S(r)\hat{p}S^\dagger(r)$$

where

$$\hat{q} := \frac{1}{\sqrt{2}}(b + b^\dagger), \qquad \hat{p} := -\frac{i}{\sqrt{2}}(b - b^\dagger).$$

Solution 6.　We obtain

$$S(r)\hat{q}S^\dagger(r) = e^{-r}\hat{q}, \qquad S(r)\hat{p}S^\dagger(r) = e^{r}\hat{p}.$$

Problem 7.　Let $\zeta \in \mathbb{C}$ and

$$\hat{G} := \frac{1}{2}(\zeta b^\dagger b^\dagger - \zeta^* bb).$$

(i) Calculate the commutators

$$[\hat{G}, b], \qquad [\hat{G}, [\hat{G}, b]].$$

(ii) Let

$$S(\zeta) := \exp\left(\frac{1}{2}(\zeta b^\dagger b^\dagger - \zeta^* bb)\right) \equiv \exp\hat{G}.$$

Find

$$S(\zeta)bS(-\zeta), \qquad S(\zeta)b^\dagger S(-\zeta).$$

Solution 7.　(i) We have

$$[\hat{G}, b] = \frac{1}{2}\zeta[b^\dagger b^\dagger, b] = -\zeta b^\dagger$$

and

$$[\hat{G}, [\hat{G}, b]] = [\hat{G}, -\zeta b^\dagger] = -\zeta[\hat{G}, b^\dagger] = |\zeta|^2 b.$$

(ii) Using the results from (i), we find

$$S(\zeta)bS(-\zeta) = (\cosh r)b - e^{i\phi}(\sinh r)b^\dagger$$

where $\zeta = re^{i\phi}$. Likewise, we find

$$S(\zeta)b^\dagger S(-\zeta) = (\cosh r)b^\dagger - e^{-i\phi}(\sinh r)b.$$

Problem 8.　Consider the unbounded linear operators

$$K_+ := \frac{1}{2}b^{\dagger 2}, \quad K_- := \frac{1}{2}b^2, \quad K_0 := \frac{1}{2}(b^\dagger b + \frac{1}{2}I)$$

$$A^\dagger := b^\dagger, \quad A := b$$

where I is the identity operator.
(i) Show that these operators form a Lie algebra.
(ii) Consider the operator

$$P := \zeta K_+ - \zeta^* K_- + \alpha A^\dagger - \alpha^* A$$

where ζ and α are complex numbers. Let

$$V = e^{\beta K_+} e^{\epsilon A^\dagger} e^{\gamma K_0} e^{\nu I} e^{\delta K_-} e^{\eta A}$$

where β, ϵ, γ, ν, δ and η are complex numbers. Let

$$e^P = V.$$

Find γ, β, δ, ϵ, η, ν as functions of ζ and α.

Solution 8. (i) For the commutators we obtain

$$[K_0, K_\pm] = \pm K_\pm, \quad [K_+, K_-] = -2K_0, \quad [K_+, A] = -A^\dagger, \quad [K_-, A^\dagger] = A$$

$$[K_0, A^\dagger] = \frac{1}{2} A^\dagger, \quad [K_0, A] = -\frac{1}{2} A, \quad [A, A^\dagger] = I.$$

This Lie algebra refers to squeezed coherent states.
(ii) We write the complex numbers ζ and α in terms of real numbers λ, μ, θ and ϕ as

$$\zeta = r e^{i\theta}, \quad \alpha = s e^{i\phi}.$$

We use the formula

$$e^A B e^{-A} = \sum_{n=0}^{\infty} \frac{[A, B]_n}{n!}$$

where the *repeated commutator* $[A, B]_n$ is defined by

$$[A, B]_n := [A, [A, B]_{n-1}]$$

with $[A, B]_0 := B$. We find

$$e^P A e^{-P} = \cosh(r) A - e^{i\theta} \sinh(r) A^\dagger + \frac{s}{r}((\cosh(r) - 1) e^{i(\theta - \phi)} - \sinh(r) e^{i\phi})$$

$$e^P A^\dagger e^{-P} = \cosh(r) A^\dagger - e^{-i\theta} \sinh(r) A + \frac{s}{r}((\cosh(r) - 1) e^{i(\phi - \theta)} - \sinh(r) e^{-i\phi}).$$

The corresponding similarity transformations, induced by the operator V, are

$$VAV^{-1} = e^{-\gamma/2}(A - \beta A^\dagger - \epsilon I)$$
$$VA^\dagger V^{-1} = (e^{\gamma/2} - \beta\delta e^{-\gamma/2}) A^\dagger + \delta e^{-\gamma/2} A + \eta I - \epsilon\delta e^{-\gamma/2} I.$$

From

$$e^P A e^{-P} = V A V^{-1}, \qquad e^P A^\dagger e^{-P} = V A^\dagger V^{-1}$$

we find, by separating out terms with $A^\dagger$, A and I, that

$$\gamma = -2\ln(\cosh(r))$$
$$\beta = e^{i\theta}\tanh(r)$$
$$\delta = -e^{-i\theta}\tanh()$$
$$\epsilon = -\frac{s}{r\cosh(r)}((\cosh(r) - 1)e^{i(\theta - \phi)} - \sinh(r)e^{i\phi})$$
$$\eta = -\frac{s}{r\cosh(r)}((\cosh(r) - 1)e^{-i(\theta - \phi)} - \sinh(r)e^{-i\phi}).$$

The coefficient ν cannot be found by this method. How can we determine ν? One finds

$$\nu = -\frac{s^2}{r^2\cosh(r)}((\cosh(r) - 1) + i\sin(\theta - 2\phi)(\sinh(r) - r\cosh(r))).$$

Problem 9.   (i) Show that the operators

$$K_+ = \frac{1}{2}b^\dagger b^\dagger, \qquad K_- = \frac{1}{2}bb, \qquad K_0 = \frac{1}{2}\left(b^\dagger b + \frac{1}{2}I\right)$$

are generators of the Lie algebra $su(1,1)$.
(ii) Let $D(\beta)$ be the displacement operator and

$$S(z) = \exp(zK_+ - z^*K_-), \qquad z \in \mathbb{C}.$$

We define the state

$$|(\beta, z)\rangle := D(\beta)S(z)|0\rangle.$$

Show that the ordering $D(\beta)S(z)$ versus $S(z)D(\beta)$ is unitarily equivalent.

Solution 9.   (i) We have

$$[K_0, K_+] = K_+, \quad [K_0, K_-] = -K_-, \quad [K_+, K_-] = -2K_0.$$

(ii) We have

$$D(\beta)S(z) = S(z)D(\gamma), \qquad \gamma = \beta\cosh r - \beta^* e^{i\theta}\sinh r$$

and $z = re^{i\theta}$.

Problem 10.   Consider the squeeze operator

$$S(\zeta) = \exp\left(\frac{1}{2}(\zeta^* b^2 - \zeta b^{\dagger 2})\right)$$

with $\zeta = |\zeta| \exp(i(\phi + \theta))$. Is

$$S(\zeta')S(\zeta'') = S(\zeta' + \zeta'') ?$$

Solution 10.   The answer is no. We find

$$S(\zeta')S(\zeta'') = \exp\left(-i\phi\left(b^\dagger b + \frac{1}{2}I\right)\right) S(\zeta) \tag{1}$$

with

$$\zeta' = |\zeta'| \exp(i\theta'), \quad \mu' = \cosh|\zeta'|, \quad \nu' = \exp(i\theta')\sinh|\zeta'|$$

and $|\mu'|^2 - |\nu'|^2 = 1$ with analogous relations for ζ'', μ'' and ν''. In (1) ζ is obtained from

$$\zeta = |\zeta| \exp(i(\phi + \theta)), \qquad |\zeta| = \tanh^{-1}\left(\frac{|\nu|}{|\mu|}\right)$$

with

$$\mu = \mu'\mu'' + \nu'^*\nu'', \qquad \nu = \mu'^*\nu'' + \nu'\mu''$$

and $|\mu|^2 - |\nu|^2 = 1$.

Problem 11.   (i) A single mode squeeze operator is defined by

$$S(z) = \exp\left(\frac{1}{2}(z(b^\dagger)^2 - z^*b^2)\right)$$

where $z \in \mathbb{C}$ with $z = re^{i\theta}$. Write $S(z)$ in the form $S(z) = \exp(iH(z))$. Find $H(z)$.
(ii) Calculate

$$R^\dagger(\phi)H(z)R(\phi)$$

where $R(\phi) = \exp(i\phi b^\dagger b)$ is a single-mode rotation. Set $\phi = \theta/2$ and thus find $H(r)$.
(iii) Use the transformation

$$b = \frac{x + ip}{\sqrt{2}}, \qquad b^\dagger = \frac{x - ip}{\sqrt{2}}$$

and express $H(r)$ in terms of x and p.
(iv) Discuss the $H(r)$ as a Hamilton function.
(v) Write down the eigenvalue equation $H\psi(x) = E\psi$ with $p = -id/dx$. Discuss the spectrum.

Solution 11.   (i) Obviously we have

$$H(z) = \frac{1}{2i}(z(b^\dagger)^2 - z^*b^2).$$

(ii) We find

$$R^\dagger(\phi)H(z)R(\phi) = H(ze^{-2i\phi}).$$

Thus for $\phi = \theta/2$ we obtain $H(r)$.
(iii) Inserting the transformation yields

$$H(r) = -\frac{r}{2}(xp + px).$$

(iv) The Hamilton function $H(r) = -rxp$ leads to the Hamilton equations of motion

$$\frac{dx}{dt} = -rx, \qquad \frac{dp}{dt} = rp.$$

This describes damping of x and the pumping of p. This is a classical picture of a squeezing process.
(v) Since

$$H(r)\psi(x) = -\frac{r}{2}(xp + px)\psi(x) = irx\frac{d\psi}{dx} + i\frac{r}{2}\psi(x)$$

it follows that the eigenvalue equation is

$$x\frac{d}{dx}\psi(x) = -\left(i\frac{E}{r} + \frac{1}{2}\right)\psi(x).$$

$H(r)$ has a purely continuous spectrum covering the whole real axis. $H(r)$ is also parity invariant $(x \to -x)$. Therefore each generalized eigenvalue $E \in \mathbb{R}$ is doubly degenerated. The eigenvectors in the sense of generalized functions are

$$\psi_\pm^E(x) = \frac{1}{\sqrt{2\pi r}}x_\pm^{-(iE/r+1/2)}$$

where

$$x_+^\lambda := \begin{cases} x^\lambda & \text{for } x \geq 0 \\ 0 & \text{for } x < 0 \end{cases}, \qquad x_-^\lambda := \begin{cases} 0 & \text{for } x \geq 0 \\ |x|^\lambda & \text{for } x < 0 \end{cases}$$

and $\lambda \in \mathbb{C}$.

Problem 12.   Consider the linear operator

$$\hat{D} := \frac{1}{2}(\hat{q}\hat{p} + \hat{p}\hat{q}).$$

We set $\hbar = 1$.
(i) Find the commutators $[\hat{D}, \hat{q}]$ and $[\hat{D}, \hat{p}]$.

(ii) We consider the linear operator

$$S_\epsilon := \exp(-i\epsilon\hat{D}), \qquad \epsilon \in \mathbb{R}.$$

We define

$$S_\epsilon^\dagger \hat{q} S_\epsilon := \exp(i\epsilon \mathrm{ad}\hat{D})\hat{q} \equiv \sum_{n=0}^{\infty} \frac{(i\epsilon)^n}{n!}(\mathrm{ad}\hat{D})^n \hat{q} \tag{1}$$

where

$$(\mathrm{ad}\hat{D})\hat{q} := [\hat{D}, \hat{q}].$$

Calculate the operators $S_\epsilon^\dagger \hat{q} S_\epsilon$ and $S_\epsilon^\dagger \hat{p} S_\epsilon$.

(iii) Let

$$b = \frac{\sqrt{m\omega}}{\sqrt{2}}\left(\hat{q} + i\frac{\hat{p}}{m\omega}\right), \qquad b^\dagger = \frac{\sqrt{m\omega}}{\sqrt{2}}\left(\hat{q} - i\frac{\hat{p}}{m\omega}\right).$$

Express $\hat{D}$ in terms of b and $b^\dagger$.

(iv) Consider the normalized state

$$|\epsilon\rangle := S_\epsilon|0\rangle.$$

Calculate the expectation values

$$\langle\epsilon|\hat{q}|\epsilon\rangle, \quad \langle\epsilon|\hat{p}|\epsilon\rangle, \quad \langle\epsilon|\hat{q}^2|\epsilon\rangle, \quad \langle\epsilon|\hat{p}^2|\epsilon\rangle.$$

Solution 12. (i) Since $[\hat{q}, \hat{p}] = iI$ we find

$$[\hat{D}, \hat{q}] = -i\hat{q}, \qquad [\hat{D}, \hat{p}] = i\hat{p}.$$

(ii) Using the result from (i) and the definition (1) we find the operator

$$S_\epsilon^\dagger \hat{q} S_\epsilon = e^\epsilon \hat{q}.$$

Using the result from (i) and the definition (1) with $\hat{q}$ replaced by $\hat{p}$, we find

$$S_\epsilon^\dagger \hat{p} S_\epsilon = e^{-\epsilon}\hat{p}.$$

(iii) First we express $\hat{q}$ and $\hat{p}$ in terms of b and $b^\dagger$. Using the commutation relation $[\hat{q}, \hat{p}] = iI$, we obtain

$$\hat{D} = 2i(b^\dagger b^\dagger - bb).$$

(iv) Using the results from (i) through (iii) we find the expectation values

$$\langle\epsilon|\hat{q}|\epsilon\rangle = \langle 0|S_\epsilon^\dagger \hat{q} S_\epsilon|0\rangle = e^\epsilon \langle 0|\hat{q}|0\rangle = 0$$

$$\langle\epsilon|\hat{p}|\epsilon\rangle = \langle 0|S_\epsilon^\dagger \hat{p} S_\epsilon|0\rangle = e^{-\epsilon}\langle 0|\hat{p}|0\rangle = 0$$

$$\langle\epsilon|\hat{q}^2|\epsilon\rangle = e^{2\epsilon}\langle 0|\hat{q}^2|0\rangle = e^{2\epsilon}\frac{1}{2m\omega} = (\Delta q)^2$$

$$\langle\epsilon|\hat{p}^2|\epsilon\rangle = e^{-2\epsilon}\langle 0|\hat{p}^2|0\rangle = e^{-2\epsilon}\frac{m\omega}{2} = (\Delta p)^2.$$

19.2   Two Mode Squeezing

Problem 13.   Consider the operator

$$U(z) := e^{zb_1^\dagger b_2 - \bar{z}b_2^\dagger b_1}$$

where $b_1^\dagger, b_2^\dagger$ are Bose creation operators and b_1, b_2 are Bose annihilation operators and $z \in \mathbb{C}$. Find the operators

$$U(z)b_1 U(z)^{-1}, \qquad U(z)b_2 U(z)^{-1}.$$

Solution 13.   We have

$$U^{-1}(z) = e^{-zb_1^\dagger b_2 + \bar{z}b_2^\dagger b_1}.$$

Thus

$$U(z)b_1 U(z)^{-1} = \cos(|z|)b_1 - \frac{z\sin(|z|)}{|z|}b_2$$

$$U(z)b_2 U(z)^{-1} = \cos(|z|)b_2 + \frac{\bar{z}\sin(|z|)}{|z|}b_1.$$

We can write

$$(U(z)b_1 U(z)^{-1}, U(z)b_2 U(z)^{-1}) = (b_1, b_2) \begin{pmatrix} \cos(|z|) & \frac{\bar{z}\sin(|z|)}{|z|} \\ -\frac{z\sin(|z|)}{|z|} & \cos(|z|) \end{pmatrix}$$

where the matrix on the right-hand side is an element of the Lie group $SU(2)$.

Problem 14.   Consider the operator

$$U(z) := e^{zb_1^\dagger b_2^\dagger - \bar{z}b_2 b_1}$$

where $b_1^\dagger$, $b_2^\dagger$ are Bose creation operators and b_1, b_2 are Bose annihilation operators and $z \in \mathbb{C}$. Find the operators

$$U(z)b_1 U(z)^{-1}, \qquad U(z)b_2^\dagger U(z)^{-1}.$$

Solution 14.   We have

$$U^{-1}(z) = e^{-zb_1^\dagger b_2^\dagger + \bar{z}b_2 b_1}.$$

Thus

$$U(z)b_1 U(z)^{-1} = \cosh(|z|)b_1 - \frac{z\sinh(|z|)}{|z|}b_2^\dagger$$

$$U(z)b_2^\dagger U(z)^{-1} = \cosh(|z|)b_2^\dagger - \frac{\bar{z}\sinh(|z|)}{|z|}b_1 .$$

We can write in matrix form

$$(U(z)b_1 U(z)^{-1}, U(z)b_2^\dagger U(z)^{-1}) = (b_1, b_2^\dagger) \begin{pmatrix} \cosh(|z|) & -\frac{\bar{z}\sinh(|z|)}{|z|} \\ -\frac{z\sinh(|z|)}{|z|} & \cosh(|z|) \end{pmatrix}$$

where the matrix on the right-hand side is an element of the Lie group $SU(1,1)$.

Problem 15. Describe how two-mode squeezed state can be generated.

Solution 15. Two mode squeezed states can be generated either by entangling two independent single-mode squeezed states via a 50:50 beam splitter or by employing the non-degenerate operation of a nonlinear medium in the presence of two incoming modes. The unitary operator describing two-mode squeezing is

$$U_{12}(\zeta) = \exp(-i(\zeta b_1 b_2 + \zeta^* b_1^\dagger b_2^\dagger)/2)$$

where $\zeta \in \mathbb{C}$ is the squeezing parameter.

Problem 16. Consider the two-mode squeezing operator

$$S_2(\zeta) = \exp(\zeta b_1^\dagger b_2^\dagger - \zeta^* b_1 b_2) .$$

Let $|0\rangle \otimes |0\rangle$ be the two-mode vacuum state. Find the normalized state

$$S_2(\zeta)(|0\rangle \otimes |0\rangle)$$

expressed in number states.

Solution 16. We obtain

$$S_2(\zeta)(|0\rangle \otimes |0\rangle) = \frac{1}{\sqrt{\mu}} \sum_{k=0}^{\infty} \left(\frac{\nu}{\mu}\right)^k |k\rangle \otimes |k\rangle$$

where

$$\zeta = re^{i\phi}, \quad \mu = \cosh(r), \quad \nu = e^{i\phi}\sinh(r) .$$

This state is known as *two mode squeezed vacuum* or *twin beam state*.

19.3   Coherent Squeezed States

Problem 17.   Let $|0\rangle$ be the vacuum state. Then we define the coherent squeezed state as

$$|\beta, \zeta\rangle := D(\beta)S(\zeta)|0\rangle$$

where $D(\beta)$ is the displacement operator and $S(\zeta)$ is the squeezing operator with $\beta, \zeta \in \mathbb{C}$.
(i) Show that the state $|\beta, \zeta\rangle$ is normalized.
(ii) Show that

$$S(\zeta)D(\beta)|0\rangle \neq |\beta, \zeta\rangle.$$

Solution 17.   (i) We have

$$\langle \beta, \zeta | \beta, \zeta \rangle = \langle 0 | S(-\zeta)D(-\beta)D(\beta)S(\zeta)|0\rangle = 1.$$

(ii) We have

$$
\begin{aligned}
S(\zeta)D(\alpha)|0\rangle &= S(\zeta)D(\beta)S(-\zeta)S(\zeta)|0\rangle \\
&= \exp(\beta(b^\dagger \cosh(r) + be^{-i\phi}\sinh(r)) \\
&\quad - \beta^*(b\cosh(r) + b^\dagger e^{i\phi}\sinh(r)))S(\zeta)|0\rangle \\
&= D(\beta \cosh(r) - \beta^* e^{i\phi}\sinh(r))S(\zeta)|0\rangle \\
&= |\beta \cosh(r) - \beta^* e^{i\phi}\sinh(r), \zeta\rangle
\end{aligned}
$$

where we used that

$$S(\zeta)D(\beta)S(-\zeta) =$$
$$\exp(\beta(b^\dagger \cosh(r) + be^{-i\phi}\sinh(r)) - \beta^*(b\cosh(r) + b^\dagger e^{i\phi}\sinh(r))).$$

Thus $|\beta, \zeta\rangle$ can be written as

$$|\beta, \zeta\rangle = S(\zeta)D(\beta \cosh(r) + \beta^* \sinh(r))|0\rangle.$$

Problem 18.   Let $\hat{n} = b^\dagger b$. Find

$$\langle \beta, \zeta | \hat{n} | \beta, \zeta \rangle.$$

Solution 18.   We have

$$
\begin{aligned}
\langle \beta, \zeta | \hat{n} | \beta, \zeta \rangle &= \langle 0 | (b^\dagger \cosh(r) - be^{-\phi}\sinh(r) + \beta^* I) \\
&\quad \times (b\cosh(r) - b^\dagger e^{i\phi}\sinh(r) + \beta I)|0\rangle \\
&= \sinh^2(r) + |\beta|^2.
\end{aligned}
$$

Problem 19.   Let $|\alpha\rangle$ be a coherent state. The *Husimi distribution* of a quantum wave function $|\psi\rangle$ is given by

$$\rho_{|\psi\rangle}^{H}(\alpha) = |\langle\psi|\alpha\rangle|^2 .$$

(i) A *coherent squeezed state* $|\gamma, \beta\rangle$ is defined as

$$|\zeta, \beta\rangle := D(\beta)S(\zeta)|0\rangle$$

where $D(\beta)$ is the displacement operator and $S(\zeta) := \exp((\zeta^* b^2 - \zeta(b^\dagger)^2)/2)$. The coherent squeezed states $|\zeta, \beta\rangle$ also minimize the uncertainty relation, however, the variance of both canonically coupled variables are not equal. The modulus g of the complex number $\zeta = g e^{2i\theta}$ determines the strength of squeezing, $s = e^g - 1$, while the angle θ orients the squeezing axis. Find the Husimi distribution

$$\rho_{|\zeta, \beta\rangle}^{H}(\alpha) = |\langle\alpha|\zeta, \beta\rangle|^2$$

of a coherent squeezed state $|\zeta, \beta\rangle$.

Solution 19.   Let $\alpha = \alpha_1 + i\alpha_2$, where $\alpha_1, \alpha_2 \in \mathbb{R}$. We obtain

$$\rho_{|\zeta, \beta\rangle}^{H}(\alpha) = |\langle\alpha|\zeta, \beta\rangle|^2 = \exp(-(\Re(\beta) - \alpha_1)^2/(s+1)^2 - (\Im(\beta) - \alpha_2)^2(s+1)^2) .$$

Problem 20.   Let $(r \in \mathbb{R})$

$$S(r) = \exp\left(\frac{1}{2}r(b^2 - b^{\dagger 2})\right), \qquad D(\beta) = \exp(\beta b^\dagger - \beta^* b) .$$

Find the operator $S^\dagger(r)D(\beta)S(r)$.

Solution 20.   We find

$$S^\dagger(r)D(\beta)S(r) = D(\Re(\beta)e^r + i\Im(\beta)e^{-r}) .$$

Problem 21.   Let $D(\beta)$ be the displacement operator

$$D(\beta) = \exp(\beta b^\dagger - \beta^* b)$$

and

$$S(r, \phi) = \exp(re^{i\phi}(b^\dagger)^2/2 - re^{-i\phi}b^2/2)$$

be the squeeze operator with the squeeze factor $r \geq 0$ and squeeze angle $\phi \in (-\pi, \pi]$. Let

$$\rho_T := \frac{1}{\bar{n}+1}\sum_{n=0}^{\infty}\left(\frac{\bar{n}}{\bar{n}+1}\right)^n |n\rangle\langle n|$$

be the *Bose-Einstein density operator* with the mean occupancy

$$\overline{n} = \left(\exp \left(\frac{\hbar \omega}{k_B T} \right) - 1 \right)^{-1}.$$

Consider the displaced squeezed thermal state

$$\rho = D(\beta) S(r, \phi) \rho_T S^{\dagger}(r, \phi) D^{\dagger}(\beta). \tag{1}$$

The *Weyl expansion* of the density operator

$$\rho = \frac{1}{\pi} \int_{\mathbb{C}} d^2 \lambda \, \chi(\lambda) D(-\lambda)$$

with $d^2 \lambda = d\Re\lambda \, d\Im\lambda$ provides the one-to-one correspondence between the density operator ρ and its characteristic function

$$\chi(\lambda) := \mathrm{tr}(\rho D(\lambda)) \equiv \sum_{n=0}^{\infty} \langle n | \rho D(\lambda) | n \rangle. \tag{2}$$

A Gaussian state has a characteristic function of the form

$$\chi(\lambda) = \exp \left(- \left(A + \frac{1}{2} \right) |\lambda|^2 - \frac{1}{2} B^* \lambda^2 - \frac{1}{2} B (\lambda^*)^2 + C^* \lambda - C \lambda^* \right)$$

with $A > 0$. Find the coefficients A, B, C for the given ρ.

Solution 21.   Calculating (2) where ρ is given by (1) and comparing coefficients yields

$$A = \left(\overline{n} + \frac{1}{2} \right) \cosh(2r) - \frac{1}{2}, \quad B = - \left(\overline{n} + \frac{1}{2} \right) e^{i\phi} \sinh(2r), \quad C = \beta.$$

Problem 22.   Let $S(\zeta)$ be the one-mode squeeze operator with $\zeta = |\zeta| e^{i\phi}$. Let $D(\beta)$ be the displacement operator. Find

$$S(\zeta) D(\beta) S^{-1}(\zeta).$$

Solution 22.   We obtain

$$S(\zeta) D(\beta) S^{-1}(\zeta) = D(\tilde{\beta}), \qquad \tilde{\beta} = \cosh(|\zeta|)\beta + e^{i\phi} \sinh(|\zeta|)\beta^*.$$

Chapter 20

Trace and Partial Trace

Let $|n\rangle$ be the number states $(n = 0, 1, \ldots)$ and $|\beta\rangle$ be a coherent state $(\beta \in \mathbb{C})$. The trace of a bounded linear operator $\hat{A}$ could be calculated using the number states (Fock states) $|n\rangle$

$$\text{tr}(\hat{A}) = \sum_{n=0}^{\infty} \langle n|\hat{A}|n\rangle$$

or using coherent states $|\beta\rangle$

$$\text{tr}(\hat{A}) = \frac{1}{\pi} \int_{\mathbb{C}} d^2\beta \langle \beta|\hat{A}|\beta\rangle.$$

The integration is over the complex plane $\mathbb{C}$. One usually introduces polar coordinates $\beta = re^{i\phi}$ with $r \geq 0$ and $\phi \in [0, 2\pi)$.

Can the squeezed state $|\zeta\rangle$ be used to calculate the trace?

Let $\mathcal{H}_1$, $\mathcal{H}_2$ be complex Hilbert spaces. Let $\mathcal{H}$ be their Hilbert tensor product. Each Hilbert space admits at least one orthonormal basis. Then tr_2 would be the the operator of taking the partial trace with respect to the Hilbert space $\mathcal{H}_2$. The linear operation tr_2 maps states in the Hilbert space $\mathcal{H}$ into states in the Hilbert space $\mathcal{H}_1$. Analogously we consider tr_1.

345

Problem 1.   Show that

$$\sum_{n=0}^{\infty} \langle n|\hat{A}|n\rangle = \frac{1}{\pi} \int_{\mathbb{C}} \langle \beta|\hat{A}|\beta\rangle\,.$$

Solution 1.   We have

$$|n\rangle = \frac{1}{\pi} \int_{\mathbb{C}} \langle \beta|n\rangle |\beta\rangle d^2\beta$$

and therefore

$$\langle n| = \frac{1}{\pi} \int_{\mathbb{C}} \overline{\langle \beta|n\rangle} \langle \beta| d^2\beta = \frac{1}{\pi} \int_{\mathbb{C}} \langle n|\beta\rangle \langle \beta| d^2\beta\,.$$

We also apply the completeness relations

$$I = \frac{1}{\pi} \int_{\mathbb{C}} |\beta\rangle \langle \beta| d^2\beta, \qquad I = \sum_{n=0}^{\infty} |n\rangle \langle n|\,.$$

Thus we have

$$\begin{aligned}
\text{tr}\hat{A} &= \sum_{n=0}^{\infty} \langle n|\hat{A}|n\rangle \\
&= \sum_{n=0}^{\infty} \langle n| \left(\frac{1}{\pi} \int_{\mathbb{C}} |\beta\rangle \langle \beta| d^2\beta \right) \hat{A}|n\rangle \\
&= \frac{1}{\pi} \int_{\mathbb{C}} \sum_{n=0}^{\infty} \langle n|\beta\rangle \langle \beta|\hat{A}|n\rangle d^2\beta \\
&= \frac{1}{\pi^3} \int_{\mathbb{C}} \sum_{n=0}^{\infty} \left(\langle n|\gamma\rangle \langle \gamma|\beta\rangle d^2\gamma \right) \langle \beta|\hat{A}| \left(\int_{\mathbb{C}} |\delta\rangle \langle \delta|n\rangle \right) d^2\delta d^2\beta \\
&= \frac{1}{\pi^3} \int_{\mathbb{C}} \int_{\mathbb{C}} \int_{\mathbb{C}} d^2\beta d^2\gamma d^2\delta \sum_{n=0}^{\infty} \langle n|\gamma\rangle \langle \gamma|\beta\rangle \langle \delta|n\rangle \langle \beta|\hat{A}|\delta\rangle \\
&= \frac{1}{\pi^3} \int_{\mathbb{C}} \int_{\mathbb{C}} \int_{\mathbb{C}} d^2\beta d^2\gamma d^2\delta \left(\langle \delta| \left(\sum_{n=0}^{\infty} |n\rangle \langle n| \right) |\gamma\rangle \right) \langle \gamma|\beta\rangle \\
&= \frac{1}{\pi^3} \int_{\mathbb{C}} \int_{\mathbb{C}} \int_{\mathbb{C}} d^2\beta d^2\gamma d^2\delta \langle \delta|\gamma\rangle \langle \delta|\gamma\rangle \langle \gamma|\beta\rangle \langle \beta|\hat{A}|\delta\rangle \\
&= \frac{1}{\pi^2} \int_{\mathbb{C}} \int_{\mathbb{C}} d^2\beta d^2\gamma \langle \delta|\beta\rangle \langle \beta|\hat{A}|\delta\rangle \\
&= \frac{1}{\pi} \int_{\mathbb{C}} d^2\delta \langle \delta|\hat{A}|\delta\rangle\,.
\end{aligned}$$

Problem 2. The *trace* of an analytic function $f(b, b^\dagger)$ can be calculated as

$$\text{tr}(f(b, b^\dagger)) = \sum_{n=0}^{\infty} \langle n| f(b, b^\dagger) |n \rangle$$

where $\{ |n\rangle \; : \; n = 0, 1, 2, \ldots \}$ are the number states. A second method consists of obtaining the normal order function of f and integrating over the complex plane

$$\text{tr}(f(b, b^\dagger)) = \frac{1}{\pi} \int_{\mathbb{C}} \overline{f}^{(n)}(\beta, \beta^*) d^2\beta \,.$$

(i) Find the trace of $e^{-\epsilon b^\dagger b}$ using this second method, where $\epsilon > 0$.
(ii) Compare with the first method.

Solution 2. (i) The normal order form of $e^{-\epsilon b^\dagger b}$ is given by

$$e^{-\epsilon b^\dagger b} = \sum_{k=0}^{\infty} \frac{1}{k!} (e^{-\epsilon} - 1)^k (b^\dagger)^k b^k \,.$$

Thus we have to calculate the integral

$$\frac{1}{\pi} \int_{\mathbb{C}} \sum_{k=0}^{\infty} \frac{1}{k!} (e^{-\epsilon} - 1)^k (\beta^*)^k \beta^k d^2\beta \,.$$

We set $\beta = re^{i\phi}$. Thus $\beta\beta^* = r^2$. Since $d^2\beta \to d\phi r dr$ with $\phi \in [0, 2\pi)$, $r \in [0, \infty)$ and

$$\int_0^{2\pi} d\phi = 2\pi, \qquad \int_0^{\infty} re^{-ar^2} dr = \frac{1}{2a}$$

we obtain

$$\text{tr}(e^{-\epsilon b^\dagger b}) = \frac{1}{1 - e^{-\epsilon}} \,.$$

(ii) Using the first method we find

$$\begin{aligned}
\text{tr}(e^{-\epsilon b^\dagger b}) &= \sum_{n=0}^{\infty} \langle n| e^{-\epsilon b^\dagger b} |n\rangle = \sum_{n=0}^{\infty} \langle n| e^{-\epsilon n} |n\rangle \\
&= \sum_{n=0}^{\infty} e^{-\epsilon n} \langle n|n\rangle = \sum_{n=0}^{\infty} e^{-\epsilon n} \\
&= \frac{1}{1 - e^{-\epsilon}} \,.
\end{aligned}$$

Thus the first method is simpler to apply for this case.

Problem 3.   The single-mode squeezed state $|\zeta\rangle$ is given by $|\zeta\rangle = S(\zeta)|0\rangle$, where

$$S(\zeta) = \exp\left(-\frac{\zeta}{2}(b^\dagger)^2 + \frac{\zeta^*}{2}b^2\right)$$

with $\zeta \in \mathbb{C}$. One sets $\zeta = re^{i\phi}$. Why does the squeezed state $|\zeta\rangle$ cannot be used to calculate the trace.

Solution 3.   The single-mode squeezed operator $S(\zeta)$ can be written as

$$S(\zeta) = \exp\left(-\frac{1}{2}(b^\dagger)^2 e^{i\phi}\tanh(r)\right)\exp\left(-\frac{1}{2}(b^\dagger b + bb^\dagger)\ln(\cosh(r))\right)$$
$$\times \exp\left(\frac{1}{2}b^2 e^{-i\phi}\tanh(r)\right).$$

From $b|0\rangle = 0$ and $|\zeta\rangle = S(\zeta)|0\rangle$ we find

$$|\zeta\rangle = \sqrt{(\operatorname{sech}(r))}\sum_{n=0}^{\infty}\frac{\sqrt{((2n)!)}}{n!}\left(-\frac{1}{2}e^{i\phi}\tanh(r)\right)^n|2n\rangle.$$

Consequently the single-mode squeezed state is a superposition only of even number states. Thus the single-mode squeezed states are not complete. Note that the coherent states are overcomplete. The coherent squeezed states $D(\beta)S(\zeta)|0\rangle$ also form an overcomplete set. One has

$$\frac{1}{\pi}\int_{\mathbb{C}}d^2\beta|\beta,\zeta\rangle\langle\beta,\zeta| = I$$

where I is the identity operator.

Problem 4.   Consider the two-mode squeezed state

$$|\psi\rangle = e^{r(b_1^\dagger b_2^\dagger - b_1 b_2)}|00\rangle$$

where $|00\rangle \equiv |0\rangle \otimes |0\rangle$ and r is the squeezing parameter. This state can also be written as

$$|\psi\rangle = \frac{1}{\cosh(r)}\sum_{n=0}^{\infty}(\tanh(r))^n|n\rangle \otimes |n\rangle.$$

This is the *Schmidt basis* for this state. The density operator ρ is given by $\rho = |\psi\rangle\langle\psi|$. Calculate the partial traces using the number states.

Solution 4.   We have

$$\rho = \frac{1}{(\cosh(r))^2}\sum_{n=0}^{\infty}(\tanh(r))^n|n\rangle \otimes |n\rangle\sum_{m=0}^{\infty}(\tanh(r))^m\langle m| \otimes \langle m|$$
$$= \frac{1}{(\cosh(r))^2}\sum_{m=0}^{\infty}\sum_{n=0}^{\infty}(\tanh(r))^n(\tanh(r))^m|n\rangle\langle m| \otimes |n\rangle\langle m|.$$

Let I be the identity operator. Using that $\langle k|n\rangle = \delta_{kn}$ and $\langle m|k\rangle = \delta_{mk}$ we have

$$\rho_1 = \sum_{k=0}^{\infty}(I \otimes \langle k|)\rho(I \otimes |k\rangle)$$

$$= \frac{1}{(\cosh(r))^2}\sum_{k=0}^{\infty}(I \otimes \langle k|)\sum_{m,n=0}^{\infty}(\tanh(r))^n(\tanh(r))^m|n\rangle\langle m| \otimes |n\rangle\langle m|(I \otimes |k\rangle)$$

$$= \frac{1}{(\cosh(r))^2}\sum_{k=0}^{\infty}\sum_{m,n=0}^{\infty}(\tanh(r))^n(\tanh(r))^m|n\rangle\langle m|\delta_{kn}\delta_{mk}$$

$$= \frac{1}{(\cosh(r))^2}\sum_{k=0}^{\infty}(\tanh(r))^{2k}|k\rangle\langle k|.$$

We obtain the same result for ρ_2.

Problem 5. Use the reduced density operators ρ_1 and ρ_2 from the previous problem and calculate the entanglement

$$E(r) := -\mathrm{tr}(\rho_1 \log_2 \rho_1) = -\mathrm{tr}(\rho_2 \log_2 \rho_2).$$

Discuss E as a function of the squeezing parameter r.

Solution 5. We have

$$E(r) = -\mathrm{tr}\left(\left(\frac{1}{\cosh^2(r)}\sum_{k=0}^{\infty}(\tanh^{2k}(r))|k\rangle\langle k|\right)\right.$$

$$\left. \times \log_2\left(\frac{1}{\cosh^2(r)}\sum_{\ell=0}^{\infty}(\tanh^{2\ell}(r))|\ell\rangle\langle\ell|\right)\right).$$

The two matrices inside the trace are diagonal matrices and thus the product is again a diagonal matrix. Thus

$$E(r) = -\frac{1}{\cosh^2(r)}\sum_{k=0}^{\infty}\tanh^{2k}(r)\log_2\left(\frac{\tanh^{2k}(r)}{\cosh^2(r)}\right).$$

Thus

$$E(r) = -\frac{1}{\cosh^2(r)}\sum_{k=0}^{\infty}\tanh^{2k}(r)\log_2\tanh^{2k}(r) + \frac{\log_2(\cosh^2(r))}{\cosh^2(r)}\sum_{k=0}^{\infty}\tanh^{2k}(r)$$

or using the property of log

$$E(r) = -\frac{\log_2(\tanh^2(r))}{\cosh^2(r)}\sum_{k=0}^{\infty}k\tanh^{2k}(r) + \frac{\log_2(\cosh^2(r))}{\cosh^2(r)}\sum_{k=0}^{\infty}\tanh^{2k}(r)$$

The identity

$$\sum_{k=0}^{\infty} \tanh^{2k}(r) \equiv \cosh^2(r)$$

follows from a geometric series. The identity

$$\sum_{k=0}^{\infty} k \tanh^{2k}(r) \equiv \sinh^2(r) \cosh^2(r)$$

can be obtained from the first identity by parameter differentiation with respect to r. Using these results we obtain

$$
\begin{aligned}
E(r) &= -\sinh^2(r)(\log_2(\sinh^2(r)) - \log_2(\cosh^2(r)) + \log_2(\cosh^2(r)) \\
&= -\sinh^2(r)\log_2(\sinh^2(r)) + (\sinh^2(r) + 1)\log_2(\cosh^2(r)) \\
&= -\sinh^2(r)\log_2(\sinh^2(r)) + \cosh^2(r)\log_2(\cosh^2(r)) \,.
\end{aligned}
$$

For $r = 0$ we have $E(r = 0) = 0$.

Chapter 21

Entanglement

In the original paper of Einstein, Podolsky and Rosen the spin version of entanglement was not used. They considered measurement of position and momentum observables for two particles in one-dimensional motion. The entangled state

$$|\psi\rangle = \int_{-\infty}^{\infty} |p\rangle \otimes |-p\rangle e^{-i\ell p} dp$$

was studied, where the first component in the tensor product refers to particle 1 and the second to particle 2. The state $|\psi\rangle$ is thus a superposition of simultaneous eigenkets of the momenta $\hat{P}_1$ and $\hat{P}_2$ of the two particles with associated eigenvalues p and $-p$, respectively. Thus $|\psi\rangle$ is itself an eigenket of

$$\hat{P}_1 \otimes I + I \otimes \hat{P}_2$$

with the eigenvalue 0. The entangled state $|\psi\rangle$ is also an eigenket of the operator $\hat{Q}_1 \otimes I + I \otimes \hat{Q}_2$, where $\hat{Q}_1$ and $\hat{Q}_2$ are the positions of the two particles. The maximally entangled state of the original EPR pair can also be written as

$$|\psi\rangle = \frac{1}{(2\pi)^{3/2}} \int d\mathbf{k} \exp(i\mathbf{k} \cdot \mathbf{r}_1) \exp(i\mathbf{k} \cdot \mathbf{r}_2) = \delta(\mathbf{r}_1 + \mathbf{r}_2)$$

where δ denotes the Dirac delta function and $\cdot$ is the scalar product. If the potential energy between the two particles is assumed to be spherically symmetric, we can assume that the vectors $\mathbf{r}_1$ and $\mathbf{r}_2$ lie in the xy-plane in the laboratory frame without loss of generality.

Problem 1.   Let

$$f(\mathbf{p}, \mathbf{q}) := \frac{2}{N} \left(\sum_{j=1}^{N} q_j \right)^2 + \frac{1}{N} \sum_{j,k=1}^{N} (p_j - p_k)^2$$

and

$$g(\mathbf{p}, \mathbf{q}) := \frac{2}{N} \left(\sum_{j=1}^{N} p_j \right)^2 + \frac{1}{N} \sum_{j,k=1}^{N} (q_j - q_k)^2.$$

The *Wigner function* of the pure entangled N-mode state is given by

$$W(\mathbf{q}, \mathbf{p}) = \left(\frac{2}{\pi} \right)^N \exp\left(-e^{-2r} f(\mathbf{p}, \mathbf{q}) - e^{2r} g(\mathbf{p}, \mathbf{q}) \right) \tag{1}$$

where $\mathbf{q} = (q_1, q_2, \ldots, q_N)$ and $\mathbf{p} = (p_1, p_2, \ldots, p_N)$ are the positions and momenta of the N modes and r is the squeezing parameter with equal squeezing in all initial modes. Consider the case $N = 2$. What happens if $r \to \infty$?

Solution 1.   The state $W(\mathbf{q}, \mathbf{p})$ is always positive, symmetric among the N modes and becomes peaked at $q_i - q_j = 0$ $(i, j = 1, 2, \ldots, N)$ and $p_1 + p_2 + \cdots + p_n = 0$ for large squeezing parameter r. From (1) we have

$$W(q_1, q_2, p_1, p_2) =$$

$$\frac{4}{\pi^2} \exp\left(-e^{-2r} \left((q_1 + q_2)^2 + (p_1 - p_2)^2 \right) - e^{2r} \left((p_1 + p_2)^2 + (q_1 - q_2)^2 \right) \right).$$

For $r \to \infty$ we find in the sense of generalized functions

$$C\delta(q_1 - q_2)\delta(p_1 + p_2)$$

where δ denotes the *Dirac delta function*. This makes a connection to the original *EPR state* of Einstein, Podolsky and Rosen. Thus for large r the function W peaks at $q_1 - q_2 = 0$ and $p_1 + p_2 = 0$.

Problem 2.   Consider the operator

$$U(r) := e^{-r(b_1^\dagger b_2^\dagger - b_1 b_2)}$$

where $b_1^\dagger$, $b_2^\dagger$ are Bose creation operators and b_1, b_2 are Bose annihilation operators and $r \in \mathbb{R}$. Thus $b_1^\dagger = b^\dagger \otimes I$, $b_2^\dagger = I \otimes b^\dagger$. Let $|0\rangle \otimes |0\rangle$ be the vacuum state, i.e.,

$$(b \otimes I)(|0\rangle \otimes |0\rangle) = 0, \qquad (I \otimes b)(|0\rangle \otimes |0\rangle) = 0.$$

(i) Calculate
$$|\psi(r)\rangle = U(r)(|0\rangle \otimes |0\rangle).$$

(ii) Let

$$\hat{X}_1 := b_1 + b_1^\dagger = b \otimes I + b^\dagger \otimes I, \qquad \hat{Y}_1 := -i(b_1 - b_1^\dagger) = -i(b \otimes I - b^\dagger \otimes I),$$

$$\hat{X}_2 := b_2 + b_2^\dagger = I \otimes b + I \otimes b^\dagger, \qquad \hat{Y}_2 := -i(b_2 - b_2^\dagger) = -i(I \otimes b - I \otimes b^\dagger).$$

Find
$$\mathrm{var}(\hat{X}_1 + \hat{X}_2), \qquad \mathrm{var}(\hat{Y}_1 - \hat{Y}_2)$$

where
$$\mathrm{var}(A) := \langle A^2 \rangle - \langle A \rangle^2$$

is the *variance*.

(iii) What happens in the limit $r \to \infty$ to the state $|\psi(r)\rangle$?

Solution 2. (i) We find

$$|\psi(r)\rangle = U(r)(|0\rangle \otimes |0\rangle) = \sqrt{(1-\lambda^2)} \sum_{n=0}^{\infty} \lambda^n |n\rangle \otimes |n\rangle$$

where $\lambda = \tanh(r)$ and therefore $\sqrt{1-\lambda^2} = 1/\cosh(r)$. The entanglement of this state can be viewed as an entanglement between quadrature phases in the two modes (EPR entanglement) or as an entanglement between number and phase in the two modes.

(ii) We find

$$\mathrm{var}(\hat{X}_1 + \hat{X}_2) = 2e^{-2r}, \qquad \mathrm{var}(\hat{Y}_1 - \hat{Y}_2) = 2e^{-2r}.$$

(iii) The state $|\psi(r)\rangle$ approaches a simultaneous eigenstate of $\hat{X}_1 + \hat{X}_2$ and $\hat{Y}_1 - \hat{Y}_2$.

Problem 3. Consider a quantum-mechanical system governed by the Hamilton operator

$$\hat{H} = \hbar\omega_1 b_1^\dagger b_1 + \hbar\omega_2 b_2^\dagger b_2 + \hbar\chi b_1^\dagger b_1 b_2^\dagger b_2$$

where b_1 and b_2 are Bose annihilation operators for two distinct harmonic oscillator modes, respectively and χ is a coupling constant. Such a Hamilton operator for optical systems describes a four-wave mixing process, when the constant χ is then proportional to the third order susceptibility. It can also be used to describe two distinct modes interaction in Bose condensate. Furthermore, it describes the effective interaction of output pump and probe fields of an optical-cavity mediated by a two-level atom, in the dispersive limit. Let

$$|\psi(t = 0)\rangle := |\beta_1\rangle \otimes |\beta_2\rangle$$

where $|\beta_1\rangle$ and $|\beta_2\rangle$ are coherent states.
(i) Find

$$U(t)|\psi(t = 0)\rangle$$

where

$$U(t) = \exp(-i\hat{H}t/\hbar).$$

(ii) Consider the special case $t = \pi/\chi$. Discuss.
(iii) Consider the four cases
  a) $\omega_1 = 2\chi$, $\quad \omega_2 = 2\chi$,
  b) $\omega_1 = 2\chi$, $\quad \omega_2 = \chi$,
  c) $\omega_1 = \chi$, $\quad \omega_2 = 2\chi$,
  d) $\omega_1 = \chi$, $\quad \omega_2 = \chi$

for $|\psi(\pi/\chi)\rangle$.

Solution 3. (i) We find

$$|\psi(t)\rangle = U(t)|\beta_1\rangle \otimes |\beta_2\rangle = e^{-|\beta_1|^2/2} \sum_{m=0}^{\infty} \frac{(\beta_1 e^{-i\omega_1 t})^m}{\sqrt{m}} |m\rangle \otimes |\beta_2 e^{-i\omega_2 t} e^{-i\chi m t}\rangle.$$

(ii) For $t = \pi/\chi$ we have

$$\exp(-i\chi m t) = \exp(-im\pi) = \begin{cases} 1 & m = \text{even} \\ -1 & m = \text{odd} \end{cases}$$

Thus

$$|\psi(\pi/\chi)\rangle = |\beta_{1+}e^{-i\pi\omega_1/\chi}\rangle \otimes |\beta_2 e^{-i\pi\omega_2/\chi}\rangle + |\beta_{1-}e^{-i\pi\omega_1/\chi}\rangle \otimes |-\beta_2 e^{-i\pi\omega_2/\chi}\rangle$$

or

$$|\psi(\pi/\chi)\rangle = |\beta_1 e^{-i\pi\omega_1/\chi}\rangle \otimes |\beta_{2+}e^{-i\pi\omega_2/\chi}\rangle + |-\beta_1 e^{-i\pi\omega_1/\chi}\rangle \otimes |\beta_{2-}e^{-i\pi\omega_2/\chi}\rangle$$

where

$$|\epsilon_{\pm}e^{-i\pi\omega_k\chi}\rangle := \frac{1}{2}(|\epsilon e^{-i\pi\omega_k/\chi}\rangle \pm |-\epsilon e^{-i\pi\omega_k/\chi}\rangle)$$

with $k = 1, 2$ and $\epsilon = \beta_1, \beta_2$. Hence the state is entangled.
(iii) For case a) we find

$$|\Phi_+\rangle = |\beta_1\rangle \otimes |\beta_{2+}\rangle + |-\beta_1\rangle \otimes |\beta_{2-}\rangle.$$

For case b) we find

$$|\Phi_-\rangle = |\beta_1\rangle \otimes |\beta_{2+}\rangle - |-\beta_1\rangle \otimes |\beta_{2-}\rangle.$$

For case c) we find

$$|\Psi_+\rangle = |\beta_1\rangle \otimes |\beta_{2-}\rangle + |-\beta_1\rangle \otimes |\beta_{2+}\rangle \,.$$

For case d) we find

$$|\Psi_-\rangle = |\beta_1\rangle \otimes |\beta_{2-}\rangle - |-\beta_1\rangle \otimes |\beta_{2+}\rangle \,.$$

These states may be considered as *Bell states*. However these states are not perfectly orthogonal, but for large-amplitude fields $|\beta_1|, |\beta_2| \ll 1$ this can be achieved approximately. Furthermore there is an asymmetry in these states.

Problem 4. Discuss the *entanglement* of the state

$$|\Psi\rangle = \sum_s \sum_i \delta(\omega_s + \omega_i - \omega_p)\delta(\mathbf{k}_s + \mathbf{k}_i - \mathbf{k}_p)b_s^\dagger(\omega(\mathbf{k}_i))b_i^\dagger(\omega(\mathbf{k}_i))|0\rangle$$

which appears with *spontaneous parametric down conversion*. Here ω_j, $\mathbf{k}_j$ $(j = s, i, p)$ are the frequencies and wave vectors of the signal (s), idler (i), and pump (p) respectively, ω_p and $\mathbf{k}_p$ can be considered as constants while $b_s^\dagger$ and $b_i^\dagger$ are the respective Bose creation operators for the signal and idler.

Solution 4. The entanglement of this state can be thought of as the superposition of an infinite number of two-photon states, corresponding to the infinite number of ways the spontaneous parametric down conversion signal-idler can satisfy the expression for energy and momentum conservation (owing to the delta functions)

$$\hbar\omega_s + \hbar\omega_i = \hbar\omega_p, \qquad \hbar\mathbf{k}_s + \hbar\mathbf{k}_i = \hbar\mathbf{k}_p \,.$$

Even if there is no precise knowledge of the momentum for either the signal or the idler, the state does give precise knowledge of the momentum correlation of the pair. In EPR's language, the momentum for neither the signal photon nor the idler photon is determined. However, if measurement on one of the photons yields a certain value, then the momentum of the other photon is determined.

Problem 5. Consider the function

$$G(x_1, x_2; r) = \frac{1}{\sqrt{2\pi}} \exp\left(-\frac{1}{4}(x_1 + x_2)^2 e^{2r} - \frac{1}{4}(x_1 - x_2)^2 e^{-2r}\right)$$

where $r > 0$ is the squeezing parameter. Find

$$\lim_{r \to \infty} (G(x_1, x_2; r), \phi(x_1, x_2))$$

in the sense of generalized functions, where $\phi \in S(\mathbb{R}^2)$. Here $S(\mathbb{R}^2)$ is the set of all infinitely-differentiable functions which decrease as $|\mathbf{x}| \to \infty$, together with all their derivatives, faster than any power of $|\mathbf{x}|^{-1}$.

Solution 5. We find

$$\lim_{r \to \infty} (G(x_1, x_2; r), \phi(x_1, x_2)) \to \phi(x_1, x_1).$$

Thus

$$\lim_{r \to \infty} G(x_1, x_2; r) \to \delta(x_1 - x_2)$$

in the sense of generalized functions, where δ is the Dirac delta function.

Problem 6. Consider the operator

$$U_\epsilon = \exp(\epsilon(b_1^\dagger b_2 - b_2^\dagger b_1)), \qquad \epsilon \in \mathbb{R}.$$

(i) Find

$$U_\epsilon^\dagger b_1 U_\epsilon, \qquad U_\epsilon^\dagger b_2 U_\epsilon.$$

(ii) Consider the special case $\epsilon = \pi/4$.
(iii) Find

$$D = U_{\pi/4}^\dagger (b_1^\dagger b_1 - b_2^\dagger b_2) U_{\pi/4}$$

(iv) Solve the eigenvalue problem $D|\delta\rangle = d|\delta\rangle$.

Solution 6. (i) Using the expansion

$$e^{\hat{A}} \hat{B} e^{-\hat{A}} = \hat{B} + [\hat{A}, \hat{B}] + \frac{1}{2}[\hat{A}, [\hat{A}, \hat{B}]] + \cdots + \frac{1}{n!}[\hat{A}, [\hat{A}, \cdots, [\hat{A}, \hat{B}] \ldots]] + \cdots$$

we find

$$U_\epsilon^\dagger b_1 U_\epsilon = b_1 \cos(\epsilon) + b_2 \sin(\epsilon), \qquad U_\epsilon^\dagger b_2 U_\epsilon = -b_1 \sin(\epsilon) + b_2 \cos(\epsilon).$$

(ii) For the special case $\epsilon = \pi/4$ we obtain

$$U_{\pi/4}^\dagger b_1 U_{\pi/4} = \frac{1}{\sqrt{2}}(b_2 + b_1), \qquad U_{\pi/4}^\dagger b_2 U_{\pi/4} = \frac{1}{\sqrt{2}}(b_2 - b_1)$$

since $\sin(\pi/4) = \cos(\pi/4) = 1/\sqrt{2}$.
(iii) From (ii) we find

$$U_\epsilon^\dagger b_1^\dagger U_\epsilon = b_1^\dagger \cos(\epsilon) + b_2^\dagger \sin(\epsilon), \qquad U_\epsilon^\dagger b_2^\dagger U_\epsilon = -b_1^\dagger \sin(\epsilon) + b_1^\dagger \cos(\epsilon).$$

Thus

$$D = b_1^\dagger b_2 + b_1 b_2^\dagger.$$

(iv) The eigenvalue problem $D|\delta\rangle = d|\delta\rangle$ can be rewritten as

$$(b_1^\dagger b_1 - b_2^\dagger b_2)|\nu\rangle = d|\nu\rangle$$

where

$$|\nu\rangle = U_{\pi/4}|\delta\rangle .$$

The eigenvalue problem can easily be solved since $b^\dagger b|n\rangle = n|n\rangle$. We find

$$|\nu^{(n)}\rangle = \begin{cases} |n+d\rangle \otimes |n\rangle & d \in \mathbb{Z}^+ \\ |n\rangle \otimes |n\rangle & d = 0 \\ |n\rangle \otimes |n-d\rangle & d \in \mathbb{Z}^- \end{cases}$$

where $\mathbb{Z}^+$ denotes the positive set of integers and $\mathbb{Z}^-$ denotes the negative set of integers. The eigenvalue d has countable degeneracy corresponding to the one-integer parameter set $|\nu^{(n)}\rangle$ of eigenstates. In order to solve for the original eigenvalues we have to compute their transformation under the action of the operator $U_{\pi/4}$, i.e.,

$$|\delta^{(n)}\rangle = U_{\pi/4}^\dagger |\nu^{(n)}\rangle .$$

We consider the *Schwinger two-bosons realization* of the $su(2)$ Lie algebra

$$J_+ := b_1 b_2^\dagger, \qquad J_- := b_1^\dagger b_2, \qquad J_3 := \frac{1}{2}(b_2^\dagger b_2 - b_1^\dagger b_1)$$

with

$$[J_+, J_-] = 2J_3, \qquad [J_3, J_\pm] = \pm J_\pm .$$

Thus

$$U_{\pi/4} = \exp\left(\frac{\pi}{4}(J_+ - J_-)\right) .$$

Using the Baker-Campbell-Hausdorff formula we find

$$\exp(\xi J_+ - \bar\xi J_-) = \exp(\eta J_+)\exp(\beta J_3)\exp(-\bar\eta J_-)$$

where

$$\eta = \frac{\xi}{|\xi|}\tan(\xi), \qquad \beta = \ln(1 + |\eta|^2) .$$

Thus

$$U_{\pi/4} = \exp(b_1 b_2^\dagger)\exp(\ln 2(b_2^\dagger b_2 - b_1^\dagger b_1))\exp(-b_1^\dagger b_2)$$

where we used that $\tan(\pi/4) = 1$.

Problem 7. Consider the Hamilton operator

$$\hat{H} = \sum_{i,j=0}^{N-1} h_{ij} b_i^\dagger b_j + \sum_{i,j,l,m=0}^{N-1} V_{ijlm} b_i^\dagger b_j^\dagger b_m b_l .$$

The operators $b_j^\dagger$ are Bose creation operators and the operators b_j are Bose annihilation operators. Show that an eigenstate $|\psi\rangle$ of $\hat{H}$ is entangled or $[\hat{H}, \hat{n}_j]|\psi\rangle = 0$, where $\hat{n}_j := b_j^\dagger b_j$ is the particle number operator of mode j for an appropriate basis.

Solution 7.   Suppose $|\psi\rangle$ is not entangled. We write $|\psi\rangle$ as

$$|\psi\rangle = |n_0\rangle \otimes \cdots \otimes |n_{N-1}\rangle$$

where $\{ |0_j\rangle, |1_j\rangle, \dots \}$ is a basis for particles in mode j with $0 \leq j < N$. We define the creation, annihilation and number operators, for particles in mode j, by

$$b_j^\dagger |n_j\rangle := \sqrt{n_j + 1}|n + 1_j\rangle, \qquad b_j|n_j\rangle := \sqrt{n_j}|n - 1_j\rangle$$

and $\hat{n}_j := b_j^\dagger b_j$. We have the eigenvalue equation for $|\psi\rangle$, namely $\hat{H}|\psi\rangle = \lambda|\psi\rangle$. Thus

$$
\begin{aligned}
[\hat{H}, \hat{n}_i]|\psi\rangle &= \hat{H}\hat{n}_i|\psi\rangle - \hat{n}_i\hat{H}|\psi\rangle = \hat{H}n_i|\psi\rangle - \lambda\hat{n}_i|\psi\rangle \\
&= n_i\hat{H}|\psi\rangle - \lambda\hat{n}_i|\psi\rangle = \lambda n_i|\psi\rangle - \lambda n_i|\psi\rangle \\
&= 0.
\end{aligned}
$$

Problem 8.   Let $|\beta\rangle$ be a coherent state. Consider the entangled coherent state

$$|\psi\rangle = C(|\beta_1\rangle \otimes |\beta_2\rangle + e^{i\phi}|-\beta_1\rangle \otimes |-\beta_2\rangle)$$

where C is the normalization factor and $\phi \in \mathbb{R}$.
(i) Find the normalization factor C.
(ii) Calculate the *partial trace* using the basis $\{ |n\rangle \otimes I : n = 0, 1, 2, \dots \}$ where $\{ |n\rangle : n = 0, 1, 2, \dots \}$ are the number states and I is the identity operator.

Solution 8.   (i) Since

$$\langle \beta|\gamma\rangle = \exp\left(-\frac{1}{2}(|\beta|^2 + |\gamma|^2) + \beta\gamma^* \right)$$

for coherent states $|\beta\rangle$ and $|\gamma\rangle$, we have

$$\langle \beta|\beta\rangle = 1, \qquad \langle \beta|-\beta\rangle = \exp(-2|\beta|^2).$$

We find from the condition $\langle \psi|\psi\rangle = 1$ that

$$1 = |C|^2(2 + 2\cos(\phi)\exp(-2|\beta_1|^2 - 2|\beta_2|^2)).$$

Thus

$$C = \frac{1}{\sqrt{2 + 2\cos(\phi)\exp(-2|\beta_1|^2 - 2|\beta_2|^2)}}\,.$$

(ii) We have to calculate

$$\mathrm{tr}_1(|\psi\rangle\langle\psi|) = C^2 \sum_{n=0}^{\infty} (((\langle n| \otimes I)(|\beta_1\rangle \otimes |\beta_2\rangle) + e^{i\phi}|-\beta_1\rangle \otimes |-\beta_2\rangle))$$
$$\times ((\langle\beta_1| \otimes \langle\beta_2| + e^{-i\phi}\langle\beta_1-| \otimes \langle\beta_2-|)(|n\rangle \otimes I))\,.$$

Thus

$$\mathrm{tr}_1(|\psi\rangle\langle\psi|) = C^2 \sum_{n=0}^{\infty} (\langle n|\beta_1\rangle\langle\beta_1|n\rangle|\beta_2\rangle\langle\beta_2| + e^{-i\phi}\langle n|\beta_1\rangle\langle\beta_1-|n\rangle|\beta_2\rangle\langle\beta_2-|$$
$$+ e^{i\phi}\langle n|-\beta_1\rangle\langle\beta_1|n\rangle|-\beta_2\rangle\langle\beta_2| + \langle n|-\beta_1\rangle\langle\beta_1-|n\rangle|-\beta_2\rangle\langle\beta_2-|)\,.$$

Using

$$\langle n|\beta_1\rangle\langle\beta_1|n\rangle = \frac{e^{-|\beta_1|^2}(|\beta_1|^2)^n}{n!}$$

$$\langle n|\beta_1\rangle\langle\beta_1-|n\rangle = \frac{e^{-|\beta_1|^2}(-|\beta_1|^2)^n}{n!}$$

$$\langle n|-\beta_1\rangle\langle\beta_1|n\rangle = \frac{e^{-|\beta_1|^2}(-|\beta_1|^2)^n}{n!}$$

$$\langle n|-\beta_1\rangle\langle\beta_1-|n\rangle = \frac{e^{-|\beta_1|^2}(|\beta_1|^2)^n}{n!}$$

and

$$\sum_{n=0}^{\infty} \frac{(|\beta|^2)^n}{n!} = e^{|\beta|^2}, \qquad \sum_{n=0}^{\infty} \frac{(-|\beta|^2)^n}{n!} = e^{-|\beta|^2}$$

we arrive at

$$\mathrm{tr}_1(|\psi\rangle\langle\psi|) = C^2(|\beta_2\rangle\langle\beta_2| + e^{i\phi}e^{-2|\beta_1|^2}|\beta_2\rangle\langle\beta_2-|$$
$$+ e^{-i\phi}e^{-2|\beta_1|^2}|-\beta_2\rangle\langle\beta_2| + |-\beta_2\rangle\langle\beta_2-|)\,.$$

Problem 9. A *beam splitter* is a simple device which can act to entangle output optical fields. The input field described by the Bose annihilation operator b_1 is superposed on the other input field with Bose annihilation operator b_2 by a lossless symmetric beam splitter with amplitude reflection and transmission coefficients r and t. The output field annihilation operators are given by

$$\tilde{b}_1 = \hat{B}b_1\hat{B}^{\dagger}, \qquad \tilde{b}_2 = \hat{B}b_2\hat{B}^{\dagger}$$

where the *beam splitter operator* is

$$\hat{B} := \exp\left(\frac{\theta}{2}(b_1^\dagger b_2 e^{i\phi} - b_1 b_2^\dagger e^{-i\phi})\right)$$

with the amplitude reflection and transmission coefficients

$$t := \cos(\theta/2), \qquad r := \sin(\theta/2).$$

The beam splitter gives the phase difference ϕ between the reflected and transmitted fields.

(i) Assume that the input states are two independent number states $|n_1\rangle \otimes |n_2\rangle$, where $n_1, n_2 = 0, 1, 2, \ldots$. Calculate the state

$$\hat{B}(|n_1\rangle \otimes |n_2\rangle).$$

(ii) Consider the special case $n_1 = 0$ and $n_2 = N$.

Solution 9.   (i) We obtain the state

$$\hat{B}(|n_1\rangle \otimes |n_2\rangle) = \sum_{m_1=0}^{\infty} \sum_{m_2=0}^{\infty} ((\langle m_1| \otimes \langle m_2|)\hat{B}(|n_1\rangle \otimes |n_2\rangle))|m_1\rangle \otimes |m_2\rangle$$

$$= \sum_{m_1=0}^{\infty} \sum_{m_2=0}^{\infty} B_{n_1 n_2}^{m_1 m_2}|m_1\rangle \otimes |m_2\rangle$$

where

$$B_{n_1 n_2}^{m_1 m_2} = e^{-i\phi(n_1-m_1)} \sum_{k=0}^{n_1} \sum_{\ell=0}^{n_2} (-1)^{n_1-k} r^{n_1+n_2-k-\ell} t^{k+\ell}$$

$$\times \frac{\sqrt{n_1! n_2! m_1! m_2!}}{k!(n_1-k)!\ell!(n_2-\ell)!} \delta_{m_1, n_2+k-\ell} \delta_{m_2, n_1-k+\ell}$$

with $\delta_{m,n}$ is the Kronecker delta. When the total number of input photons is $N = n_1 + n_2$, the output state becomes an $(N+1)$-dimensional entangled state.

(ii) We obtain from the results of (i)

$$B(|0\rangle \otimes |N\rangle) = \sum_{k=0}^{N} c_k^N |k\rangle \otimes |N-k\rangle$$

where the expansion coefficients are given by

$$c_k^N = \binom{N}{k}^{1/2} r^k t^{N-k} e^{ik\phi}.$$

Problem 10. The *beam splitter operator* is given by

$$\hat{B} = \exp\left(\frac{\theta}{2}(b_1^\dagger b_2 e^{i\phi} - b_1 b_2^\dagger e^{-i\phi})\right).$$

The input field described by the Bose operator b_1 is superposed with another input field with Bose operator b_2 by a lossless symmetric beam splitter with amplitude reflection and transmission coefficients r and t, i.e.

$$t := \cos(\theta/2), \qquad r := \sin(\theta/2).$$

The output-field Bose annihilation operators are given by

$$\widetilde{b}_1 = \hat{B} b_1 \hat{B}^\dagger, \qquad \widetilde{b}_2 = \hat{B} b_2 \hat{B}^\dagger.$$

Consider the input state (product state of two independent Fock states)

$$|n_1, n_2\rangle \equiv |n_1\rangle |n_2\rangle \equiv |n_1\rangle \otimes |n_2\rangle$$

where $n_1, n_2 = 0, 1, \dots$. Calculate $\hat{B}|n_1, n_2\rangle$. Is this state entangled?

Solution 10. We obtain

$$\hat{B}|n_1, n_2\rangle = \sum_{N_1=0}^{\infty} \sum_{N_2=0}^{\infty} |N_1, N_2\rangle\langle N_1, N_2|\hat{B}|n_1, n_2\rangle = \sum_{N_1=0}^{\infty} \sum_{N_2=0}^{\infty} B_{n_1 n_2}^{N_1 N_2}|N_1, N_2\rangle$$

where

$$B_{n_1 n_2}^{N_1 N_2} = e^{-i\phi(n_1 - N_1)} \sum_{k=0}^{n_1} \sum_{l=0}^{n_2} (-1)^{n_1 - k} r^{n_1 + n_2 - k - l} t^{k+l}$$

$$\times \frac{\sqrt{n_1! n_2! N_1! N_2!}}{k!(n_1 - k)! l!(n_2 - l)!} \delta_{N_1, n_2 + k - l} \delta_{N_2, n_1 - k + l}$$

where δ is the Kronecker delta function. When the total number of input photons is $N = n_1 + n_2$ with $N \geq 1$, the output state is an $(N+1)$ entangled state.

Problem 11. Consider the product Hilbert space $\mathbb{C}^2 \otimes \mathcal{H}$ where $\mathcal{H}$ denotes an arbitrary Hilbert space. For the one-Bose system we would set $\mathcal{H} = l_2(\mathbb{N})$. An arbitrary pure state in this product Hilbert space can be written as

$$|\psi\rangle := |0\rangle \otimes |\phi_0\rangle + |1\rangle \otimes |\phi_1\rangle$$

where $|\phi_0\rangle, |\phi_1\rangle \in \mathcal{H}$ and $\{|0\rangle, |1\rangle\}$ forms an orthonormal basis in $\mathbb{C}^2$. The condition for the state $|\psi\rangle$ to be normalized, i.e. $\langle \psi|\psi\rangle = 1$, leads to the constraint

$$\langle \phi_0|\phi_0\rangle + \langle \phi_1|\phi_1\rangle = 1.$$

If we assume that $|\phi_0\rangle$ and $|\phi_1\rangle$ have identical norms, then $|\psi\rangle$ takes the form

$$|\psi\rangle = \frac{1}{\sqrt{2}}(|0\rangle \otimes |\varphi_0\rangle + |1\rangle \otimes |\varphi_1\rangle)$$

where $|\phi_0\rangle = \frac{1}{\sqrt{2}}|\varphi_0\rangle$, $|\phi_1\rangle = \frac{1}{\sqrt{2}}|\varphi_1\rangle$ and $|\varphi_0\rangle$, $|\varphi_1\rangle$ are normalized. Defining the reduced density matrices (using the partial trace)

$$\rho_1 := \mathrm{tr}_{\mathbb{C}^2}(|\psi\rangle\langle\psi|), \qquad \rho_2 := \mathrm{tr}_{\mathcal{H}}(|\psi\rangle\langle\psi|)$$

the entanglement of $|\psi\rangle$ is given by

$$E(|\psi\rangle) := -\mathrm{tr}(\rho_1 \log_2 \rho_1) = -\mathrm{tr}(\rho_2 \log_2 \rho_2).$$

Describe the entanglement.

Solution 11.   Straightforward calculation yields

$$\rho_1 = |\phi_0\rangle\langle\phi_0| + |\phi_1\rangle\langle\phi_1|$$

and

$$\rho_2 = \langle\phi_0|\phi_0\rangle|0\rangle\langle0| + \langle\phi_1|\phi_0\rangle|0\rangle\langle1| + \langle\phi_0|\phi_1\rangle|1\rangle\langle0| + \langle\phi_1|\phi_1\rangle|1\rangle\langle1|.$$

Applying the constraint we find that the non-zero eigenvalues of ρ_1 and ρ_2 are given by

$$\lambda(\langle\phi_0|\phi_0\rangle, |\langle\phi_0|\phi_1\rangle|^2) := \frac{1}{2}\left(1 + \sqrt{(1 - 2\langle\phi_0|\phi_0\rangle)^2 + 4|\langle\phi_0|\phi_1\rangle|^2}\right)$$

and $1 - \lambda$. Thus

$$E(|\psi\rangle) = -\lambda \log_2 \lambda - (1 - \lambda) \log_2(1 - \lambda).$$

The entanglement is described exclusively by $\langle\phi_0|\phi_0\rangle$ and $|\langle\phi_0|\phi_1\rangle|^2$. Furthermore we have the inequality

$$|\langle\phi_0|\phi_1\rangle|^2 \le \langle\phi_0|\phi_0\rangle - \langle\phi_0|\phi_0\rangle^2 \le \frac{1}{4}.$$

Problem 12.   Consider the superposition (*macroscopic quantum superposition states*) $|\phi_0\rangle = c_0(|\alpha\rangle + |-\alpha\rangle)$ and $|\phi_1\rangle = c_1(|\beta\rangle + |-\beta\rangle)$ where $c_0, c_1 \in \mathbb{C}$ and $|\alpha\rangle, |-\alpha\rangle, |\beta\rangle, |-\beta\rangle$, are coherent states, i.e.,

$$|\psi\rangle = c_0|0\rangle \otimes (|\alpha\rangle + |-\alpha\rangle) + c_1|1\rangle \otimes (|\beta\rangle + |-\beta\rangle).$$

Discuss the entanglement.

Solution 12. The conditions for entanglement from the previous problem can be applied. In this case the normalization condition yield

$$2|c_0|^2(1 + e^{-2|\alpha|^2}) + 2|c_1|^2(1 + e^{-2|\beta|^2}) = 1.$$

Consequently $\langle \phi_0|\phi_0 \rangle = 2|c_0|^2(1 + e^{-2|\alpha|^2})$ and

$$|\langle \phi_0|\phi_1 \rangle|^2 = \langle \phi_0|\phi_0 \rangle (1 - \langle \phi_0|\phi_0 \rangle) \frac{(e^{-\frac{1}{2}|\alpha-\beta|^2} + e^{-\frac{1}{2}|\alpha+\beta|^2})^2}{(1 + e^{-2|\alpha|^2})(1 + e^{-2|\beta|^2})}.$$

It is convenient to define the real valued quantities

$$p_{00} := \langle \phi_0|\phi_0 \rangle = 2|c_0|^2(1 + e^{-2|\alpha|^2}), \qquad p_{01} := \frac{(e^{-\frac{1}{2}|\alpha-\beta|^2} + e^{-\frac{1}{2}|\alpha+\beta|^2})^2}{(1 + e^{-2|\alpha|^2})(1 + e^{-2|\beta|^2})}.$$

Thus we obtain

$$|\langle \phi_0|\phi_1 \rangle|^2 = p_{00}(1 - p_{00})p_{01}.$$

The maximum entanglement occurs when $\langle \phi_0|\phi_1 \rangle = 0$ and $\langle \phi_0|\phi_0 \rangle = \frac{1}{2}$. Since the above equation implies $\langle \phi_0|\phi_1 \rangle \neq 0$ for $\langle \phi_0|\phi_0 \rangle = \frac{1}{2}$, the maximum entanglement is approached asymptotically for $\alpha = 0$, $|\beta| \to \infty$ or $\beta = 0$, $|\alpha| \to \infty$. This is due to

$$|\alpha - \beta|^2 = |\alpha|^2 + |\beta|^2 - 2\Re(\alpha\overline{\beta}), \qquad |\alpha + \beta|^2 = |\alpha|^2 + |\beta|^2 + 2\Re(\alpha\overline{\beta}).$$

In other words, for $\alpha, \beta \neq 0$ one term shrinking in the numerator of p_{01} implies the other is growing. To find the entanglement we first determine the eigenvalues of ρ_1 and ρ_2 which are now given by

$$\lambda = \frac{1}{2}\left(1 + \sqrt{1 - 4(1 - p_{01})p_{00}(1 - p_{00})}\right).$$

Problem 13. Consider the case when $|\phi_0\rangle$ is described by a number state and $|\phi_1\rangle$ is described by a coherent state, i.e.

$$|\psi\rangle = c_0|0\rangle \otimes |n\rangle + c_1|1\rangle \otimes |\alpha\rangle.$$

The scalar product between a number state $|n\rangle$ and a coherent state $|\alpha\rangle$ is given by

$$\langle n|\alpha \rangle = e^{-\frac{1}{2}|\alpha|^2} \frac{\alpha^n}{\sqrt{n!}}.$$

Discuss the entanglement.

Solution 13. The normalization condition for this case gives

$$|c_0|^2 + |c_1|^2 = 1.$$

Consequently $\langle\phi_0|\phi_0\rangle = |c_0|^2$, and

$$|\langle\phi_0|\phi_1\rangle|^2 = |c_0|^2(1 - |c_0|^2)e^{-|\alpha|^2}\frac{|\alpha|^{2n}}{n!}.$$

It is again convenient to define the quantities p_{00} and p_{01} as

$$p_{00} := \langle\phi_0|\phi_0\rangle = |c_0|^2, \qquad p_{01} := e^{-|\alpha|^2}\frac{|\alpha|^{2n}}{n!}.$$

Thus we obtain $|\langle\phi_0|\phi_1\rangle|^2 = p_{00}(1 - p_{00})p_{01}$. The entanglement can again be determined from p_{00} and p_{01} and proceeds as described in the previous problem.

Problem 14.   Let $|\beta\rangle$ be a coherent state. Consider the entangled state

$$|\psi\rangle = \frac{1}{\sqrt{N_\beta}}(|\beta\rangle \otimes |\beta\rangle - |-\beta\rangle \otimes |-\beta\rangle).$$

(i) Find the normalization N_β.
(ii) Express the entangled state $|\psi\rangle$ using Fock states $|n\rangle \otimes |m\rangle$ with $n, m \in \mathbb{N}_0$.

Solution 14.   Using $\langle\beta|-\beta\rangle = \exp(-2|\beta|^2)$ we obtain

$$N_\beta = 2 - 2\exp(-4|\beta|^2).$$

(ii) We obtain

$$|\psi\rangle = \frac{2\exp(-|\beta|^2)}{\sqrt{N_\beta}} \sum_{n,m\,|\,n+m \text{ odd}}^{\infty} \frac{\beta^{n+m}}{\sqrt{n!m!}}|n\rangle \otimes |m\rangle.$$

Thus the total number of photons is always odd.

Problem 15.   Let $|\beta\rangle$, $|\gamma\rangle$ be coherent states. Consider the balanced entangled coherent state

$$|\psi\rangle = \frac{1}{\sqrt{2}}(|\beta\rangle \otimes |\gamma\rangle + e^{i\phi}|-\beta\rangle \otimes |-\gamma\rangle)$$

and $\phi = \pi/2$. Consider the operator

$$\Pi = \exp(i\pi(b^\dagger b \otimes I + I \otimes b^\dagger b)).$$

Find $\Pi|\psi\rangle$ and $\Pi^2|\psi\rangle$. Discuss.

Solution 15. Since $e^{i\pi} = -1$ and $e^{i\pi/2} = i$ and

$$\exp(i\phi b^\dagger b)|\beta\rangle = |\beta e^{i\phi}\rangle$$

we find the entangled state

$$\Pi|\psi\rangle = \frac{i}{\sqrt{2}}(|\beta\rangle \otimes |\gamma\rangle - i| - \beta\rangle \otimes | - \gamma\rangle).$$

Applying the operator Π again yields

$$\Pi^2|\psi\rangle = |\psi\rangle.$$

Note that

$$e^{i\pi b^\dagger b}be^{-i\pi b^\dagger b} = -b.$$

Problem 16. Consider the unitary evolution operator for the *beam splitter*

$$U_{BS} = \exp(\theta(b_1^\dagger b_2 e^{i\phi} - b_1 b_2^\dagger e^{-i\phi}))$$

where the real angular parameter θ determines the transmission and reflection coefficients via $T = t^2 = \cos^2\theta$ and $R = r^2 = \sin^2\theta$. The internal phase shift ϕ between the reflected and transmitted modes is given by the beam splitter itself. To control ϕ we can place a phase shifter in one of the output channels.

(i) Let $|\psi_{in}\rangle = |0\rangle \otimes |1\rangle$, i.e. the one input is a one-photon state and the other the vacuum state. Calculate $|\psi_{out}\rangle = U_{BS}|\psi_{in}\rangle$.

(ii) To test quantum non locality of the state $|\psi_{out}\rangle$ we apply the *displaced parity operator* based on joint parity measurements

$$\hat{\Pi}_{12}(\beta_1, \beta_2) := D_1(\beta_1)D_2(\beta_2)\exp(i\pi(\hat{n}_1 + \hat{n}_2))D_1^\dagger(\beta_1)D_2^\dagger(\beta_2)$$

where $D_1(\beta_1)$ and $D_2(\beta_2)$ are the unitary displacement operators. Calculate

$$\Pi_{12}(\beta_1, \beta_2) := \langle\psi_{out}|\hat{\Pi}_{12}(\beta_1, \beta_2)|\psi_{out}\rangle.$$

(iii) The *two mode Bell function* $B(\beta_1, \beta_2)$ can be written as

$$B(\beta_1, \beta_2) = \Pi_{12}(0,0) + \Pi_{12}(\beta_1, 0) + \Pi_{12}(0, \beta_2) - \Pi_{12}(\beta_1, \beta_2).$$

For local realistic theory $B(\beta_1, \beta_2)$ should satisfy the *Bell-CHSH inequality*

$$|B(\beta_1, \beta_2)| \leq 2.$$

The violation of this inequality indicates quantum non locality of the single photon entangled state. Calculate $B(\beta_1, \beta_2)$ and discuss the case where $|\beta_1|^2 = |\beta_2|^2$.

Solution 16.   (i) We find

$$|\psi_{out}\rangle = t|0\rangle \otimes |1\rangle + re^{i\phi}|1\rangle \otimes |0\rangle \,.$$

(ii) We obtain

$$\Pi(\beta_1, \beta_2) = \langle \psi_{out}|\hat{\Pi}(\beta_1, \beta_2)|\psi_{out}\rangle$$
$$= (4|re^{-i\phi}\beta_1 + t\beta_2|^2 - 1)\exp(-2(|\beta_1|^2 + |\beta_2|^2)) \,.$$

(iii) Let $|\beta_1|^2 = |\beta_2|^2 = J$ and let γ_{12} be an arbitrary phase space difference between the two coherent displacements β_1 and β_2. Then we can write $\beta_2 = \beta_1 e^{i\gamma_{12}}$. Thus the two-mode Bell function is given by

$$B(\beta_1, \beta_2) = -1 + (4J - 2)e^{-2J} - (4J - 1)e^{-4J} - 8rtJe^{-4J}\cos\Delta$$

where $\Delta := \gamma_{12} + \phi$. When $\gamma_{12} = -\phi$ we obtain the maximal value $|B|_{max}$ of the two mode Bell function

$$|B|_{max} = 1 + (4J - 1)e^{-4J} + 8rtJe^{-4J} - (4J - 2)e^{-2J} \,.$$

Problem 17.   There are various ways in which photons can be entangled. The photon is a spin-1 particle. One has a) polarization entanglement, b) momentum (direction) entanglement c) time-energy entanglement d) orbital angular momentum states entanglement. Describe the different types of entanglement for photons. Parametric down conversion can produce photons that are entangled both in polarization and in space.

Solution 17.   Polarization entanglement. The highest contrast in experiments can be achieved for polarization-entangled states created by parametric down-conversion. Type-II sources can produce polarization entanglement directly. Parametric down-conversion or spontaneous parametric fluorescence is the spontaneous reverse process of second-harmonic generation, or more generally speaking three-wave mixing in nonlinear optical media (for example Beta-Barium Borate crystal). In nonlinear optics the polarization **P** depends nonlinear on the electric field **E**. The nonlinearity is given by a power series expansion of the polarization vector (summation convention is used)

$$P_i = \chi_{ij}^{(1)}E_j + \chi_{ijk}^{(2)}E_jE_k + \chi_{ijkl}^{(3)}E_jE_kE_l + \cdots$$

where $\chi_{ij}^{(1)}$ describes the normal refractive properties of a material including any kind of birefringence. $\chi_{ijk}^{(2)}$ is the coefficient tensor for three-wave, because two E terms can lead to another P term, mixing in strongly nonlinear material. The $\chi_{ijkl}^{(3)}$ term describes effects that occur at even higher intensities, e.g. Kerr-lensing or phase conjugation.

In down-conversion one has a high-frequency pump field and two lower frequency down-converted fields. Let ω be the frequency. Energy conservation yields

$$\hbar\omega_p = \hbar\omega_1 + \hbar\omega_2$$

whereas phase matching is given by

$$\hbar\mathbf{k}_p = \hbar\mathbf{k}_1 + \hbar\mathbf{k}_2$$

where $\mathbf{k}$ is the wave vector. As most nonlinear media are birefringent the second criterion can be satisfied by choosing an appropriate cut, thus managing the refractive indices and therefore wave velocities such that light can be emitted in specific directions. In type-II spontaneous parametric down-conversion a pump beam is incident on a nonlinear optical crystal in which pump photons spontaneously split with a low probability into two orthogonally polarized photons called signal and idler. From energy and momentum conversation one finds that the wavelength and emission directions of the down conversion photons are correlated. They depend on the pump wavelength as well as on the angle between optical axis of the crystal and the pump beam. In the degenerate case (signal and idler having the same wavelength) the photons leave the crystal symmetrically with respect to the pump beam along two cones. For certain orientations of the optical crystal, the two emission cones intersect and the photons emerging along the intersection directions can not be assigned to one of the two orthogonally polarized cones anymore and thus form a polarization entangled pair. The polarization entangled state (Bell state) is described by

$$\frac{1}{\sqrt{2}}(|h\rangle \otimes |v\rangle + e^{i\phi}|v\rangle \otimes |h\rangle)$$

where h and v denote horizontal and vertical polarizations of light. The state is fully entangled (Bell state). By using only standard optical elements in one of the two output beams, one can transform any one of the Bell states into any of the other.

The energy-entangled states from down-conversion photons are the most universal. They are present for any pair of photons. Since there are many ways to partition the energy of the pump photon, each daughter photon has a broad spectrum, and hence a narrow wave packet in time. The sum of the two daughter photons energies is well-defined. They must add up to the energy of the monochromatic pump laser photon. This correlation is given by the energy entangled state

$$|\psi\rangle = \int_0^{E_p} dE A(E)|E\rangle_s \otimes |E_p - E\rangle_i$$

where each ket describes the energy of one of the photons, s and i denotes the signal and idler, respectively, and $A(E)$ is the spectral distribution of the collected down conversion light.

Another common entanglement from the parametric down-conversion process is the momentum direction entanglement. From the emission of a parametric down-conversion source two pairs of spatial (momentum direction) modes are extracted by pinholes. Photon pairs are emitted such that whenever a photon is emitted into one of the inner two modes its partner will be found in the opposite outer mode due to the phase matching in the crystal. The superposition of the two inner and the two outer modes on the beam-splitter serves to measure coincidence rates in various superposition of the initial spatial modes. After the beam-splitters we cannot distinguish the upper two modes from the lower two and therefore interference will be observed in the various coincidence rates.

In a discrete version (time-bin entanglement) of energy-time entangled states sources one sends a double pulse through the down-conversion crystal. If the delay between the two pump pulses equals the time difference between the short and the long arms of the Mach-Zehnder interferometer then again there are two indistinguishable ways of obtaining a coincidence detection.

Chapter 22

Teleportation

Quantum continuous variables provide a new approach to quantum information processing and quantum communication. They describe highly excited quantum systems such as multi-photon fields of light. Continuous variables offer additional advantages over the single-photon system. They involve the use of highly efficient telecommunication photodiodes. The coherent sources of continuous entanglement are also orders of magnitude more efficient than the spontaneous sources of discrete entanglement. Teleportation schemes can be demonstrated involving bright light sources.

Problem 1. Consider the numbers states $|n\rangle$ with $n = 0, 1, 2, \ldots$. Let

$$|\Psi_{in}\rangle = \alpha|0\rangle_1 + \beta|1\rangle_1 + \gamma|2\rangle_1 \tag{1}$$

be an input quantum state in mode 1. An ancilla quantum state

$$|\Psi_{ancilla}\rangle = |0\rangle_2|0\rangle_3 = |0\rangle \otimes |0\rangle$$

in the vacuum state is also given. The modes 2 and 3 pass through a first parametric amplifier whose transformation is given by the operator ($\theta_1 \in \mathbb{R}$)

$$U_{23}(\theta_1) = \exp(\theta_1(b_2^\dagger b_3^\dagger - b_2 b_3))$$

(i) Calculate

$$|\Psi'_{ancilla}\rangle = U_{23}(\theta_1)|\Psi_{ancilla}\rangle = U_{23}(\theta_1)|0\rangle_2|0\rangle_3$$

and set $\gamma_1 = \tanh^2 \theta_1$.

(ii) Consider the product state

$$|\Psi_{in}\rangle|\Psi'_{ancilla}\rangle \equiv |\Psi_{in}\rangle \otimes |\Psi_{ancilla}\rangle .$$

The output mode 2 of the first parametric amplifier and the mode 1 are used as the input modes of the second parametric amplifier ($\theta_2 \in \mathbb{R}$)

$$U_{12}(\theta_2) = \exp(\theta_2(b_1^\dagger b_2^\dagger - b_1 b_2)) \otimes I_3$$

Calculate

$$|\Psi_{out}\rangle = U_{12}(\theta_2)|\Psi_{in}\rangle|\Psi'_{ancilla}\rangle$$

which contains only the state $|1\rangle_1|1\rangle_2$ and set $\gamma_2 = \tanh^2 \theta_2$.

(iii) Find the projection

$$(_1\langle 1|_2\langle 1| \otimes I_3)|\Psi_{out}\rangle .$$

(iv) Assume that we want to transform the input state (1) into

$$|\Psi_{out}\rangle = \alpha|0\rangle_3 + \beta|1\rangle_3 - \gamma|2\rangle_3 .$$

What is the relation between the coefficients γ_1 and γ_2?

Solution 1. (i) Since

$$[b_2^\dagger b_3^\dagger, b_2 b_3] = -b_2^\dagger b_2 - b_3^\dagger b_3 - I$$

and

$$[b_2^\dagger b_2, b_2^\dagger b_3^\dagger] = b_2^\dagger b_3^\dagger, \qquad [b_3^\dagger b_3, b_2^\dagger b_3^\dagger] = b_2^\dagger b_3^\dagger$$

$$[b_2^\dagger b_2, b_2 b_3] = -b_2 b_3, \qquad [b_3^\dagger b_3, b_2 b_3] = -b_2 b_3$$

we have a Lie algebra with the basis

$$b_2^\dagger b_3^\dagger, \quad b_2 b_3, \quad b_2^\dagger b_2, \quad b_3^\dagger b_3, \quad I .$$

Thus we can *disentangle* the operator $U_{23}(\theta_1)$ as

$$U_{23}(\theta_1) = \exp(\sqrt{\gamma_1} b_2^\dagger b_3^\dagger) \exp(\ln(1 - \gamma_1)(b_2^\dagger b_2 + b_3^\dagger b_3 + I)/2) \exp(-\sqrt{\gamma_1} b_2 b_3) .$$

Now we have

$$\exp(-\sqrt{\gamma_1} b_2 b_3)|0\rangle_2|0\rangle_3 = |0\rangle_2|0\rangle_3$$

$$\exp(\ln(1 - \gamma_1)(b_2^\dagger b_2 + b_3^\dagger b_3 + I)/2)|0\rangle_2|0\rangle_3 = \exp(\ln((1 - \gamma_1)^{1/2})|0\rangle_2|0\rangle_3$$

$$= \sqrt{1 - \gamma_1}|0\rangle_2|0\rangle_3$$

and

$$\exp(\sqrt{\gamma_1} b_2^\dagger b_3^\dagger)|0\rangle_2|0\rangle_3 = \sum_{n=0}^{\infty} \gamma_1^{n/2}|n\rangle_2|n\rangle_3 .$$

Thus

$$U_{23}(\theta_1)|\Psi_{ancilla}\rangle = \sqrt{1-\gamma_1}\sum_{n=0}^{\infty}\gamma_1^{n/2}|n\rangle_2|n\rangle_3\,.$$

(ii) Using the result from (i) since the structure of the operator U_{12} is the same as U_{23} we obtain

$$|\Psi_{out}\rangle = \sqrt{(1-\gamma_1)(1-\gamma_2)}|1\rangle_1|1\rangle_2(\sqrt{\gamma_2}\alpha|0\rangle_3 + \sqrt{\gamma_1}(1-2\gamma_2)\beta|1\rangle_3$$
$$+\gamma_1\sqrt{\gamma_2}(3\gamma_2-2)\gamma|2\rangle_3 + \Psi_{others}$$

(iii) The projection yields the state

$$|\Psi_{out}\rangle = \sqrt{(1-\gamma_1)(1-\gamma_2)}(\sqrt{\gamma_2}\alpha|0\rangle_3 + \sqrt{\gamma_1}(1-2\gamma_2)\beta|1\rangle_3$$
$$+\gamma_1\sqrt{\gamma_2}(3\gamma_2-2)\gamma|2\rangle_3\,.$$

(iv) The relation for the parameters is

$$\sqrt{\gamma_2} = \sqrt{\gamma_1}(1-2\gamma_2) = -\gamma_1\sqrt{\gamma_2}(3\gamma_2-2)$$

with the solution

$$\gamma_1 = \frac{21-7\sqrt{2}}{9+4\sqrt{2}} \approx 0.757, \qquad \gamma_2 = \frac{3-\sqrt{2}}{7} \approx 0.226\,.$$

Problem 2. Let $|\beta\rangle$ be a coherent state. Let b and $b^{\dagger}$ be Bose annihilation and creation operators, respectively. Let $D(\mu)$ be the displacement operator ($\mu \in \mathbb{C}$). Consider the product state

$$|\psi\rangle := \frac{1}{\pi}\int_{\mathbb{C}}d^2\beta|\beta\rangle \otimes |\beta^*\rangle\,.$$

This is a maximally entangled continuous-variable state. The state is not normalized. For teleportation we assume the unknown state $|\phi\rangle$ to be in mode 1, the sender's part of the quantum channel to be in mode 2, and the receiver's part in mode 3. Calculate

$$(_{12}\langle\psi| \otimes I_3)(D_1^{\dagger}(\mu) \otimes I_2 \otimes I_3)(|\phi\rangle_1 \otimes |\psi\rangle_{23})$$

where I_2 is the identity operator acting on mode 2, I_3 the identity operator acting on mode 3, and $D_1^{\dagger}$ indicates that the operator acts on mode 1.

Solution 2. Using the completeness relation of coherent states

$$\frac{1}{\pi}\int_{\mathbb{C}}d^2\beta|\beta\rangle\langle\beta| = I$$

yields

$$|\gamma\rangle = \frac{1}{\pi} \int_C d^2\beta |\beta\rangle\langle\beta|\gamma\rangle$$

where we used $I|\beta\rangle = |\beta\rangle$. Applying this expansion and the identity

$$\langle\gamma|\beta\rangle = \langle\beta^*|\gamma^*\rangle$$

we find

$$(_{12}\langle\psi| \otimes I_3)(D_1^\dagger(\mu) \otimes I_2 \otimes I_3)(|\phi\rangle_1 \otimes |\psi\rangle_{23})$$

$$= \frac{1}{\pi^2} \int_C \int_C d^2\beta d^2\gamma \langle\gamma|D^\dagger(\mu)|\phi\rangle\langle\gamma^*|\beta\rangle|\beta^*\rangle_3$$

$$= \frac{1}{\pi^2} \int_C \int_C d^2\beta d^2\gamma \langle\beta^*|\gamma\rangle\langle\gamma|D^\dagger(\mu)|\phi\rangle|\beta^*\rangle_3$$

$$= \frac{1}{\pi} \int_C d^2\beta \langle\beta^*|D^\dagger(\mu)|\phi\rangle|\beta^*\rangle_3$$

$$= \frac{1}{\pi} \int_C d^2\beta |\beta^*\rangle\langle\beta^*|D_3^\dagger(\mu)|\phi\rangle_3$$

$$= D_3^\dagger(\mu)|\phi\rangle_3$$

where we used the identity

$$\frac{1}{\pi} \int_C d^2\gamma \langle\beta^*|\gamma\rangle\langle\gamma|D^\dagger(\mu)|\phi\rangle = \langle\beta^*|D^\dagger(\mu)|\phi\rangle .$$

We conclude that after the joint measurement, the sender's state is projected onto the state which is a unitarily transformed unknown state. Upon receiving the measurement outcome μ, the receiver recovers the unknown state by using the appropriate unitary transformation $D(\mu)$.

Chapter 23

Swapping and Cloning

Swapping and cloning need not only be studied for finite-dimensional systems but also for continuous variables. We can therefore investigate whether coherent states $|\beta\rangle$ and squeezed states $|\zeta\rangle$ can be swapped or cloned.

Problem 1. Can two coherent states be swapped, i.e., can we find a unitary transformation (*swap operator*) such that

$$U_{swap}(|\beta_1\rangle \otimes |\beta_2\rangle) = |\beta_2\rangle \otimes |\beta_1\rangle$$

holds? Consider the unitary operator

$$U(z) := e^{z b_1^\dagger b_2 - z^* b_1 b_2^\dagger}, \qquad z \in \mathbb{C}.$$

Solution 1. Yes, we can find a swap operator. From the unitary operator given above we find $U(z)(|0\rangle \otimes |0\rangle) = |0\rangle \otimes |0\rangle$. Now we have

$$U(z)(|\beta_1\rangle \otimes |\beta_2\rangle) = U(z)(D(\beta_1) \otimes D(\beta_2))|0\rangle \otimes |0\rangle$$
$$= U(z)D_1(\beta_1)D_2(\beta_2)U^{-1}(z)|0\rangle \otimes |0\rangle$$

where $D(\beta)$ is the displacement operator. Thus

$$U(z)D_1(\beta_1)D_2(\beta_2)U^{-1}(z) = U(z)\exp(\beta_1 b_1^\dagger - \beta_1^* b_1 + \beta_2 b_2^\dagger - \beta_2^* b_2)U(z)^{-1}$$

and therefore

$$U(z)D_1(\beta_1)D_2(\beta_2)U^{-1}(z) = \exp(\beta_1(U(z)b_1 U(z)^{-1})^\dagger - \beta_1^*(U(z)b_1 U(z)^{-1})$$
$$+ \beta_2(U(z)b_2 U(z)^{-1})^\dagger - \beta_2^*(U(z)b_2 U(z)^{-1}))$$
$$\equiv \exp(X).$$

Calculating the unitary transformations in the exponent, we find

$$X = \left(\cos(|z|)\beta_1 + \frac{z\sin(|z|)}{|z|}\beta_2\right)b_1^\dagger - \left(\cos(|z|)\beta_1^* + \frac{z^*\sin(|z|)}{|z|}\beta_2^*\right)b_1$$
$$+ \left(\cos(|z|)\beta_2 - \frac{z^*\sin(|z|)}{|z|}\beta_1\right)b_2^\dagger - \left(\cos(|z|)\beta_2^* - \frac{z\sin(|z|)}{|z|}\beta_1^*\right)b_2\,.$$

Thus

$$\exp(X) = D_1\left(\cos(|z|)\beta_1 + \frac{z\sin(|z|)}{|z|}\beta_2\right)D_2\left(\cos(|z|)\beta_2 - \frac{z^*\sin(|z|)}{|z|}\beta_1\right)$$
$$= D\left(\cos(|z|)\beta_1 + \frac{z\sin(|z|)}{|z|}\beta_2\right) \otimes D\left(\cos(|z|)\beta_2 - \frac{z^*\sin(|z|)}{|z|}\beta_1\right)\,.$$

Therefore, we have

$$|\beta_1\rangle \otimes |\beta_2\rangle \to |\cos(|z|)\beta_1 + \frac{z\sin(|z|)}{|z|}\beta_2\rangle \otimes |\cos(|z|)\beta_2 - \frac{z^*\sin(|z|)}{|z|}\beta_1\rangle\,.$$

If we write $z = |z|e^{i\delta}$, then we can write

$$|\beta_1\rangle \otimes |\beta_2\rangle \to |\cos(|z|)\beta_1 + e^{i\delta}\sin(|z|)\beta_2\rangle \otimes |\cos(|z|)\beta_2 - e^{-i\delta}\sin(|z|)\beta_1\rangle\,.$$

Choosing $\sin(|z|) = 1$ yields

$$|\beta_1\rangle \otimes |\beta_2\rangle \to |e^{i\delta}\beta_2\rangle \otimes |-e^{-i\delta}\beta_1\rangle = |e^{i\delta}\beta_2\rangle \otimes |e^{-i(\delta+\pi)}\beta_1\rangle\,. \qquad (1)$$

Applying the unitary operator

$$V = e^{-i\delta b_1^\dagger b_1}e^{i(\delta+\pi)b_2^\dagger b_2} = e^{-i\delta b^\dagger b} \otimes e^{i(\delta+\pi)b^\dagger b}$$

from the left, we find $|\beta_1\rangle \otimes |\beta_2\rangle \to |\beta_2\rangle \otimes |\beta_1\rangle$. If we set $\beta_1 = \beta$ and $\beta_2 = 0$ in (1) we obtain

$$|\beta\rangle \otimes |0\rangle = |\cos(|z|)\beta\rangle \otimes |-e^{-i\delta}\sin(|z|)\beta\rangle = |\cos(|z|)\beta\rangle \otimes |e^{-i(\delta+\pi)}\sin(|z|)\beta\rangle\,.$$

Problem 2.   We cannot *clone* coherent states, i.e., we cannot find a unitary operator which maps

$$|\beta\rangle \otimes |0\rangle \to |\beta\rangle \otimes |\beta\rangle\,.$$

Use the result from problem 1 (equation (1))

$$|\beta\rangle \otimes |0\rangle \to |\cos(|z|)\beta\rangle \otimes |e^{-i(\delta+\pi)}\sin(|z|)\beta\rangle \qquad (1)$$

to find an approximation.

Solution 2. Applying the operator $I \otimes e^{i(\delta+\pi)b^{\dagger}b}$ to the right-hand side of (1), we obtain

$$| \cos(|z|)\beta\rangle \otimes |\sin(|z|)\beta\rangle \,.$$

If we set $|z| = \pi/4$ we obtain

$$|\beta\rangle \otimes |0\rangle \to \left|\frac{\beta}{\sqrt{2}}\right\rangle \otimes \left|\frac{\beta}{\sqrt{2}}\right\rangle \,.$$

This is called *imperfect cloning*.

Problem 3. We consider three infinite-dimensional Hilbert spaces $\mathcal{H}_1$, $\mathcal{H}_2$, $\mathcal{H}_3$ and the product Hilbert space $\mathcal{H}_3 \otimes \mathcal{H}_1 \otimes \mathcal{H}_2$ with $\mathcal{H}_1 = \mathcal{H}_2 = \mathcal{H}_3$. Consider the heterodyne-current operator $Z := b_1 + b_2^{\dagger}$, where the Bose annihilation operator acts in the Hilbert space $\mathcal{H}_1$ and the Bose creation operator $b_2^{\dagger}$ acts on the Hilbert space $\mathcal{H}_2$. We have $[Z, Z^{\dagger}] = 0$ and the eigenvalue equation $Z|z\rangle\rangle_{12} = z|z\rangle\rangle_{12}$ with $z \in \mathbb{C}$. The eigenstates $|z\rangle\rangle$ are given by

$$|z\rangle\rangle_{12} = D_1(z)|0\rangle\rangle_{12} = D_2(z^*)|0\rangle\rangle_{12}$$

where D_1 denotes the displacement operator for mode 1, D_2 the displacement operator for mode 2 and

$$|0\rangle\rangle_{12} := \frac{1}{\sqrt{\pi}} \sum_{n=0}^{\infty} (-1)^n |n\rangle_1 \otimes |n\rangle_2$$

in the Fock basis (number basis). The expression

$$_{32}\langle\langle z|z'\rangle\rangle_{12} = \frac{1}{\pi} D_1(z') T_{13} D_3^{\dagger}(z)$$

where

$$T_{13} := \sum_{n=0}^{\infty} |n\rangle_{13}\langle n| \equiv \sum_{n=0}^{\infty} (|n\rangle \otimes I \otimes I)(I \otimes I \otimes \langle n|)$$

denotes the transfer operator which obviously satisfies $T_{13}|\psi\rangle_3 = |\psi\rangle_1$ for any vector $|\psi\rangle$. For a cloning operation consider the input state in the product Hilbert space $\mathcal{H}_3 \otimes \mathcal{H}_1 \otimes \mathcal{H}_2$

$$|\psi\rangle = |\phi\rangle_3 \otimes \int_{\mathbb{C}} d^2z f(z, z^*) |z\rangle\rangle_{12} \,.$$

where $|\phi\rangle_3$ is the original state in the Hilbert space $\mathcal{H}_3$ to be cloned in $\mathcal{H}_3$ itself and $\mathcal{H}_1$ and $\mathcal{H}_2$ is an ancillary Hilbert space. The cloning transformation is realized by the unitary operator

$$U = \exp\left(\left(\frac{1}{2}(b_3 + b_3^{\dagger}) + \frac{1}{2}(b_3 - b_3^{\dagger})\right) Z^{\dagger} - \left(\frac{1}{2}(b_3 + b_3^{\dagger}) - \frac{1}{2}(b_3 - b_3^{\dagger})\right) Z\right)$$

where $Z = b_1 + b_2^\dagger = b \otimes I + I \otimes b^\dagger$. Let $|\psi\rangle_{out} = U|\psi\rangle$.
(i) Calculate the commutator $[b_3 b_1^\dagger + b_3 b_2, b_3^\dagger b_1 + b_3^\dagger b_2^\dagger]$ and discuss.
(ii) Evaluate the one-site restricted density matrix ρ_3 corresponding to the state $|\psi\rangle_{out}$ for the Hilbert space $\mathcal{H}_3$.
(iii) Evaluate the one-site restricted density matrix ρ_1 corresponding to the state $|\psi\rangle_{out}$ for the Hilbert space $\mathcal{H}_1$.
(iv) Compare the two density matrices.

Solution 3.   (i) Using $[b_j, b_k^\dagger] = \delta_{jk} I$ we find

$$[b_3 b_1^\dagger + b_3 b_2, b_3^\dagger b_1 + b_3^\dagger b_2^\dagger] = I + b_1^\dagger b_1 + b_2^\dagger b_2 + b_1^\dagger b_2^\dagger + b_1 b_2 \,.$$

The right-hand side does not depend on b_3 and $b_3^\dagger$.
(ii) Let $|w\rangle\rangle_{12}$ be an eigenstate of the operator Z. We have $\rho = |\psi\rangle\langle\psi|$. Thus for the partial trace we have to calculate

$$\rho_3 = \int_{\mathbb{C}} d^2 w \int_{\mathbb{C}} d^2 z \int_{\mathbb{C}} d^2 z' f(z, z^*) f^*(z', z'^*) A$$

where $A \equiv {}_{12}\langle\langle w|D_3^\dagger(z)|\phi\rangle_{33}\langle\phi|D_3(z') \otimes |z\rangle\rangle_{12} {}_{12}\langle\langle z'|w\rangle\rangle_{12}$. Using the completeness and orthogonality of the eigenstates $|w\rangle\rangle_{12}$ of the operator Z we find

$$\rho_3 = \int_{\mathbb{C}} d^2 z |f(z, z^*)|^2 D_3^\dagger(z)|\phi\rangle_{33}\langle\phi|D_3(z) \,.$$

(iii) For ρ_1 we have to calculate

$$\rho_1 = \int_{\mathbb{C}} d^2 w \int_{\mathbb{C}} \frac{d^2 z}{\pi} \int_{\mathbb{C}} \frac{d^2 z'}{\pi} f(z, z^*) f^*(z', z'^*)$$
$$\times D_1(z) T_{13}(D_3^\dagger(w) D_3^\dagger(z)|\phi\rangle_{33}\langle\phi|D_3(z') D_3(w)) T_{31} D_3^\dagger(z') \,.$$

Using the completeness and orthogonality of the eigenstates $|w\rangle\rangle$ of the operator Z we find after integration over z and z'

$$\rho_1 = \int_{\mathbb{C}} d^2 w |\widetilde{f}(w, w^*)|^2 D_1^\dagger(w)|\phi\rangle_{11}\langle\phi|D_1(z)$$

where $\widetilde{f}(w, w^*)$ denotes the *Fourier transform* over the complex plane

$$\widetilde{f}(w, w^*) = \frac{1}{\pi} \int_{\mathbb{C}} d^2 z e^{wz^* - w^* z} f(z, z^*) \,.$$

(iv) For

$$f(z, z^*) = \sqrt{\frac{2}{\pi}} e^{-|z|^2}$$

one has two identical clones, i.e., $\rho_3 = \rho_1$ which are given by the original state $|\psi\rangle$ degraded by Gaussian noise.

Chapter 24

Homodyne Detection

In optical homodyne detection one mixes a local oscillator field (for example a coherent state or squeezed state of light) with a signal field at a balanced (50/50) *beam splitter* (unitary operator)

$$U_{BS} = e^{i\pi(b_S^\dagger \otimes b_{LO} + b_S \otimes b_{LO}^\dagger)/4}$$

where the signal field mode is represented by the Bose operator b_S and the local oscillator is represented by the Bose operator b_{LO}. Then

$$U_{BS} \begin{pmatrix} b_S \otimes I \\ I \otimes b_{LO} \end{pmatrix} U_{BS}^\dagger = \begin{pmatrix} U_{BS}(b_S \otimes I)U_{BS}^\dagger \\ U_{BS}(I \otimes b_{LO}U_{BS}^\dagger \end{pmatrix} = \frac{1}{\sqrt{2}} \begin{pmatrix} b_S \otimes I - iI \otimes b_{LO} \\ I \otimes b_{LO} - ib_S \otimes I \end{pmatrix}.$$

Thus the signal wave is overlapped on a beam splitter with a relatively strong local oscillator wave in the matching optical mode. The two fields emerging from the beam splitter are incident on two high-efficiency photodiodes whose output photocurrents are subtracted. The photocurrent difference is proportional to the value of the electric field operator. Thus optical homodyne detection corresponds to the difference photon counting of the two output electric fields from the beam splitter.

In heterodyne detection the signal electric field and another electric field, the auxiliary electric field, feed the same port of a beam splitter. The local electric field oscillator enters the other port of the beam splitter. The frequencies of the signal, auxiliary and local oscillator fields are different. One has $\omega_S + \omega_A = 2\omega_L$, $\omega_S - \omega_A = 2\omega$ with $\omega_L \gg \omega$.

Problem 1.   A *homodyne detector* is constructed by placing a photo-counter in each arm after the beam splitter and then considering the difference photocurrent between the two modes

$$\hat{D} := \widetilde{b}_1^\dagger \widetilde{b}_1 - \widetilde{b}_2^\dagger \widetilde{b}_2 \,.$$

Express the homodyne photocurrent in terms of the input modes b_1, b_2.

Solution 1.   Straightforward calculation yields

$$\hat{D} = (2\tau - 1)(b_1^\dagger b_1 - b_2^\dagger b_2) + 2\sqrt{\tau(1-\tau)}(b_1^\dagger b_2 + b_1 b_2^\dagger) \,.$$

This expression reduces to

$$\hat{D} = b_1^\dagger b_2 + b_1 b_2^\dagger \equiv b^\dagger \otimes b + b \otimes b^\dagger$$

for a balanced ($\tau = 1/2$) beam splitter.

Problem 2.   Let b_1, b_2 be Bose annihilation operators. Consider the operators

$$d_\pm = \frac{1}{\sqrt{2}}(b_1 \pm b_2 e^{i\theta})$$

where θ is a phase shift. Let

$$I = d_+^\dagger d_+ - d_-^\dagger d_- \,.$$

Find I in terms of the original operators b_1, b_2.

Solution 2.   We have

$$
\begin{aligned}
I &= d_+^\dagger d_+ - d_-^\dagger d_- \\
&= \frac{1}{2}(b_1^\dagger + b_2^\dagger e^{-i\theta})(b_1 + b_2 e^{i\theta}) - \frac{1}{2}(b_1^\dagger - b_2^\dagger e^{-i\theta})(b_1 - b_2 e^{i\theta}) \\
&= b_2^\dagger b_1 e^{-i\theta} + b_1^\dagger b_2 e^{i\theta} \,.
\end{aligned}
$$

This plays a role in homodyne measurement, where b_1 describes the signal field and b_2 describes the local oscillator field.

Problem 3.   Let b_S, b_{LO}, b_P be the Bose annihilation operators corresponding to the signal, the local oscillator and the photodetector input field, respectively. Let $0 < \epsilon < 1$. By shifting the b_P field with a constant phase, we have

$$b_P = \sqrt{\epsilon} b_S + i\sqrt{1-\epsilon} b_{LO} \,.$$

Let
$$b_S = \hat{x}_S + i\hat{y}_S, \quad b_{LO} = \hat{x}_{LO} + i\hat{y}_{LO}, \quad b_P = \hat{x}_P + i\hat{y}_P$$
where $\hat{x}$, $\hat{y}$ are the (hermitian) quadrature operators. Let
$$\hat{n}_S = b_S^\dagger b_S, \quad \hat{n}_{LO} = b_{LO}^\dagger b_{LO}, \quad \hat{n}_P = b_P^\dagger b_P$$
be the photon number operators and let $\langle\,\rangle$ be the average with respect to a quantum state. Find the mean photodetector output $\langle \hat{n}_P \rangle$.

Solution 3. We obtain
$$\langle \hat{n}_P \rangle = \epsilon \hat{n} + (1 - \epsilon)\hat{n} - 2(\epsilon(1 - \epsilon))^{1/2} \langle \hat{x}_S \hat{y}_{LO} - \hat{x}_{LO} \hat{y}_S \rangle \,.$$

Problem 4. Suppose that $b_1^\dagger, b_2^\dagger$ are Bose creation operators and b_1, b_2 are Bose annihilation operators and I is the identity operator. Consider the linear operator
$$Z := b \otimes I + I \otimes b^\dagger$$
where $b_1 := b \otimes I$ and $b_2^\dagger := I \otimes b^\dagger$. Thus $Z = b_1 + b_2^\dagger$. The operator is called the *heterodyne-current operator*. One also finds the notation
$$Z = b_S + b_I$$
where the subscripts S and I, respectively denote the signal mode at frequency $\omega_0 + \Delta\omega$ and the imaging mode at frequency $\omega_0 - \Delta\omega$ ($\Delta\omega \ll \omega$).
(i) Calculate the commutator $[Z, Z^\dagger]$.
(ii) Find the state
$$Z(|0\rangle \otimes |0\rangle), \quad Z^\dagger(|0\rangle \otimes |0\rangle)\,.$$
(iii) Find the state
$$Z^2(|0\rangle \otimes |0\rangle)\,.$$

Solution 4. (i) We have
$$Z^\dagger = b_1^\dagger + b_2 \equiv b^\dagger \otimes I + I \otimes b$$
and
$$\begin{aligned}
[Z, Z^\dagger] &= (b \otimes I + I \otimes b^\dagger)(b^\dagger \otimes I + I \otimes b) \\
&\quad - (b^\dagger \otimes I + I \otimes b)(b \otimes I + I \otimes b^\dagger) \\
&= bb^\dagger \otimes I + b \otimes b + b^\dagger \otimes b^\dagger + I \otimes b^\dagger b \\
&\quad - b^\dagger b \otimes I - b^\dagger \otimes b^\dagger - b \otimes b - I \otimes bb^\dagger \\
&= (bb^\dagger - b^\dagger b) \otimes I + I \otimes (b^\dagger b - bb^\dagger) \\
&= I \otimes I - I \otimes I = 0\,.
\end{aligned}$$

(ii) We have

$$
\begin{aligned}
Z(|0\rangle \otimes |0\rangle) &= (b \otimes I + I \otimes b^\dagger)|0\rangle \otimes |0\rangle \\
&= (b \otimes I)(|0\rangle \otimes |0\rangle) + (I \otimes b^\dagger)(|0\rangle \otimes |0\rangle) \\
&= |0\rangle \otimes b^\dagger |0\rangle = |0\rangle \otimes |1\rangle
\end{aligned}
$$

since $b|0\rangle = 0$. Analogously

$$
Z^\dagger |0\rangle \otimes |0\rangle = b^\dagger |0\rangle \otimes |0\rangle = |1\rangle \otimes |0\rangle \,.
$$

(iii) We find

$$
Z^2(|0\rangle \otimes |0\rangle) = \sqrt{2}(|0\rangle \otimes |2\rangle).
$$

Chapter 25

Hamilton Operators

Most experimental realizations of quantum logic gates (Hadamard gate, quantum phase gate, controlled-NOT gate) involve several qubits and number states (Fock states). A Hamilton operator $\hat{H}$ must describe the interaction. Thus in quantum computing we are faced with two problems. One is to determine the Hamilton operator $\hat{H}$ for the system such that the time-evolution $\exp(-i\hat{H}t/\hbar)$ represents the execution of the computation. The other one is to build the hardware described by this Hamilton operator.

Problem 1. Consider a single continuous variable corresponding to a linear operator X. Let P be the operator of the conjugate variable, i.e.,

$$[X, P] = iI. \tag{1}$$

Consider the *Kerr-Hamilton operator*

$$K = H^2 = (X^2 + P^2)^2.$$

The *Kerr-Hamilton operator* corresponds to a χ^3 process in nonlinear optics. The linear unbounded operators X and P could correspond to quadrature amplitudes of a mode of the electromagnetic field. The quadrature amplitudes are the real and imaginary parts of the complex electric field. Let

$$S := \frac{1}{2}(XP + PX).$$

Calculate the commutators

$$[K, X], \quad [K, P], \quad [X, [K, S]], \quad [P, [K, S]].$$

Discuss.

Solution 1.   Since

$$P^2X^2 = X^2P^2 - 4iXP - 2I, \qquad P^2X = XP^2 - 2iP$$

we find

$$[K, X] = \frac{i}{2}(X^2P + PX^2 + 2P^3)$$
$$[K, P] = -\frac{i}{2}(P^2X + XP^2 + 2X^3)$$
$$[X, [K, S]] = P^3$$
$$[P, [K, S]] = X^3.$$

Thus the algebra generated by X, P, H, S, K by calculating commutators includes all third order polynomials in X and P. We can construct Hamilton operators that are arbitrary hermitian polynomials in any order of X and P. We have

$$[P^3, P^m X^n] = iP^{m+2}X^{n-1} + \text{lower order terms}$$

and

$$[X^3, P^m X^n] = iP^{m-1}X^{n+2} + \text{lower order terms}.$$

Problem 2.   A nonlinear four-wave mixing device can be described by the Hamilton operator

$$\hat{H} = \frac{1}{4}\hbar\omega(b_1^\dagger b_2 + b_1 b_2^\dagger)^2 \equiv \frac{1}{4}\hbar\omega(b^\dagger \otimes b + b \otimes b^\dagger)^2.$$

(i) Let $|n\rangle$ be a Fock state. Consider the normalized input state $|n\rangle \otimes |0\rangle$. Find the normalized state

$$\exp(-i\hat{H}t/\hbar)(|n\rangle \otimes |0\rangle)$$

for the interaction time $t = \pi/\omega$.
(ii) Consider the normalized input state $|n\rangle \otimes |0\rangle$. Find the state

$$\exp(-i\hat{H}t/\hbar)(|n\rangle \otimes |0\rangle)$$

for the interaction time $t = \pi/(2\omega)$.

Solution 2.   (i) For the interaction time $t = \pi/\omega$ the output state is $i|0\rangle \otimes |n\rangle$ (for n even) and $\exp(-i\pi/4)|n\rangle \otimes |0\rangle$ for n odd. Thus the device acts as an even-odd filter, switching the even numbers from one mode to the other. Under these operating conditions it can be used as a device to measure parity without counting the photon number. It is sufficient to detect any photons in either of the output channels.

(ii) If the interaction time is $t = \pi/(2\omega)$ the output state will have the form

$$\frac{1}{\sqrt{2}}(|n\rangle \otimes |0\rangle + e^{-i(n+1)\pi/2}|0\rangle \otimes |n\rangle)$$

which is a maximally entangled state for n photons.

Problem 3. *Cross phase modulation* is described by the Hamilton operator

$$\hat{H} = -\hbar\omega b_1^\dagger b_1 b_2^\dagger b_2 = -\hbar\omega(b^\dagger b \otimes b^\dagger b)$$

where ω is a function of the third order susceptibility $\chi^{(3)}$. Consider the two-mode Fock state $|m\rangle \otimes |n\rangle$, i.e. $m, n = 0, 1, 2, \ldots$. Find the state $\exp(-i\hat{H}t/\hbar)(|m\rangle \otimes |n\rangle)$.

Solution 3. Since $b_1 = b \otimes I$ and $b_2 = I \otimes b$ we have

$$b_1^\dagger b_1 |m\rangle \otimes |n\rangle = m|m\rangle \otimes |n\rangle, \quad b_2^\dagger b_2 |m\rangle \otimes |n\rangle = n|m\rangle \otimes |n\rangle.$$

Thus

$$\exp(-i\hat{H}t/\hbar)|m\rangle \otimes |n\rangle = e^{i\omega tmn}|m\rangle \otimes |n\rangle.$$

Problem 4. A *Kerr medium* is nonlinear in the sense that its refractive index n has a component which varies with the intensity of the propagating electric field $\mathbf{E}$, that is

$$n = n_0 + n_2|\mathbf{E}|^2$$

where n_0 and n_2 are constants. For a single-mode field, described by Bose creation and annihilation operators $b^\dagger$ and b, propagating through a low-loss Kerr media, the interaction Hamilton operator can be written as

$$\hat{H} = \chi b^{\dagger 2} b^2.$$

(i) Show that the *number state (Fock state)* $|n\rangle$ is an eigenstate.

(ii) Assume that the initial state is a coherent state $|\beta\rangle$. Find $|\beta(t)\rangle$.

(iii) Let $\chi t = \pi r/s$ where r and s are mutually prime with $r < s$. Write $\exp(-i\pi rn^2/s)$ as a discrete Fourier transform. Express $|\beta(t)\rangle$ using this expansion.

Solution 4. (i) Since

$$b^{\dagger 2} b^2 \equiv b^\dagger b (b^\dagger b - I)$$

and $b^\dagger b |n\rangle = n|n\rangle$, we have

$$\hat{H}|n\rangle = \chi(n^2 - n)|n\rangle \,.$$

Thus the eigenvalues are $\chi(n^2 - n)$.

(ii) The solution of the *time dependent Schrödinger equation* ($\hbar = 1$)

$$i\frac{d|\beta\rangle}{dt} = \hat{H}|\beta\rangle$$

is given by

$$|\beta(t)\rangle = \exp(-i\hat{H}t)|\beta\rangle \,.$$

Using the result from (i) we find

$$|\beta(t)\rangle = \sum_{n=0}^{\infty} c_n e^{-i\chi t(n^2 - n)}|n\rangle$$

where

$$c_n := \exp(-|\beta|^2/2)\frac{\beta^n}{\sqrt{n!}} \,.$$

Since $n^2 - n$ is always an even number, the system will revive whenever χt is a multiple of π.

(iii) Let $\chi t = \pi r/s$ where r, s are mutually prime with $r < s$. Then we can write the quadratic (in n) phase in terms of linear phases using the discrete Fourier transform

$$\exp(-i\pi n^2 r/s) = \sum_{p=0}^{\ell-1} a_p^{(r,s)} \exp(-2\pi i p n/\ell)$$

where

$$\ell = \begin{cases} s & \text{if } r \text{ is odd, } s \text{ is even or vice-versa} \\ 2s & \text{if both } r \text{ and } s \text{ are odd} \end{cases} \,.$$

Thus

$$a_p^{(r,s)} = \frac{1}{\ell} \sum_{k=0}^{\ell-1} \exp(-i\pi r k^2/s + 2\pi i p k/\ell)$$

and

$$|\beta(t)\rangle = \sum_{p=0}^{\ell-1} a_p^{(r,s)} |\beta \exp(i\pi(r/s - 2p/\ell))\rangle \,.$$

Problem 5. The Hamilton operator for the *second harmonic generation* can be written as

$$\hat{H} = i\hbar\frac{\kappa}{2}(b^{\dagger 2}b_{sh} - b^2 b_{sh}^\dagger)$$

where b is the fundamental cavity mode Bose operator, b_{sh} is the second-harmonic mode Bose operator and κ is the nonlinear coupling. Using the Heisenberg equation of motion find the time evolution of b and b_{sh}.

Solution 5. The *Heisenberg equation of motion* of an operator $\hat{A}$ is given by

$$i\hbar \frac{d\hat{A}}{dt} = [\hat{A}, \hat{H}](t).$$

The commutation relations are given by

$$[b, b^\dagger] = I, \quad [b_{sh}, b_{sh}^\dagger] = I$$

$$[b, b] = [b_{sh}, b_{sh}] = [b, b_{sh}] = [b, b_{sh}^\dagger] = 0.$$

Thus we find the operator-valued nonlinear differential equations

$$\frac{db}{dt} = \kappa b^\dagger b_{sh}, \qquad \frac{db_{sh}}{dt} = -\frac{\kappa}{2} b^2.$$

In a more realistic model, cavity photon losses must be taken into account.

Problem 6. A single spin-$\frac{1}{2}$ particle is placed on a cantilever tip. The tip can oscillate only in the z-direction. A ferromagnetic particle, whose magnetic moment points in the positive z-direction, produces a non-uniform magnetic field at the spin. A uniform magnetic field, $\mathbf{B}_0$, oriented in the positive z-direction, determines the ground state of the spin. A rotating magnetic field, $\mathbf{B}_1(t)$, induces transitions between the ground state and excited states of the spin. It is given by

$$B_x(t) = B_1 \cos(\omega t + \phi(t)), \qquad B_y(t) = -B_1 \sin(\omega t + \phi(t)), \qquad B_z(t) = 0$$

where $\phi(t)$ describes a smooth change in phase required for a cyclic adiabatic inversion of the spin

$$|d\phi(t)/dt| \ll \omega.$$

In the reference frame rotating with $\mathbf{B}_1(t)$, the time-dependent Hamilton operator is given by

$$\hat{H}(t) = \frac{P_z^2}{2m_c^*} + \frac{m_c^* \omega_c^2 Z^2}{2} - \hbar\left(\omega_L - \omega - \frac{d\phi}{dt}\right) S_z - \hbar\omega_1 S_x - g\mu \frac{\partial B_z}{\partial Z} Z S_z$$

where Z is the coordinate of the oscillator which describes the dynamics of the quasi-classical cantilever tip, P_z is its momentum, m_c^* and ω_c are the effective mass and the frequency of the cantilever, S_z and S_x are the $z-$ and the $x-$ component of the spin,

$$S_x = \frac{1}{2}\begin{pmatrix} 0 & 1 \\ 1 & 0 \end{pmatrix}, \qquad S_z = \frac{1}{2}\begin{pmatrix} 1 & 0 \\ 0 & -1 \end{pmatrix},$$

ω_L is its Larmor frequency, ω_1 is the *Rabi frequency* (the frequency of the spin precession around the magnetic field $\mathbf{B}_1(t)$ at the resonance condition $\omega = \omega_L$, $d\phi/dt = 0$), g and μ are the g-factors and the magnetic moment of the spin and we defined $m_c^* = m_c/4$ as the effective cantilever mass. The operator acts in the product Hilbert space $L_2(\mathbb{R}) \otimes \mathbb{C}^2$. One sets

$$\omega_c = (k_c/m_c^*)^{1/2}, \quad \omega_L = \gamma B_z, \quad \omega_1 = \gamma B_1$$

where $\gamma = g\mu/\hbar$ is the gyromagnetic ratio of the spin, m_c and k_c are the mass and the force constant of the cantilever, B_z includes the uniform magnetic field B_0 and the magnetic field produced by the ferromagnetic particle.

(i) Cast the Hamilton operator in dimensionless form $\hat{H}/\hbar\omega_c \to \hat{K}$ by introducing the quantities

$$E_0 := \hbar\omega_c, \quad F_0 := \sqrt{k_c E_0}, \quad Z_0 := \sqrt{E_0/k_c}, \quad P_0 := \hbar/Z_0$$

with $\omega = \omega_L$ and using the dimensionless time $\tau := \omega_c t$.

(ii) The dimensionless time-dependent Schrödinger equation

$$i\frac{\partial\Psi}{\partial\tau} = \hat{K}\Psi$$

where

$$\Psi(\tau, z) = \begin{pmatrix} \Psi_1(\tau, z) \\ \Psi_2(\tau, z) \end{pmatrix}$$

can be solved using the expansions

$$\Psi_1(\tau, z) = \sum_{n=0}^{\infty} A_n(\tau)|n\rangle, \qquad \Psi_2(\tau, z) = \sum_{n=0}^{\infty} B_n(\tau)|n\rangle$$

$$|n\rangle = \pi^{1/4} 2^{n/2} (n!)^{1/2} e^{-z^2/2} H_n(z)$$

where $\{\,|n\rangle\ :\ n = 0, 1, \dots\}$ are number states. Here $H_n(z)$ are the Hermitian polynomials. Find the time evolution of the complex expansion coefficients A_n and B_n.

(iii) What would be an initial state closest to the classical limit?

Solution 6. (i) Since $\hat{H}/(\hbar\omega_c) \to \hat{K}$ we find

$$\hat{K} = \frac{\hat{H}}{\hbar\omega_c} = \frac{1}{2}(p_z^2 + z^2) + S_z\frac{d\phi}{d\tau} - \epsilon S_x - 2\eta z S_z$$

where we used $\omega_L = \omega$ and

$$p_z := \frac{P_z}{P_0}, \quad z := \frac{Z}{Z_0}, \quad \epsilon := \frac{\omega_1}{\omega_c}, \quad \eta := \frac{g\mu}{2F_c}\frac{\partial B_z}{\partial Z}, \quad \omega_c dt = d\tau\,.$$

(ii) Inserting the series expansions into the dimensionless Schrödinger equation we find the system of linear differential equations with time-dependent coefficients for the complex amplitudes $A_n(\tau)$ and $B_n(\tau)$

$$i\frac{dA_n}{d\tau} = \left(n + \frac{1}{2} + \frac{1}{2}\frac{d\phi}{d\tau}\right)A_n - \frac{\eta}{\sqrt{2}}\left(\sqrt{n}A_{n-1} + \sqrt{n+1}A_{n+1}\right) - \frac{\epsilon}{2}B_n$$

$$i\frac{dB_n}{d\tau} = \left(n + \frac{1}{2} + \frac{1}{2}\frac{d\phi}{d\tau}\right)B_n + \frac{\eta}{\sqrt{2}}\left(\sqrt{n}B_{n-1} + \sqrt{n+1}B_{n+1}\right) - \frac{\epsilon}{2}A_n$$

where we used the Bose operators b and $b^\dagger$ defined by

$$b|n\rangle = \sqrt{n}|n-1\rangle, \qquad b^\dagger|n\rangle = \sqrt{n+1}|n+1\rangle$$

and

$$\frac{1}{2}\left(p_z^2 + z^2\right)|n\rangle = \left(n + \frac{1}{2}\right)|n\rangle$$

with

$$z := \frac{1}{\sqrt{2}}(b^\dagger + b), \quad p_z := \frac{i}{\sqrt{2}}(b^\dagger - b), \quad [b, b^\dagger] = I.$$

(iii) We can choose the coherent state

$$\Psi_1(z, 0) = \sum_{n=0}^{\infty} A_n(0)|n\rangle, \quad \Psi_2(z, 0) = 0, \quad A_n(0) = \frac{\beta^n}{\sqrt{n!}}\exp(-|\beta|^2/2).$$

Problem 7. Consider two *Bose-Einstein condensates* which both occupy the ground-state of their respective traps. They are described by the atom Bose annihilation (creation) operators b_1 ($b_1^\dagger$) and b_2 ($b_2^\dagger$). Atoms are released from each trap with momenta (wave vectors) $\mathbf{k}_1$ and $\mathbf{k}_2$, respectively, producing an interference pattern which enables a relative phase to be measured. The *intensity* $I(\mathbf{x}, t)$ of the atomic field is given by

$$I(\mathbf{x}, t) = I_0\langle\psi|(b_1^\dagger(t)e^{i\mathbf{k}_1\cdot\mathbf{x}} + b_2^\dagger(t)e^{i\mathbf{k}_2\cdot\mathbf{x}})(b_1(t)e^{-i\mathbf{k}_1\cdot\mathbf{x}} + b_2(t)e^{-i\mathbf{k}_2\cdot\mathbf{x}})|\psi\rangle \quad (1)$$

where I_0 is the single atom intensity. Atoms within each condensate collide. This can be described using the Hamilton operator

$$\hat{H} = \frac{1}{2}\hbar\chi((b_1^\dagger b_1)^2 + (b_2^\dagger b_2)^2) \quad (2)$$

where χ is the collision rate between the atoms within each condensate. Cross-collisions between the two condensates, described by the term $b_1^\dagger b_1 b_2^\dagger b_2$ could also be included. Using the Hamilton operator given by (2) the intensity $I(\mathbf{x}, t)$ is given by

$$I(\mathbf{x}, t) = I_0(\langle\psi|b_1^\dagger b_1|\psi\rangle + \langle\psi|b_2^\dagger b_2|\psi\rangle + \langle\psi|b_1^\dagger \exp(i\chi t(b_1^\dagger b_1 - b_2^\dagger b_2))b_2|\psi\rangle e^{-i\phi(\mathbf{x})}$$
$$+ \langle\psi|b_2^\dagger \exp(-i\chi t(b_1^\dagger b_1 - b_2^\dagger b_2))b_1|\psi\rangle e^{i\phi(\mathbf{x})})$$

where $\phi(\mathbf{x}) := (\mathbf{k}_2 - \mathbf{k}_1) \cdot \mathbf{x}$. Calculate $I(\mathbf{x}, t)$ for the product state

$$|\psi\rangle = |\beta_1\rangle \otimes |\beta_2\rangle$$

where $|\beta_1\rangle$ and $|\beta_2\rangle$ are coherent states.

Solution 7.   Since $b|\beta\rangle = \beta|\beta\rangle$, $\langle\beta|b^\dagger = \langle\beta|\beta^*$, and

$$e^{i\chi t b^\dagger b}|\beta\rangle = \sum_{j=0}^{\infty} \frac{1}{j!}(e^{i\chi t} - 1)(b^\dagger)^j b^j|\beta\rangle = \sum_{j=0}^{\infty} \frac{1}{j!}(e^{i\chi t} - 1)(b^\dagger)^j \beta^j$$

we have

$$\langle\beta|e^{i\chi t b^\dagger b}|\beta\rangle = \exp((e^{i\chi t} - 1)\beta^*\beta).$$

We also have

$$\langle\beta|e^{i\chi t b^\dagger b}b|\beta\rangle = \beta\exp((e^{i\chi t}-1)\beta^*\beta), \quad \langle\beta|b^\dagger e^{i\chi t b^\dagger b}|\beta\rangle = \beta^*\exp((e^{i\chi t}-1)\beta^*\beta).$$

Since $b_1^\dagger b_1 = b^\dagger b \otimes I$ and $b_2^\dagger b_2 = I \otimes b^\dagger b$ we have

$$
\begin{aligned}
e^{i\chi t(b_1^\dagger b_1 - b_2^\dagger b_2)} &= e^{i\chi t b_1^\dagger b_1} e^{-i\chi t b_2^\dagger b_2} = e^{i\chi t(b^\dagger b \otimes I)} e^{-i\chi t(I \otimes b^\dagger b)} \\
&= (e^{i\chi t b^\dagger b} \otimes I)(I \otimes e^{-i\chi t b^\dagger b}) = e^{i\chi t b^\dagger b} \otimes e^{-i\chi t b^\dagger b}.
\end{aligned}
$$

Thus

$$\langle\psi|b_1^\dagger b_1|\psi\rangle = \beta_1^*\beta_1, \qquad \langle\psi|b_2^\dagger b_2|\psi\rangle = \beta_2^*\beta_2$$

and

$$
\begin{aligned}
\langle\psi|b_1^\dagger e^{i\chi t(b_1^\dagger b_1 - b_2^\dagger b_2)}b_2|\psi\rangle e^{-i\phi(\mathbf{x})} &= \beta_1^*\beta_2 \exp((e^{i\chi t} - 1)\beta_1^*\beta_1) \\
&\times \exp((e^{-i\chi t} - 1)\beta_2^*\beta_2)e^{-i\phi(\mathbf{x})}
\end{aligned}
$$

$$
\begin{aligned}
\langle\psi|b_2^\dagger e^{-i\chi t(b_1^\dagger b_1 - b_2^\dagger b_2)}b_1|\psi\rangle e^{i\phi(\mathbf{x})} &= \beta_1\beta_2^* \exp((e^{-i\chi t} - 1)\beta_1^*\beta_1) \\
&\times \exp((e^{i\chi t} - 1)\beta_2^*\beta_2)e^{i\phi(\mathbf{x})}.
\end{aligned}
$$

Problem 8.   Consider the Hamilton operator $\hat{H}$ of a coupled one Bose one Fermi system

$$\hat{H} = \omega b^\dagger b \otimes I_F + J I_B \otimes c^\dagger c + \alpha(b^\dagger \otimes c + b \otimes c^\dagger)$$

where ω, J, α are positive quantities, I_B is the identity operator in the space of the Bose operators and I_F is the identity operator in the space of the Fermi operators. Here b and $b^\dagger$ are Bose annihilation and creation operators, respectively, and c and $c^\dagger$ are Fermi annihilation and creation operators, respectively. The commutation and anticommutation relations are

$$[b, b^\dagger] = I_B, \quad [b, b] = [b^\dagger, b^\dagger] = 0$$

$$[c, c^\dagger]_+ = I_F, \quad [c, c]_+ = [c^\dagger, c^\dagger]_+ = 0.$$

Find the eigenvalues of the Hamilton operator $\hat{H}$. Find the eigenvectors. Use the matrix representations for $c^\dagger$, c, $b^\dagger$, b, i.e.

$$c^\dagger = \begin{pmatrix} 0 & 0 \\ 1 & 0 \end{pmatrix}, \quad c = \begin{pmatrix} 0 & 1 \\ 0 & 0 \end{pmatrix}.$$

For the Fermi states we set

$$|0\rangle_F = \begin{pmatrix} 1 \\ 0 \end{pmatrix}.$$

Thus

$$|1\rangle_F = c^\dagger |0\rangle_F = \begin{pmatrix} 0 \\ 1 \end{pmatrix}.$$

For the Bose operators we have

$$b^\dagger = \begin{pmatrix} 0 & 0 & 0 & 0 & \ldots \\ 1 & 0 & 0 & 0 & \ldots \\ 0 & \sqrt{2} & 0 & 0 & \ldots \\ 0 & 0 & \sqrt{3} & 0 & \ldots \\ & & \ldots & & \end{pmatrix}, \quad b = \begin{pmatrix} 0 & 1 & 0 & 0 & 0 & \ldots \\ 0 & 0 & \sqrt{2} & 0 & 0 & \ldots \\ 0 & 0 & 0 & \sqrt{3} & 0 & \ldots \\ 0 & 0 & 0 & 0 & \sqrt{4} & \ldots \\ & & \ldots & & & \end{pmatrix}.$$

For the states of the Bose operators we have

$$|n\rangle_B = \frac{1}{\sqrt{n!}} b^{\dagger n} |0\rangle_B$$

where $n = 0, 1, 2, \ldots..$.

Solution 8. Using the matrix representation for the number operators $c^\dagger c$ and $b^\dagger b$ given above we find

$$c^\dagger c = \begin{pmatrix} 0 & 0 \\ 0 & 1 \end{pmatrix}$$

and

$$b^\dagger b = \text{diag}(0, 1, 2, \ldots).$$

Thus it follows that

$$b^\dagger b \otimes I_F = \text{diag}(0, 1, 2, \ldots) \otimes \begin{pmatrix} 1 & 0 \\ 0 & 1 \end{pmatrix} = \text{diag}(0, 0, 1, 1, 2, 2, \ldots)$$

and

$$I_B \otimes c^\dagger c = \text{diag}(1, 1, 1, \ldots) \otimes \begin{pmatrix} 0 & 0 \\ 0 & 1 \end{pmatrix} = \text{diag}(0, 1, 0, 1, \ldots).$$

For the interacting terms we find

$$b \otimes c^\dagger = (0) \oplus \begin{pmatrix} 0 & 1 \\ 0 & 0 \end{pmatrix} \oplus \begin{pmatrix} 0 & \sqrt{2} \\ 0 & 0 \end{pmatrix} \oplus \begin{pmatrix} 0 & \sqrt{3} \\ 0 & 0 \end{pmatrix} \oplus \cdots$$

$$b^\dagger \otimes c = (0) \oplus \begin{pmatrix} 0 & 0 \\ 1 & 0 \end{pmatrix} \oplus \begin{pmatrix} 0 & 0 \\ \sqrt{2} & 0 \end{pmatrix} \oplus \begin{pmatrix} 0 & 0 \\ \sqrt{3} & 0 \end{pmatrix} \oplus \cdots$$

where $\oplus$ denotes the direct sum of matrices. Adding up the matrices we obtain the matrix representation of the Hamilton operator

$$\hat{H} = (0) \oplus \begin{pmatrix} J & \alpha \\ \alpha & \omega \end{pmatrix} \oplus \begin{pmatrix} \omega + J & \sqrt{2}\alpha \\ \sqrt{2}\alpha & 2\omega \end{pmatrix} \oplus \begin{pmatrix} 2\omega + J & \sqrt{3}\alpha \\ \sqrt{3}\alpha & 3\omega \end{pmatrix} \oplus \cdots$$

$$\oplus \begin{pmatrix} n\omega + J & \sqrt{n+1}\alpha \\ \sqrt{n+1}\alpha & (n+1)\omega \end{pmatrix} \oplus \cdots .$$

Thus one eigenvalue is 0. The eigenvector for the eigenvalue 0 is given by $|0\rangle_B \otimes |0\rangle_F$. Thus this state is not entangled. To find the other eigenvalues we have to find the eigenvalues of the 2×2 matrices

$$\begin{pmatrix} n\omega + J & \sqrt{n+1}\alpha \\ \sqrt{n+1}\alpha & (n+1)\omega \end{pmatrix}$$

where $n = 0, 1, 2, \ldots$. We find

$$E_\pm(n) = \frac{J}{2} + \frac{\omega}{2} + n\omega \pm \sqrt{\frac{1}{4}(J - \omega)^2 + (n+1)\alpha^2}.$$

Whether 0 is the lowest eigenvalue depends on the values of J, ω and α. Consider the case with $n = 0$. Then

$$E_-(n = 0) = \frac{J}{2} + \frac{\omega}{2} - \sqrt{\frac{1}{4}(J - \omega)^2 + \alpha^2}$$

The eigenvector for the eigenvalue $E_-(n = 0)$ is given by $(\alpha \neq 0)$

$$|0\rangle_B \otimes |1\rangle_F + \frac{1}{\alpha}\left(-\sqrt{\frac{1}{4}(J - \omega)^2 + \alpha^2} + \frac{\omega}{2} - \frac{J}{2}\right)|1\rangle_B \otimes |0\rangle_F .$$

Thus this state is entangled except if $-\sqrt{(J - \omega)^2/4 + \alpha^2} + \omega/2 - J/2 = 0$. The condition $E_-(n = 0) = 0$ yields $\alpha^2 = J\omega$. Then we find the state

$$|0\rangle_B \otimes |1\rangle_F - \frac{J}{\alpha}|1\rangle_B \otimes |0\rangle_F .$$

This state is also entangled if $J \neq 0$. Thus we have an unentangled state with eigenvalue 0 and for given parameter values of J, ω and α we can have an entangled state with this eigenvalue.

Problem 9. Let b, $b^\dagger$ be Bose annihilation and creation operators. Let

$$\sigma_+ := \frac{1}{2}(\sigma_x + i\sigma_y) = \begin{pmatrix} 0 & 1 \\ 0 & 0 \end{pmatrix}, \quad \sigma_- := \frac{1}{2}(\sigma_x - i\sigma_y) = \begin{pmatrix} 0 & 0 \\ 1 & 0 \end{pmatrix}.$$

Consider the Hamilton operator

$$\hat{H} = b^\dagger \otimes \sigma_- + b \otimes \sigma_+$$

which describes a single atom coupled to a single mode of an electromagnetic field. $\sigma_\pm$ act on the atom and b, $b^\dagger$ act on the field.
(i) Calculate $\hat{H}^2$.
(ii) Calculate the commutator

$$[b^\dagger \otimes \sigma_-, b \otimes \sigma_+].$$

(iii) Let $|n\rangle$ be a number state and $|\beta\rangle$ be a coherent state. Calculate

$$A(n, \beta) := (\langle n| \otimes I_2)e^{i\theta \hat{H}}(|\beta\rangle \otimes I_2)$$

where $\theta \in \mathbb{R}$ and I_2 is 2×2 unit matrix.

Solution 9. (i) Since $\sigma_-\sigma_- = 0$, $\sigma_+\sigma_+ = 0$ and

$$\sigma_+\sigma_- = \begin{pmatrix} 1 & 0 \\ 0 & 0 \end{pmatrix}, \quad \sigma_-\sigma_+ = \begin{pmatrix} 0 & 0 \\ 0 & 1 \end{pmatrix}$$

we find

$$(b^\dagger \otimes \sigma_- + b \otimes \sigma_+)(b^\dagger \otimes \sigma_- + b \otimes \sigma_+) = bb^\dagger \otimes \sigma_+\sigma_- + b^\dagger b \otimes \sigma_-\sigma_+.$$

Thus we can write in matrix notation

$$\begin{pmatrix} bb^\dagger & 0 \\ 0 & b^\dagger b \end{pmatrix}.$$

(ii) We have

$$[b^\dagger \otimes \sigma_-, b \otimes \sigma_+] = b^\dagger b \otimes \sigma_-\sigma_+ - bb^\dagger \otimes \sigma_+\sigma_-.$$

(iii) Since $b|\beta\rangle = \beta|\beta\rangle$ and $b|n\rangle = \sqrt{n}|n-1\rangle$ we obtain

$$A(n, \alpha) = \exp(-|\beta|^2)\frac{|\beta|^2}{n!} \begin{pmatrix} \cos(\theta\sqrt{n}) & \frac{i\sqrt{n}}{\beta}\sin(\theta\sqrt{n}) \\ \frac{in}{\sqrt{n+1}}\sin(\theta\sqrt{n+1}) & \cos(\theta\sqrt{n+1}) \end{pmatrix}.$$

Problem 10.   Consider the model Hamilton operator for ions trapped inside an optical cavity

$$\hat{H} := \hat{H}_0 + \hat{V}$$

where

$$\hat{H}_0 = \left(\hbar\nu a^\dagger a + \frac{1}{2}I_a\right) \otimes I_b \otimes I_2 + I_a \otimes \hbar\omega_c b^\dagger b \otimes I_2 + I_a \otimes I_b \otimes \frac{\hbar\omega_0}{2}\sigma_z$$

and

$$\hat{V} = \hbar\Omega(\exp(i\eta_L(a^\dagger + a) - i(\omega_L t + \phi)I_a) \otimes I_b \otimes \sigma_+ + h.c.)$$
$$+ \hbar g \sin(\eta_c(a^\dagger + a)) \otimes (b^\dagger + b) \otimes (\sigma_+ + \sigma_-).$$

Here $a^\dagger$ (a) and $b^\dagger$ (b) are Bose creation (annihilation) operators for the vibrational phonon and the cavity field photon, respectively and ω_0 is the transition frequency of the two-level ion. The ion-phonon and ion-cavity coupling constants are Ω and g, and σ_k $(k = z, +, -)$ are the Pauli operators describing the internal state of the ion. Thus we consider a two-level ion radiated by the single mode cavity field of frequency ω_c and an external laser field of frequency ω_L. The operators I_a, I_b and I_2 are the identity operators in their respective Hilbert spaces, where I_2 is the 2×2 unit matrix. Thus we have a tripartite system. The parameters η_L and η_c are the *Lamb-Dicke parameters*.
(i) Consider the unitary operator

$$U_0(t) = \exp(-i\hat{H}_0 t/\hbar).$$

Calculate (interaction picture)

$$H_I(t) = U_0^\dagger(t)\hat{V}U_0(t) \equiv \exp(i\hat{H}_0 t/\hbar)\hat{V}\exp(-i\hat{H}_0 t/\hbar).$$

(ii) Discuss how a Hadamard gate can be realized.

Solution 10.   (i) Straightforward calculation yields

$$\hat{H}_I(t) = \hbar\Omega\left(\hat{O}_0^L \exp(i(\delta_{0L} t - \phi))\right) \otimes I_b \otimes \sigma_+$$

$$+ \hbar\Omega\left(\sum_{k=1}^\infty (i\eta_L)^k \hat{O}_k^L a^k \exp(i((\delta_{0L} - k\nu)t - \phi))\right) \otimes I_b \otimes \sigma_+$$

$$+ \hbar\Omega\left(\sum_{k=1}^\infty (i\eta_L)^k a^{\dagger k} \hat{O}_k^L \exp(i((\delta_{0L} + k\nu)t - \phi))\right) \otimes I_b \otimes \sigma_+$$

$$+ \hbar g\left(\sum_{k=1,3,\ldots}^\infty (i^{k-1}\eta_c^k) a^{\dagger k} \hat{O}_k^c \exp(i(\delta_{0c} + k\nu + 2\omega_c)t)\right) \otimes b^\dagger \otimes \sigma_+$$

$$+\hbar g \left(\sum_{k=1,3,\ldots}^{\infty} (i^{k-1}\eta_c^k)\hat{O}_k^c a^k \exp(i(\delta_{0c} - k\nu + 2\omega_c)t) \right) \otimes b^\dagger \otimes \sigma_+$$

$$+\hbar g \left(\sum_{k=1,3,\ldots}^{\infty} (i^{k-1}\eta_c^k)\hat{O}_k^c a^k \exp(i(\delta_{0c} - k\nu)t) \right) \otimes b \otimes \sigma_+$$

$$+\hbar g \left(\sum_{k=1,3,\ldots}^{\infty} (i^{k-1}\eta_c^k)a^{\dagger k}\hat{O}_k^c \exp(i(\delta_{0c} + k\nu)t) \right) \otimes b \otimes \sigma_+$$

$$+h.c.$$

where

$$\hat{O}_k^L := \exp\left(\frac{-\eta_L^2}{2}\right) \sum_{p=0}^{\infty} \frac{(i\eta_L)^{2p} a^{\dagger p} a^p}{p!(p+k)!}, \quad \hat{O}_k^c := \exp\left(\frac{-\eta_c^2}{2}\right) \sum_{p=0}^{\infty} \frac{(i\eta_c)^{2p} a^{\dagger p} a^p}{p!(p+k)!}$$

and

$$\delta_{0L} := \omega_0 - \omega_L, \qquad \delta_{0c} := \omega_0 - \omega_c.$$

(ii) A basis is

$$|m\rangle \otimes |n\rangle \otimes |g\rangle, \qquad |m\rangle \otimes |n\rangle \otimes |e\rangle$$

where $m = 0, 1, \ldots, \infty$ denotes the state of ionic vibrational motion, $n = 0, 1, \ldots, \infty$ denotes the state of the quantized cavity field and $|g\rangle$ and $|e\rangle$ denote the ground state and excited state, respectively for the two-level ion. Using appropriate values for the parameters and the time we can find an implementation of the Hadamard gate

$$|m\rangle \otimes |0\rangle \otimes |g\rangle \to \frac{1}{2}(|m\rangle \otimes |0\rangle \otimes |g\rangle + |m\rangle \otimes |0\rangle \otimes |e\rangle)$$

$$|m\rangle \otimes |0\rangle \otimes |e\rangle \to \frac{1}{2}(|m\rangle \otimes |0\rangle \otimes |g\rangle - |m\rangle \otimes |0\rangle \otimes |e\rangle).$$

Problem 11. Consider a square lattice with lattice constant a and periodic boundary conditions. Let $c_{\mathbf{j}}^\dagger$ ($c_{\mathbf{j}}$) denotes the creation (annihilation) operator for an electron in the Wannier state at the lattice site $\mathbf{j}$. The Hamilton operator $\hat{H}$ of spinless tight-binding electrons in the presence of a magnetic field can be written as

$$\hat{H} = \sum_{\langle \mathbf{j}_1, \mathbf{j}_2 \rangle} t_{\mathbf{j}_1 \mathbf{j}_2} c_{\mathbf{j}_1}^\dagger c_{\mathbf{j}_2}$$

with

$$t_{j_1 j_2} = -t \exp\left(-i\frac{e}{\hbar} \int_{\mathbf{j}_1}^{\mathbf{j}_2} \mathbf{A} \cdot d\mathbf{r}\right)$$

The summation $\langle \mathbf{j}_1, \mathbf{j}_2 \rangle$ runs over the nearest neighbour site on the square lattice. The uniform magnetic field $\mathbf{B}$ is applied in z-direction. Choosing the *Landau gauge*

$$\mathbf{A} = B(0, x, 0)$$

the line integral in $t_{\mathbf{j}_1 \mathbf{j}_2}$ can be written as

$$\frac{e}{\hbar} \int_{\mathbf{j}_1}^{\mathbf{j}_2} \mathbf{A} \cdot d\mathbf{r} = \begin{cases} 0 & \mathbf{j}_1 = (m, n) \quad \mathbf{j}_2 = (m+1, n) \\ -2\pi m \Phi / \Phi_0 & \mathbf{j}_1 = (m, n) \quad \mathbf{j}_2 = (m, n+1) \end{cases}$$

where the integers m and n refer to the $x = 1$, $y = 2$ coordinates of the square lattice sites. $\Phi = Ba^2$ is the magnetic flux through a unit plaquette. Φ_0 stands for the magnetic flux quantum h/e. Find the Hamilton operator in the *Bloch representation*. The Fourier transform is given by

$$c_{\mathbf{k}} = \frac{1}{\sqrt{N}} \sum_{\mathbf{j}} e^{i(k_1 j_1 + k_2 j_2)} c_{\mathbf{j}}$$

with the inverse

$$c_{\mathbf{j}} = \frac{1}{\sqrt{N}} \sum_{\mathbf{k}} e^{-i(k_1 j_1 + k_2 j_2)} c_{\mathbf{k}}$$

where $\mathbf{k}$ runs over the first Brillouin zone.

Solution 11.   We obtain the Hamilton operator in Bloch representation

$$\hat{H} = -2t \sum_{\mathbf{k}} \cos(k_1 a) c^{\dagger}(\mathbf{k}) c(\mathbf{k})$$

$$-t \sum_{\mathbf{k}} \left(\exp(-ik_2 a) c^{\dagger}(k_1 + 2\frac{\pi \Phi}{a\Phi_0}, k_2) c(k_1, k_2) \right.$$

$$\left. + \exp(ik_2 a) c^{\dagger}(k_1 - 2\frac{\pi \Phi}{a\Phi_0}, k_2) c(k_1, k_2) \right)$$

where $c^{\dagger}(\mathbf{k})$ creates an electron in the Block state with wave vector $\mathbf{k}$ and $c(\mathbf{k})$ annihilates an electron in the Block state with wave vector $\mathbf{k}$. If $\Phi / \Phi_0 = p/q$ is rational, the magnetic Brillouin zone can be reduced to $0 \leq k_1 \leq 2\pi/a$ and $0 \leq k_2 \leq 2\pi/(qa)$. We obtain

$$c\left(k_1 + 2\frac{\pi \Phi}{a\Phi_0}\ell, k_2\right) = c\left(k_1 + 2\frac{\pi \Phi}{a\Phi_0}(\ell + q), k_2\right)$$

with ℓ are integers.

Bibliography

Books

Barnett S. M. and Radmore P. M.
Methods in Theoretical Quantum Optics, Oxford University Press (1997)

Bell J. S.
Speakable and unspeakable in quantum mechanics, Cambridge University Press (1989)

Benenti G., Casati G. and Strini G.
Principles of Quantum Computing and Information, Volume I: Basic Concepts, World Scientific (2007)

Blank J., Exner P. and Havlíček M.
Hilbert Space Operators in Quantum Physics, second edition, Springer (2008)

Bohm A.
Quantum Mechanics, Third edition, Springer (1994)

Braunstein S. L. and Pati A. K.
Quantum Information with Continuous Variables, Kluwer (2003)

Cerf N. J., Leuchs G. and Polzik E. S.
Quantum Information with continuous variables of atoms and light, Imperial College Press (2007)

Constantinescu F. and Magyari E.
Problems in Quantum Mechanics, Pergamon Press (1971)

Dirac P. A. M.
The Principle of Quantum Mechanics, Clarendon Press (1958)

Flügge S.
Practical Quantum Mechanics, Springer (1974)

Gasiorowicz S.
Quantum Physics, Wiley (1974)

Gilmore R.
Lie Groups, Lie Algebras, and Some of Their Applications, Wiley (1974)

Glimm J. and Jaffe A.
Quantum Physics, Springer (1981)

Hardy Y. and Steeb W.-H.
Classical and Quantum Computing with C++ and Java Simulations, Birkhauser-Verlag (2002)

Hirvensalo M.
Quantum Computing, second edition, Springer (2004)

Kaye P., Laflamme R. and Mosca M.
An Introduction to Quantum Computing, Oxford University Press (2007)

Kim Y. S. and Noz M. E.
Phase Space Picture of Quantum Mechanics, World Scientific (1991)

Kitaev A. Yu., Shen A. H. and Vyalyi M. N.,
Classical and Quantum Computation, American Mathematical Society (2002)

Klauder J. R. and Skagerstam B.
Coherent States - Applications in Physics and Mathematical Physics, World Scientific (1985)

Kowalski K. and Steeb W.-H.
Nonlinear Dynamical Systems and Carleman Linearization, World Scientific (1991)

Lewand R. E.
Cryptological Mathematics, Mathematical Association of America (2000)

Louisell W. H.
Quantum Statistical Properties of Radiation, Wiley (1973)

Mermin N. D.
Quantum Computer Science, Cambridge University Press (2007)

Merzbacher E.
Quantum Mechanics, 2nd edition, Wiley (1970)

Nielsen M. A. and Chuang I. L.
Quantum Computation and Quantum Information, Cambridge University Press (2000)

Prugovećki E.
Quantum Mechanics in Hilbert Space, second edition, Academic Press (1981)

Sakurai J. J.
Modern Quantum Mechanics, Addison-Wesley (1994)

Scully M. O. and Zubairy M. S.
Quantum Optics, Cambridge University Press (1997)

Steeb W.-H. and Hardy Y.
Matrix Calculus and Kronecker Product: A Practical Approach to Linear and Multilinear Algebra, second edition, World Scientific (2010)

Steeb W.-H.
Continuous Symmetries, Lie Algebras, Differential Equations and Computer Algebra, second edition, World Scientific (2007)

Steeb W.-H.
Hilbert Spaces, Wavelets, Generalized Functions and Quantum Mechanics, Kluwer Academic Publishers (1998)

Steeb W.-H. and Hardy Y.
Quantum Mechanics using Computer Algebra, second edition, World Scientific (2010)

Steeb W.-H.
Problems and Solutions in Theoretical and Mathematical Physics, Third Edition, Volume I: Introductory Level, World Scientific (2010)

Steeb W.-H.
Problems and Solutions in Theoretical and Mathematical Physics, Third Edition, Volume II: Advanced Level, World Scientific (2010)

Stolze J. and Suter D.
Quantum Computing: A short course from theory to experiment, Wiley (2008)

von Neumann J.
Mathematical Foundations of Quantum Mechanics, Princeton University
Press (1955)

Walls D. F. and Milburn G. J.
Quantum Optics, Springer (1994)

Papers

Part I

Alber G., Delgado A., Gisin N. and Jex I., *Efficient bipartite quantum state
purification in arbitrary dimensional Hilbert spaces*, J. Phys. A: Math. Gen.
34, 8821 (2001)

Bennett C. H., DiVincenzo D. P., Smolin J. A. and Wootters W. K., *Mixed
State Entanglement and Quantum Error Correction*, Phys. Rev. A **54**,
3824 (1996)

Bennett C. H., Brassard G., Popescu S., Schumacher B., Smolin J. A. and
Wootters W.K., *Purification of Noisy Entanglement and Faithful Teleportation via Noisy Channels*, Phys. Rev. Lett. **77**, 722 (1996)

Benioff P., *Models of quantum Turing machines*, Fortsch. Phys. **46**, 423
(1997)

Berman G. P., Borgonovi F., Chapline G., Gurvitz S. A., Hammel P. C.,
Pelekhov D. V., Suter A. and Tsifrinovich V. I., *Application of Magnetic
Resonance Force Microscopy Cyclic Adiabatic Inversion for a Single-Spin
Measurement*, J. Phys. A : Math. Gen. **36**, 4417 (2003)

Brandt H. E., *Riemannian Geometry of Quantum Computation*, Nonlinear
Analysis: Theory, Methods and Applications **71**, e474 (2009)

Brassard G., Braunstein S. L. and Cleve R. *Teleportation as a quantum
computation*, Physica D **120**, 43 (1998)

Brassard G., Cleve R. and Tapp A., *The cost of Exactly Simulating Quantum Entanglement with Classical Communication*, Phys. Rev. Lett. **83**,
1874 (1999)

Braunstein S. L., Caves C. M., Jozsa R., Linden N., Popescu S. and Schack
R., *Separability of Very Noisy Mixed States and Implications for NMR
Quantum Computing*, Phys. Rev. Lett. **83**, 1054 (1999)

Braunstein S. L., Mann A. and Revzen M., *Maximal Violation of Bell Inequalities for Mixed States*, Phys. Rev. Lett. **68**, 3259 (1992)

Brukner C., Pan J.-W., Simon C., Weihs G. and Zeilinger A., *Probabilistic Instantaneous Quantum Computation*, Phys. Rev. A **67**, 034304 (2003)

Bruß D., *Characterizing Entanglement*, J. Math. Phys. **43**, 4237 (2002)

Buhrman H., Cleve R. and van Dam W., *Quantum Entanglement and Communication Complexity*, SIAM Journ. on Computing **30**, 1829–1841 (2001)

Childs A. M., Leung D., Verstraete F. and Vidal G., *Asymptotic entanglement capacity of the Ising and anisotropic Heisenberg interactions*, Quantum Information and Computation **3**, 97 (2003)

Chuang I. L. and Yamamoto Y., *A Simple Quantum Computer*, Phys. Rev. A **52**, 3489 (1995)

Cleve R. and Buhrman H., *Substituting Quantum Entanglement for Communication*, Phys. Rev. A **56**, 1201 (1997)

Cleve R., Ekert A., Macchiavello C. and Mosca M., *Quantum Algorithms Revisited*, Proc. Roy. Soc. Lond. A **454**, 339 (1998)

Dehaene J. and De Moor B., *Clifford group, stabilizer states, and linear and quadratic operations over GF(2)*, Phys. Rev. A **68**, 042318 (2003)

Deutsch D., *Quantum Computation Networks*, Proc. R. Soc. London A **425**, 73 (1989)

Deutsch D., Ekert A., Josza R., Macchiavello C., Popescu S. and Sanpera A., *Quantum Privacy Amplification and the Security of Quantum Cryptography over Noisy Channels*, Phys. Rev. Lett. **77**, 2818 (1996)

Deutsch D. and Jozsa R., *Rapid solutions of problems by quantum computation*, Proc. R. Soc. London **439**, 553 (1992)

Ekert A., *Quantum Cryptography Based on Bell's Theorem*, Phys. Rev. Lett. **67**, 661 (1991)

Fan H., Matsumoto K. and Imai H., *Quantify entanglement by concurrence hierarchy*, J. Phys. A: Math. Gen. **36**, 4151 (2003)

Gottesman D., *Grand Mathematical Challenge for the Twenty-First Century and the Millennium*, edited by S. J. Lomonaco Jr., (American Mathematical Society, Providence, Rhode Island), 221 (2002)

Gottesman D., *A Class of Quantum Error-Correcting Codes Saturating the Quantum Hamming Bound*, Phys. Rev. A **54**, 1862 (1996)

Gottesman D., *An Introduction to Quantum Error Corrections and Fault-Tolerant Quantum Computation*, in "Quantum Information Science and Its Contributions to Mathematics", Proceedings of Symposia in Applied Mathematics, **68**, 13 (2010)

Grover L., *Quantum Mechanics helps searching for a needle in a haystack*, Phys. Rev. Lett. **79**, 325 (1997)

Grudka A. and Wojcik A., *How to encode the states of two non-entangled qubits in one qutrit*, Phys. Lett. A **314**, 350 (2003)

Hardy Y., Steeb W.-H. and Stoop R., *Fully Entangled Quantum States in C^{N^2} and Bell Measurement*, Int. J. Theor. Phys. **42**, 2314 (2003)

Hardy Y., Steeb W.-H. and Stoop R., *Entanglement, Disentanglement and Wigner Functions*, Physica Scripta **69**, 166 (2004)

Hardy Y. and Steeb W.-H., *Unitary Operators, Entanglement, and Gram-Schmidt Orthogonalization*, Int. J. of Mod. Phys. C **20**, 891 (2009)

Henderson L. and Vedral V., *Information, Relative Entropy of Entanglement and Irreversibility*, Physical Review Letters, **84**, 2263 (2000)

Horodecki M., P. Horodecki and R. Horodecki, *Separability of Mixed States: necessary and sufficient conditions*, Physics Letters A, **223**, 1 (1996)

Jordan T. F., *Quantum mysteries explored*, Am. J. Phys. **62**, 874 (1994)

Jozsa R., *Quantum Algorithms and the Fourier transform*, Proc. Roy. Soc. London A **454**, 323 (1998)

Kimura G., *The Bloch vector for N-level systems*, Phys. Lett. A **314**, 319 (2003)

Korbelář M. and Tolar J., *Symmetries of the finite Heisenberg group for composite systems*, J. Phys. A: Math. Theor. **43**, 375302 (2010)

Kuzmich A., Walmsley A. and Mandel L., *Violation of a Bell-type inequality in the homodyne measurement of light in an Einstein-Podolsky-Rosen state*, Phys. Rev. A **64**, 063804 (2001)

Laflamme L., Miquel C., Paz J. P. and Zurek W. H., *Perfect Quantum Error Correction Code*, Phys. Rev. Lett. **77**, 198 (1996)

Mermin N. D., *From Classical State-Swapping to Quantum Teleportation*, Phys. Rev. A **65**, 012320 (2002)

Nielsen M. A., *Quantum information theory*, Ph.D. thesis, University of New Mexico (1998)

Nielsen M. A., C. M. Dawson, J. L. Dodd, A. Gilchrist, D. Mortimer, T. J. Osborne, M. J. Bremner, A. W. Harrow and A. Hines, *Quantum dynamics as a physical resource*, Physical Review A, **67**, 052301 (2003)

Nielsen M. A. and Chuang L., *Programmable Quantum Gate Arrays*, Phys. Rev. Lett. **79**, 321 (1997)

Pan J.-W., Bouwmeester D., Daniell M., Weinfurter H. and Zeilinger A., *Experimental test of quantum nonlocality in three-photon Greenberger-Horne-Zeilinger entanglement*, Nature **403**, 515 (2000)

Planat M. and Jorrand P., *Group theory for quantum gates and quantum coherence*, J. Phys. A: Math. Theor. **41**, 182001 (2008)

Sharma S. S., *Tripartite GHZ state generation with trapped ion in optical cavity*, Phys. Lett. A **311**, 111 (2003)

Shor P. W., *Scheme of reducing decoherence in quantum computer memory*, Phys. Rev. A **52**, 2493 (1995)

Shor P. W., *Polynomial-time algorithm for prime factorization and discrete logarithms on a quantum computer*, Soc. Ind. Appl. Math. J. Comp. 1484 (1997)

Steeb W.-H. and Hardy Y., *Entangled Quantum States and a C++ Implementation*, Int. J. Mod. Phys. C **11**, 69 (2000)

Steeb W.-H. and Hardy Y., *Quantum Computing and SymbolicC++ Simulations*, Int. J. Mod. Phys. C **11**, 323 (2000)

Steeb W.-H. and Hardy Y., *Energy Eigenvalue Level Motion and a SymbolicC++ Implementation*, Int. J. Mod. Phys. C **11**, 1347 (2000)

Steeb W.-H. and Hardy Y., *Fermi Systems, Hubbard Model and a SymbolicC++ Implementation*, Int. J. Mod. Phys. C **12**, 235 (2001)

Steeb W.-H. and Hardy Y., *Energy Eigenvalue Level Motion with Two Parameters*, Z. Naturforsch. **56 a**, 565 (2001)

Steeb W.-H. and Hardy Y., *Entangled Quantum States and the Kronecker Product*, Z. Naturforsch. **57 a**, 689 (2002)

Steeb W.-H., Hardy Y., Stoop R., *Discrete wavelets and perturbation theory*, J. Phys A: Math. Gen. **36**, 6807 (2003)

Steeb W.-H. and Hardy Y., *A hierarchy of Hamilton operators and entanglement*, Central European Journal of Physics, **7**, 854 (2009)

Steeb W.-H., *Diabolic Points and Entanglement*, Physica Scripta **81**, 025012 (2010)

Thomsen M. K., Glück R. and Axelsen H. B., *Reversible arithmetic logic unit for quantum arithmetic*, J. Phys. A: Math. Theor. **43**, 382002 (2010)

Trifonov A., Tsegaye T., Björk G., Söderholm J., Goobar E., Atatüre M. and Sergienko A. V., *Experimental demonstration of the relative phase operator*, J. Opt. B: Quantum Semiclass. Opt., **2**, 105 (2000)

Tyson J., *Operator-Schmidt decomposition of the quantum Fourier transform on $C^{N_1} \otimes C^{N_2}$*, J. Phys. A: Math. Gen. **36** 6813 (2003)

van Loock P. and Braunstein S. L., *Greenberger-Horne-Zeilinger non-locality in phase space*, Phys. Rev. A **63**, 022106 (2001)

Wang X. and Sanders B. C., *Entanglement capability of self-inverse Hamiltonian evolution*, Phys. Rev. A **68**, 014301 (2003)

Werner R. F., *Quantum states with Einstein-Podolsky-Rosen correlations admitting a hidden-variable model*, Phys. Rev. A **40**, 4277 (1989)

Yang G., Hung W. N. N., Song X. and Perkowski M., *Majority based reversible logic gates*, Theoretical Computer Science **334**, 259 (2005)

Yura F., *Entanglement cost of three-level antisymmetric states*, J. Phys. A: Math. Gen. **36**, L237 (2003)

Part II

Ban M., *The phase operator in quantum information processing*, J. Phys. A: Math. Gen. **35**, L193 (2002)

Ban M., *Phase-space approach to continuous variable quantum teleportation*, Phys. Rev. A **69**, 054304 (2004)

Banerji J., *Non-linear wave packet dynamics of coherent states*, PRAMANA J. Phys. **56**, 267 (2001)

Barak R. and Ben-Aryeh Y., *Photon statistics and entanglement in coherent-squeezed linear Mach-Zehnder and Michelson interferometers*, J. Opt. Soc. Am. B **25**, 361 (2008)

Bartlett S. D., Sanders B. C., Braunstein S. L. and Nemoto K., *Efficient Classical Simulations of Continuous Variable Quantum Information Processes*, Phys. Rev. Lett. **88**, 097904 (2002)

Barnett S. M., Jeffers J. and Gatti A., *Quantum optics of lossy beam splitters*, Phys. Rev. A **57**, 2134 (1998)

Braunstein S. L. and Kimble H. J., *Teleportation of Continuous Quantum Variables*, Phys. Rev. Lett. **80**, 869 (1998)

Braunstein S. L., Fuchs C. A., Kimble H. J. and van Loock P., *Quantum versus classical domains for teleportation with continuous variables*, Phys. Rev. A **64**, 022321 (2001)

Cerf N. J., Adami C. and Kwiat P. G., *Optical Simulation of Quantum Logic*, Phys. Rev. A **57**, 1477 (1998)

Cirac J. I., *Quantum Information: Entanglement, Purification, Error Correction, and Quantum Optical Implementations*, in Fundamentals of Quantum Information, Heiss D. (Ed.), Springer, Berlin (2002)

Chruściński D., *Spectral Properties of the Squeeze Operator*, Phys. Lett. A **327**, 290-295 (2004)

D'Ariano G. M., De Martin F. and Sacchi M. F., *Continuous Variable Cloning via Network of Parametric Gates*, Phys. Rev. Lett. **86**, 914 (2001)

DasGupta A., *Disentanglement formulas: An alternative derivation and some applications to squeezed coherent states*, Am. J. Phys. **64**, 1422 (1996)

de Oliveira M. C. and Milburn G. J., *Discrete teleportation protocol of continuum spectra field states*, Phys. Rev. A **65**, 032304 (2002)

Dell'Anno F., De Siena S. and Illuminati F., *Multiphoton quantum optics and quantum state engineering*, Physics Reports, **428**, 53 (2006)

Dodonov V. V., Man'ko V. I. and Nikonov D. E., *Even and odd coherent states for multimode parametric systems*, Phys. Rev. A **51**, 3328 (1995)

Einstein A., Podolsky B. and Rosen N., *Can Quantum-Mechanical Description of Physical Reality Be Considered Complete?*, Phys. Rev. **47**, 777 (1935)

Eisert J., *Optimizing Linear Optics Quantum Gates*, Phys. Rev. Lett. **95**, 040502 (2005)

Ferraro A., Olivares S. and Paris M. G. A., *Gaussian states in a continuous variable quantum information*, arXiv:quant-ph/0503237v1, 2005

Fujii K., *Basic Properties of Coherent and Generalized Coherent Operators Revisited*, Mod. Phys. Lett. A **16**, 1277 (2001)

Hardy Y. and Steeb W.-H., *Fermi-Bose Systems, Macroscopic Quantum Superposition States and Entanglement*, Int. J. Theor. Phys. **43**, 2207 (2004)

Hong-Yi F. and Klauder J. R., *Canonical coherent state representation of some squeeze operators*, J. Phys. A: Math. Gen. **21**, L725-L730 (1988)

Imoto N., Haus H. A. and Yamamoto Y., *Quantum nondemolition measurement of the photon number via the optical Kerr effect*, Phys. Rev. A **32**, 2287 (1985)

Jeffers J., Barnett S. M. and Pegg D., *Retrodiction as a tool for micromaser field measurement*, J. Mod. Opt. **49**, 925 (2002)

Knill E., Laflamme R. and Milburn G. J., *A scheme for efficient quantum computation with linear optics*, Nature **409**, 46 (2001)

Kok P., Munro W. J., Nemoto K., Ralph T. C., Dowling J. P. and Milburn G. J., *Linear optical quantum computing with photonic qubits*, Rev. Mod. Phys. **79**, 1 (2007)

Koniorczyk M., Buzek V. and Janszky J., *Wigner-function description of quantum teleportation in arbitrary dimensions and a continuous limit*, Phys. Rev. A **64**, 034301 (2001)

Kok P., Munro W. J., Nemoto K., Ralph T. C., Dowling J. P. and Milburn G. J., *Linear Optical Quantum Computing*, arXiv:quant-ph/0512071v2

Kok P. and Braunstein S. L., *Multi-dimensional Hermite polynomials in quantum optics*, J. Phys. A: Math. Gen. **34**, 6185 (2001)

Lee J. and Kim M. S., *Quantum Teleportation Via Mixed Two-Mode Squeezed States in the Coherent-State Representation*, J. Kor. Phys. Soc. **42**, 457 (2003)

Li Shang-Bin and Xu Jing-Bo, *Quantum probabilistic teleportation via entangled coherent states*, Phys. Lett. A **309**, 321 (2003)

Lloyd S. and Braunstein S. L., *Quantum Computing over Continuous Variables*, Phys. Rev. Lett. **82**, 1784 (1999)

Lo C. F., *Eigenfunctions and eigenvalues of squeeze operators*, Phys. Rev. A **42**, 752 (1990)

Milburn G. J. and Braunstein S. L., *Quantum teleportation with squeezed states*, Phys. Rev. A **60**, 937 (1999)

Pan Jian-Wei, Daniell M., Gasparoni S., Weihs G., and Zeilinger A., *Experimental Demonstration of Four-Photon Entanglement and High-Fidelity Teleportation*, Phys. Rev. Lett. **86**, 4435 (2001)

Paris M. G. A., *Generation of mesoscopic quantum superpositions through Kerr-stimulated degenerate downconversion*, J. Opt. B: Quantum Semiclass. Opt. **1**, 662 (1999)

Pegg D. T., Barnett S. M. and Jeffers J., *Quantum theory of preparation and measurement*, J. Mod. Opt. **49**, 913 (2002)

Schrödinger E., *Die gegenwärtige Situation in der Quantenmechanik*, Naturwissenschaften **23**, pp. 807-812; 823-828; 844-849 (1935)

Serafini A., De Siena S., Illuminati F. and Paris M. G. A., *Minimum decoherence cat-like states in Gaussian noisy channels*, arXiv:quant-ph/0310005v2

Sharma S. S. and Vidiella-Barranco A., *Fast quantum logic gates with trapped ions interacting with external laser and quantized cavity field beyond Lamb-Dicke regime*, Phys. Lett. A **309** 345 (2003)

Slosser J. J. and Milburn G. J., *Creating Metastable Schrödinger Cat States*, Phys. Rev. Lett. **75**, 418 (1995)

Steeb W.-H. and Hardy Y., *Quantum Optics Networks, Unitary Operators and Computer Algebra*, Int. J. of Mod. Phys. C **19**, 1069 (2008)

Thamm R., Kolley E. and Kolley W., *Green's Functions for Tight-Binding Electrons on a Square Lattice in a Magnetic Field*, phys. stat. sol. (b) **170**, 163 (1992)

van Enk S. J. and Hirota O., *Entangled coherent states: Teleportation and decoherence*, Phys. Rev. A **64**, 022313 (2001)

van Loock P. and Braunstein S. L., *Multipartite Entanglement for Continuous Variables: A Quantum Teleportation Network*, **84**, 3482 (2000)

Vaziri A., Mair A., Weihs G. and Zeilinger A., *Entanglement of the Angular Orbital Momentum States of the Photons*, Nature **412**, 313 (2001)

Yurke B. and Stoler D., *Generating quantum mechanical superpositions of macroscopically distinguishable states via amplitude dispersion*, Phys. Rev. Lett. **57**, 13 (1986)

Zou Xu-Bo, Pahlke K. and Mathis W., *The non-deterministic quantum logic operation and the teleportation of the superposition of the vacuum state and the single-photon state via parametric amplifiers*, Phys. Lett. A **311** 271 (2003)

Index